AF775398

Selected Titles in This Series

204 **A. M. Vinogradov,** Cohomological analysis of partial differential equations and Secondary Calculus, 2001
203 **Te Sun Han and Kingo Kobayashi,** Mathematics of information and coding, 2001
202 **V. P. Maslov and G. A. Omel′yanov,** Geometric asymptotics for nonlinear PDE. I, 2001
201 **Shigeyuki Morita,** Geometry of differential forms, 2001
200 **V. V. Prasolov and V. M. Tikhomirov,** Geometry, 2001
199 **Shigeyuki Morita,** Geometry of characteristic classes, 2001
198 **V. A. Smirnov,** Simplicial and operad methods in algebraic topology, 2001
197 **Kenji Ueno,** Algebraic geometry 2: Sheaves and cohomology, 2001
196 **Yu. N. Lin′kov,** Asymptotic statistical methods for stochastic processes, 2001
195 **Minoru Wakimoto,** Infinite-dimensional Lie algebras, 2001
194 **Valery B. Nevzorov,** Records: Mathematical theory, 2001
193 **Toshio Nishino,** Function theory in several complex variables, 2001
192 **Yu. P. Solovyov and E. V. Troitsky,** C^*-algebras and elliptic operators in differential topology, 2001
191 **Shun-ichi Amari and Hiroshi Nagaoka,** Methods of information geometry, 2000
190 **Alexander N. Starkov,** Dynamical systems on homogeneous spaces, 2000
189 **Mitsuru Ikawa,** Hyperbolic partial differential equations and wave phenomena, 2000
188 **V. V. Buldygin and Yu. V. Kozachenko,** Metric characterization of random variables and random processes, 2000
187 **A. V. Fursikov,** Optimal control of distributed systems. Theory and applications, 2000
186 **Kazuya Kato, Nobushige Kurokawa, and Takeshi Saito,** Number theory 1: Fermat's dream, 2000
185 **Kenji Ueno,** Algebraic Geometry 1: From algebraic varieties to schemes, 1999
184 **A. V. Mel′nikov,** Financial markets, 1999
183 **Hajime Sato,** Algebraic topology: an intuitive approach, 1999
182 **I. S. Krasil′shchik and A. M. Vinogradov, Editors,** Symmetries and conservation laws for differential equations of mathematical physics, 1999
181 **Ya. G. Berkovich and E. M. Zhmud′,** Characters of finite groups. Part 2, 1999
180 **A. A. Milyutin and N. P. Osmolovskii,** Calculus of variations and optimal control, 1998
179 **V. E. Voskresenskiĭ,** Algebraic groups and their birational invariants, 1998
178 **Mitsuo Morimoto,** Analytic functionals on the sphere, 1998
177 **Satoru Igari,** Real analysis—with an introduction to wavelet theory, 1998
176 **L. M. Lerman and Ya. L. Umanskiy,** Four-dimensional integrable Hamiltonian systems with simple singular points (topological aspects), 1998
175 **S. K. Godunov,** Modern aspects of linear algebra, 1998
174 **Ya-Zhe Chen and Lan-Cheng Wu,** Second order elliptic equations and elliptic systems, 1998
173 **Yu. A. Davydov, M. A. Lifshits, and N. V. Smorodina,** Local properties of distributions of stochastic functionals, 1998
172 **Ya. G. Berkovich and E. M. Zhmud′,** Characters of finite groups. Part 1, 1998
171 **E. M. Landis,** Second order equations of elliptic and parabolic type, 1998
170 **Viktor Prasolov and Yuri Solovyev,** Elliptic functions and elliptic integrals, 1997

For a complete list of titles in this series, visit the
AMS Bookstore at **www.ams.org/bookstore/**.

Translations of

MATHEMATICAL MONOGRAPHS

Volume 204

Cohomological Analysis of Partial Differential Equations and Secondary Calculus

A. M. Vinogradov

American Mathematical Society
Providence, Rhode Island

EDITORIAL COMMITTEE

AMS Subcommittee
Robert D. MacPherson
Grigorii A. Margulis
James D. Stasheff (Chair)
ASL Subcommittee Steffen Lempp (Chair)
IMS Subcommittee Mark I. Freidlin (Chair)

А. М. Виноградов

Когомологический анализ дифференциальных уравнений в частных производных и вторичное исчисление

Translated from the original Russian manuscript by Joseph Krasil′shchik

2000 *Mathematics Subject Classification.* Primary 35A30, 37K10;
Secondary 37Jxx, 58J10.

ABSTRACT. This book deals with the principles of a new theory, which plays the same role in general (nonlinear) systems of partial differential equations as algebraic geometry does in systems of algebraic equations. Starting with the fundamentals of the classical geometry of partial differential equations in the spirit of S. Lie, rewritten in terms of modern theory of jet spaces, and incorporating some important impulses from the theory of integrable systems, the exposition eventually arrives at first elements of Secondary Calculus. A systematic use of the coordinate-free language of differential calculus over commutative algebras allows us to reveal basic underlying structures of the developed theory and is a characteristic feature of the presented approach.

The book is intended for mathematicians (pure and applied), specialists in mathematical and theoretical physics, mechanics of continuous media, etc., interested in modern approaches to partial differential equations. The text is also comprehensible to advanced graduate students in these fields.

Library of Congress Cataloging-in-Publication Data

Vinogradov, A. M. (Aleksandr Mikhailovich)

Cohomological analysis of partial differential equations and secondary calculus / A.M. Vinogradov.

p. cm. — (Translations of mathematical monographs, ISSN 0065-9282 ; v. 204)

Translated from the original Russian manuscript.

Includes bibliographical references and index.

ISBN 0-8218-2922-X (alk. paper)

1. Differential equations, Nonlinear. 2. Geometry, Differential. 3. Homology theory. I. Title. II. Series.

QA377.V54 2001
515′.353—dc21 2001046087

Copying and reprinting. Individual readers of this publication, and nonprofit libraries acting for them, are permitted to make fair use of the material, such as to copy a chapter for use in teaching or research. Permission is granted to quote brief passages from this publication in reviews, provided the customary acknowledgment of the source is given.

Republication, systematic copying, or multiple reproduction of any material in this publication is permitted only under license from the American Mathematical Society. Requests for such permission should be addressed to the Assistant to the Publisher, American Mathematical Society, P. O. Box 6248, Providence, Rhode Island 02940-6248. Requests can also be made by e-mail to `reprint-permission@ams.org`.

© 2001 by the American Mathematical Society. All rights reserved.
The American Mathematical Society retains all rights
except those granted to the United States Government.
Printed in the United States of America.

∞ The paper used in this book is acid-free and falls within the guidelines
established to ensure permanence and durability.
Visit the AMS home page at URL: `http://www.ams.org/`

10 9 8 7 6 5 4 3 2 1 06 05 04 03 02 01

Contents

Foreword vii

Introduction ix

Chapter 0. From Symmetries of Partial Differential Equations to Secondary Calculus 1

0.1. What are symmetries of partial differential equations, and what are partial differential equations themselves? 2

0.2. Jets 3

0.3. Higher-order contact structures 5

0.4. Differential equations are diffieties 6

0.5. What are symmetries of partial differential equations? 8

0.6. Infinitesimal symmetries of partial differential equations are secondary quantized vector fields 10

0.7. Digression: on symmetries of partial differential equations 12

0.8. Secondary ("quantized") functions 14

0.9. Higher-order scalar secondary ("quantized") operators 16

0.10. Secondary ("quantized") differential forms and $\mathcal{C}$-spectral sequences 19

0.11. How does the $\mathcal{C}$-spectral sequence work? 21

Chapter 1. Elements of Differential Calculus in Commutative Algebras 25

1.1. General preliminaries 25

1.2. Adjoint operators 31

1.3. Spencer complexes and the Green formula 38

1.4. Quadratic Lagrangians and the Euler operator 44

1.5. Conservation laws in the linear theory 47

1.6. Automorphisms and the linear Noether theorem 51

1.7. Concluding remarks 55

Chapter 2. Geometry of Finite-Order Contact Structures and the Classical Theory of Symmetries of Partial Differential Equations 57

2.1. Necessary facts from the geometry of jet spaces 57

2.2. The structure of U-transformations 66

2.3. Infinitesimal automorphisms of the Cartan distribution 72

2.4. The structure of automorphisms of the Cartan distribution on the manifolds $J^k(E,m)$, $k<\infty$ 80

2.5. Classical theory of symmetries of partial differential equations 86

2.6. Concluding remarks 93

Chapter 3. Geometry of Infinitely Prolonged Differential Equations and Higher Symmetries 95
3.1. Geometry of infinitely prolonged equations, and related differential calculus 95
3.2. Horizontalization operation, the structure of Cartan submodules and of $\mathcal{C}$-differential operators 100
3.3. Higher infinitesimal symmetries of differential equations 107
3.4. The structure of $\mathcal{C}$-transformations 113
3.5. Examples of diffieties 122
3.6. Concluding remarks 125

Chapter 4. $\mathcal{C}$-spectral Sequence and Some Applications 127
4.1. Spencer complexes and the Green formula in the $\mathcal{C}$-theory 127
4.2. Nonlinear Lagrangian formalism 133
4.3. The $\mathcal{C}$-spectral sequence 137
4.4. The $\mathcal{C}$-spectral sequence of infinitely prolonged equations 155
4.5. Applications to the Lagrangian formalism with constraints and to the theory of conservation laws 172
4.6. Characteristic classes 182
4.7. Concluding remarks 187

Chapter 5. Introduction to Secondary Calculus 189
5.1. First idea of Secondary Calculus 189
5.2. Secondary linear algebra 196
5.3. Secondary modules 203
5.4. Secondary linear differential operators 210
5.5. $\mathcal{C}_\nabla$-spectral sequence 213
5.6. Secondary multivector-valued differential forms 221
5.7. The secondary Frölicher–Nijenhuis formalism 227
5.8. Concluding remarks 233

Bibliography 237

Foreword

This book originates from a text I wrote at the beginning of the 1980s, when the existence of Secondary Calculus became evident to me. The subsequent exploration of a possibility to publish a polished version of this text as a book in my Soviet homeland showed that it would be hardly possible in a reasonably near future. This is why the text was divided into several parts published as separate papers. The fact that they are parts of one construction was lost to a greater extent during this operation.

After my move to Italy in 1990, I was strongly and permanently pressed by Natasha, my wife, to publish the old manuscript in English. Current research prevented it for a long time, till the moment when Joseph Krasil′shchik became her ally and proposed to translate the text into English. After the translation was finished and I wrote the necessary new chapter containing recent developments in the field, it has become rather clear that the whole text must be reelaborated. This is finally done, and I am very grateful to Joseph for his patience during this not very pleasant process of repeated modifications which required his additional efforts to keep the text smooth.

My pleasant obligation is to acknowledge that this work was partially supported by the Italian *Ministero dell'Universitá e della Ricerca Scientifica e Tecnologica* and by *Istituto Nazionale di Fisica Nucleare*. Both the author and the translator were also supported by the INTAS grant 96-0793.

Introduction

1. This book deals with principles of a theory playing the same role for general systems of (nonlinear) partial differential equations as algebraic geometry does for algebraic equations. The objects studied in algebraic geometry are varieties of all solutions to a given system of algebraic equations. The history, from Descartes to Grothendieck's schemes, of what should be called the "variety of all solutions" shows that it is not a very simple problem. It becomes even more complicated if one tries to answer the question how the "variety of all solutions to a given system of partial differential equations" should be formalized properly. A specific feature of this problem is to understand how to match, in a reasonable way, solutions with different domains. This aspect is of crucial importance in physics, where a possibility to localize the theory must be guaranteed. It is quite clear that these requirements cannot be satisfied by constructions built with standard means of functional analysis, topology, etc. One of the goals of this book is to introduce the concept of a *diffiety* (*differential variety*), a new mathematical object, which plays the same role in the theory of (nonlinear) partial differential equations as *affine algebraic varieties* do in the theory of algebraic equations. Rather lengthy preparations are needed to organize properly preliminary material coming from various areas, but after that the results proved bring inevitably to this notion.

Diffieties are, generally infinite-dimensional, manifolds carrying a structure that may be called the *infinite-order contact structure*, and are locally equivalent to infinite prolongations of differential equations. Due to this structure, a very special kind of differential calculus, called *Secondary Calculus*, can be developed on a diffiety. Various natural characteristics of a diffiety and, consequently, of the corresponding systems of partial differential equations, are expressed in terms of Secondary Calculus and vice versa. Secondary Calculus is distinguished by its completely cohomological nature. In other words, it operates with quantities that are certain cohomology classes of differential complexes together with the operators connecting them. From the point of view of category theory, the objects in Secondary Calculus are homotopy types of special differential complexes over diffieties while the morphisms are suitable homotopy classes of differential cochain mappings. From the technical point of view, the principal ingredients of Secondary Calculus are a rather specific mixture of commutative and homological algebra with differential geometry. At first glance, it looks rather surprising that a natural language one needs to deal with systems of nonlinear partial differential equations is of this kind. However, now we can be sure that quantum physics and, above all, quantum field theory and its generalizations find their natural mathematical background in Secondary Calculus.

We use the adjective "secondary" to emphasize that the Secondary Calculus is a natural language programmed in terms of the classical, i.e., primary Calculus. Having in mind its relationship with quantum physics, it could be also called "Quantized", or "Quantum", Calculus. These words, however, were too abused recently to be mentioned once again in a completely different context. Nevertheless, we shall do it just once to state that the main goal of this book is to build a provisional bridge between classical and quantum mathematics. The author recognizes all deficiencies of the presented construction but hopes that it has at least one merit: to serve as a mean for reasonably courageous adventures in crossing the rifts.

2. This book consists of the main text preceded by introductory Chapter 0. The basic text is written in a standard mathematical style, which leaves no room for motivations and informal arguments. But this is hardly justified in the situation the book deals with and a compensation is given in the introductory chapter.

We do not touch two important aspects of Secondary Calculus in the book. One of them concerns *nonlocal quantities*, i.e., the quantities that depend not only on original unknown functions and their arbitrary derivatives, but also on variables obtained by applying inversions of differential operations (for instance, integrals). An introduction to this topic is given in the survey article by Krasil′shchik and the author *Nonlocal trends in the geometry of differential equations* [**74**] (see also [**19**]). The central concept of the nonlocal theory is that of a *covering* in the category of differential equations. The notion was introduced by the author in [**122**] and is discussed quite briefly in Chapter 5 of this book. Another aspect is the graded (or "super") version of Secondary Calculus. The first results in this direction were obtained by Krasil′shchik and are presented in his joint book with Kersten [**70**].

The theory presented in this book has a wide range of applications. The simplest of them are related to computations and the subsequent use of higher symmetries and conservation laws of a given system of partial differential equations. A relatively short introduction to this subject can be found in [**127**], while the collection [**139**] shows the corresponding computational techniques "in action". The book *Symmetries and Conservation Laws for Differential Equations of Mathematical Physics* [**19**] contains a systematic exposition of these matters and may be regarded as a collection of examples illustrating the general theory exposed in our book. It is strongly recommended to the readers interested in seeing how this theory is used in actual applications.

It should be stressed from the very beginning that both differential operators and all natural constructions relating them are treated in the main text in an algebraic coordinate-free manner. This simplifies computations considerably and makes it possible to single out some fundamental structures of the theory, which could hardly be observed in the framework of the standard approach. Although adaptation to this nontraditional approach will probably require some additional efforts from the readers, they will be rewarded for their work eventually.

Our exposition is self-contained in the sense that all preliminary facts and constructions are introduced. Further details concerning the preliminary material can be found in the book [**71**], while the book [**19**], as already mentioned, can be viewed as a collection of particular examples illustrating the general theory. An elementary introduction to differential calculus over commutative algebras can be found in [**95**].

3. The text is organized as follows.

Chapter 0 is an informal introduction to the basic concepts of the modern theory of partial differential equations. Our main goal there is to show why these fundamentals are unavoidable from both mathematical and physical points of view.

In the first chapter all necessary prerequisites of differential calculus in commutative algebras are presented. Further exposition is based on this language, which is absolutely indispensable to reveal basic structures of the theory. This is exactly the point at which our approach starts to diverge rapidly from the one commonly adopted in mathematical physics.

Fundamentals of linear Lagrangian formalism are then developed using this language. This allows us to discover a natural cohomological structure of the formalism itself and some related constructions. Among the latter, conservation laws are to be mentioned first. The related cohomology comes from some multianalogs of Diff-Spencer complexes. A version of this kind of cohomology will play an important role in Chapters 4 and 5 in the context of Secondary Calculus.

Naturalness of a purely algebraic framework for Lagrangian formalism given in this chapter is well confirmed by the fact that it is generalized straightforwardly to any reasonably noncommutative situation (see [**116**]).

By contrast, the second chapter is of purely geometric nature and deals with finite-order jet spaces and general systems of (nonlinear) partial differential equations interpreted as submanifolds in these spaces. Exposition is concentrated there around various kinds of automorphisms of higher-order contact structures naturally interpreted as symmetries of partial differential equations. It is shown that the symmetry group is not enlarged when the order of jets, i.e., of involved derivatives, increases. This means that the classical symmetry theory, originated by Lie and developed later by Bäcklund and others, has no chance to incorporate new "experimental data" emerging in the theory of integrable systems. At the same time, this compels us to pass to infinite jets in search for a suitable extension for the concept of a symmetry of nonlinear partial differential equations.

This step is rewarded later in Chapter 3 where, among other constructions, the theory of *higher symmetries* for general systems of partial differential equations is developed. This is done by studying the geometry of infinite-order jet spaces and infinitely prolonged differential equations. These are the simplest examples of diffieties. It turns out that infinite-order infinitesimal contact transformations form a much larger Lie algebra than the classical ones. The group of infinite-order contact transformations also becomes larger, but to much smaller extent than infinitesimal transformations. Even more, in some situations it does not change at all. This, at first sight disappointing, discrepancy, which was mysterious for a long time, has a natural explanation: infinite-order infinitesimal contact transformations are, by their nature, cohomology classes of vector fields. These cohomology classes are exactly the *secondary vector fields* as explained in Chapter 5. But, similarly to quantum particles, secondary vector fields do not have trajectories. This is the main philosophical lesson of Chapter 3.

In this chapter, elements of the differential calculus agree with the infinite-order contact structure ($\mathcal{C}$-differential operators, etc.) are also elaborated. The advantages of the algebraic setting for differential calculus given in Chapter 1 become evident when we pass to the infinite jet space. It allows us to avoid difficulties typical for infinite-dimensional situations, and also improper temptations.

Foundations of the higher symmetry theory, as presented in Chapter 3, were developed by the author in 1976–77 (see [**121, 137, 124, 142**]) and appeared later in a number of publications (see, for instance, [**54, 96, 146**]) presented in a coordinatewise way. Computational aspects of this theory were, for the first time, tested extensively in the collection [**139**]. The book [**19**] contains an updated and detailed exposition of the basic theory together with the recent ramifications and applications. It is very impressive that the first attempt to interpret infinite-order contact transformations as a natural extension of Lie's theory was done by Bäcklund in [**11**] about 100 years ago.

The central cohomological construction for the theory of partial differential equations, the *$\mathcal{C}$-spectral sequence* associated to a given system of equations, is introduced earlier in Chapter 4. The first term of this sequence is most important, since Lagrangians (actions), conservation laws, symplectic structures in the field theory, etc., are parts of it. It allows one to answer such questions as what is the complete list of conservation laws admitted by a given system of partial differential equations, or could such a system be obtained from variational principles, and so on. When the number of independent variables becomes equal to zero, the $\mathcal{C}$-spectral sequence reduces to the standard de Rham complex. This is one of the reasons why later, in Chapter 5, we call its elements *secondary differential forms.*

It is more difficult to characterize the second term of the $\mathcal{C}$-spectral sequence in the standard language. For instance, its elements might be called characteristic classes of cobordisms composed of solutions of the equation in question. In particular, various well-known characteristic classes can be obtained in this way, i.e., as elements of the second term of the $\mathcal{C}$-spectral sequence associated with some *universal* equations. Alternatively, they may be called *secondary de Rham cohomology classes*, since in the case when the number of independent variables equals zero, the construction gives the standard de Rham cohomology. Classical ("primary") mathematics gives no hints on how to call higher terms of the $\mathcal{C}$-spectral sequence.

The main technical problem discussed in Chapter 4 is how to compute the first term of the $\mathcal{C}$-spectral sequence. A general method based on a direct use of the Spencer type cohomology is developed there. It gives an exhaustive answer for infinite-order jet spaces, and a rather efficient computational algorithm for infinitely prolonged equations is constructed on its basis. In particular, it allows one to describe the domain in the standard (p, q)-diagram where all nontrivial terms are located (the 2- and p-Line Theorems). A formally equivalent but much more manageable method is discussed in Chapter 5.

As regards the second term of the $\mathcal{C}$-spectral sequence, an analogy with usual vector fields and differential forms allows one to suggest some simple homotopy techniques that work well for a large class of equations. On the other hand, this analogy leads to the first idea of Secondary Calculus as is explained in Chapter 5.

Some applications to the theory of conservation laws, Lagrangian formalism with constraints and other problems are discussed at the end of the chapter. Even in standard situations, the results obtained there reveal new important features. For instance, the method of "generating functions" developed with computational purposes shows that higher symmetries and conservation laws of a (nonlinear) system of partial equations are solutions of two mutually adjoint $\mathcal{C}$-differential equations. These equations coincide for Euler–Lagrange equations. This clarifies the nature of the Noether theorem and allows us to generalize it to a much larger class of equations. The computational techniques for conservation laws of concrete equations can be found in [**127**]. The book [**19**] is once again strongly recommended to the readers interested in these aspects.

The results contained in Chapter 4 were obtained by the author mainly in 1976–77 and published in two short notes [**129**] and [**121**]. However, it turned out to be impossible to publish a detailed exposition of these results in the USSR then. They appeared later, with almost six year delay, in the USA [**134**]. Meanwhile Tsujishita published a detailed description of a simplified version of the $\mathcal{C}$-spectral sequence called the *variational bicomplex* [**112**]. Tsujishita's approach requires a fibered structure of the equation in question, which furnishes the second differential coming from the corresponding Leray–Serre spectral sequence. For this reason, problems such as that of minimal surfaces, etc., cannot be treated directly with this method. Moreover, his approach appeals to a coordinate description rather than the algebraic conceptual language and thus hides basic natural structures of the theory and, in particular, the perspective of Secondary Calculus. For the same reasons, Tsujishita's approach becomes more popular among specialists in mathematical physics. On the other hand, in the same paper Tsujishita discovered important links to the theory of characteristic classes. Perhaps, the difficulties related to a systematic use of the natural algebraic language of differential calculus in the context under consideration gave rise to some works of a coordinate-oriented nature (see, for instance, books [**3**] and [**146**]). The area of mathematics that was formed around the variational bicomplex was called by Tsujishita [**110**] "formal geometry of differential equations". Since then, a flavor of formality of the subject was maintained by several authors. Up to a certain extent, this point of view can be justified, but for Secondary Calculus in general (see Chapter 5) it is no longer true and, even more, would be counterproductive in a strategic perspective.

The concluding Chapter 5 is of a synthetic nature. Its main goal is, first of all, to show how the results of the preceding chapters are self-organized naturally into a new Calculus called Secondary, to present further structures of Secondary Calculus and to provide some perspectives. Proofs of the results reported in this chapter are, as a rule, omitted.

The notion of a diffiety introduced earlier in Chapter 3 is used systematically. Infinitely prolonged equations are the simplest examples of this concept. Diffieties may also be called "secondary manifolds" and as such are the stages where the performances of Secondary Calculus take place. Then we pass to "secondary vector bundles" and "secondary modules" over "secondary smooth function algebras" on which "secondary differential operators" act. All these objects are homotopy types of special differential complexes over the base diffiety $\mathcal{O}$. For example, the secondary smooth function algebra on $\mathcal{O}$ is the homotopy type of the algebra of graded differential operators acting on the horizontal de Rham complex over $\mathcal{O}$. Homotopy classes of the columns of the zero term of the associated $\mathcal{C}$-spectral sequence are secondary differential forms. Secondary differential operators are defined as suitable homotopy classes of differential cochain mappings of complexes representing secondary modules, etc.

When we take for the base diffiety all coverings over an infinitely prolonged equation, the cohomology of the corresponding complexes contains exhaustive information concerning the original equation. This is why Secondary Calculus is much more than just a formal theory (see above). Thus, the problem of efficient computations of the cohomology becomes central in Secondary Calculus. It is remarkable that the $\mathcal{C}$-spectral sequence method can be extended quite naturally to general secondary modules. This generalization is described together with a very important improvement of the original author's method (Chapter 4) due to the works by M. Marvan, T. Tsujishita, D. Gessler, and A. Verbovetsky, where the *compatibility complex* was used instead of the related Spencer type cohomology.

Another topic to which considerable attention is paid in Chapter 5 is the "secondarization problem", i.e., the problem of finding secondary counterparts for all ingredients of the standard (primary) Calculus. It is a rather delicate problem as one could see from the preceding discussion, and of rather great importance as a whole. In spite of the fact that an almost algorithmic procedure can be suggested to solve it now, a detailed elaboration of nonstandard cohomological techniques and, not infrequently, remaking of well-known pieces of the standard Calculus in order to bring them to a *secondarizable* form is required. Such an enterprise goes far beyond the bounds of this book. For this reason we restrict ourselves to a description of secondary differential operators, some elements of the secondary Hamiltonian formalism and, in more details, of secondary multivector-valued differential forms including the secondary Schouten–Nijenhuis and Frölicher–Nijenhuis bracket formalism. These concrete constructions are sufficiently instructive and lead to numerous interesting applications as well. As an illustration, we mention here that the Euler operator, associating to a given Lagrangian the corresponding Euler–Lagrange equations, turns out to be a first-order secondary differential operator while its formal coordinate description presents it as an infinite-order operator.

It should also be stressed that Secondary Calculus has various levels corresponding to the second and higher order terms of the $\mathcal{C}$-spectral sequence. We do not touch upon these aspects here.

4. Some general remarks concerning the actual status of the theory presented in this book are to be added.

Algebraic geometry, i.e., the general theory of systems of (nonlinear) algebraic equations, is one of the most respected and "noble" areas of the modern pure mathematics. On the contrary, according to the 1991 AMS Subject Classification, a similar theory concerning (nonlinear) partial differential equations does not exist at all. Moreover, the common opinion is that a substantive general theory of (nonlinear) partial differential equations even cannot exist. Thus, the study of a concrete system of (nonlinear) differential equations of a particular applied interest is left to experts in the corresponding special fields of mechanics, mathematical or theoretical physics, etc.

It is also commonly forgotten that at the end of the 19th century and the beginning of the 20th the situation was completely opposite. The great symphony by Sophus Lie, who laid the first foundation stones in the building of the general theory of nonlinear partial differential equations, was highly recognized as a "noble" part of pure mathematics. It attracted attention of many distinguished mathematicians of that epoch. But, quite surprisingly, at least at the first glance, this glorious period suddenly ended after the World War I, which seemingly completely destroyed the great nonlinear culture of old masters. From Lie's symphony, only the scores of Lie groups and Lie algebras were extracted, subsequently elaborated and developed in thousands of works. Dynamics and causes of these events are yet to be analyzed. Two of them are clear, however. First, a very attractive temptation to linearize real life was satisfied by functional analysis and became dominant in almost everything that concerns partial differential equations. Second, a consistent interaction of many mathematical structures is indispensable when studying such equations, even linear. A lot of these structures, and especially those related to the general homological culture have not even been discovered then. The resulting underestimation and underdevelopment of the general theory of nonlinear partial differential equations is in the origin of a difficult and tortuous progress in many key areas of mathematical natural sciences.

Secondary Calculus is a natural continuation of the opera by Sophus Lie and his followers, enriched by the results of the formal theory of partial differential equations. It is worth emphasizing that cohomological methods in the general theory of differential equations appeared for the first time exactly in the context of the formal theory due to D. Spencer and Goldschmidt (see, for instance, [**44, 45, 103**]) as Spencer's resolution, Spencer's cohomology, etc. These authors discovered a deep cohomological nature of the old formal theory developed by P. Riquier, M. Janet and others. Since the beginning of the 1970s the Moscow school in geometry of partial differential equations has brought the above theories to a natural synthesis on the basis of the algebraic "perestroika" of Calculus [**136, 133**], and this revealed the first structures of Secondary Calculus at the beginning of the 1980s.

At present, it is clear that with Secondary Calculus the general theory of (nonlinear) partial differential equations finds its solid natural basis. Even more, Secondary Calculus showed itself to be a unifying theory, whose potential applications range from arithmetic algebraic geometry to modern theories of elementary particles. This features could be hardly expected from the very beginning. Of course, a substantial systematic preliminary work has to be done to get a realization of the opened perspectives. Thus, the author would be very satisfied if this book drew the interest of young mathematicians and maybe physicists towards this new nonlinear mathematics.

Since the works by Elie Cartan summarized in his book [**29**] (see also [**24**] for a modern formulation of Cartan's approach), there exists a tradition to treat partial differential equations as exterior differential systems (EDS) [**79, 99**]. This point of view is useful for a variety of problems. EDS are geometric structures very close to PDE, and for this reason many topics considered in this book can be brought into the context of EDS. The reader will find an instructive example of this kind in the papers by Bryant and Griffiths [**25, 26**], as well as in [**27, 28**].

CHAPTER 0

From Symmetries of Partial Differential Equations to Secondary Calculus

An ordinary man wonders marvelous things; a wise man wonders the common places.

Confucius

Since the subject of this book is a rather new and perhaps not very standard area of pure contemporary mathematics, a preliminary glimpse trip through its basic ideas and results would help the reader to capture from the very beginning the necessary polarization of mind to enter it. This is the goal of the introductory chapter, which reproduces, in a very straightened and idealized manner, the author's experience in searching for *natural* answers to the following two strongly related very general questions:

(i) what are the basic structures of the theory of nonlinear partial differential equations,

and

(ii) what is the natural mathematical background for quantum field theory and its recent generalizations?

It may seem that these questions are too general to be the object of a serious study. In fact, to follow this direction, one must wonder what are partial differential equations, what are their solutions, how to "observe" these solutions and many other similar "absolutely idiotic" things. Such questions are, however, neither ad hoc, nor metaphysical ones. In this chapter we show informally in what context they arise very naturally and what kind of results one could get by answering them in a systematic way.

Notice also that when one has to face a sufficiently complicated new scientific area, the best way to proceed, and this is one of the most important common places to respect, is to start with a study of symmetric models. In the context we are interested in this means to study symmetries of general systems of partial differential equations. This is why it is chosen as the departure point in our subsequent analysis.

The topics of this chapter will be discussed later with all necessary details.

0.1. What are symmetries of partial differential equations, and what are partial differential equations themselves?

To answer conceptually these two questions is not so easy as it might appear at first glance. In fact, consider a partial differential equation, say,

$$F(x, u, \dots, u_{(k)}) = 0, \tag{0.1}$$

where $x = (x_1, \dots, x_n)$, $u = (u^1, \dots, u^m)$ and $u_{(l)}$ stands for the totality of all l-th order derivatives of the dependent variables u with respect to the independent variables x, and ask what the symmetries of (0.1) are. If one treats (0.1) to be a relation between two different groups of variables, dependent and independent ones, then it looks natural to define a symmetry of (0.1) to be a transformation of the dependent variables, and separately, of the independent ones, that preserves this relation. More exactly, it is suggested to call symmetries of (0.1) transformations of the form

$$(x, u) \mapsto (\tilde{x}(x), \tilde{u}(u)) \tag{0.2}$$

which, being extended canonically to derivatives, preserve (0.1).

This historically first proposal was followed by the next two, which enlarge the group of transformations (0.2), first, to

$$(x, u) \mapsto (\tilde{x}(x), \tilde{u}(x, u))$$

and then to

$$(x, u) \mapsto (\tilde{x}(x, u), \tilde{u}(x, u)). \tag{0.3}$$

Unfortunately, all these definitions are based on an *ad hoc* choice of transformations to be taken as symmetries. So, going this way one can never be sure that a new "Ansatz", like (0.2) or (0.3), is the "true" final definition. For example, the transformations (0.3) were, historically, followed by the famous contact transformations discovered by Sophus Lie. Namely, he suggested to regard as symmetries of (0.1) transformations of the form

$$(x, u, u_x) \mapsto (\tilde{x}, \tilde{u}, \tilde{u}_x),$$

where $\tilde{x} = \tilde{x}(x, u, u_x)$, $\tilde{u} = \tilde{u}(x, u, u_x)$, $\tilde{u}_x = \tilde{u}_x(x, u, u_x)$, which preserve (0.1) together with the equation $du - \sum u_{x_i} dx_i = 0$.

But why cannot somebody find something else? Lie himself did not answer this question. But he understood deeply that an expression of the form (0.1) is not a sovereign object to be transformed directly, but only a label of it. In particular, his discovery of contact transformations was based on a geometric interpretation of what stands behind labels of the form $f(x, u, u_x) = 0$ (for a more detailed discussion see [**127**]).

In general, it is clear that symmetries of a mathematical object are to be its invertible morphisms ("transformations") to itself. Therefore, one can see from the above discussion that the observed uncertainty in defining symmetries of PDE's comes from the fact that we do not really know what, properly, partial differential equations are.

The last question is, in fact, neither too absurd nor so innocent as it might appear at first glance. On the contrary, having answered it, we shall gain much more than just the true concept of symmetry for PDE's.

In what follows, we sometimes call *labels* what is commonly called partial differential equations, i.e., expressions of the form (0.1), in order to distinguish the common and the conceptual use of this term.

Now we pass to the necessary preliminaries.

0.2. Jets

Note that labels of differential equations are "algebraic" relations between independent and dependent variables and derivatives of the latter up to a certain order. Thus, each label defines a submanifold in the manifold, one of the possible local charts of which is formed by all these variables and derivatives. These manifolds are called jet spaces (or manifolds), and we are interested in them to reinterpret labels in an invariant coordinate-free form. In what follows, the necessary definitions and elementary facts on jets are collected.

Let us fix an $(n+m)$-dimensional manifold E, $m \geq 0$, and a nonnegative integer n. This is done to emphasize our intention to study n-dimensional submanifolds of E. If L, L' are two of them and $a \in L \cap L'$, we say that they have the same k-th order jet at $a \in E$ if they are tangent to one another up to order k at this point. Then k-jets are equivalence classes of n-dimensional submanifolds of E with respect to the k-th order tangency. We denote by $[L]_a^k$ the k-th order jet of the n-dimensional submanifold $L \subset E$ at $a \in L$. The totality of k-th order jets of all possible n-dimensional submanifolds $L \subset E$ at all points $a \in E$ is denoted by $J^k(E,n)$. Supplied with the natural structure of a smooth manifold, it is called the k-th order jet space of n-dimensional submanifolds of E.

REMARK. If E has a fibered structure, say, $\pi \colon E \to M$, $\dim M = n$, then one can consider a special class of n-dimensional submanifolds of E that are graphs of local sections of π. The k-th order jets of these graphs constitute an open and everywhere dense subset of $J^k(E,n)$ called the k-th order jet space of sections of π and denoted by $J^k(\pi) \subset J^k(E,n)$.

EXAMPLE ($k = 0$). Evidently, every two n-dimensional submanifolds of E passing through a point $a \in E$ are tangent to one another with order zero at a. Therefore, there exists only one 0-jet at a and one can identify $J^0(E,n)$ with E.

EXAMPLE ($k = 1$). Two n-dimensional submanifolds of E are tangent to one another with order one at a point a if and only if they have the same tangent space at a. Hence first-order jets at $a \in E$ can be identified with n-dimensional linear subspaces of the tangent space T_aE of E at a.

We emphasize that all these definitions are also valid for $k = \infty$. In particular, $J^\infty(E,n)$ is well defined as a set. However, to introduce a smooth structure in it requires some care. This will be done below.

Natural mappings

$$\alpha_{k,l} \colon J^k(E,n) \to J^l(E,n), \quad [L]_a^k \mapsto [L]_a^l,$$

with $\infty \geq k \geq l \geq 0$, combine jet spaces of different order into a family. In particular, they form the sequence

$$E = J^0(E,n) \xleftarrow{\alpha_{1,0}} J^1(E,n) \xleftarrow{\alpha_{2,1}} \cdots \xleftarrow{\alpha_{k,k-1}} J^k(E,n) \xleftarrow{\alpha_{k+1,k}} \cdots \tag{0.4}$$

whose inverse limit coincides with $J^\infty(E,n)$.

For a given $L \subset E$, $\dim L = n$, we have the mapping

$$j_k(L)\colon L \to J^k(E,n), \quad L \ni a \mapsto [L]_a^k \in J^k(E,n).$$

A function φ on $J^k(E,n)$ is said to be smooth if and only if $\varphi \circ j_k(L) \in C^\infty(L)$ for every n-dimensional $L \subset E$. In this way we define the smooth function algebra on $J^k(E,n)$ and, therefore, a smooth manifold structure on it. Let us note that this definition works for $k = \infty$ as well.

It follows directly from the definitions that $j_k(L)$ is a smooth mapping and

$$L_{(k)} = \operatorname{im} j_k(L) \subset J^k(E,n)$$

is an n-dimensional smooth submanifold of $J^k(E,n)$. In addition, the identities

$$j_l(L) = \alpha_{k,l} \circ j_l(L), \quad k \geq l,$$

show that $\alpha_{k,l}$ is a smooth surjection and

$$\alpha_{k,l}^*\colon C^\infty(J^l(E,n)) \to C^\infty(J^k(E,n))$$

is an imbedding. For the smooth function algebra of $J^\infty(E,n)$ we take the direct limit of the following sequence of monomorphisms:

$$C^\infty(E) \xrightarrow{\alpha_{1,0}^*} C^\infty(J^1(E,n)) \xrightarrow{\alpha_{2,1}^*} \cdots \xrightarrow{\alpha_{k,k-1}^*} C^\infty(J^k(E,n)) \xrightarrow{\alpha_{k+1,k}^*} \cdots.$$

We use the standard notation $C^\infty(J^\infty(E,n))$ for this algebra.

It is convenient to shorten the notation as follows:

$$\mathcal{F}_k = \mathcal{F}_k(E,n) \stackrel{\text{def}}{=} C^\infty(J^k(E,n)), \quad 0 \leq k < \infty,$$
$$\mathcal{F} = \mathcal{F}(E,n) \stackrel{\text{def}}{=} C^\infty(J^\infty(E,n)).$$

We get the filtration

$$\mathcal{F}_0 \subset \mathcal{F}_1 \subset \cdots \subset \mathcal{F}_k \subset \cdots \subset \mathcal{F}$$

of the algebra $\mathcal{F}$ by identifying $\mathcal{F}_k$ with its image in $\mathcal{F}$ under the monomorphism $\alpha_{k,\infty}^*$. The differential calculus over the filtered algebra $\mathcal{F} = \{\mathcal{F}_k\}$ gives the necessary rigorous foundations for all our subsequent constructions (see [**71**]). In particular, it enables us to treat $J^\infty(E,n)$ in many aspects as a usual finite-dimensional smooth manifold.

A local chart on E is said to be *divided* if the corresponding coordinate functions are divided into two parts consisting of n and m functions, respectively. The elements of the first, say $x_1, \ldots, x_n$, are called "*independent*" and those of the second, say $u^1, \ldots, u^m$, "*dependent*" variables.

A divided chart on E generates a local chart on $J^k(E,n)$, $0 \leq k \leq \infty$, which consists of the functions

$$x_1, \ldots, x_n, u^1, \ldots, u^m, \ldots, u_\sigma^i, \quad |\sigma| \leq k,$$

where σ stands for a multiindex, say $\sigma = (i_1, \ldots, i_n)$, and $|\sigma| = i_1 + \cdots + i_n$. The function u_σ^i is defined by the condition

$$u_\sigma^i \circ j_k(L) = \frac{\partial^{|\sigma|} f^i}{\partial x_1^{i_1} \cdots \partial x_n^{i_n}}, \quad \forall L \subset E, \ \dim L = n,$$

where

$$u^i = f^i(x_1, \ldots, x_n), \quad i = 1, \ldots, m, \tag{0.5}$$

are local equations of L. We see that the jet manifolds $J^k(E,n)$ (or $J^k\pi$) are exactly those that carry natural coordinate systems composed of independent and dependent variables together with their derivatives.

A standard label of a system of partial differential equations looks as

$$F_j(x, u, \dots, u^i, \dots) = 0, \quad j = 1, \dots, l, \tag{0.6}$$

with $x = (x_1, \dots, x_n)$, $u = (u^1, \dots, u^m)$. It is natural to interpret (0.6) as local equations of a submanifold $\mathcal{Y} \subset J^k(E,n)$ for a suitable E. For instance, $u_t = u_{xx} + uu_x$ defines the hypersurface

$$u_{2,0} + uu_{1,0} - u_{0,1} = 0$$

in $J^3(\mathbb{R}^3, 2)$, where $\mathbb{R}^3 = \{(x,t,u)\}$ and $x_1 = x$, $x_2 = t$, $u^1 = u$ is a divided coordinate system in it.

This motivates the following coordinate-free version of the standard definition of partial differential equations.

DEFINITION. A submanifold $\mathcal{Y} \subset J^k(E,n)$ is called the *label of a system of partial differential equations* imposed on n-dimensional submanifolds of the given manifold E.

Let a submanifold $L \subset E$ be given by (0.5). Then the functions $f^i(x)$, $i = 1, \dots, m$, satisfy system (0.6) if and only if $L_{(k)} \subset \mathcal{Y}$. Thus, the manifold $L_{(k)}$ can be taken as the coordinate-free version of the notion of solution.

We remark that $J^k(\pi)$ can replace $J^k(E,n)$ in all previous considerations.

0.3. Higher-order contact structures

Jet spaces possess primordially a natural geometric structure, namely, the so-called *k-th order contact structure*, or the *Cartan distribution*. This means that to each point $\theta \in J^k(E,n)$ a linear subspace, say $\mathcal{C}_\theta$, of the tangent space $T_\theta(J^k(E,n))$ is assigned.

The subspace $\mathcal{C}_\theta$ can be defined as follows. First, let us introduce some special n-dimensional subspaces of $T_\theta(J^k(E,n))$ called *R-planes*. By definition, an R-plane at $\theta \in J^k(E,n)$ is a subspace of the form $T_\theta L_{(k)}$, where that $\theta = [L]^k_a$. It is important to emphasize that not every n-dimensional subspace of $T_\theta(J^k(E,n))$ is an R-plane, and more than one R-plane pass through θ if $k < \infty$ and $mn > 0$. Second, put

$$\mathcal{C}_\theta = \{\text{the linear span of all } R\text{-planes at } \theta\}.$$

A simple computation shows that

$$\dim J^k(E,n) = \dim T_\theta(J^k(E,n)) = m\binom{n+k}{k} + n, \quad 0 \le k < \infty,$$

and

$$\dim \mathcal{C}_\theta = m\binom{n+k-1}{k} + n, \quad 0 \le k < \infty.$$

In particular, $\dim \mathcal{C}_\theta \to \infty$ with $k \to \infty$.

EXAMPLE (classical contact structure). Consider the jet manifold $J^1(\mathbb{R}^{n+1}, n)$. In this case $m = k = 1$, $\dim J^1(\mathbb{R}^{n+1}, n) = 2n+1$, $\dim \mathcal{C}_\theta = 2n$, i.e., $\mathcal{C}_\theta$ is a hyperplane in $T_\theta(J^1(\mathbb{R}^{n+1}, n))$. Dividing standard Cartesian coordinates $(x_1, \dots, x_{n+1})$ in the space $\mathbb{R}^{n+1}$ into $x = (x_1, \dots, x_n)$ and $u = x_{n+1}$, we get the local chart

$(x_1, \ldots, x_n, u, p_1, \ldots, p_n)$ in the manifold $J^1(\mathbb{R}^{n+1}, n)$, where $p_i = u_{x_i}$. In this chart, $\mathcal{C}_\theta$ is given by the equation

$$du - \sum_{i=1}^{n} p_i \, dx_i = 0,$$

which is just the classical (= first-order) contact structure discovered by S. Lie.

In the general case, $\mathcal{C}_\theta$ is given by the system

$$du_\sigma^i - \sum_{j=1}^{n} u_{\sigma+1_j}^i \, dx_j = 0, \quad 1 \le i \le m, \; 0 \le |\sigma| < k, \tag{0.7}$$

where $\sigma + 1_j = (i_1, \ldots, i_j + 1, \ldots, i_n)$ for $\sigma = (i_1, \ldots, i_n)$.

The common feature of all k-th order contact structures, $0 < k < \infty$, is that they all are, in a sense, "completely nonintegrable" distributions. On the other hand, the infinite-order contact structure on $J^\infty(E, n)$ is completely integrable. This is easily seen from the fact that the Pfaff system (0.7) satisfies the Frobenius complete integrability conditions if $k = \infty$. Moreover, the Cartan distribution on infinite-order jet spaces is finite-dimensional, in spite of the fact that it is the inverse limit of the Cartan distributions on finite-order jet spaces whose dimension grows infinitely with the order of jets. In fact, the dimension of the infinite-order distribution is equal to n, i.e., $\dim \mathcal{C}_\theta = n$ for $\theta \in J^\infty(E, n)$. This means that for any $\theta \in J^\infty(E, n)$ there exists only one R-plane that coincides with $\mathcal{C}_\theta$.

All these remarks are given to emphasize that infinite-order jet spaces differ substantially from finite-order ones not only because they are infinite-dimensional. It looks paradoxical but the former are simpler than the latter when we study the properties of the basic structures they naturally carry.

0.4. Differential equations are diffieties

Having at our disposal infinite jet spaces, now we can answer the question: what are differential equations?

Let

$$D_i = \frac{\partial}{\partial x_i} + \sum_{k,\sigma} u_{\sigma+1_i}^k \frac{\partial}{\partial u_\sigma^k} \tag{0.8}$$

be the so-called *total derivative operator* with respect to x_i. Then the system of equations

$$F_j = 0, \; D_i F_j = 0, \quad 1 \le j \le l, \; 1 \le i \le n, \tag{0.9}$$

defines the first prolongation of (0.6). Evidently, system (0.9) is equivalent to (0.6) in the sense that both have the same set of solutions. System (0.9) is, in fact, a local coordinate description of a new label $\mathcal{Y}_{(1)} \subset J^{k+1}(E, n)$, which can be defined, purely geometrically, in terms of the contact structure on $J^k(E, n)$ only. More generally, one can define the s-th prolongation $\mathcal{Y}_{(s)} \subset J^{k+s}(E, n)$ of the label $\mathcal{Y} \subset J^k(E, n)$ by means of the system

$$F_j = 0, \; \ldots, \; D_\sigma F_j = 0, \ldots, \quad |\sigma| \le s, \tag{0.10}$$

with $D_\sigma = D_1^{i_1} \circ \cdots \circ D_n^{i_n}$ for $\sigma = (i_1, \ldots, i_n)$. In this definition, s can equal infinity. In this case we write $\mathcal{Y}_\infty$ instead of $\mathcal{Y}_{(\infty)}$. Clearly, $\mathcal{Y}_\infty \subset J^\infty(E, n)$. It is not difficult to see that

$$\alpha_{k+s,k+l}(\mathcal{Y}_{(s)}) \subset \mathcal{Y}_{(l)}$$

for every $s \geq t$. In particular, we have the following sequence:

$$\mathcal{Y} = \mathcal{Y}_{(0)} \xleftarrow{\alpha_{k+1,k}} \mathcal{Y}_{(1)} \xleftarrow{\alpha_{k+2,k+1}} \cdots \leftarrow \mathcal{Y}_{(s)} \xleftarrow{\alpha_{k+s+1,k+s}} \cdots,$$

whose inverse limit coincides with $\mathcal{Y}_\infty$.

It is useful to note that the ideal of the submanifold $\mathcal{Y}_\infty \subset J^\infty(E,n)$ is obtained from the functions F_i in system (0.6) by applying all arithmetical operations together with differentiations. In other words, this ideal is *differentially generated by the functions* F_i, similarly to the ideal of an algebraic manifold *algebraically generated* by the left-hand sides of the initial system of algebraic equations. This shows that infinite prolongations of differential equations are analogs of affine algebraic manifolds in the geometric theory of partial differential equations.

We define the smooth function algebra $\mathcal{F}(\mathcal{Y}_\infty)$ on $\mathcal{Y}_\infty$ to be the restriction of the smooth function algebra on $J^\infty(E,n)$ to $\mathcal{Y}_\infty$:

$$\mathcal{F}(\mathcal{Y}_\infty) \stackrel{\text{def}}{=} \mathcal{F}(E,n)|_{\mathcal{Y}_\infty}.$$

This algebra is filtered naturally by the subalgebras $\mathcal{F}_s(\mathcal{Y}_\infty)$, $s = 0,1,2,\dots$, where $\mathcal{F}_s(\mathcal{Y}_\infty)$ is the pullback of the algebra $C^\infty(\mathcal{Y}_{(s)})$ under the mapping $\alpha_{\infty,k+s}\colon \mathcal{Y}_\infty \to \mathcal{Y}_{(s)}$,

$$\mathcal{F}_0(\mathcal{Y}_\infty) \subset \mathcal{F}_1(\mathcal{Y}_\infty) \subset \cdots \subset \mathcal{F}_s(\mathcal{Y}_\infty) \subset \cdots \subset \mathcal{F}(\mathcal{Y}_\infty).$$

This enables us to develop the necessary differential calculus over $\mathcal{Y}_\infty$ as the calculus over the filtered commutative algebra $\mathcal{F}(\mathcal{Y}_\infty) = \{\mathcal{F}_s(\mathcal{Y}_\infty)\}$ (see [**71**] for details).

In addition, $\mathcal{Y}_\infty$ inherits the infinite-order contact structure from the ambient space $J^\infty(E,n)$. More precisely, if $\mathcal{Y}$ is a formally integrable system and $\theta \in \mathcal{Y}_\infty$, then $\mathcal{C}_\theta \subset T_\theta(\mathcal{Y}_\infty)$. Thus, $\mathcal{Y}_\infty$ is supplied canonically with a "contact structure", which is an n-dimensional distribution on it satisfying the Frobenius complete integrability conditions. Manifolds of the form $\mathcal{Y}_\infty$ considered together with the contact structure described above are local forms of the objects which we call *diffieties.*

DEFINITION. A manifold $\mathcal{O}$ supplied with an n-dimensional distribution is called a *diffiety* if it is locally of the form $\mathcal{Y}_\infty$.

This definition of a diffiety will be extended in the sequel (see the end of Section 0.7 and Section 5.1).

The number n is called the *diffiety dimension* of $\mathcal{O}$ and is denoted by $\operatorname{Dim}\mathcal{O}$. Of course, in general it differs from the ordinary dimension of $\mathcal{O}$, which is usually infinite.

Now we take as the leading principle that, conceptually,

systems of partial differential equations are diffieties.

In particular, $\mathcal{Y}_\infty$ is the object which stands behind the label $\mathcal{Y} \subset J^k(E,n)$ (or behind (0.6)).

It would be too naive to try to change the long-time established terminology by using the words "differential equations" in their new meaning, i.e., as a synonym of the word "diffiety". For this reason we retain their traditional meaning, keeping in mind, however, that it simply refers to a label. Thus, these words are to be replaced by "diffiety" when treating a conceptual problem. For example, the question: What are the symmetries of partial differential equations?, is to be formulated as

what are the automorphisms of diffieties?

A little later we shall see that it is not difficult to answer this question. But before that we must add some details to the general picture.

First of all, we remark that starting from (0.6), one can produce many new labels by transforming, for instance, dependent and independent variables, or passing to the corresponding first-order system, or prolonging it, etc. If $\mathcal{Y} \subset J^k(E, n)$ and $\mathcal{Y}' \subset J^{k'}(E', n)$ are two labels related to one another in this way, then $\mathcal{Y}_\infty = \mathcal{Y}'_\infty$. This shows clearly that $\mathcal{Y}$ and $\mathcal{Y}'$ are, in fact, different "labels" of the same object.

Now we have to interpret in terms of diffieties what are solutions of partial differential equations. The concept of integral submanifold of a diffiety gives us the answer. More precisely, let $\mathcal{O}$ be a diffiety and $\operatorname{Dim} \mathcal{O} = n$, i.e., $\dim \mathcal{C}_\theta = n$ for every $\theta \in \mathcal{O}$, where $\theta \mapsto \mathcal{C}_\theta \subset T_\theta \mathcal{O}$ is the distribution with which $\mathcal{O}$ is supplied by definition. A submanifold $W \subset \mathcal{O}$ is said to be *integral* if $T_\theta W = \mathcal{C}_\theta$ for every $\theta \in W$. Evidently, $\dim W = n$. It can be proved that every integral submanifold of the diffiety $\mathcal{O} = \mathcal{Y}_\infty$ is locally of the form $L_{(\infty)}$, where $L \subset E$ is a solution of $\mathcal{Y}$ in the usual sense of this word, i.e., local equations (0.5) of L satisfy equations (0.6), which define $\mathcal{Y}$ locally (see [**71**]). This justifies the above interpretation of solutions as integral submanifolds.

What was said before enables us to informally regard a diffiety as a shelf on which all solutions of the corresponding differential equation are stored. Also, it leads to an important generalization of the notion of a solution to arbitrary nonlinear systems of partial differential equations (see [**71, 84, 86, 83, 128, 130, 123, 124**]).

The following types of diffieties are to be mentioned as examples. Let N be a foliated manifold with n-dimensional leaves. For $\theta \in N$, we define $\mathcal{C}_\theta$ to be the tangent space to the leaf passing through θ. The manifold N equipped with the n-dimensional distribution $\theta \mapsto \mathcal{C}_\theta$ is a diffiety. In fact, $N = \mathcal{Y}_\infty$, where $\mathcal{Y}$ is the system of partial differential equations corresponding to the Pfaff system associated with the distribution $\{\mathcal{C}_\theta\}$. Thus, foliated manifolds are diffieties. The diffiety dimension of such a diffiety is equal to the dimension of the leaves.

Every n-dimensional manifold, say M, can also be viewed as a foliated manifold consisting of only one leaf (in this case $\mathcal{C}_\theta = T_\theta M$), and therefore, as a diffiety. Evidently, $\operatorname{Dim} M = \dim M$ in this case.

REMARK. There are two different natural ways to treat finite-dimensional manifolds as diffieties. One of them was presented above. We get the second one by supplying M with the zero-dimensional distribution $\theta \mapsto \mathcal{C}_\theta = \{0\} \subset T_\theta M$. In this case, $\operatorname{Dim} M = 0$. These two ways are, in a sense, dual to each other, and therefore lead to a kind of duality in the theory of differential equations.

These simple observations show that the traditional differential mathematics growing on smooth finite-dimensional manifolds is just the zero-**D**imensional (not zero-**d**imensional!) part of the diffiety theory. In other words, it is just the top of a very big iceberg whose underwater bulk is, obviously, worth a systematic exploration, independent of any possible application.

0.5. What are symmetries of partial differential equations?

As we have already noted, this question is answered naturally by substituting "diffiety" for "differential equation" in it. This is done below.

Recalling that diffieties are manifolds equipped with a geometric structure (the Cartan distribution), we see that their symmetries should be diffeomorphisms preserving this structure. More exactly we have:

DEFINITION. A mapping $\Phi\colon \mathcal{O} \to \mathcal{O}$ is called a *symmetry* of the diffiety $\mathcal{O}$ if

(i) Φ is a diffeomorphism, i.e., $\Phi^*(C^\infty(\mathcal{O})) = C^\infty(\mathcal{O})$;

(ii) $d_\theta\Phi(\mathcal{C}_\theta) = \mathcal{C}_{\Phi(\theta)}$.

Here $\mathcal{C}_\theta$ denotes, as before, the contact "plane" at θ and $d_\theta\Phi\colon T_\theta \to T_{\Phi(\theta)}$ stands for the differential of Φ at θ.

Specializing this definition to the case $\mathcal{O} = \mathcal{Y}_\infty$, we get the definition of symmetry for a concrete system of partial differential equations given by its label $\mathcal{Y}$, i.e., by (0.6).

Of course, the symmetries of diffieties should constitute a special class of their morphisms. The general morphisms are called *smaps* and their definition is as follows.

DEFINITION. A mapping $\Phi\colon \mathcal{O}_1 \to \mathcal{O}_2$ of a diffiety $\mathcal{O}_1$ into a diffiety $\mathcal{O}_2$ is called a *smap* if

(i) Φ is smooth, i.e., $f \circ \Phi \in C^\infty(\mathcal{O}_1)$ for every $f \in C^\infty(\mathcal{O}_2)$;

(ii) $d_\theta\Phi(\mathcal{C}_\theta) \subset \mathcal{C}_{\Phi(\theta)}$ for every $\theta \in \mathcal{O}_1$.

It can be shown that smaps can be identified locally with differential operators, generally nonlinear. Smaps are analogs of smooth maps connecting finite-dimensional manifolds, i.e., smooth mappings of finite-dimensional manifolds may be viewed as smaps of zero-dimensional diffieties.

Diffieties and smaps are the objects and morphisms, respectively, of a category called the *category of differential equations*.

Now we have to introduce the infinitesimal version of diffiety symmetries. Let $\mathcal{O}$ be a diffiety. A vector field X on $\mathcal{O}$ is called a *trivial contact field* (on $\mathcal{O}$) if $X_\theta \in \mathcal{C}_\theta$ for every $\theta \in \mathcal{O}$. Here X_θ denotes the value of the field X at the point θ.

DEFINITION. A vector field X on $\mathcal{O}$ is called a *contact* field or a *$\mathcal{C}$-field* if for every trivial contact field Y the commutator $[X, Y]$ is a trivial contact field.

It results from the complete integrability of the contact structure on $\mathcal{O}$ that commutators of trivial contact vector fields are also trivial contact fields. In other words, the set $\mathcal{C}D(\mathcal{O})$ of all trivial contact fields on $\mathcal{O}$ is a Lie algebra.

It is now easily seen from the definition and the Jacobi identity that the $\mathcal{C}$-fields on $\mathcal{O}$ form a Lie algebra with respect to the standard commutator operation and that $\mathcal{C}D(\mathcal{O})$ is an ideal of this Lie algebra. We denote this algebra by $D_\mathcal{C}(\mathcal{O})$. The quotient Lie algebra

$$\operatorname{Sym}\mathcal{O} = D_\mathcal{C}(\mathcal{O})/\mathcal{C}D(\mathcal{O})$$

is called the *symmetry algebra* of $\mathcal{O}$. Specializing this definition to the case $\mathcal{O} = \mathcal{Y}_\infty$, we arrive at the (higher) symmetry algebra of a system of partial differential equations given by its label $\mathcal{Y}$ (or by (0.6)).

It might seem more natural to call the symmetry algebra the whole algebra of $\mathcal{C}$-fields $D_\mathcal{C}(\mathcal{O})$. The reasons why we must pass to quotients are explained in the next section. For now, we just notice this delicate conceptual point.

Elements of the algebra $\operatorname{Sym}\mathcal{Y}$ are called (*higher*) *infinitesimal symmetries* of the system of partial differential equations given by the label $\mathcal{Y}$.

It turns out that $\operatorname{Sym}\mathcal{O}$ is the cohomology of a differential complex and the quotienting in question complies with this fact properly.

0.6. Infinitesimal symmetries of partial differential equations are secondary quantized vector fields

A description of the algebra $\operatorname{Sym}\mathcal{O}$ in local coordinates will be needed to justify this assertion. We can restrict ourselves to the case $\mathcal{O} = \mathcal{Y}_\infty$ because every diffiety $\mathcal{O}$ is locally of this form.

First of all, we consider the simplest situation where $\mathcal{Y}$ is the "empty" equation $\{0 = 0\}$. In this case, $\mathcal{Y}_\infty = J^\infty(E, n)$ and we can use a local chart of the form $(x, u, \dots, u^i_\sigma, \dots)$ described above.

Coordinate expressions of trivial $\mathcal{C}$-fields in these coordinates look as follows:

$$Y = \sum_{i=1}^{n} a_i D_i, \quad a_i \in \mathcal{F}(E, n),$$

where D_i denotes the total derivative operator (0.8).

Now, let $\varphi = (\varphi_1, \dots, \varphi_m)$, $\varphi_i \in \mathcal{F}(E, n)$, be an $\mathbb{R}^m$-valued function on $J^\infty(E, n)$. It determines the so-called *evolutionary derivation operator*

$$\text{(0.11)} \qquad \mathcal{Э}_\varphi = \sum_{\sigma,i} D_\sigma(\varphi_i) \frac{\partial}{\partial u^i_\sigma},$$

which is a vector field on $J^\infty(E, n)$. In fact, $\mathcal{Э}_\varphi$ is a $\mathcal{C}$-field and φ is called the *generating function* of it. Let us note that the identification of generating functions with m-tuples of ordinary functions, as was done above, is possible only locally.

The following result gives us the necessary local description of $\mathcal{C}$-fields.

PROPOSITION. *Every $\mathcal{C}$-field X on $J^\infty(E, n)$ can be uniquely presented in the form*

$$\text{(0.12)} \qquad X = \mathcal{Э}_\varphi + Y,$$

where Y is a trivial $\mathcal{C}$-field.

We also have

$$[\mathcal{Э}_\varphi, \mathcal{Э}_\psi] = \mathcal{Э}_{\{\varphi,\psi\}},$$

where

$$\text{(0.13)} \qquad \{\varphi, \psi\} = \mathcal{Э}_\varphi(\psi) - \mathcal{Э}_\psi(\varphi).$$

This bracket operation endows the linear space of all smooth $\mathbb{R}^m$-valued functions defined on the local chart under consideration, with a Lie algebra structure. Then the above proposition shows that this Lie algebra is locally isomorphic to the Lie algebra $\operatorname{Sym} J^\infty(E, n)$.

If $\xi \in \operatorname{Sym} J^\infty(E, n)$ and $\xi = \mathcal{Э}_\varphi \bmod \mathcal{C}D(J^\infty(E, n))$, φ is also called the *generating function* of ξ (with respect to the chosen coordinate system).

REMARK. Note that the bracket (0.13) coincides with the standard Poisson bracket for functions $\varphi, \psi \in \mathcal{F}_1(E, n)$ in the situation when $m = 1$ and φ, ψ do not depend on u.

Let us observe now that every symmetry $\Phi\colon \mathcal{Y}_\infty \to \mathcal{Y}_\infty$ generates a mapping $\tilde{\Phi}\colon \operatorname{Sol}\mathcal{Y} \to \operatorname{Sol}\mathcal{Y}$ of the "space" of all local solutions of $\mathcal{Y}$. In fact, it transforms an integral submanifold of $\mathcal{Y}_\infty$ into another, as is clearly seen from the definitions. Therefore, identifying integral submanifolds of $\mathcal{Y}_\infty$ with local solutions of $\mathcal{Y}$, as was explained before, we get the mapping $\tilde{\Phi}$.

When $\mathcal{Y}$ is the "empty" equation $\{0 = 0\}$, we have $\mathcal{Y}_\infty = J^\infty(E, n)$ and the set $\operatorname{Sol}\mathcal{Y}$ is canonically identified with the set of all n-dimensional submanifolds of E. Therefore, every symmetry $\Phi\colon J^\infty(E,n) \to J^\infty(E,n)$ generates a transformation $\tilde{\Phi}$ of the "space" of all n-dimensional submanifolds of E.

For the same reason every infinitesimal symmetry of $\mathcal{Y}_\infty$, i.e., every $\mathcal{C}$-field, say X, on $\mathcal{Y}_\infty$ generates a (virtual) flow on the "space" $\operatorname{Sol}\mathcal{Y}$. The velocity field of this flow is given by the formula

$$\frac{\partial u^i}{\partial t} = \varphi_i(x, u, \dots, u^\alpha_\sigma, \dots), \quad i = 1, \dots, m, \tag{0.14}$$

in which t is a new independent variable (the "evolution time") and $X = \mathbin{\raisebox{\depth}{\scalebox{1}[-1]{∂}}}_\varphi + Y$, $\varphi = (\varphi_1, \dots, \varphi_m)$, according to the above proposition.

It is seen from (0.14) that the flow on $\operatorname{Sol}\mathcal{Y}$ generated by X does not depend on its trivial part Y. In other words, $\mathcal{C}$-fields belonging to the same coset modulo $\mathcal{C}D(\mathcal{Y}_\infty)$ generate the same flow on $\operatorname{Sol}\mathcal{Y}$. This is why the Lie algebra $D_{\mathcal{C}}(\mathcal{O})$ of all contact fields must be quotiented by trivial ones in order to obtain true symmetries. The meaning of generating functions becomes clear from (0.14).

Now we note that the generating function occurring in (0.14) is not arbitrary unless $\mathcal{Y}_\infty = J^\infty(E, n)$. In fact, it must satisfy the following equation, supposing that $\mathcal{Y}$ is given by (0.6):

$$\bar{\ell}_F \bar{\varphi} = 0.$$

Here $F = (F_1, \dots, F_l)$ and

$$\ell_F = \begin{pmatrix} \sum_\sigma \frac{\partial F_1}{\partial u^1_\sigma} D_\sigma & \dots & \sum_\sigma \frac{\partial F_1}{\partial u^m_\sigma} D_\sigma \\ \dots & \dots & \dots \\ \sum_\sigma \frac{\partial F_l}{\partial u^1_\sigma} D_\sigma & \dots & \sum_\sigma \frac{\partial F_l}{\partial u^m_\sigma} D_\sigma \end{pmatrix}$$

is an $(l \times m)$-matrix differential operator on $J^\infty(E, n)$ and the bars over ℓ and φ indicate the restrictions to $\mathcal{Y}_\infty$.

It is now natural to ask the following question: what do the notions just introduced mean when applied to zero-dimensional diffieties, i.e., to usual manifolds (see the end of Section 0.4). In this case, the contact distribution is zero-dimensional, and therefore, $\mathcal{C}D(M) = 0$ for a manifold M viewed as a zero-dimensional diffiety. For the same reasons, every vector field on M is contact and we see that $\operatorname{Sym} M = D(M)$, where $D(M)$ stands for the Lie algebra of all vector fields on M.

Since $n = 0$ in the situation in question, every local chart on M, say $u^1, \dots, u^m$, can be regarded as a divided one. This shows that the standard coordinate expression $X = \sum_i \varphi_i(u)\partial/\partial u^i$, $u = (u^1, \dots, u^m)$, for a vector field X on M is a particular case of (0.11). Namely, we have $X = \mathbin{\raisebox{\depth}{\scalebox{1}[-1]{∂}}}_\varphi$ for $\varphi = (\varphi_1(u), \dots, \varphi_m(u))$, and system (0.14) in this case looks as follows:

$$\frac{du^i}{dt} = \varphi_i(u^1, \dots, u^m), \quad i = 1, \dots, m. \tag{0.15}$$

But in (0.15) we recognize the ordinary differential equations of characteristics for the first-order partial differential operator $X = \sum \varphi_i \partial/\partial u^i$. Now the analogy $X \mapsto \mathcal{Э}_\varphi$, (0.15) $\mapsto$ (0.14) motivates the following principal statement:

> *If $\varphi = (\varphi_1, \ldots, \varphi_m)$ is the generating function of a symmetry $\xi \in \operatorname{Sym} \mathcal{Y}_\infty$, then system (0.14) of partial differential equations can be regarded naturally as the characteristic system corresponding to the operator ξ.*

This gives the desired answer to the main question—*What are secondary quantized differential equations?*—for first-order operators:

> *Higher symmetries of partial differential equations are first-order secondary quantized differential operators.*

This is our starting point when we look for secondary quantized differential operators of higher orders. Before, however, we shall discuss some aspects of the symmetry theory for partial differential equations because of its potential importance for the future Secondary Calculus.

0.7. Digression: on symmetries of partial differential equations

Besides the aim just mentioned, in the first part of this section we shall collect some brief historical remarks in order to illustrate the previous discussion and also for terminological reasons.

The symmetry theory for differential equations was founded by S. Lie, who discovered both the conceptual structure of this theory and the main technical tools to work with (see [**81**]). Unfortunately, only the two simplest, in a sense, aspects of his work, namely Lie groups and Lie algebras, were mainly assimilated by the mathematical community and then developed in thousands of works which have nothing in common with differential equations. As we have already mentioned, the symmetries of a differential equation, according to Lie, are transformations of dependent and independent variables, or the first-order contact transformations, discovered by Lie himself, that leave the chosen label of the equation in question invariant. We remark that the so-defined symmetry group or algebra actually depends on the choice of a label of the equation in question. Also, the generating functions of these classical infinitesimal symmetries depend only on the derivatives of the first order. This explains the relations between the classical and modern symmetry theories.

The first systematic attempts to apply Lie's theory to mechanics of continuous media were made by L. Ovsiannikov and his collaborators about 70–80 years after Lie's original work (see [**97**]). These authors merely elaborated some technical aspects of Lie's theory, while its conceptual content was not touched. Probably, Ovsiannikov was the first who recognized Lie's theory behind many special methods and tricks used in mechanics. The so-called dimensional analysis by Sedov [**101**] and Birkhoff [**16**] is perhaps the most remarkable example of this kind.

It looks surprising that the first attempts to generalize the classical notion of symmetry were made almost immediately after Lie's work, first by A. Bäcklund and then by E. Noether. For example, in Bäcklund's paper [**11**] one can find a coordinatewise definition of $\mathcal{C}$-fields on $J^\infty(\mathbb{R}^{n+1}, n)$. Of course, these attempts of Bäcklund and Noether could not be well founded at that time for many reasons, and in particular, it was not clear how to treat infinite-dimensional manifolds. For example, Bäcklund considered all vector fields on infinite jets to be integrable.

However, one can find the same deficiency in some recent works (for instance, in [**6**]).

New times for the symmetry theory arrived with the discovery of nonlinear equations integrable via the so-called inverse scattering transform method. The first examples of nonclassical ("higher") symmetries were discovered at that time. For instance, it turned out that the higher analogs of the Korteweg–de Vries equation are, in fact, its nonclassical symmetries.

Apparently, the rigorous nonclassical symmetry theory starts with Kupershmidt's work [**78**] in which all $\mathcal{C}$-fields on $J^\infty(\pi)$ were completely described. Subsequently, the author introduced all necessary basic notions, constructions and formulas of the theory [**121, 137, 142**], some of which were sketched above.

We call these new nonclassical symmetries "higher" to emphasize that their generating functions may depend on the derivatives of arbitrary order, in contrast to the classical ones, which may depend only on the derivatives of order not greater than one. In the current literature the terms "generalized symmetries" (for instance, [**96**]) and "Lie–Bäcklund transformations" (for instance, [**54**]) are also used in the same sense.

First rigorous and systematic computations of higher symmetries for concrete equations with more than one independent variable, i.e., beyond the scope of integrable systems, were performed within the Pereslavl-Zalesski program in 1984–85 (see [**139**]).

Nowadays, the foundations of the higher symmetry theory as well as the basic computational algorithms are well established. Thus, the main general problem in this field is to enlarge the area of possible applications by developing and elaborating computational techniques. Another interesting related problem is to determine what are the obstructions for partial differential equations to be symmetric. It is remarkable that this problem admits a natural solution in terms of Secondary Calculus, but we cannot touch upon this topic here.

One can imagine naturally a generalization of the higher symmetry theory in which generating functions depend not only on the derivatives of arbitrary order but also on "nonlocal" variables such as $\int f\,dx$, $f \in \mathcal{F}$. In fact, there exist a number of problems suggesting such a generalization, and the first few steps in this direction have already been taken; see [**17, 36, 60, 74, 80, 92**].

We can apply once again the philosophy of Section 0.1 in searching the desired concept of nonlocal symmetry. This means answering the question "what are differential equations" in a manner different from that in Section 0.1. Below we shall indicate very briefly how it can be done (see [**74**] for more details and motivations).

Let $\mathcal{O}$ and $\mathcal{O}'$ be diffieties, $\operatorname{Dim}\mathcal{O} = \operatorname{Dim}\mathcal{O}'$. A smap $\Phi\colon \mathcal{O}' \to \mathcal{O}$ is called a *covering* of $\mathcal{O}$ by $\mathcal{O}'$ if $\Phi(\mathcal{O}') = \mathcal{O}$ and $d_\theta\Phi(\mathcal{C}_\theta) = \mathcal{C}_{\Phi(\theta)}$, for all $\theta \in \mathcal{O}'$. A symmetry $\xi \in \operatorname{Sym}\mathcal{O}'$ is called a *nonlocal (infinitesimal) symmetry* or, more exactly, a Φ-*symmetry* of $\mathcal{O}$. We emphasize that nonlocal symmetries of $\mathcal{O}$ are pairs of the form (Φ, ξ), where Φ is a covering of $\mathcal{O}$ and ξ is a higher symmetry of the covering diffiety $\mathcal{O}'$. Roughly speaking, the reason to call such symmetries nonlocal is that functions on $\mathcal{O}'$ (in particular, the generating ones) are seen by an "observer" on $\mathcal{O}$ as depending on some new ("nonlocal") variables.

Example. Consider the Burgers equation $\mathcal{Y} = \{u_t = uu_x + u_{xx}\}$ and the heat equation $\tilde{\mathcal{Y}} = \{v_t = v_{xx}\}$. The Cole–Hopf substitution $u = 2v_x/v$ connecting these equations is, in fact, the label of a covering $\Phi\colon \tilde{\mathcal{Y}}_\infty \to \mathcal{Y}_\infty$. Therefore, higher symmetries of the heat equation can be regarded as nonlocal symmetries of the Burgers equation. The heat equation is linear and thus every solution $a(x,t)$ of it can be treated also as its symmetry with the generating function $a(x,t)$. On the other hand, the generating function of the corresponding nonlocal symmetry of the Burgers equation is

$$(2a_x - au)\exp\left(-\frac{1}{2}\int u\,dx\right).$$

It involves a nonlocal variable, namely, $\int u\,dx$.

The so-defined nonlocal symmetries can be used in the standard applications exactly in the same way as the classical or higher ones. However, the fact that different nonlocal symmetries live, generally, on different coverings leads to some nonstandard and surprising features of nonlocal theory. One of them is the following.

Evidently, for a fixed $\Phi\colon \mathcal{O}' \to \mathcal{O}$ all Φ-symmetries of $\mathcal{O}$ form a Lie algebra which coincides with $\operatorname{Sym}\mathcal{O}'$. But, at first glance, it seems absurd to look for the commutator of two nonlocal symmetries defined on two different coverings. However, it turns out to be possible to find the desired commutator on a suitable third covering. Therefore, in order to organize all nonlocal symmetries of $\mathcal{O}$ into something like a Lie algebra one must take into consideration all coverings of $\mathcal{O}$ simultaneously.

All coverings of a given diffiety constitute in a natural way a category which is called a *cobweb* (see [**74**]). Now the philosophy of Section 0.1 forces us to answer the question what are differential equations by saying that they are cobwebs. This answer implies many important consequences for Secondary Calculus.

We remark also that, while diffieties are analogs of affine varieties of algebraic geometry, cobwebs are analogs of fields of rational functions on them.

0.8. Secondary ("quantized") functions

It seems natural to define higher-order secondary quantized differential operators as compositions of first-order ones. But following this way we immediately encounter the following difficulty.

Recall that first-order secondary differential operators are elements of the Lie algebra $\operatorname{Sym}\mathcal{O}$. On the other hand, they are not proper differential operators but cosets (equivalence classes) of them. So the question arises: how to compose these cosets? We leave to the interested reader the task to verify that a direct attack to this problem fails.

Another aspect of the problem can be extracted from a similar question: on what kind of objects do secondary differential operators act? No doubt, secondary differential operators should be proper operators, i.e., they should act on some kind of objects. The usual functions cannot be taken as such. One can see this by trying to define an action of the algebra $\operatorname{Sym}\mathcal{O}$ on $C^\infty(\mathcal{O})$. The only natural way to do this would be to put $\xi(f) = X(f)$ for $\xi \in \operatorname{Sym}\mathcal{O}$, $X \in D_{\mathcal{C}}(\mathcal{O})$, $\xi = X \bmod \mathcal{C}D(\mathcal{O})$ and $f \in C^\infty(\mathcal{O})$. But this definition is clearly not correct. Namely, if $X_1, X_2 \in D_{\mathcal{C}}(\mathcal{O})$ and $X_1 = X_2 \bmod \mathcal{C}D(\mathcal{O})$, then, generally, $X_1(f) \neq X_2(f)$ if $X_1 \neq X_2$.

However, it is clear that, linguistically, secondary operators should act on "secondary functions" and we shift this question by asking what are the latter. The following analogy will help us to answer it.

Let $\Lambda^i(M)$ denote the space of i-th degree differential forms on the manifold M. The mapping

$$C^\infty(M) \xrightarrow{d} \Lambda^1(M) \tag{0.16}$$

(the standard de Rham differential) provides extremal problems on smooth functions on M with the "universal solution". Treating smooth manifolds as zero-dimensional diffieties we see that the analog of (0.16) should be a mapping that provides variational problems for multiple integrals with the universal solution. But this is the well-known Euler–Lagrange mapping:

$$\boxed{\begin{array}{c}\text{variational functionals,}\\ \text{or "actions"}\end{array}} \xrightarrow{\mathcal{E}} \boxed{\begin{array}{c}\text{differential}\\ \text{operators}\end{array}} \tag{0.17}$$

i.e., $\mathcal{E}$ associates with an "action" $\int_\Omega L\,dx_1\cdots dx_n$ the left-hand side of the corresponding Euler–Lagrange equation. Therefore, this analogy between (0.16) and (0.17) suggests to adopt "actions" as secondary (or "quantized") functions. This idea is to be corrected because "actions", as understood in the standard way, contain a feature that is parasitic for our aim and should be eliminated. This feature is the exact reference to the domain of integration Ω. Thus, our next problem is to find a meaning for hieroglyphs of the form $\int L\,dx_1\cdots dx_n$ (without Ω!). We shall solve it by interpreting them as some kind of cohomology classes (for more motivations and details see [**47**]). But before that we need some preliminaries.

Let $\mathcal{O}$ be a diffiety, $\operatorname{Dim}\mathcal{O} = n$, and let $\Lambda^i(\mathcal{O})$ denote the $C^\infty(\mathcal{O})$-module of all i-th degree differential forms on $\mathcal{O}$. We denote by $\mathcal{C}\Lambda^i(\mathcal{O})$ the submodule of $\Lambda^i(\mathcal{O})$ consisting of forms whose restrictions to the contact distribution of $\mathcal{O}$ vanish. In other words,

$$\mathcal{C}\Lambda^i(\mathcal{O}) \ni \omega \Leftrightarrow \omega(Y_1,\dots,Y_i) = 0, \quad \forall Y_1,\dots,Y_i \in \mathcal{C}D(\mathcal{O}).$$

We put

$$\bar\Lambda^i(\mathcal{O}) = \Lambda^i(\mathcal{O})/\mathcal{C}\Lambda^i(\mathcal{O}).$$

Elements of $\bar\Lambda^i(\mathcal{O})$ are called *horizontal differential forms* on $\mathcal{O}$. Evidently, $\bar\Lambda^i(\mathcal{O}) = 0$ if $i > n$.

It is easy to see that $d(\mathcal{C}\Lambda^i(\mathcal{O})) \subset \mathcal{C}\Lambda^{i+1}(\mathcal{O})$ and, therefore, the standard differential $d\colon \Lambda^i(\mathcal{O}) \to \Lambda^{i+1}(\mathcal{O})$ induces the *horizontal differential*

$$\bar d\colon \bar\Lambda^i(\mathcal{O}) \to \bar\Lambda^{i+1}(\mathcal{O}).$$

Of course, $\bar d^2 = 0$ and this enables us to introduce the horizontal de Rham complex of $\mathcal{O}$:

$$0 \to \bar\Lambda^0(\mathcal{O}) = C^\infty(\mathcal{O}) \xrightarrow{\bar d} \bar\Lambda^1(\mathcal{O}) \xrightarrow{\bar d} \cdots \xrightarrow{\bar d} \bar\Lambda^n(\mathcal{O}) \to 0.$$

The cohomology of this complex is called *horizontal de Rham cohomology* of $\mathcal{O}$. They are denoted by $\bar H^i(\mathcal{O})$, $i = 0, 1, \dots, n$.

Finally, we accept the following basic interpretation:

> *Secondary (or "quantized") functions on $\mathcal{O}$ are elements of the cohomology group $\bar H^n(\mathcal{O})$.*

In other words, we consider the cohomology group $\bar H^n(\mathcal{O})$, $\operatorname{Dim}\mathcal{O} = n$, to be an analog of the smooth function algebras in Secondary Calculus. Actually, it would be more correct to take $\bar H^*(\mathcal{O})$ for such an analog. The reason will become clearer in Section 5.1, after the corresponding theory is elaborated.

To justify the choice we made, we describe the given "horizontal" constructions for $\mathcal{O} = J^\infty(E, n)$ in coordinates. First of all, we observe that the coset of a differential form $\omega \in \Lambda^i(J^\infty(E, n))$ modulo $\mathcal{C}\Lambda^i(J^\infty(E, n))$ contains only one element of the form

$$\rho = \sum_{1<k_1<\cdots<k_i<n} a_{k_1,\ldots,k_i}(x, u, \ldots, u^j_\sigma, \ldots)\, dx_{k_1} \wedge \cdots \wedge dx_{k_i},$$

where $a_{k_1,\ldots,k_i} \in C^\infty(J^\infty(E, n))$. The characteristic feature of such a form is that the differentials $du^j, \ldots du^j_\sigma, \ldots$ are absent in its coordinate expression. Therefore, the module $\bar{\Lambda}^i(J^\infty(E, n))$ can be identified locally with the module of forms of this type.

Under the identification made, the horizontal de Rham differential $\bar{d}$ looks as

$$\bar{d}\rho = \sum_{s,k_1,\ldots,k_i} D_s(a_{k_1,\ldots,k_i})\, dx_s \wedge dx_{k_1} \wedge \cdots \wedge dx_{k_i}.$$

In particular, every horizontal $(n-1)$-form can be represented uniquely as

$$\rho = \sum_i (-1)^i a_i\, dx_1 \wedge \cdots \wedge \widehat{dx_i} \wedge \cdots \wedge dx_n, \quad a_i \in C^\infty(J^\infty(E, n)),$$

and

$$\bar{d}\rho = \operatorname{div}(A) dx_1 \cdots dx_n,$$

where $A = (a_1, \ldots, a_n)$ and $\operatorname{div} A = \sum_i D_i(a_i)$. Also, horizontal n-forms look as follows:

$$L(x, u, \ldots, u^i, \ldots)\, dx_1 \wedge \cdots \wedge dx_n, \quad L \in C^\infty(J^\infty(E, n)),$$

and one can recognize in them Lagrangian densities. Therefore, we see that the horizontal cohomology $\bar{H}^n(J^\infty(E, n))$ can be identified locally with the linear space of equivalence classes of Lagrangian densities on $J^\infty(E, n)$ with respect to the following relation:

$$L_1 \sim L_2 \Leftrightarrow L_1 - L_2 = \operatorname{div} A \text{ for some } A.$$

On the other hand, actions $\int_\Omega L_i\, dx_1 \cdots dx_n$, $i = 1, 2$, are equivalent in the sense that they lead to the identical Euler–Lagrange equations if and only if $L_1 \sim L_2$. This is independent of the choice of Ω. For these reasons, it is natural to identify the hieroglyphs $\int L\, dx_1 \cdots dx_n$ with n-dimensional horizontal cohomology classes.

We conclude this section by noting that the above arguments remain valid for arbitrary diffieties as well.

0.9. Higher-order scalar secondary ("quantized") operators

First of all, we must justify the above definition of secondary functions by demonstrating that, indeed, first-order secondary differential operators act naturally on them. In other words, we must look for a natural action of the algebra $\operatorname{Sym} \mathcal{O}$ on the space $\bar{H}^n(\mathcal{O})$. This can be done straightforwardly.

First, note that if $X \in D_\mathcal{C}(\mathcal{O})$, $\omega \in \mathcal{C}\Lambda^i(\mathcal{O})$ and L_X denotes the Lie derivative along X, then $L_X(\omega) \in \mathcal{C}\Lambda^i(\mathcal{O})$ as it follows from the definitions and from the fact that

$$\mathcal{C}\Lambda^i(\mathcal{O}) = \mathcal{C}\Lambda^1(\mathcal{O}) \wedge \Lambda^{i-1}(\mathcal{O}), \quad i > 1.$$

This allows us to define the Lie derivative on horizontal forms by passing to quotients:

$$L_X \colon \bar{\Lambda}^i(\mathcal{O}) \to \bar{\Lambda}^i(\mathcal{O}).$$

Next, let

$$\xi \in \operatorname{Sym}\mathcal{O}, \qquad \xi = X \bmod \mathcal{C}D(\mathcal{O}) \;\text{ for }\; X \in D_{\mathcal{C}}(\mathcal{O}),$$
$$\theta \in \bar{H}^n(\mathcal{O}), \qquad \theta = \omega \bmod \bar{d}\bar{\Lambda}^{n-1}(\mathcal{O}) \;\text{ for }\; \omega \in \bar{\Lambda}^n(\mathcal{O}).$$

Let us define now an action of ξ on θ by putting

$$\xi(\theta) \stackrel{\text{def}}{=} \{L_X(\omega) \bmod \bar{d}\bar{\Lambda}^{n-1}(\mathcal{O})\} \in \bar{H}^n(\mathcal{O}).$$

The correctness of this definition is deduced directly from the following two facts:

(i) $L_Y(\omega) \in \bar{d}\bar{\Lambda}^{n-1}(\mathcal{O})$ if $\omega \in \bar{\Lambda}^n(\mathcal{O})$ and $Y \in \mathcal{C}D(\mathcal{O})$,
(ii) $L_X \circ \bar{d} = \bar{d} \circ L_X$.

They both are direct consequences of definitions.

Now we see that the above definition of secondary functions agrees well with other "secondary" constructions and, therefore, may serve as a model when dealing with more complicated "secondary" notions. For example, let us observe that we have succeeded to define a well-posed action of one quotient (namely, $\operatorname{Sym}\mathcal{O} = D_{\mathcal{C}}(\mathcal{C})/\mathcal{C}D(\mathcal{O})$) on another (namely, $\bar{H}^n(\mathcal{O}) = \bar{\Lambda}^n(\mathcal{O})/\bar{d}\bar{\Lambda}^{n-1}(\mathcal{O})$) because

1. $D_{\mathcal{C}}(\mathcal{O})$ consists of first-order differential operators which act on the $C^\infty(\mathcal{O})$-module $\bar{\Lambda}^n(\mathcal{O})$ leaving $\bar{d}\bar{\Lambda}^{n-1}(\mathcal{O})$ invariant;
2. the image of $\bar{\Lambda}^n(\mathcal{O})$ under the action of a first-order operator belonging to $\mathcal{C}D(\mathcal{O})$ is contained in $\bar{d}\bar{\Lambda}^{n-1}(\mathcal{O})$.

We get the necessary generalization to higher-order secondary differential operators simply by replacing the words "first-order" by "k-th order" in 1 and 2 above. More precisely, let $\operatorname{Diff}_k(\bar{\Lambda}^n(\mathcal{O}))$ denote the $C^\infty(\mathcal{O})$-module of all ("usual") differential operators of order $\leq k$ acting on $\bar{\Lambda}^n(\mathcal{O})$ and put

$$\overline{\operatorname{Diff}}_k(\mathcal{O}) = \{\Delta \in \operatorname{Diff}_k(\bar{\Lambda}^n(\mathcal{O})) \mid \Delta(\bar{d}\bar{\Lambda}^{n-1}(\mathcal{O})) \subset \bar{d}\bar{\Lambda}^{n-1}(\mathcal{O})\},$$
$$\underline{\operatorname{Diff}}_k(\mathcal{O}) = \{\Delta \in \operatorname{Diff}_k(\bar{\Lambda}^n(\mathcal{O})) \mid \Delta(\bar{\Lambda}^n(\mathcal{O})) \subset \bar{d}\bar{\Lambda}^{n-1}(\mathcal{O})\}.$$

Then the space of all *scalar secondary* ("*quantized*") *differential operators* of order $\leq k$ on $\mathcal{O}$ is defined to be the quotient

$$\mathfrak{Diff}_k(\mathcal{O}) = \overline{\operatorname{Diff}}_k(\mathcal{O})/\underline{\operatorname{Diff}}_k(\mathcal{O}). \tag{0.18}$$

Of course, every secondary operator $\Delta \in \mathfrak{Diff}_k(\mathcal{O})$ can be understood as an actual operator

$$\Delta \colon \bar{H}^n(\mathcal{O}) \to \bar{H}^n(\mathcal{O})$$

acting on secondary functions. Indeed, if

$$\Delta = \delta \bmod \underline{\operatorname{Diff}}_k(\mathcal{O}), \qquad \text{for } \delta \in \overline{\operatorname{Diff}}_k(\mathcal{O}),$$
$$\Theta = \omega \bmod \bar{d}\bar{\Lambda}^{n-1}(\mathcal{O}), \qquad \text{for } \omega \in \bar{\Lambda}^n(\mathcal{O}),$$

then the horizontal cohomology class

$$\Delta(\Theta) \stackrel{\text{def}}{=} \{\, \delta(\omega) \bmod \bar{d}\bar{\Lambda}^{n-1}(\mathcal{O}) \,\}$$

is well defined, i.e., does not depend on the choice of representatives δ and ω.

For $\mathcal{O} = J^\infty(E,n)$, the secondary differential operators thus defined admit the following coordinate description. Operators of the form

$$\sum_{s=1}^{k} \sum_{\substack{i_1,\dots,i_s\\ \sigma_1,\dots,\sigma_s}} a^{i_1,\dots,i_s}_{\sigma_1,\dots,\sigma_s} \frac{\partial^s}{\partial u^{i_1}_{\sigma_1} \cdots \partial u^{i_s}_{\sigma_s}} + \text{const},$$

where $\sigma_1, \dots, \sigma_s$ are multiindices, are called *vertical* (with respect to the chosen coordinate system). Then it can be proved that every coset

$$\Delta = \{\, \delta \bmod \underline{\mathrm{Diff}}_k(J^\infty(E,n)) \,\} \in \mathfrak{Diff}(J^\infty(E,n))$$

for $\delta \in \overline{\mathrm{Diff}}_k(J^\infty(E,n))$ contains only one vertical operator. Thus, the quotient (0.18) representing secondary differential operators can be identified locally with the set of all vertical secondary operators. These operators of order $\le k$ can be presented in the form

$$\partial_\Delta \overset{\mathrm{def}}{=} \sum_{i=1}^{m} \sum_{\sigma} \mathcal{L}(\Delta^i) \circ \frac{\partial}{\partial u^i_\sigma},$$

where $\Delta = (\Delta^1, \dots, \Delta^m)$, $\Delta^i \in \mathrm{Diff}_{k-1} \mathcal{F}(E,n)$, are arbitrary vertical operators and

$$\mathcal{L}_\sigma(\Delta^i) = [D_{i_1}, \dots, [D_{i_n}, \Delta^i] \dots],$$

where $\sigma = (i_1, \dots, i_n)$. The *generating operator* Δ is a higher-order analog of the generating functions for evolutionary derivations but, unlike the latter, it is not defined uniquely if $k > 1$.

Secondary operators of order > 1 are not reduced to compositions of first-order secondary operators. This fact is instructive in connection with the discussion at the beginning of Section 0.8.

We note also that an explicit description of secondary differential operators on arbitrary diffieties is a much more complicated problem.

Further details, results and alternative views concerning secondary differential operators can be found in [**47**].

Finally, turning back to question on p. 12, we can exhibit the simplest k-th order linear secondary (quantized) differential equations as

$$\Delta(H) = 0, \quad \Delta \in \mathfrak{Diff}_k(\mathcal{O}), \ H \in \bar{H}^n(\mathcal{O}).$$

It must be emphasized, however, that these equations form only a very special class of secondary differential equations. For instance, differentials $d_1^{p,q} = d_1^{p,q}(\mathcal{O})$ of the $\mathcal{C}$-spectral sequence (see Sections 0.10 and 0.11) give us other examples of secondary quantized differential operators and, therefore, secondary differential equations. One of them looks as

$$\mathcal{E}\left(\int L\,dx\right) = 0,$$

where $\int L\,dx \in \bar{H}^n(\mathcal{O})$ and $\mathcal{E}$ is the Euler operator assigning to an action $\int L\,dx \in \bar{H}^n(\mathcal{O})$ the corresponding Euler–Lagrange equation. This is due to the fact that $\mathcal{E} = d_1^{0,n}$ (see Section 0.11). We note also that secondary operators $d_1^{p,q}$ are of the first order while their coordinate description shows them to be "usual" infinite-order operators.

0.10. Secondary ("quantized") differential forms and $\mathcal{C}$-spectral sequences

In this section, we shall consider another aspect of Secondary Calculus, namely, secondary ("quantized") differential forms. What are they? This is a more difficult question than the one concerning secondary differential operators we have already discussed. For this and some other reasons, we shall omit preliminary motivations showing, as before, how to arrive at the exact definitions. Instead, some *a posteriori* justifications will be given.

Let $\mathcal{O}$ be a diffiety. Adopting the notation of Section 0.9, we consider the algebra

$$\Lambda^*(\mathcal{O}) = \sum_{i\geq 0} \Lambda^i(\mathcal{O})$$

of all differential forms on $\mathcal{O}$ and the ideal

$$\mathcal{C}\Lambda^*(\mathcal{O}) = \sum_{i\geq 0} \mathcal{C}\Lambda^i(\mathcal{O}).$$

Denote by $\mathcal{C}^k\Lambda^*(\mathcal{O})$ the k-th power of this ideal, which, by definition, is the linear subspace of $\mathcal{C}\Lambda^*(\mathcal{O})$ generated by all products of the form

$$\omega_1 \wedge \omega_2 \wedge \cdots \wedge \omega_n, \quad \omega_i \in \mathcal{C}\Lambda^*(\mathcal{O}).$$

The ideal $\mathcal{C}\Lambda^*(\mathcal{O})$ and, therefore, all ideals $\mathcal{C}^k\Lambda^*(\mathcal{O})$ are stable with respect to the exterior differential d. Thus, we get the filtration

$$\Lambda^*(\mathcal{O}) \supset \mathcal{C}\Lambda^*(\mathcal{O}) \supset \mathcal{C}^2\Lambda^*(\mathcal{O}) \supset \cdots \supset \mathcal{C}^k\Lambda^*(\mathcal{O}) \supset \cdots \tag{0.19}$$

of the de Rham complex of $\mathcal{O}$ by the subcomplexes $\{\mathcal{C}^k\Lambda^*(\mathcal{O}), d\}$. The ideals $\mathcal{C}^k\Lambda^*(\mathcal{O})$ are naturally graded,

$$\mathcal{C}^k\Lambda^*(\mathcal{O}) = \sum_{s\geq 0} \mathcal{C}^k\Lambda^{k+s}(\mathcal{O}),$$

where

$$\mathcal{C}^k\Lambda^{k+s}(\mathcal{O}) = \mathcal{C}^k\Lambda^*(\mathcal{O}) \cap \Lambda^{k+s}(\mathcal{O}).$$

Now we shall achieve our goal by making use of standard spectral sequence arguments applied to the filtration (0.19) (see, for instance, [**91**] for the general theory). We start with the quotients

$$E_0^{p,q}(\mathcal{O}) \stackrel{\text{def}}{=} \mathcal{C}^p\Lambda^{p+q}(\mathcal{O})/\mathcal{C}^{p+1}\Lambda^{p+q}(\mathcal{O})$$

connected by the differentials

$$d_0^{p,q} \colon E_0^{p,q} \to E_0^{p,q+1},$$

which are natural quotients of the exterior differential d. They all constitute a complex, namely,

$$E_0(\mathcal{O}) = \sum_{p,q} E_0^{p,q}(\mathcal{O}), \quad d_0 = \sum_{p,q} d_0^{p,q}, \ d_0 \colon E_0(\mathcal{O}) \to E_0(\mathcal{O}),$$

which is called the *zero term* of the spectral sequence corresponding to the filtration (0.19). Its *first term* is then defined to be the cohomology of the zero term.

More precisely, we put

$$E_1^{p,q} \stackrel{\text{def}}{=} \ker d_0^{p,q} / \operatorname{im} d_0^{p,q}$$

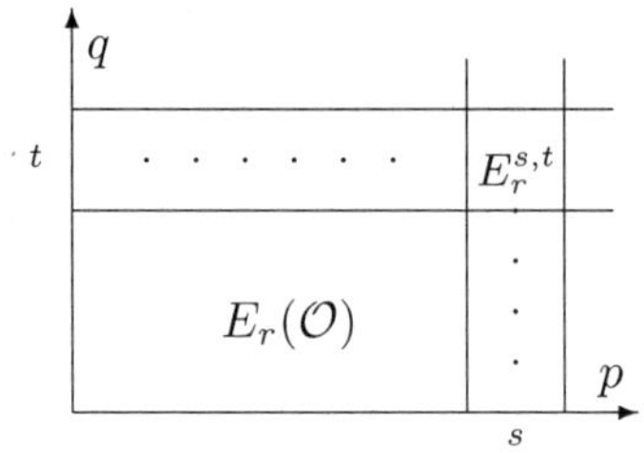

FIGURE 0.1

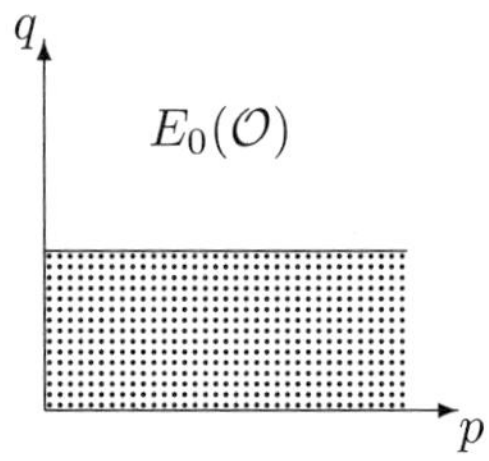

FIGURE 0.2

and note that the exterior differential d generates, by passing to quotients, the differentials

$$d_1^{p,q}\colon E_1^{p,q} \to E_1^{p+1,q}.$$

The first term is now defined to be the complex

$$E_1(\mathcal{O}) = \sum_{p,q} E_1^{p,q}(\mathcal{O}), \quad d_1 = \sum_{p,q} d_1^{p,q},\ d_1\colon E_1(\mathcal{O}) \to E_1(\mathcal{O}).$$

Continuing this procedure, one can define the *r-th term*

$$E_r(\mathcal{O}) = \sum_{p,q} E_r^{p,q}(\mathcal{O}), \quad d_r = \sum_{p,q} d_r^{p,q},\ d_r\colon E_r^{p,q}(\mathcal{O}) \to E_r^{p+r,q-r+1}(\mathcal{O}),$$

as the cohomology of the $(r-1)$-st term.

The system $\{E_r(\mathcal{O}), d_r\}$ of complexes defined in such a way is called the $\mathcal{C}$-*spectral sequence* of the diffiety $\mathcal{O}$. It is often convenient to display the terms of a spectral sequence in a diagram as it is shown in Figure 0.1. For instance, the shaded part of the diagram in Figure 0.2 shows where all nontrivial terms $E_0^{p,q}(\mathcal{O})$ are situated if $\operatorname{Dim}\mathcal{O} = n$. Since the terms $E_r^{p,q}(\mathcal{O})$, $r > 0$, are consecutive cohomology of the preceding ones, the nontrivial part of $E_t(\mathcal{O})$ lies in the same part of the (p,q)-diagram.

DEFINITION. Elements of $E_1(\mathcal{O})$ are called *secondary* (*"quantized"*) *differential forms* on the diffiety $\mathcal{O}$.

Some arguments in favor of this interpretation are as follows. Let M be a finite-dimensional manifold considered as a zero-dimensional diffiety (see Section 0.4). Then the diagram for the first term of its $\mathcal{C}$-spectral sequence looks as is shown in Figure 0.3. Moreover, $d_1^{p,0} = d$. In other words, we see that the de Rham complex of M coincides with the nontrivial part of the first term of the corresponding $\mathcal{C}$-spectral sequence.

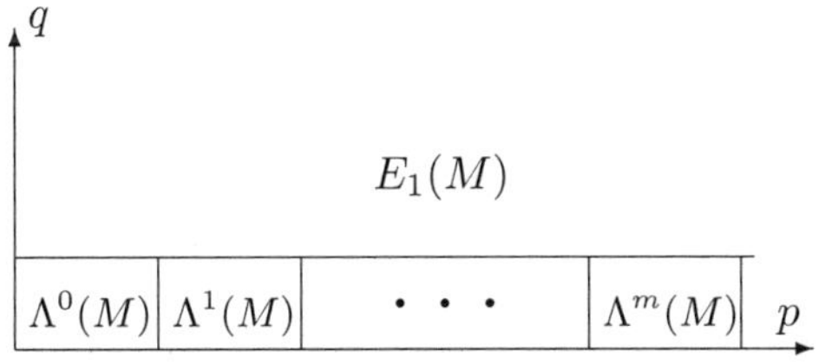

FIGURE 0.3

We can observe that standard constructions and formulas connecting "usual" vector fields and differential forms remain valid for their secondary ("quantized") analogs. For instance, the insertion operator of a secondary vector field ("symmetry") into a secondary differential form as well as the corresponding Lie derivative is well defined. Moreover, they are connected by the secondary analog of the infinitesimal Stokes formula

$$L_X = i_X \circ d + d \circ i_X,$$

in which the exterior differential d is to be replaced by its secondary analog $d_1(\mathcal{O})$.

Finally, we note that secondary differential forms are bigraded objects unlike the "usual" ones which are 1-graded. This is an illustration of the fact that secondary objects are richer and more complicated structures than their "primary" analogs. The same idea can be expressed differently by saying that the "usual" (or "primary") mathematical objects are degenerate forms of the secondary ones. This statement can be also viewed as the following general mathematical paraphrase of the Bohr correspondence principle:

$$\boxed{\text{Secondary Calculus}} \xrightarrow{\text{Dim}\to 0} \boxed{\text{Calculus}}$$

It seems plausible that in the cobweb theory (see the end of Section 0.7) a rigorous meaning to "Dim $\to$ 0" can be given on the basis of a more delicate $\mathbb{R}$-valued concept of diffiety dimension.

0.11. How does the $\mathcal{C}$-spectral sequence work?

In this section, we collect some results that demonstrate secondary differential forms "in action". The following two diagrams illustrate how the term $E_1(\mathcal{O})$, i.e., secondary differential forms, reflects the structure of $\mathcal{O}$. They make the general estimate of the term $E_1(\mathcal{O})$ given on the diagram in Figure 0.4 more precise. In the general case, the number of nontrivial rows in the diagram of $E_1(\mathcal{Y}_\infty)$, $\mathcal{Y} = \{F = O\}$, is equal to the highest order of nontrivial Spencer cohomology groups of the universal linearization operator ℓ_F (see Section 0.6).

Now we want to discuss the meaning of some terms $E_1^{p,q}(\mathcal{O})$ of particular interest. First of all, we have

$$E_1^{0,i}(\mathcal{O}) = \bar{H}^i(\mathcal{O})$$

(see Section 0.8) as it is easily seen from the definitions. In particular, elements of the term $E_1^{0,n}(\mathcal{Y}_\infty)$ are "actions" of variational problems constrained by the

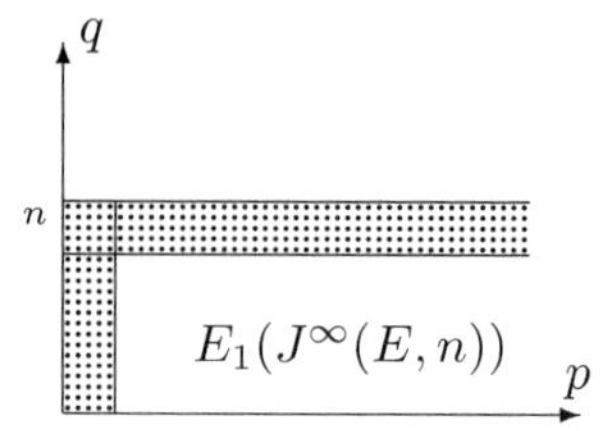

FIGURE 0.4

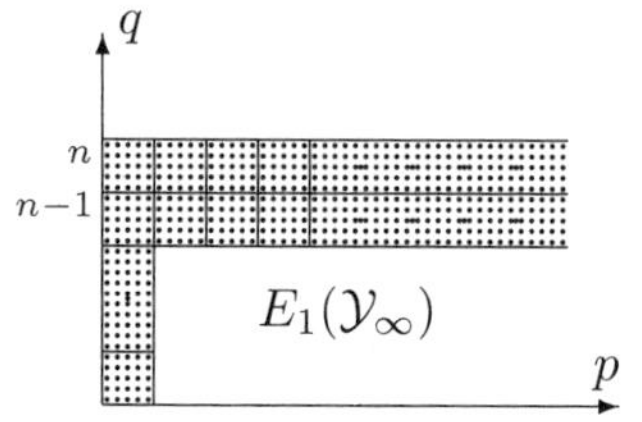

FIGURE 0.5

equation $\mathcal{Y}$ (see Section 0.8). Moreover, if $\mathcal{L} \in E_1^{0,n}(\mathcal{Y}_\infty)$ is an action, then

$$d_1^{0,n}(\mathcal{L}) = 0 \tag{0.20}$$

is the corresponding (constrained) Euler–Lagrange equation. This gives a solution of the "direct" problem of the calculus of variations in the general case of local constraints, i.e., constraints given by differential equations. In particular, equation (0.20) incorporates automatically the theory of Lagrange multipliers.

The standard spectral sequence arguments applied to diagrams of the above type allow one to solve immediately the so-called *triviality problem for Lagrangians* both in local and in global settings. This problem is to describe those "actions", or Lagrangians, to which trivial Euler–Lagrange equations correspond. For example, the answer for free (=nonconstrained) problems is that globally trivial (in this sense) Lagrangian densities are of the form $\omega + \bar{d}\rho$, where ω is a closed differential form on $J^1(E,n)$ and $\bar{d}\rho$ is a full divergence term (see Section 0.8).

Now let $\mathcal{Y} \subset J^k(E,n)$ be the Euler–Lagrange equations corresponding to a Lagrangian $\mathcal{L} \in \bar{H}^n(J^\infty(E,n)) = E_1^{0,n}(J^\infty(E,n))$. Then $E_2^{0,n}(\mathcal{Y}_\infty)$ consists of all Lagrangians *alternative* to $\mathcal{L}$. To compute all alternative Lagrangians one can use spectral sequence arguments as follows.

Let $\mathcal{Y}$ satisfy the conditions indicated on the diagram in Figure 0.5. We also assume that $\mathcal{Y}$ is formally integrable. Then the fact that the diagram for $E_1(\mathcal{Y}_\infty)$ consists of two nontrivial rows leads directly to the following exact sequence:

$$\begin{aligned} 0 &\to E_2^{1,n-1}(\mathcal{Y}_\infty) \to H^n(\mathcal{Y}) \to E_2^{0,n}(\mathcal{Y}_\infty) \xrightarrow{d_2^{0,n}} E_2^{2,n-1}(\mathcal{Y}_\infty) \\ &\to H^{n+1}(\mathcal{Y}) \to \cdots \to E_2^{k,n}(\mathcal{Y}_\infty) \xrightarrow{d_2^{k,n}} E_2^{k+2,n-1}(\mathcal{Y}_\infty) \to H^{n+k+1}(\mathcal{Y}) \to \cdots, \end{aligned} \tag{0.21}$$

where $H^i(\mathcal{Y})$ stands for the i-th de Rham cohomology group of $\mathcal{Y}$. The terms $E_2^{p,n-1}(\mathcal{Y}_\infty)$, $p > 0$, of this sequence can be evaluated or even computed exactly by using Spencer cohomology type techniques. This gives a solution of the alternative Lagrangian problem for a given $\mathcal{Y}$ as well as of a number of similar problems. The most famous of them is the *inverse problem* in the calculus of variations. This is the problem of detecting Euler–Lagrange equations in the case when the constraints are given by $\mathcal{Y}$. Its solution is equivalent to the computation of $E_2^{1,n}(\mathcal{Y}_\infty)$, which is another term of (0.21). The description of symplectic structures for field theories constrained by $\mathcal{Y}$ is also reduced to computation of the term $E_2^{2,n}$ of (0.21).

Coming back to the first term of the $\mathcal{C}$-spectral sequence, we note that the term $E_1^{0,n-1}(\mathcal{Y}_\infty)$ can be interpreted as the space of all *conservation laws* for solutions of $\mathcal{Y}$. This observation allows one to develop a consistent theory of conservation laws (="conserved currents"="integrals of motion") independently of any symmetry considerations, which works equally well in situations where the latter cannot even be applied. This theory is sketched below for formally integrable equations $\mathcal{Y}$ satisfying assumptions that guarantee the two-row diagram for $E_1(\mathcal{Y}_\infty)$ (Figure 0.3).

First of all, the diagram in Figure 0.3 shows that

$$H^{n-1}(\mathcal{Y}) = E_2^{0,n-1}(\mathcal{Y}_\infty) \subset E_1^{0,n-1}(\mathcal{Y}_\infty).$$

Thus, cohomology classes $\Omega \in H^{n-1}(\mathcal{Y})$ can be interpreted as conservation laws of $\mathcal{Y}$. We call them *rigid*. In addition, we see that the kernel of the differential

$$d_1^{0,n-1}\colon E_1^{0,n-1}(\mathcal{Y}_\infty) \to E_1^{1,n-1}(\mathcal{Y}_\infty)$$

consists of these rigid conservation laws. They cannot distinguish between two solutions of $\mathcal{Y}$ if one can be deformed into another. For this and some other reasons, we may ignore them. On the other hand, a conservation law is uniquely characterized up to a rigid one by its image under the differential $d_1^{0,n-1}$. This motivates introduction of the following basic notion: the image $d_1^{0,n-1}(\Omega) \in E_1^{1,n-1}(\mathcal{Y}_\infty)$ of a conservation law $\Omega \in E_1^{1,n-1}(\mathcal{Y}_\infty)$ is called the *generating function* of Ω.

The following isomorphism is fundamental in finding generating functions:

$$E_1^{1,n-1}(\mathcal{Y}_\infty) = \ker \bar{\ell}_F^*,$$

where $\mathcal{Y} = \{F = 0\}$, $\bar{\ell}_F = \ell_F|_{\mathcal{Y}_\infty}$ (see Section 0.6) and "$*$" stands for the formal adjoint. So, generating functions of conservation laws can be found by solving the equation

$$\bar{\ell}_F^*(\psi) = 0. \tag{0.22}$$

This is the most efficient general method of finding conservation laws for concrete equations (see [**139**] and [**19**]).

REMARK. In fact, not all solutions of (0.22) are generating functions of conservation laws. However, one can make use of the differential $d_1^{1,n-1}$ to throw away unnecessary (in this sense) solutions of (0.22).

Now we note that generating functions of symmetries (see Section 0.7) and of conservation laws satisfy mutually adjoint equations $\bar{\ell}_F\varphi = 0$ and $\bar{\ell}_F^*\psi = 0$. We also have $\ell_F = \ell_F^*$ for Euler–Lagrange equations. This demonstrates clearly the nature of the intimate relations between symmetries and conservation laws for Euler–Lagrange equations as given by the classical theorem of Noether. But now we see that the same relations hold for a much larger class of equations, which can be called conformally selfadjoint, i.e., such that

$$\bar{\ell}_F^* = A \circ \bar{\ell}_F,$$

where A is an invertible operator on $\mathcal{Y}_\infty$. The equation $u_x = u_y$ is a simple example of this kind with $A = -1$.

An interpretation of the term $E_2(\mathcal{Y}_\infty)$ is that it consists of characteristic classes of bordisms composed of solutions of $\mathcal{Y}$. The standard "differential characteristic classes" theories can be obtained in this way under a suitable choice of $\mathcal{Y}$ (see [**131**]). This approach leads, however, to finer characteristic classes, for instance, *special characteristic classes.* We illustrate this for solutions of the (vacuum) Einstein equations, or Einstein manifolds. Let $\mathcal{Y}$ be the Einstein system on a manifold M. Then it is possible to show that

$$\mathcal{Y}_\infty/\mathrm{Diffeo}\,(M) = \mathcal{Y}'_\infty,$$

where Diffeo (M) is the diffeomorphism group of M acting naturally on $\mathcal{Y}_\infty$, and $\mathcal{Y}'$ is a universal system of partial differential equations. In fact, $\mathcal{Y}'$ does not depend on M and, therefore, its solutions are just diffeomorphism classes of Einstein manifolds. The corresponding special characteristic classes are elements of $E_2(\mathcal{Y}'_\infty)$.

It is interesting to note a direct relation between the $\mathcal{C}$-spectral sequence and recent facts in field theory, the so-called *anti-field formalism* [**10**]. As it happens, anti-fields are the quantities arising in the first column of the $\mathcal{C}$-spectral sequence for the Yang–Mills equations. Computations one finds in physical literature implicitly simulate computations of this column.

The $\mathcal{C}$-spectral sequence machinery was discovered by the author while searching for a conceptual definition of the Euler operator.

CHAPTER 1

Elements of Differential Calculus in Commutative Algebras

The not yet commonly known fact that differential calculus is part of commutative algebra (see [**64, 71, 136, 133, 124**]), is of fundamental importance, at least for our purposes. A coordinate-free algebraic language is a consequence of this fact and allows one to discover a natural cohomological nature of the Lagrangian formalism in field theory, or, better to say, of the calculus of variations. Our exposition follows the lines sketched for the first time in [**129**] (details were published later in [**134**]).

For convenience of exposiion, the algebraic construction of the adjoint operator, Green's formula, etc., is given in this chapter in a formal axiomatic manner. A conceptual cohomological approach is postponed to Chapter 5.

It is worth noting that the algebraic construction of the Lagrangian formalism as presented below generalizes very naturally to both the graded and noncommutative situations. See [**117**] for a detailed discussion of these topics.

1.1. General preliminaries

In this section, we collected the necessary basic facts of the theory of differential operators over commutative algebras. The main ideas of the theory were exposed by the author in [**136, 133**]. The proofs of the results given here can be found in [**71**].

1.1.1. Let M be a smooth ($= C^\infty$) manifold and let $\mathbb{F}$ be equal to $\mathbb{R}$ or $\mathbb{C}$. The $\mathbb{F}$-algebra of smooth $\mathbb{F}$-valued functions on M is denoted by $C^\infty_{\mathbb{F}}(M)$ or simply by $C^\infty(M)$. The $C^\infty(M)$-module of smooth differential forms of degree i on M is denoted by $\Lambda^i(M) = \Lambda^i_{\mathbb{F}}(M)$. We also use the notation $\Lambda^*(M) = \sum_i \Lambda^i(M)$. We denote by $D(M) = D_{\mathbb{F}}(M)$ the $C^\infty(M)$-module of vector fields on M and identify these fields with derivations of the algebra $C^\infty_{\mathbb{F}}(M)$. We use the standard notation of the calculus of differential forms and vector fields on manifolds (see, for example, [**20, 43, 104**]). The only exception is the Lie derivative: if $X \in D(M)$ is a vector field and ω is an object, then the Lie derivative of ω along X is denoted by $X(\omega)$.

In the sequel, all the constructions and operations related to smooth manifolds are supposed to be smooth, i.e., of the class C^∞. All tensor products are taken over the basic algebra A (mainly, $A = C^\infty(M)$) unless otherwise stated explicitly.

1.1.2. Let K be a commutative ring with unity and A a unitary commutative K-algebra. A pair (a, Δ), $a \in A$, $\Delta \in \mathrm{Hom}_K(P, Q)$, where P and Q are A-modules, determines an element $\delta_a(\Delta) \in \mathrm{Hom}_K(P, Q)$:

$$\delta_a(\Delta)(p) = \Delta(ap) - a\Delta(p), \quad p \in P.$$

The operator $\delta_a\colon \operatorname{Hom}_K(P,Q) \to \operatorname{Hom}_K(P,Q)$, $\Delta \mapsto \delta_a(\Delta)$, is a K-homomorphism and $\delta_a \circ \delta_b = \delta_b \circ \delta_a$. Set

$$\delta_{a_0,a_1,\dots,a_s} = \delta_{a_0} \circ \delta_{a_1} \circ \dots \circ \delta_{a_s}.$$

DEFINITION. A K-homomorphism $\Delta \in \operatorname{Hom}_K(P,Q)$ is called a *K-differential operator* (DO) *of order* $\le s$ over A, if for any elements $a_0,\dots,a_s \in A$ one has $\delta_{a_0,a_1,\dots,a_s}(\Delta) = 0$.

Let $K = \mathbb{F}$, $A = C^\infty_{\mathbb{F}}(M)$, $P = \Gamma(\xi)$, and $Q = \Gamma(\eta)$, where ξ, η are smooth $\mathbb{F}$-vector bundles over M and $\Gamma(\alpha)$ denotes the set of all smooth sections of the bundle α. Such a situation will be called *classical.*

PROPOSITION. *In the classical situation the "usual" definition of differential operator coincides with the given above.*

1.1.3. The set of all DO of order $\le s$ acting from P to Q can be endowed with a structure of left (resp., right) A-module by setting $a\Delta = a \circ \Delta$, i.e., $(a\Delta)(p) = a\Delta(p)$ (resp., $\Delta a = \Delta \circ a$, i.e., $(\Delta a)(p) = \Delta(ap)$). The module defined in such a way is denoted by $\operatorname{Diff}_s(P,Q)$ (resp., $\operatorname{Diff}^+_s(P,Q)$). We also denote by $\operatorname{Diff}^{(+)}_s(P,Q)$ the bimodule endowed with both structures. Obviously, one has the embeddings $\operatorname{Diff}^{(+)}_s(P,Q) \subset \operatorname{Diff}^{(+)}_t(P,Q)$, $s \le t$. Let $\operatorname{Diff}^{(+)}(P,Q) = \bigcup_{s\ge 0} \operatorname{Diff}^{(+)}_s(P,Q)$.

Let P, Q, and R be A-modules. Then for any $\Delta_1 \in \operatorname{Hom}_K(P,Q)$ and $\Delta_2 \in \operatorname{Hom}_K(Q,R)$ one has

$$\delta_a(\Delta_2 \circ \Delta_1) = \delta_a(\Delta_2) \circ \Delta_1 + \Delta_2 \circ \delta_a(\Delta_1),$$

from which it follows that $\Delta_2 \circ \Delta_1 \in \operatorname{Diff}_{s+t}(P,R)$ for any $\Delta_1 \in \operatorname{Diff}_s(P,Q)$, $\Delta_2 \in \operatorname{Diff}_t(Q,R)$. Thus, composition of DO turns $\operatorname{Diff}(P,P)$ into an associative algebra and allows one to regard $\operatorname{Diff}(P,Q)$ as a left $\operatorname{Diff}(Q,Q)$- and a right $\operatorname{Diff}(P,P)$-module.

Let $\operatorname{Diff}^{(+)}_s Q = \operatorname{Diff}^{(+)}_s(A,Q)$ and $\operatorname{Diff}^{(+)} Q = \operatorname{Diff}^{(+)}(A,Q)$. Define a mapping

$$\mathcal{D}^{(+)} = \mathcal{D}^{(+)}(Q)\colon \operatorname{Diff}^{(+)} Q \to Q$$

by $\mathcal{D}(\Delta) = \Delta(1)$ and set $\mathcal{D}^{(+)}_s = \mathcal{D}^{(+)}|_{\operatorname{Diff}^{(+)}_s Q}$. Then $\mathcal{D}_s$ is an A-module homomorphism, while $\mathcal{D}^+_s$ is a DO of order $\le s$. Note also that the mappings $i_+\colon \operatorname{Diff}_s(P,Q) \to \operatorname{Diff}^+_s(P,Q)$ and $i^+\colon \operatorname{Diff}^+_s(P,Q) \to \operatorname{Diff}_s(P,Q)$ induced by the identity mapping of the set $\operatorname{Hom}_K(P,Q)$ are s-th order differential operators.

One can see that to any $\Delta \in \operatorname{Diff}_s(P,Q)$ there corresponds a homomorphism $h_\Delta \in \operatorname{Hom}_A(P, \operatorname{Diff}^+_s Q)$ defined by $h_\Delta(p)(a) = \Delta(ap)$.

PROPOSITION. *The correspondences $\Delta \mapsto h_\Delta$ and $h \mapsto \mathcal{D}_s \circ h$ define mutually inverse isomorphisms between the A-modules* $\operatorname{Diff}_s(P,Q)$ *and* $\operatorname{Hom}^\bullet_A(P, \operatorname{Diff}^+_s Q)$, *where* $\operatorname{Hom}^\bullet$ *means that the set of all A-homomorphisms from P to* $\operatorname{Diff}^+_s Q$ *is endowed with an A-module structure by means of the left module structure in* $\operatorname{Diff}^{(+)}_s Q$.

Let $\Delta \in \operatorname{Diff}_t(P,Q)$ be a differential operator. Consider the mapping

$$\Delta^{(s)} = \Delta \circ \mathcal{D}^+_s(P)$$

and

$$h^s_\Delta = h_{\Delta^{(s)}}\colon \operatorname{Diff}^+_s P \to \operatorname{Diff}^+_{s+t} Q.$$

Passing to the limit as $s \to \infty$, we get a homomorphism $h^*_\Delta\colon \operatorname{Diff}^+ P \to \operatorname{Diff}^+ Q$ of A-modules which is a $\operatorname{Diff} A$-module homomorphism as well. For the operator

$\Delta = \mathcal{I}_t^+$, the corresponding homomorphism $h_\Delta^s\colon \mathrm{Diff}_s^+(\mathrm{Diff}_t^+ P) \to \mathrm{Diff}_{s+t}^+ Q$ is called the *gluing operator* and is denoted by $c_{s,t}$.

Note that $\mathrm{Diff}_0(P,Q) = \mathrm{Hom}_A(P,Q)$ and $\mathrm{Diff}_0 P = P$. Due to the latter, the representation $\Delta = (\Delta - \Delta(1)) + \Delta(1)$ determines a splitting of the A-module $\mathrm{Diff}_s P$ into the direct sum $\mathrm{Diff}_s P = \widetilde{\mathrm{Diff}_s P} \oplus P$, where $\widetilde{\mathrm{Diff}_s P} = \{\Delta \in \mathrm{Diff}_s P \mid \Delta(1) = 0\}$. In a similar way, $\mathrm{Diff}\, P = \widetilde{\mathrm{Diff}\, P} \oplus P$.

1.1.4. The module $D(P) = \widetilde{\mathrm{Diff}_1 P}$ consists of all K-derivations of the K-algebra A with values in the A-module P. Hence, in the classical situation one has $D(C^\infty(M)) = D(M)$.

A function $\varphi\colon (a_1, \dots, a_s) \mapsto \varphi(a_1, \dots, a_s)$ will be called a *skew-symmetric multiderivation* if it is skew-symmetric and is a K-derivation with respect to each argument. The set of all such multiderivations is an A-module, denoted by $D_s(P)$. Obviously, $D_1(P) = D(P)$. We also set $D_0(P) = P$.

The modules $D_s(P)$ can be defined in a more functorial manner as follows. Let $(\mathrm{Diff}_1^+)^i(P) = \mathrm{Diff}_1^+((\mathrm{Diff}_1^+)^{i-1}(P))$. For any *subset* Q_1 of an A-module Q let us define the module $D(Q_1 \subset Q) = \{\nabla \in D(Q) \mid \mathrm{im}\, \nabla \subset Q_1\}$. Then the modules $D_i(P)$ are inductively defined by the equality

$$D_i(P) = D(D_{i-1}(P) \subset (\mathrm{Diff}_1^+)^{i-1}(P)).$$

The embedding $D_i(P) \subset (\mathrm{Diff}_1^+)^i(P)$, which is not an A-module homomorphism, is the composition of the following natural embeddings:

$$D_i(P) = D(D_{i-1}(P) \subset (\mathrm{Diff}_1^+)^{i-1}(P)) \subset D((\mathrm{Diff}_1^+)^{i-1}(P)) \subset (\mathrm{Diff}_1^+)^i(P).$$

In the sequel, we denote by $D_s^\bullet(\mathrm{Diff}_t^+ Q)$ the set of multiderivations $D_s(\mathrm{Diff}_t^+ Q)$ endowed with the structure of an A-module induced by the left module structure in $\mathrm{Diff}_t^{(+)} Q$. In other words, if $\Delta \in D_s^\bullet(\mathrm{Diff}_t^+ Q)$ and $a \in A$, then $(a\Delta)(a_1, \dots, a_s) = a \cdot \Delta(a_1, \dots, a_s)$.

1.1.5. Let $\mathcal{M}_A$ be the category of all A-modules and $\mathcal{F}$ be one of the functors Diff_s, Diff_s^+, $\mathrm{Diff}_s^{(+)}$, etc. or a composition of these functors (e.g., $D_i(\mathrm{Diff}_s^+)$). Then the correspondence $P \Rightarrow \mathcal{F}(P)$ is a functor on the category $\mathcal{M}_A$ taking values either in the same category, or in the category $\mathcal{M}_A^k$ of A-polymodules (e.g., $\mathrm{Diff}^{(+)} P \in \mathrm{Ob}\, \mathcal{M}_A^2$). If Φ is an A-module such that the functor $P \Rightarrow \mathrm{Hom}_A(\Phi, P)$ is equivalent to the functor $P \Rightarrow \mathcal{F}(P)$, then Φ is called the *representing object* for $\mathcal{F}$.

Functors $\mathcal{F}$ of the above type are examples of *differential calculus functors.* All of them possess representing objects. In particular, the representing object for $\mathcal{F} = D_i$ is the module of *i-differential forms* $\Lambda^i = \Lambda^i(A)$. The functor $\mathcal{F} = \mathrm{Diff}_s$ is represented by the module $\mathcal{J}^s = \mathcal{J}^s(A)$ of s-jets. The representing object for the functor $Q \Rightarrow \mathrm{Diff}_s(P,Q)$ is denoted by $\mathcal{J}^s(P)$ and is called the *module of s-jets of the module P*.

Let $Q = \mathcal{J}^s(P)$. Then $\mathrm{Diff}_s(P, \mathcal{J}^s(P)) = \mathrm{Hom}_A(\mathcal{J}^s(P), \mathcal{J}^s(P))$. The s-th order operator $j_s = j_s(P)\colon P \to \mathcal{J}^s(P)$ corresponding to $\mathrm{id}_{\mathcal{J}^s(P)}$ is called the *s-jet operator*. Consider the homomorphism $\varphi = \varphi_\square \in \mathrm{Hom}_A(\mathcal{J}^s(P), Q)$ corresponding to $\square \in \mathrm{Diff}_s(P,Q)$ under the natural A-module isomorphism $\mathrm{Diff}_s(P,Q) = \mathrm{Hom}_A(\mathcal{J}^s(P), Q)$. Then $\square = \varphi \circ j_s$.

Note that, as usual, $\Lambda^i = \Lambda^1 \wedge \cdots \wedge \Lambda^1$ (i factors).

1.1.6. Natural transformations of the differential calculus functors define, via duality, the operators connecting corresponding representing objects. Here are three examples we will need later.

1. The natural embedding $\mathrm{Diff}_{k-1} \hookrightarrow \mathrm{Diff}_k$ induces an A-module homomorphism $\nu_{k,k-1}\colon \mathcal{J}^k(P) \to \mathcal{J}^{k-1}(P)$ of the *jet order reduction.*

2. The representing object for the functor D_iD_j is $\Lambda^i \otimes_A \Lambda^j$. Then the natural embedding $D_{i+j} \hookrightarrow D_iD_j$ induces a homomorphism $\Lambda^i \otimes_A \Lambda^j \to \Lambda^{i+j}$ which is the *exterior product of differential forms.*

3. The representing object for the functor $D_s^\bullet(\mathrm{Diff}_1^+)$ is $\mathcal{J}^1(\Lambda^s)$. Using the inductive definition of D_s, one can easily construct natural embeddings $D_s \hookrightarrow D_s^\bullet(\mathrm{Diff}_1^+)$ starting with $D \hookrightarrow \mathrm{Diff}_1$. Passing to representing objects, we get a homomorphism $\mathcal{J}^1(\Lambda^{s-1}) \to \Lambda^s$. The operator $d\colon \Lambda^{s-1} \to \Lambda^s$ defined as the composition $\Lambda^{s-1} \xrightarrow{j_s} \mathcal{J}^1(\Lambda^{s-1}) \to \Lambda^s$ is a first-order operator called the *exterior differentiation operator.* Thus we get the sequence

$$0 \to A = \Lambda^0 \xrightarrow{d} \Lambda^1 \xrightarrow{d} \Lambda^2 \xrightarrow{d} \cdots$$

which is a complex called the *de Rham complex* of the algebra A.

1.1.7. Let $\mathcal{K}$ be a subcategory of the category $\mathcal{M}_A$. Consider the set $\{\mathcal{F}\}$ of differential calculus functors of the above type on $\mathcal{M}_A$, and for any natural transformation $T\colon \mathcal{F}_1 \Rightarrow \mathcal{F}_2$ denote by $T_P \in \mathrm{Mor}(\mathcal{F}_1(P), \mathcal{F}_2(P))$ the corresponding homomorphism $\mathcal{F}_1(P) \to \mathcal{F}_2(P)$. We assume that $\mathcal{K}$ is such that $P \in \mathrm{Ob}\,\mathcal{K}$ implies both $\mathcal{F}(P) \in \mathrm{Ob}\,\mathcal{K}$ and $T_P \in \mathrm{Mor}_{\mathcal{K}}(\mathcal{F}_1(P), \mathcal{F}_2(P))$. In this situation, it is reasonable to look for representing objects in $\mathcal{K}$. If such objects exist, we denote them in the same way as in Subsection 1.1.4 by adding the subscript $\mathcal{K}$: e.g., $\mathcal{J}^s_{\mathcal{K}}$ or $d_{\mathcal{K}}\colon \Lambda^{i-1}_{\mathcal{K}} \to \Lambda^i_{\mathcal{K}}$.

One of the most important examples of such a subcategory is the category $\mathcal{MG}_A$ of *geometric modules.* A module $Q \in \mathrm{Ob}\,\mathcal{M}_A$ is said to be geometric if $\tilde{Q} = \bigcap_I IQ = 0$, $I \in \mathrm{Spec}\,A$. One can define the *geometrization functor* $\mathcal{G}\colon \mathcal{M}_A \Rightarrow \mathcal{MG}_A$ by setting $Q \Rightarrow Q/\tilde{Q}$.

PROPOSITION. *The differential calculus functors $\mathcal{F}$ are representable in $\mathcal{MG}_A$, and the geometrization functor takes the representing objects in $\mathcal{M}_A$ to the corresponding representing objects in $\mathcal{MG}_A$.*

Consider the classical case (see Subsection 1.1.2). Let $\mathcal{K} = \mathcal{MG}_A$. Then $\Lambda^i_{\mathcal{K}} = \Lambda^i(M)$ and $\mathcal{J}^s_{\mathcal{K}}(P) = \Gamma(\xi_s)$, where $\xi_s\colon J^s(\xi) \to M$ is the bundle of s-jets corresponding to the bundle ξ over M (see Subsection 2.2.1). Similar statements are valid for other standard constructions related to differential calculus on manifolds.

A category $\mathcal{K}$ is said to be *smooth* if the module $\Lambda^1_{\mathcal{K}}$ is projective and of finite type. For example, $\mathcal{MG}_A$ is a smooth category in the classical situation.

1.1.8. If $a \in A$, $\Delta \in \mathrm{Diff}_k^{(+)}(P,Q)$, then

$$a \cdot \Delta = \Delta \cdot a \bmod \mathrm{Diff}_{k-1}^{(+)}(P,Q)$$

due to the fact that $a\cdot\Delta - \Delta\cdot a = \delta_a(\Delta)$ is a DO of order $\le k-1$. So, the quotient modules

$$\mathrm{smbl}_k(P,Q) = \frac{\mathrm{Diff}_k(P,Q)}{\mathrm{Diff}_{k-1}(P,Q)} = \frac{\mathrm{Diff}_k^+(P,Q)}{\mathrm{Diff}_{k-1}^+(P,Q)}$$

coincide. Let us set

$$\mathrm{smbl}(P,Q) = \sum_k \mathrm{smbl}_k(P,Q)$$

and call this module the *module of symbols*. For $P = A$, we write $\mathrm{smbl}_k(Q)$ instead of $\mathrm{smbl}_k(A,Q)$ and $\mathrm{smbl}(Q)$ instead of $\mathrm{smbl}(A,Q)$. If an operator Δ lies in $\mathrm{Diff}_k^{(+)}(P,Q)$, then its class in $\mathrm{smbl}_k(P,Q)$ is denoted by $\mathrm{smbl}_k \Delta$ and called the *symbol* of the operator Δ.

If $\Delta \in \mathrm{Diff}_k(P,Q)$ and $\nabla \in \mathrm{Diff}_l(S,P)$, then the product

$$(\mathrm{smbl}_k \Delta) \cdot (\mathrm{smbl}_l \nabla) = \mathrm{smbl}_{k+l}(\Delta \circ \nabla)\, \mathrm{smbl}_{k+l}(S,Q)$$

is well defined. With this product, $\mathrm{smbl}(P,P)$ becomes an associative algebra. This algebra is commutative in the case $P = A$ and is called the *algebra of symbols*. By the same reason, $\mathrm{smbl}(P,Q)$ is a left $\mathrm{smbl}(Q,Q)$-module and a right $\mathrm{smbl}(P,P)$-module. In particular, $\mathrm{smbl}\, P$ is a $\mathrm{smbl}\, A$-module.

When passing to quotient modules, the homomorphisms h_Δ^s (see Subsection 1.1.3) induce the homomorphisms $\mathrm{smbl}_s P \to \mathrm{smbl}_{s+t} Q$, $s \geq 0$; their direct sum is denoted by

$$\mathrm{smbl}\, \Delta \colon \mathrm{smbl}\, P \to \mathrm{smbl}\, Q$$

and is also called the *symbol* of Δ. Note that $\mathrm{smbl}(\nabla \circ \Delta) = \mathrm{smbl}\, \nabla \circ \mathrm{smbl}\, \Delta$.

The isomorphism $\mathrm{Diff}_s(P,Q) = \mathrm{Hom}_A^{\bullet}(P, \mathrm{Diff}_s^+ Q)$ (see Proposition 1.1.3) generates in an obvious way an A-homomorphism $\mathrm{smbl}_s(P,Q) \to \mathrm{Hom}_A(P, \mathrm{smbl}_s Q)$, which is an isomorphism if P is projective. Putting together these homomorphisms for all $s = 0, 1, \dots$, one gets a $\mathrm{smbl}\, A$-module homomorphism

$$\mathrm{smbl}(P,Q) \to \mathrm{Hom}_A(P, \mathrm{smbl}\, P),$$

which is an isomorphism for a projective P. If Q is also projective, then we have the isomorphisms

$$\begin{aligned} \mathrm{smbl}(P,Q) &= \mathrm{Hom}_A(P, \mathrm{smbl}\, Q) = \mathrm{Hom}_A(P, \mathrm{smbl}\, A \otimes_A Q) \\ &= \mathrm{smbl}\, A \otimes_A \mathrm{Hom}_A(P,Q). \end{aligned}$$

In particular, if $P = A$, then

$$\mathrm{smbl}\, Q = \mathrm{smbl}\, A \otimes_A Q.$$

Denote by $D_{\mathrm{sym}}^k(P)$ the A-module of all symmetric multiderivations with values in P depending on k arguments. Any $\Delta \in \mathrm{Diff}_k(P,Q)$ determines a multiderivation $\mathrm{sm}_k(\Delta) \in D_{\mathrm{sym}}^k(\mathrm{Hom}_A(P,Q))$ by the equality

$$\mathrm{sm}_k(\Delta)(a_1, \dots, a_k) = \delta_{a_1,\dots,a_k}(\Delta).$$

If $\Delta \in \mathrm{Diff}_{k-1}(P,Q)$, then $\mathrm{sm}_k(\Delta) = 0$. Therefore, the correspondence $\Delta \mapsto \mathrm{sm}_k(\Delta)$ generates an A-module homomorphism

$$\chi_k \colon \mathrm{smbl}_k(P,Q) \longrightarrow D_{\mathrm{sym}}^k(\mathrm{Hom}_A(P,Q)).$$

Let a multiderivation $[\nabla_1, \dots, \nabla_k] \in D_{\mathrm{sym}}^k(A)$, $\nabla_i \subset D(A)$, be defined by

$$(a_1, \dots, a_k) \mapsto \frac{1}{k!} \sum_\sigma \nabla_1(a_{\sigma(1)}) \cdots \nabla_k(a_{\sigma(k)}),$$

where the sum is taken over all permutations of the set $\{1, \dots, k\}$. Then one has the homomorphism

$$\gamma_k \colon S^k(D) \otimes \mathrm{Hom}_A(P, Q) \longrightarrow D^k_{\mathrm{sym}}(\mathrm{Hom}_A(P, Q)),$$

where $S^k(D)$ denotes the k-th symmetric power of the A-module $D = D(A)$ and

$$\gamma_k(\nabla_1 \cdots \nabla_k \otimes h)(a_1, \dots, a_k) = [\nabla_1, \dots, \nabla_k](a_1, \dots, a_k) \otimes h$$

with $h \in \mathrm{Hom}_A(P, Q)$.

PROPOSITION. *If $\mathcal{K}$ is a smooth category, $A \in \mathrm{Ob}\,\mathcal{K}$, and P is a projective A-module of finite type, then χ_k and γ_k are isomorphisms.*

Hence in the classical situation, the modules $\mathrm{smbl}_k(P, Q)$, $D^k_{\mathrm{sym}}(\mathrm{Hom}_A(P, Q))$, and the tensor product $S^k(D) \otimes \mathrm{Hom}_A(P, Q)$ can be identified. Under the assumptions made above one can also identify $\ker \nu_{k,k-1}$ with $S^k(\Lambda^1_{\mathcal{K}}) \otimes P$, where, as before, S^k denotes the k-th symmetric power.

1.1.9. A geometric interpretation of the algebra of symbols is as follows (see [**141**] for details). Consider $\mathrm{Spec}_{\mathbb{F}}(\mathrm{smbl}\, A)$, the $\mathbb{F}$-spectrum of the $\mathbb{F}$-algebra $A = C^\infty_{\mathbb{F}}(M)$. Its elements are homomorphisms $h \colon \mathrm{smbl}\, A \to \mathbb{F}$ of unitary $\mathbb{F}$-algebras, i.e., $h(1_{\mathrm{smbl}\, A}) = 1_{\mathbb{F}}$. There exists a canonical identification

$$\mathrm{Spec}_{\mathbb{F}}(\mathrm{smbl}\, A) = T^*M$$

under which elements of the algebra $\mathrm{smbl}\, A$ are identified with smooth functions on T^*M that are polynomial along the fibers of the projection $\tau \colon T^*M \to M$.

If $\Delta_i \in \mathrm{Diff}_{k_i} A$, $i = 1, 2$, then $[\Delta_1, \Delta_2] \in \mathrm{Diff}_{k_1+k_2-1} A$. Denote by f_i the functions on T^*M corresponding to $\mathrm{smbl}_{k_i} \Delta_i$ under the above identification. Then the function corresponding to $\mathrm{smbl}_{k_1+k_2-1}[\Delta_1, \Delta_2]$ is the canonical *Poisson bracket* $\{f_1, f_2\}$ of two functions on T^*M.

If $P = \Gamma(\xi)$ with $\xi \colon E_\xi \to M$ being a vector bundle over M, then the module $\mathrm{smbl}\, P$ is identified with the $\mathrm{smbl}\, A$-module $\Gamma_{\mathrm{poly}}(\tau^*(\xi))$ of sections of the pullback $\tau^*(\xi) \colon \tau^*(E_\xi) \to T^*M$ polynomial along fibers of τ. Suppose that $Q = \Gamma(\eta)$ for a vector bundle $\eta \colon E_\eta \to M$. Then the canonical isomorphism

$$\mathrm{smbl}(P, Q) = \mathrm{smbl}\, A \otimes_A \mathrm{Hom}(P, Q)$$

allows one to identify the $(\mathrm{smbl}\, A)$-module $\mathrm{smbl}(P, Q)$ with the $(\mathrm{smbl}\, A)$-module $\mathrm{Hom}^{\mathrm{poly}}(\tau^*(\xi), \tau^*(\eta))$ of vector bundle homomorphisms from $\tau^*(\xi)$ to $\tau^*(\eta)$ that are polynomial along fibers of τ. In particular, if $\Delta \in \mathrm{Diff}_s(P, Q)$, then $\mathrm{smbl}\, \Delta$ may be viewed as a fiber bundle homomorphism

$$s_\Delta \colon \tau^*(\xi) \to \tau^*(\eta).$$

A covector $\theta \in T^*_x M \subset T^*M$ is said to be *characteristic* for Δ if the evaluation $s_{\Delta,\theta}$ of s_Δ at θ taking the fiber of $\tau^*(\xi)$ at θ to that of $\tau^*(\eta)$ has a nontrivial cokernel. Put

$$\mathrm{char}\, \Delta = \{\theta \in T^*M \mid \theta \text{ is characteristic for } \Delta\}.$$

Obviously, $\mathrm{char}\, \Delta$ is a closed subset in T^*M.

1.1.10. To an A-module P and a number $k \in \mathbb{N}$ one associates the complex denoted by $\mathcal{J}_k P$ and called the *k-th Spencer jet-complex* of P:

$$0 \to P \xrightarrow{j_k} \mathcal{J}^k(P) \xrightarrow{\Sigma_{0,k}} \mathcal{J}^{k-1}(P) \otimes \Lambda^1 \xrightarrow{\Sigma_{1,k-1}} \cdots \xrightarrow{\Sigma_{s-1,k-s+1}} \mathcal{J}^{k-s}(P) \otimes \Lambda^s \xrightarrow{\Sigma_{s,k-s}} \cdots .$$

The operators $\Sigma = \Sigma_{s,t}$ are defined by the formula

$$\Sigma(j_t(p) \otimes \omega) = j_{t-1}(p) \otimes d\omega, \quad p \in P, \ \omega \in \Lambda^s.$$

Evidently, Σ is a first-order operator. Note that the case $k = \infty$ can be considered as well.

1.1.11. In the sequel the theory of differential operators on smooth manifolds is understood as differential calculus in the category $\mathcal{K} = \mathcal{MG}_A$ with $A = C^\infty_{\mathbb{F}}(M)$.

1.2. Adjoint operators

In this section, we construct the theory of adjoint operators in a way which allows one to apply it in situations where the usual coordinate approach is not efficient. We fix a manifold M, $\dim M = n$, and use the notations $\Lambda^i = \Lambda^i_{\mathbb{F}}(M)$, $A = C^\infty_{\mathbb{F}}(M)$.

1.2.1. Action of operators on Λ^n. Define the action $\Delta[\omega] \in \Lambda^n$ of operators $\Delta \in \operatorname{Diff} A$ on differential forms $\omega \in \Lambda^n$ by the following rules:

1. $(\Delta_1 + \Delta_2)[\omega] = \Delta_1[\omega] + \Delta_2[\omega]$, $\Delta_1, \Delta_2 \in \operatorname{Diff} A$.
2. $f[\omega] = f\omega$, $f \in \operatorname{Diff}_0 A = A$.
3. $X[\omega] = -X(\omega)$, $X \in D(M)$.
4. $(\Delta_1 \circ \Delta_2)[\omega] = \Delta_2[\Delta_1[\omega]]$.

PROPOSITION. *The action satisfying the above rules* 1–4 *exists and is unique.*

PROOF. Any operator $\Delta \in \operatorname{Diff} A$ can be presented as the sum of a "scalar" $f = \Delta(1)$ and monomials of the form $X_1 \circ \cdots \circ X_s$, $X_i \in D(M)$. But due to rules 3 and 4,

$$(X_1 \circ \cdots \circ X_s)[\omega] = (-1)^s X_s(\dots(X_1(\omega))\dots).$$

Therefore, $\Delta[\omega] = f\omega + \sum(-1)^s X_s(\dots(X_1(\omega)))\dots)$ which proves the uniqueness.

To prove the existence, we show first that

(1) $(fX)[\omega] = (X \circ f)[\omega]$

and

(2) $(X \circ Y)[\omega] = (Y \circ X)[\omega] + ([X,Y])[\omega]$.

To check the first equality, note that $Z(\omega) = Z \lrcorner\, d\omega + d(Z \lrcorner\, \omega) = d(Z \lrcorner\, \omega)$, since $d\omega = 0$, and that

$$0 = Y \lrcorner\, (df \wedge \omega) = Y(f)\omega - df \wedge (Y \lrcorner\, \omega),$$

i.e., $Y(f)\omega = df \wedge (Y \lrcorner\, \omega)$. Hence,

$$(fX)(\omega) = d(fX \lrcorner\, \omega) = fd(X \lrcorner\, \omega) + df \wedge (X \lrcorner\, \omega) - fX(\omega) + X(f)\omega.$$

This shows that $(fX)[\omega] = -fX(\omega) - X(f)\omega$. On the other hand,

$$(X \circ f)[\omega] - X(f)[\omega] = f[X[\omega]] - X(f)[\omega] = -fX(\omega) - X(f)\omega.$$

The second equality immediately follows from the well-known property of the Lie derivative: $X(Y(\omega)) - Y(X(\omega)) = [X, Y](\omega)$.

Let us represent the operator Δ in the form $\Delta = f + \sum_i X_1^i \circ \cdots \circ X_{s(i)}^i$, where $X_j^i \in D(M)$, and set

$$\Delta[\omega] = f\omega + \sum_i (-1)^{s(i)} X_{s(i)}^i(\ldots(X_1^i(\omega))\ldots).$$

This formula gives the desired action if we prove that it is well defined. To prove the latter, we need to show that the expression above vanishes when $\Delta = 0$. Consider a "monomial" $\Delta = X_1 \circ \cdots \circ X_s$. From (1) and (2) it follows that the definition of $\Delta[\omega]$ is invariant under the following elementary operations:

(a) for $X_{i+1} = fY$, one can replace $X_i \circ X_{i+1}$ by $fX_i \circ Y - X_i(f)Y$;

(b) $X_i \circ X_{i+1}$ can be replaced by $X_{i+1} \circ X_i + [X_i, X_{i+1}]$.

It can be easily seen that the expression $f + \sum X_1^i \circ \cdots \circ X_{s(i)}^i = 0$ can be reduced to zero by these elementary operations. □

1.2.2. Adjoint operator. Suppose at first that $\Delta \in \operatorname{Diff} \Lambda^n$ and M is an orientable manifold. Let $\omega \in \Lambda^n$ be the volume form, i.e., $\omega_x \neq 0$ for all $x \in M$. Let us define $\Delta_\omega \in \operatorname{Diff} A$ by setting $\Delta(f) = \Delta_\omega(f)\omega$.

DEFINITION. An operator $\Delta^* \colon A \to \Lambda^n$ is called *adjoint* to $\Delta \in \operatorname{Diff} \Lambda^n$ if $\Delta^*(f) = \Delta_\omega[f\omega]$.

Let $\omega' = g\omega$ be another volume form. Then $\Delta_\omega = g\Delta_{\omega'}$ and

$$\Delta_\omega[f\omega] = (g \circ \Delta_{\omega'})[f\omega] = \Delta_{\omega'}[fg\omega] = \Delta_{\omega'}[f\omega']$$

which means that the adjoint operator is well defined.

If M is nonorientable, then in the same manner one can construct the adjoint operator Δ_U^* in any orientable domain $U \subset M$. Then, due to the uniqueness of the action $\Delta[\cdot]$, one has $\Delta_U^*|_{U\cap V} = \Delta_{U\cap V}^* = \Delta_V^*|_{U\cap V}$. Hence, the system of operators $\{\Delta_U^*\}$ produces a well-defined operator Δ^* on the entire manifold M.

PROPOSITION. *We have*

1. *If* $\Delta \in \operatorname{Diff}_s \Lambda^n$, *then* $\Delta^* \in \operatorname{Diff}_s \Lambda^n$.
2. *If* $\omega \in \Lambda^n = \operatorname{Diff}_0 \Lambda^n$, *then* $\omega^* = \omega$.
3. *If* $X \in D(\Lambda^n)$, *then* $X + X^* \in \operatorname{Diff}_0 \Lambda^n = \Lambda^n$.
4. *One has* $\operatorname{smbl}_s \Delta^* = (-1)^s \operatorname{smbl}_s \Delta$ *for any* $\Delta \in \operatorname{Diff}_s \Lambda^n$.
5. $(\Delta^*)^* = \Delta$ *for any* $\Delta \in \operatorname{Diff} \Lambda^n$.

PROOF. (1) Let $f, g \in A$. Then

$$\begin{aligned}\delta_g(\Delta^*)(f) &= \Delta^*(gf) - g\Delta^*(f) = \Delta_\omega[gf\omega] - g\Delta_\omega[f\omega] \\ &= -\delta_g(\Delta_\omega)[f\omega] = -\delta_g(\Delta)_\omega[f\omega] = -\delta_g(\Delta)^*(f),\end{aligned}$$

i.e., $\delta_g(\Delta^*) = -\delta_g(\Delta)^*$. Hence,

$$0 = [\delta_{g_0,\ldots,g_s}(\Delta)]^* = (-1)^{s+1}\delta_{g_0,\ldots,g_s}(\Delta^*)$$

for all $g_0, \ldots, g_s \in A$.

(2) is obvious.

(3) Let $X \in D(\Lambda^n)$. Then $X_\omega \in D(M)$ and

$$X^*(f) = -X_\omega(f\omega) = -X_\omega(f)\omega - fX_\omega(\omega).$$

Therefore, $X^* = -X - X_\omega(\omega)$ and $X + X^* = -X_\omega(\omega)$.

(4) Let us start with Δ such that $\Delta_\omega = X_1 \circ \cdots \circ X_k$, where $X_i \in D(M)$, $k \le s$. Then

$$\Delta^*(f) = (-1)^k X_k(\dots(X_1(f\omega))\dots) = (-1)^k (X_k \circ \cdots \circ X_1)(f) \cdot \omega + \Delta'(f),$$

where $\Delta' \in \mathrm{Diff}_{k-1}\Lambda^n$. But $X_k \circ \cdots \circ X_1 = X_1 \circ \cdots \circ X_k + \Delta''$, where $\Delta'' \in \mathrm{Diff}_{k-1}A$, from which it follows that $\mathrm{smbl}_s \Delta^* = \mathrm{smbl}_s \Delta = 0$ if $k < s$. Hence, for $k = s$ one has $\mathrm{smbl}_s \Delta^* = \Delta^* \bmod \mathrm{Diff}_{s-1}\Lambda^n = (-1)^s \mathrm{smbl}_s(X_1 \circ \cdots \circ X_s) \cdot \omega = (-1)^s \mathrm{smbl}_s \Delta$.

The general case is reduced to the above one, since Δ_ω can be represented in the form $\Delta(1) + \sum_i X_1^i \circ \cdots \circ X_{s(i)}^i$, where $s(i) \le s$ and $X_j^i \in D(M)$.

(5) Let us show at first that $(\Delta \circ X)^* = -\bar{X} \circ \Delta^*$ and $(\bar{X} \circ \Box)^* = -\Box^* \circ X$, where $\Delta, \Box \in \mathrm{Diff}\,\Lambda^n$, $X \in D(M)$, and $\bar{X} \in \mathrm{Diff}_1(\Lambda^n, \Lambda^n)$ is defined by $\bar{X}(\rho) = X(\rho)$ for any $\rho \in \Lambda^n$. Note that $(\Delta \circ X)_\omega = \Delta_\omega \circ X$. Hence,

$$(\Delta \circ X)^*(f) = -X(\Delta_\omega[f\omega]) = -X(\Delta^*(f)) = -(\bar{X} \circ \Delta^*)(f).$$

On the other hand,

$$(\bar{X} \circ \Box)(f) = X(\Box_\omega(f)\omega) = (X \circ \Box_\omega + g\Box_\omega)(f) \cdot \omega,$$

where $g\omega = X(\omega)$. Therefore, $(\bar{X} \circ \Box)_\omega = (X + g) \circ \Box_\omega$ and hence $(\bar{X} \circ \Box)^*(f) = ((X+g) \circ \Box_\omega)[f\omega] = \Box_\omega[(X+g)[f\omega]] = -\Box_\omega[X(f)\omega] = -\Box^*(X(f))$.

Now let $\Delta = \Box \circ X$, where $\Delta, \Box \in \mathrm{Diff}\,\Lambda^n$, $X \in D(M)$. Then from the above equalities it follows that $\Delta^{**} = \Box^{**} \circ X$. But since any $\Delta \in \mathrm{Diff}_s\Lambda^n$ can be represented in the form $\Delta = \sum_i \Box_i \circ X_i + \Delta(1)$, where $\Box_i \in \mathrm{Diff}_{s-1}\Lambda^n$, $X_i \in D(M)$, the last equality allows us to prove the desired fact by induction on $\deg \Delta$. $\Box$

REMARK. If $X \in D(\Lambda^n)$, then the form $\operatorname{div} X = X + X^* = -X_\omega(\omega) \in \Lambda^n$ is natural to be called the *divergence* of the "field" X.

1.2.3. Let P, Q be A-modules and $\hat{Q} = \mathrm{Hom}_A(Q, \Lambda^n)$. Now we construct the adjoint operator in the general case for any $\Delta \in \mathrm{Diff}(P, \hat{Q})$. To do this, define a family of operators $\Delta(p,q) \in \mathrm{Diff}\,\Lambda^n$, $p \in P$, $q \in Q$, by setting $\Delta(p,q)(f) = \Delta(fp)(q)$, $f \in A$. Obviously, Δ is uniquely determined by the family $\Delta(p,q)$. The following result is also evident.

LEMMA. *Consider a family of operators $\Delta[p,q] \in \mathrm{Diff}\,\Lambda^n$, $p \in P$, $q \in Q$. Then $\Delta[p,q] = \Delta(p,q)$ for some operator $\Delta \in \mathrm{Diff}(P, \hat{Q})$ if and only if*

$$\Delta[p, \sum_i f_i q_i] = \sum_i f_i \Delta[p, q_i], \quad \Delta[\sum_i f_i p_i, q] = \sum_i f_i \Delta[p_i, q_i]$$

for any $f_i \in A$, $p, p_i \in P$, $q, q_i \in Q$.

According to the lemma, the family $\Delta^*(q,p) = \Delta(p,q)^*$ uniquely determines an operator $\Delta^* \in \mathrm{Hom}_A(Q, \hat{P})$ which will be called the *adjoint operator* of Δ.

In more explicit terms, the adjoint operator is defined by

$$\langle \Delta^*(q), p \rangle = \Delta(p,q)^*(1), \tag{1.1}$$

where the bracket $\langle \cdot, \cdot \rangle$ stands for a natural pairing $\hat{P} \times P \to \Lambda^n$.

A more convenient description of operators $\Delta(p,q)$ is as follows. Let $p \in P$. Let the homomorphism $f \mapsto fp$ of A to P be also denoted by p. The corresponding adjoint operator $p^* \colon \hat{P} \to \Lambda^n$ is given by

$$p^*(p) = \langle \hat{p}, p \rangle \tag{1.2}$$

as it results directly from the definition and Proposition 1.2.2 (2). Then, obviously,

$$\Delta(p,q) = q^* \circ \Delta \circ p, \quad p^* \circ \Delta^* \circ q = (q^* \circ \Delta \circ p)^*. \tag{1.3}$$

PROPOSITION. *Let P, Q, R be projective A-modules of finite type. Let us identify $\hat{\hat{P}}$ with P, etc. Then*

1. *If $\Delta\colon P \to \hat{Q}$ is an A-homomorphism, then for $\Delta^*\colon Q \to \hat{P}$,*
$$\langle \Delta(p), q\rangle = \langle \Delta^*(q), p\rangle.$$
2. *For any $\Delta \in \mathrm{Diff}(P, \hat{Q})$ one has $\Delta^{**} = \Delta$.*
3. *For any $\Delta_1 \in \mathrm{Diff}(P, \hat{Q})$, $\Delta_2 \in \mathrm{Diff}(\hat{Q}, R)$ one has $(\Delta_2 \circ \Delta_1)^* = \Delta_1^* \circ \Delta_2^*$.*

PROOF. (1) If Δ is a homomorphism, then, obviously, $\Delta \circ p = \Delta(p)\colon A \to \hat{P}$ and $q^* \circ \Delta \circ p = q^* \circ \Delta(p) = \langle \Delta(p), q\rangle$ (the last is due to (1.2)). Thus, Proposition 1.2.2 (2) implies that

$$(q^* \circ \Delta \circ p)^* = \langle \Delta(p), q\rangle^* = \langle \Delta(p), q\rangle \in \Lambda^n.$$

Therefore, $\langle \Delta^*(q), p\rangle = \Delta(p,q)^*(1) = (q^* \circ \Delta \circ p)^*(1) = \langle \Delta(p), q\rangle$.

(2) We have $\Delta^{**}(p,q) = (\Delta^*(p,q))^* = \Delta(p,q)^{**}$. Since $\Delta(p,q) \in \mathrm{Diff}\,\Lambda^n$, Proposition 1.2.2 (5) implies that $\Delta(p,q)^{**} = \Delta(p,q)$.

(3) We start with the particular case $A \xrightarrow{\Delta_1} A \xrightarrow{\Delta_2} \Lambda^n$. Represent Δ_1 in the form

$$\Delta_1 = \Delta_1(1) + \sum X_1^i \circ \cdots \circ X_{s(i)}^i, \quad X_j^i \in D(M).$$

From the formula $(\Box \circ X)^* = -\bar{X} \circ \Box^*$ proved above it is easy to see that $(\Delta_2 \circ \Delta_1)^* = \tilde{\Delta}_1 \circ \Delta_2^*$, where $\tilde{\Delta}_1\colon \Lambda^n \to \Lambda^n$ is defined by $\tilde{\Delta}_1(\omega) = \Delta_1[\omega]$. Directly from the definition it follows that $\tilde{\Box} = \Box^*$ for any $\Box \in \mathrm{Diff}\,A$. Hence, in the case under consideration one has $(\Delta_2 \circ \Delta_1)^* = \Delta_1^* \circ \Delta_2^*$.

Note also that the formula $(\Delta_2 \circ \Delta_1)^* = \Delta_1^* \circ \Delta_2^*$ holds if at least one the operators Δ_1, Δ_2 is an A-homomorphism. For instance, let Δ_1 be a homomorphism. Then, due to Proposition 1.2.3 (1),

$$(p^* \circ \Delta_1^*)(q) = p^*(\Delta_1^*(q)) = \langle \Delta_1^*(q), p\rangle = \langle \Delta_1(p), q\rangle = \Delta_1(p)^*(q),$$

which shows that

$$p^* \circ \Delta_1^* = \Delta_1(p)^*. \tag{1.4}$$

On the other hand, taking into account (1.3) and (1.4), we have

$$\begin{aligned} p^* \circ (\Delta_2 \circ \Delta_1)^* \circ r &= (r^* \circ \Delta_2 \circ \Delta_1 \circ p)^* = (r^* \circ \Delta_2 \circ \Delta_1(p))^* \\ &= \Delta_1(p)^* \circ \Delta_2^* \circ r = p^* \circ \Delta_1^* \circ \Delta_2^* \circ r. \end{aligned}$$

This proves the desired assertion. If Δ_2 is an A-homomorphism, the proof is similar and is based on the formula

$$r^* \circ \Delta_2 = (\Delta_2^*(r))^*,$$

which reduces to (1.4) by setting $\Delta_1 = \Delta_2^*$ and taking into account Proposition 1.2.3 (2).

Now we can prove the desired formula in the situation $A \xrightarrow{\Delta_1} \hat{Q} \xrightarrow{\Delta_2} \Lambda^n$ with $\hat{Q}$ being a free A-module of finite type. Consider a splitting of $\hat{Q}$:

$$\hat{Q} \xrightarrow{\beta_i} Q \xrightarrow{\alpha_i} \hat{Q} \quad \text{with} \quad \mathrm{id}_{\hat{Q}} = \bigoplus_{i=1}^{m} \alpha_i \circ \beta_i.$$

Since α_i and β_i are homomorphisms, one has

$$(\beta_i \circ \Delta_1)^* = \Delta_1^* \circ \beta_i^*, \quad (\Delta_2 \circ \alpha_i)^* = \alpha_i^* \circ \Delta_2^*.$$

Moreover, $\beta_i \circ \Delta_1 \colon A \to A$ and $\Delta_2 \circ \alpha_i \colon A \to \Lambda^n$. In such a situation, as was proved earlier,

$$(\Delta_2 \circ \alpha_i \circ \beta_i \circ \Delta_1)^* = (\beta_i \circ \Delta_1)^* \circ (\Delta_2 \circ \alpha_i).$$

So, putting these facts together, one finds that

$$\begin{aligned}(\Delta_2 \circ \Delta_1)^* &= \Big(\Delta_2 \circ (\bigoplus_{i=1}^{m} \alpha_i \circ \beta_i) \circ \Delta_1\Big)^* = \Big(\bigoplus_{i=1}^{m} (\Delta_2 \circ \alpha_i) \circ (\beta_i \circ \Delta_1)\Big)^* \\ &= \bigoplus_{i=1}^{m} (\beta_i \circ \Delta_1)^* \circ (\Delta_2 \circ \alpha_i)^* = \bigoplus_{i=1}^{m} \Delta_1^* \circ \beta_i^* \circ \alpha_i^* \circ \Delta_2^* \\ &= \Delta_1^* \circ \bigoplus_{i=1}^{m} (\alpha_i \circ \beta_i)^* \circ \Delta_2^* = \Delta_1^* \circ \mathrm{id}_{\hat{Q}}^* \circ \Delta_2^* = \Delta_1^* \circ \Delta_2^*.\end{aligned}$$

Now we are in a position to consider the situation $A \xrightarrow{\Delta_1} \hat{Q} \to \Lambda^n$ with a projective module $\hat{Q}$. Recall that such a $\hat{Q}$ is a direct summand of a free module T. Denote by $i \colon \hat{Q} \to T$ and $\pi \colon T \to \hat{Q}$ the corresponding embedding and projection, $\pi \circ i = \mathrm{id}_{\hat{Q}}$. Then $\Delta_2 \circ \Delta_1 = (\Delta_2 \circ \pi) \circ (i \circ \Delta_1)$ and $i \circ \Delta_1 \colon A \to T$, $\Delta_2 \circ \pi \colon T \to \Lambda^n$. So, as was just proved,

$$\begin{aligned}((\Delta_2 \circ \pi) \circ (i \circ \Delta_1))^* &= (i \circ \Delta_1)^* \circ (\Delta_2 \circ \pi)^* = \Delta_1^* \circ i^* \circ \pi^* \circ \Delta_2^* \\ &= \Delta_1^* \circ (\pi \circ i)^* \circ \Delta_2^* = \Delta_1^* \circ \Delta_2^*.\end{aligned}$$

This proves that $(\Delta_2 \circ \Delta_1)^* = \Delta_1^* \circ \Delta_2^*$ for $A \xrightarrow{\Delta_1} \hat{Q} \xrightarrow{\Delta_2} \Lambda^n$ with a projective Q.

Finally, the general case reduces to the one considered above since

$$\begin{aligned}(\Delta_2 \circ \Delta_1)^*(r, p) &= (r^* \circ \Delta_2 \circ \Delta_1 \circ p)^* \\ &= (\Delta_1 \circ p)^* \circ (r^* \circ \Delta_2)^* = p^* \circ \Delta_1^* \circ \Delta_2^* \circ r.\end{aligned}$$

□

1.2.4. Coordinates. Let $\Delta \in \mathrm{Diff}_s A$. Consider $\tilde{\Delta} \colon \Lambda^n \to \Lambda^n$ such that $\tilde{\Delta}(\omega) = \Delta[\omega]$. Then $\delta_f(\tilde{\Delta}) = -\delta_f(\Delta)$. This shows that $\tilde{\Delta} \in \mathrm{Diff}_s(\Lambda^n, \Lambda^n)$. In particular, $\tilde{\Delta}$ is a local operator and hence $(\tilde{\Delta}|_U)(\omega|_U) = \tilde{\Delta}(\omega)|_U$, where $U \subset M$ is an open domain in M. In other words, $(\Delta|_U)[\omega|_U] = \Delta[\omega]|_U$, i.e., the action of operators $\Delta \in \mathrm{Diff}\, A$ on n-forms is local. From the above it also follows that $(\Delta|_U)^* = \Delta^*|_U$. This allows us to compute expressions $\Delta[\omega]|_U$ and $\Delta^*|_U$ in local coordinates in the form $(\Delta|_U)[\omega|_U]$ and $(\Delta|_U)^*$, respectively, where U is the corresponding local chart.

Let $(x_1, \dots, x_n)$ be local coordinates in U and $\omega = dx_1 \wedge \dots \wedge dx_n$. Denote $\partial^k/\partial x_{i_1} \cdots \partial x_{i_k}$ by $\partial^{|\sigma|}/\partial x_\sigma$, where $\sigma = (i_1, \dots, i_k)$ is an unordered set of integers $1 \le i_j \le n$, $|\sigma| = k$. Since $(\partial/\partial x_i)(f\omega) = (\partial f/\partial x_i)\omega$, one has $(a\partial^{|\sigma|}/\partial x_\sigma)[\omega] = (-1)^{|\sigma|}(\partial^{|\sigma|} a/\partial x_\sigma)\omega$. Hence, for an operator $\Delta \in \mathrm{Diff}\, \Lambda^n$ such that $\Delta_\omega = \sum_\sigma a_\sigma \partial^{|\sigma|}/\partial x_\sigma$ we get

$$(\Delta^*)_\omega = \sum_\sigma (-1)^{|\sigma|} \frac{\partial^{|\sigma|}}{\partial x_\sigma} \circ a_\sigma. \tag{1.5}$$

Note also that $\Delta = \omega \circ \Delta_\omega$. Therefore, due to Proposition 1.2.2 (2), $\Delta^* = (\omega \circ \Delta_\omega)^* = (\Delta_\omega)^* \circ \omega^* = (\Delta_\omega)^* \circ \omega$. This shows that

$$(\Delta_\omega)^* \circ \omega = \omega \circ (\Delta^*)_\omega.$$

Let $\alpha_i \colon E_i \to M$ be vector bundles, $i = 1, 2$, and let $P = \Gamma(\alpha_1)$, $Q = \Gamma(\alpha_2)$ be the modules of their sections. Consider an operator $\Delta \in \operatorname{Diff}(P, \hat{Q})$ and suppose that a coordinate neighborhood U is such that the bundles α_i possess bases over U, say, $e_1, \dots, e_r$ and $f_1, \dots, f_s$, respectively. Let $\hat{e}_i \in \hat{P}_U$, $\hat{f}_i \in \hat{Q}_U$ be such that $\hat{e}_i(e_j) = \delta_{i,j}\omega$, $\hat{f}_i(f_j) = \delta_{i,j}\omega$. Then for $p \in P_U$ one has $p = \sum_i p_i e_i$ and, as is easy to see,

$$\Delta(p) = \sum_{i,j} (\Delta_{i,j})_\omega (p_j) \hat{f}_i \quad \text{with} \quad \Delta_{i,j} = \Delta(e_j, f_i) = f_i^* \circ \Delta \circ e_i.$$

Similarly, if $q = \sum_l q_l f_l \in Q_U$, then

$$\Delta^*(q) = \sum_{k,l} ((\Delta^*)_{k,l})_\omega (q_l) \hat{e}_k$$

with $(\Delta^*)_{k,l} = \Delta^*(f_l, e_k) = \Delta(e_k, f_l)^* = (\Delta_{l,k})^*$. Hence, $\big((\Delta^*)_{k,l}\big)_\omega = \big((\Delta_{l,k})^*\big)_\omega$. In other words, if the operator Δ is represented by the matrix $C = \|(\Delta_{ij})_\omega\|$ which acts on columns $(p_1, \dots, p_r)^t$, then Δ^* is represented by the matrix $\|c_{kl}^*\|$ with $c_{kl}^* = \big((\Delta_{lk})^*\big)_\omega$ acting on columns $(q_1, \dots, q_s)^t$. Since $\Delta_{l,k} \in \operatorname{Diff} \Lambda^n$, the operator $\big((\Delta_{l,k})^*\big)_\omega$ is described by formula (1.5).

1.2.5. Transformations. To finish this section, we describe transformation properties of the above-described operations. This is done in two versions, finite and infinitesimal ones. The latter is understood in the following sense.

Consider the classical situation and let F_t be a one-parameter family of transformations. Then the action $X(z) = (d(F_t(z)/dt))|_{t=0}$ is naturally called an *infinitesimal transformation* of the object z. In the general algebraic situation, infinitesimal transformations will be understood as operation of this type. For example, let $F \colon A_1 \to A_2$ be an isomorphism of commutative K-algebras and F_Λ (or simply F) the corresponding mapping of differential forms. Let $\Delta \in \operatorname{Diff} \Lambda^k(A_1)$. Then we set $F(\Delta) = F_\Lambda \circ \Delta \circ F^{-1} \in \operatorname{Diff} \Lambda^k(A_2)$. When $A_1 = A_2 = A$ and $X \in D(A)$, the object $X(\Delta)$ should be defined by "differentiating" $(F_t)_\Lambda \circ \Delta \circ F_t^{-1}$ and setting $X = (dF_t/dt)|_{t=0}$. In this way, one gets the definition $X(\Delta) = X_\Lambda \circ \Delta - \Delta \circ X$, or $X(\Delta) = [X, \Delta]$. Here $X_\Lambda(\omega)$ denotes the Lie derivative of the form ω. This system of notation will be used below to indicate the nature of the object under transformation. For example, $F_{\operatorname{Diff} \Lambda^k} \colon \operatorname{Diff} \Lambda^k(A_1) \to \operatorname{Diff} \Lambda^k(A_2)$ or $X_{\operatorname{Diff} \Lambda^k}(\Delta) = [X, \Delta]$. We emphasize that in the classical situation the role of A_i is played by $C^\infty(M_i)$, while $F = f^*$ for a smooth mapping $f \colon M_2 \to M_1$.

The action $[\,\cdot\,]$ defined over A_1 is transformed into an action $[\,\cdot\,]_F$ over A_2 according to the formula

$$\Delta[\omega]_F = F_\Lambda\big(F^{-1}(\Delta)[F_\Lambda^{-1}(\omega)]\big).$$

Its infinitesimal analog is denoted by $[\,\cdot\,]_X$, $X \in D(A)$, and is expressed by

$$\Delta[\omega]_X = X_\Lambda(\Delta[\omega]) - [X, \Delta][\omega] - \Delta[X_\Lambda(\omega)].$$

In accordance with Proposition 1.2.2, which states the uniqueness of the action, one has

$$\Delta[\omega] = F_\Lambda\bigl(F^{-1}(\Delta)[F_\Lambda(\omega)]\bigr). \tag{1.6}$$

The infinitesimal version of (1.6) is

$$\Delta[\omega]|_X = 0, \quad \text{or} \quad X(\Delta[\omega]) = [X,\Delta][\omega] + \Delta[X(\omega)]. \tag{1.7}$$

To prove the latter, it suffices to use the definition of the commutator together with the axioms of action.

Now we show that

$$* \circ F_{\mathrm{Diff}\,\Lambda^n} = F_{\mathrm{Diff}\,\Lambda^n} \circ *, \quad \text{or} \quad F(\Delta)^* = F(\Delta^*), \ \ \Delta \in \mathrm{Diff}\,\Lambda^n. \tag{1.8}$$

First, let us check that $F(\Delta)_{F(\omega)} = F(\Delta_\omega)$. Indeed, for $a \in A_2$ one has

$$\begin{aligned}(F(\Delta)_{F(\omega)}(a))F(\omega) &= F(\Delta)(a) = F(\Delta(F^{-1}(a))) \\ &= F(\Delta_\omega(F^{-1}(a))\omega) = (F \circ \Delta_\omega \circ F^{-1})(a)(\omega).\end{aligned}$$

On the other hand,

$$F(\Delta)^*(a) = (F(\Delta)_{F(\omega)})[aF(\omega)] = F(\Delta_\omega)[F(F^{-1}(a)\omega)].$$

According to equality (1.6) applied to the homomorphism F^{-1}, the last expression equals $F(\Delta_\omega[F^{-1}(a)\omega]) = F(\Delta^*(F^{-1}(a))) = F(\Delta^*)(a)$.

Obviously, the infinitesimal form of (1.8) is

$$* \circ X_{\mathrm{Diff}\,\Lambda^n} = X_{\mathrm{Diff}\,\Lambda^n} \circ *, \quad \text{or} \quad [X,\Delta]^* = [X,\Delta^*]. \tag{1.9}$$

Let A_i be K-algebras, $i = 1,2$, and P_i be A_i-modules. We say that a K-linear mapping $F_P\colon P_1 \to P_2$ *covers* a homomorphism $F\colon A_1 \to A_2$ if $F_P(ap) = F(a)F_P(p)$, $a \in A_1$, $p \in P_1$. In a similar way, $X_P \in \mathrm{Diff}_1(P,P)$ *covers* $X \in D(A)$ if $X_P(ap) = X(a)p + aX_P(p)$. Let

$$\mathrm{Der}\,P = \{(X_P, X) \mid X_P \in \mathrm{Diff}_1(P,P) \ \text{ covers } \ X \in D(A)\}.$$

The set $\mathrm{Der}\,P$ carries an obvious A-module structure. The identity $\delta_a(X_P) = X(a)$ shows that X is uniquely determined by X_P and one has an embedding $\mathrm{Der}\,P \subset \mathrm{Diff}_1(P,P)$. If, in the notations of the previous subsection, $P = \Gamma(\alpha_1)$, then $X_P \in \mathrm{Der}\,P$ if and only if $(X_P)_U = XI_r + B$, where I_r is the identity $(r \times r)$-matrix and $B = \|b_{ij}\|$, $b_{ij} \in C^\infty(U)$, $X \in D(U)$, $i,j = 1,\dots,r$.

Let $F_P\colon P_1 \to P_2$, $F_Q\colon Q_1 \to Q_2$ be bijections covering F. Then one can define $F_{\mathrm{Diff}}\colon \mathrm{Diff}(P_1,Q_1) \to \mathrm{Diff}(P_2,Q_2)$ by setting $F_{\mathrm{Diff}}(\Delta) = F_Q \circ \Delta \circ F_P^{-1}$. Similarly, if both $X_P \in \mathrm{Der}\,P$ and $X_Q \in \mathrm{Der}\,Q$ cover the same derivation $X \in D(A)$, we put

$$X_{\mathrm{Diff}}(\Delta) = X_Q \circ \Delta - \Delta \circ X_P = [X,\Delta] \in \mathrm{Diff}(P,Q).$$

The following formulas are valid in this situation:

$$* \circ F_{\mathrm{Diff}} = F_{\mathrm{Diff}} \circ *, \qquad * \circ X_{\mathrm{Diff}} = X_{\mathrm{Diff}} \circ *.$$

To prove the first one (the second is obvious), note that

$$F_P^{-1} \circ F_P(p_1) = p_1 \circ F^{-1}. \tag{1.10}$$

Indeed, if $a_2 \in A$, then

$$\bigl(F_P^{-1} \circ F_P(p_1)\bigr)(a_2) = F_P^{-1}\bigl(a_2F_P(p_1)\bigr) = F^{-1}(a_2) \circ p_1 = (p_1 \circ F^{-1})(a_2).$$

Similarly, one has

$$\hat{q}_2^* \circ F_Q = F \circ F_{\hat{Q}}^{-1}(q_2)^*. \tag{1.11}$$

Formulas (1.10) and (1.11) imply

$$F(\square)(F_P(p_1), \hat{q}_2) = F(\square(p_1, F_{\hat{Q}}(\hat{q}_2))), \quad p_1 \in P_1, \ \hat{q}_2 \in \hat{Q}_2. \tag{1.12}$$

Namely,

$$\begin{aligned} F(\square)\big(F_P(p_1), \hat{q}_2\big) &= \hat{q}_2^* \circ F_Q \circ \square \circ F_P^{-1} \circ F_P(p_1) = F \circ F_{\hat{Q}}^{-1}(\hat{q}_2)^* \circ \square \circ p_1 \circ F^{-1} \\ &= F \circ \square(p_1, F_{\hat{Q}})^{-1}(\hat{q}_2) \circ F^{-1} = F\big(\square(p_1, F_{\hat{Q}}^{-1}(\hat{q}_2))\big). \end{aligned}$$

Therefore, for $\square = \Delta^*$ one has

$$F(\Delta^*)(\hat{q}_2, p_2) = F\big(\Delta^*(F_{\hat{Q}}^{-1}\hat{q}_2, F_P^{-1}p_2)\big) = F\big(\Delta(F_P^{-1}p_2, F_{\hat{Q}}^{-1}\hat{q}_2)^*\big).$$

But the last expression, due to (1.8) and (1.12), equals

$$[F\big(\Delta(F_P^{-1}p_2, F_{\hat{Q}}^{-1}\hat{q}_2)\big)]^* = [F(\Delta)(p_2, \hat{q}_2)]^* = F(\Delta)^*(\hat{q}_2, p_2)$$

for any $\hat{q}_2 \in \hat{Q}_2$, $p_2 \in P_2$. Hence, $F(\Delta^*) = F(\Delta)^*$.

1.3. Spencer complexes and the Green formula

The goal of this section is to establish some natural relations between the operator $*$ and the Spencer Diff-complexes. The classical Green formula is obtained as a consequence of these relations.

1.3.1. Spencer Diff-complexes. Recall the scheme of the Spencer Diff-complex construction (see [**136**] for details and also [**71, 124**]). Consider the composition

$$D_i^\bullet(\mathrm{Diff}_s^+) \xrightarrow{\alpha} D_{i-1}^\bullet(\mathrm{Diff}_1^+(\mathrm{Diff}_s^+ P)) \xrightarrow{\beta} D_{i-1}^\bullet(\mathrm{Diff}_{s+1}^+ P),$$

where α is induced by the functor transformation $D_i \Rightarrow D_{i-1}^\bullet(\mathrm{Diff}_1^+)$ (see Subsection 1.1.5) while β is due to the gluing operator

$$c_{1,s}\colon \mathrm{Diff}_1^+(\mathrm{Diff}_s^+ P) \to \mathrm{Diff}_{s+1}^+ P.$$

The operators

$$S = S_{i,s} = \beta \circ \alpha\colon D_i^\bullet(\mathrm{Diff}_s^+ P) \to D_{i-1}^\bullet(\mathrm{Diff}_{s+1}^+ P)$$

are called the *Spencer* Diff-*operators*. They form the *Spencer* Diff-*complex* $S_k P$:

$$\begin{aligned} 0 \leftarrow P \xleftarrow{\text{Д}_k} \mathrm{Diff}_k P &\xleftarrow{S_{1,k-1}} D^\bullet(\mathrm{Diff}_{k-1}^+ P) \\ &\xleftarrow{S_{2,k-2}} \cdots \xleftarrow{S_{i,k-i}} D_i^\bullet(\mathrm{Diff}_{k-i}^+) \leftarrow \cdots . \end{aligned}$$

For $k = \infty$ we get the "stable" version SP of this complex,

$$0 \leftarrow P \xleftarrow{\text{Д}} \mathrm{Diff}\, P \xleftarrow{S_1} D^\bullet(\mathrm{Diff}^+ P) \xleftarrow{S_2} \cdots \xleftarrow{S_i} D^\bullet(\mathrm{Diff}^+) \leftarrow \cdots,$$

which is the direct limit with respect to the natural embeddings $\cdots \subset S_k P \subset S_{k+1} P \subset \cdots$. Note that the operators $S_{i,j}$ are A-homomorphisms.

Below we shall need an explicit description of the Spencer operator based on the interpretation of elements of $D_i(P)$ as multiderivations. Denote by $X_1 \wedge \cdots \wedge X_i \otimes p \in D_i(P)$ the multiderivation

$$(f_1, \dots, f_i) \mapsto [(X_1 \wedge \cdots \wedge X_i) \lrcorner (df_1 \wedge \cdots \wedge df_i)]p,$$

where $X_j \in D(A)$, $f_j \in A$, $p \in P$. If $P = \mathrm{Diff}_s^+ Q$, then $\Delta \otimes X_1 \wedge \cdots \wedge X_i$ denotes the multiderivation

$$(f_1, \ldots, f_i) \mapsto \Delta \circ [(X_1 \wedge \cdots \wedge X_i) \lrcorner (df_1 \wedge \cdots \wedge df_i)], \quad \Delta \in \mathrm{Diff}_s Q,$$

belonging to $D_i(\mathrm{Diff}_s^+ Q)$. In this notation,

$$S_{i,s}(\Delta \otimes X_1 \wedge \cdots \wedge X_i) = \sum_j (-1)^{j-1} \Delta \circ X^{(j)} \otimes X_j,$$

where $X^{(j)} = X_1 \wedge \cdots \wedge \hat{X}_j \wedge \cdots \wedge X^i$ and the multiderivation $\Delta \circ X^{(j)} \otimes X_j$ acts according to the rule $(f_1, \ldots, f_{i-1}) \mapsto \Delta \circ X^{(j)}(f_1, \ldots, f_{i-1}) X_j$. Here

$$X^{(j)}(f_1, \ldots, f_{i-1}) = (X_1 \wedge \cdots \wedge \hat{X}_j \wedge \cdots \wedge X^i) \lrcorner (df_1 \wedge \cdots \wedge df_{i-1}).$$

Note that the multiderivations of the form $\Delta \otimes X_1 \wedge \cdots \wedge X_i$ generate the module $D_i^\bullet(\mathrm{Diff}_s^+ P)$.

We shall also need the natural isomorphism

$$\Pi = \Pi_i \colon D_i(\Lambda^n) \to \Lambda^{n-i}$$

(the Poincaré duality) defined in the following way. Let $Z = X_1 \wedge \cdots \wedge X_i \otimes \omega \in D_i(\Lambda^n)$, where $\omega \in \Lambda^n$ is a (local) volume form. Then $\Pi(Z) = X_1 \lrcorner (\ldots (X_i \lrcorner \omega) \ldots)$. This definition is obviously independent of the choice of ω and by linearity extends to the entire module $D_i(\Lambda^n)$.

1.3.2. Consider the diagram

$$
\begin{array}{ccccccccccc}
0 \longleftarrow & \Lambda^n & \xleftarrow{\text{Д}} & \mathrm{Diff}_k^+ \Lambda^n & \xleftarrow{S} & D^\bullet(\mathrm{Diff}_{k-1}^+ \Lambda^n) & \xleftarrow{S} & \cdots & \xleftarrow{S} & D_n^\bullet(\mathrm{Diff}_{k-n}^+ \Lambda^n) & \longleftarrow 0 \\
(1.13) \quad \scriptstyle{\mathrm{id}} & \big\downarrow & & \scriptstyle{\psi}\big\downarrow & & \scriptstyle{\psi}\big\downarrow & & & & \scriptstyle{\psi}\big\downarrow & \\
0 \longleftarrow & \Lambda^n & \xleftarrow{-\mu} & \mathrm{Diff}_k^+ \Lambda^n & \xleftarrow{\mathcal{S}} & \mathrm{Diff}_{k-1}^+ \Lambda^{n-1} & \xleftarrow{\mathcal{S}} & \cdots & \xleftarrow{\mathcal{S}} & \mathrm{Diff}_{k-n}^+ A & \longleftarrow 0
\end{array}
$$

whose upper row is the Spencer complex. The A-homomorphisms $\mathcal{S}$ in the lower row are defined by $\mathcal{S}(\nabla) = d \circ \nabla \in \mathrm{Diff}_{i+1} \Lambda^{s+1}$, $\nabla \in \mathrm{Diff}_i^+ \Lambda^s$. Finally, $\mu(\Delta) = \Delta^*(1)$, $\Delta \in \mathrm{Diff}_k \Lambda^n$, and ψ is the composition of the following isomorphisms:

$$
\begin{aligned}
D_s^\bullet(\mathrm{Diff}_{k-s}^+ \Lambda^n) &\xrightarrow{D_s^\bullet(*)} D_s^\bullet(\mathrm{Diff}_{k-s} \Lambda^n) \\
&\longrightarrow \mathrm{Diff}_{k-s}(D_s(\Lambda^n)) \xrightarrow{\mathrm{Diff}_{k-s}(\Pi)} \mathrm{Diff}_{k-s} \Lambda^{n-s} \xrightarrow{i_+} \mathrm{Diff}_{k-s}^+ \Lambda^{n-s},
\end{aligned}
$$

where the middle isomorphism is the natural transposition of functors $D_s(\mathrm{Diff}_{k-s}) \Rightarrow \mathrm{Diff}_{k-s}(D_s)$ and i_+ was defined in Subsection 1.1.3.

THEOREM. *For any $k = 1, \ldots, \infty$, all the squares in* (1.13) *are anti-commutative.*

PROOF. Anti-commutativity of the left square is evident, since $\psi = *$ in this case. Let $Z = \Delta \otimes X_1 \wedge \cdots \wedge X_s \in D_s(\mathrm{Diff}_{k-s}^+ \Lambda^n)$. Then, due to Subsection 1.3.1,

$Z' = (D_{s-1}(*) \circ S)(Z)$ is the following multiderivation:

$$\begin{aligned}(f_1,\dots,f_{s-1}) \mapsto &\sum_j (-1)^{j-1}\{\Delta \circ X^{(j)}(f_1,\dots,f_{s-1}) \circ X_j\}^* \\ &= \sum_j (-1)^j \bar{X}_j \circ X^{(j)}(f_1,\dots,f_{s-1}) \circ \Delta^* \\ &= \sum_j (-1)^j [X^{(j)}(f_1,\dots,f_{s-1})\bar{X}_j] \circ \Delta^* \\ &\quad + \sum_j (-1)^j X_j(X^{(j)}(f_1,\dots,f_{s-1}))\Delta^* \\ &= \sum_j (-1)^j X^{(j)}(f_1,\dots,f_{s-1})\bar{X}_j \circ \Delta^* \\ &\quad + \sum_{i<j} (-1)^{i+j} X^{(i,j)}(f_1,\dots,f_{s-1})\Delta^* \\ &\in D_{s-1}(\mathrm{Diff}_{k+s-1}\Lambda^n),\end{aligned}$$

where $X^{(i,j)} = [X_i, X_j] \wedge X_1 \wedge \dots \wedge \hat{X}_i \wedge \dots \wedge \hat{X}_j \wedge \dots \wedge X_s$. The transposition of functors $D_{s-1}(\mathrm{Diff}_{k-s+1}) \Rightarrow \mathrm{Diff}_{k-s+1}(D_{s-1})$ transforms Z' to the differential operator

$$f \mapsto \sum_j (-1)^j X^{(j)} \otimes X_j(\Delta^*(f)) + \sum_{i<j} (-1)^{i+j} X^{(i,j)} \otimes \Delta^*(f).$$

Hence, for $\Box = (\psi \circ S)(Z) \in \mathrm{Diff}_{k-s}\Lambda^{n-s}$ one has

$$\Box(f) = \sum_j (-1)^j X^{(j)} \lrcorner\, X_j(\Delta^*(f)) + \sum_{i<j} (-1)^{i+j} X^{(i,j)} \lrcorner\, \Delta^*(f),$$

where for any element $W = Y_1 \wedge \dots \wedge Y_l$ the contraction $W \lrcorner\, \rho$ is understood as $Y_1 \lrcorner\, (\dots(Y_l \lrcorner\, \rho)\dots)$.

On the other hand, $\psi(Z)(f) = X_1 \lrcorner\, (\dots(X_s \lrcorner\, \Delta^*(f))\dots)$. This shows that $\Box'(f) = d(X_1 \lrcorner\, \dots \lrcorner\, (X_s \lrcorner\, \Delta^*(f))\dots)$ for $\Box' = \mathcal{S}(\psi(Z))$. The desired result is a consequence of the lemma below together with the relations $d\Delta^*(f) = 0$ and $X_j(\Delta^*(f)) = d(X_j \lrcorner\, \Delta^*(f))$. □

LEMMA. *If* $\rho \in \Lambda^k$ *and* $X_i \in D(M)$, *then*

$$\begin{aligned}d(X_1 \lrcorner\, \dots \lrcorner\, X_s \lrcorner\, \rho) = \sum_i (-1)^{i-1} X^{(i)} \lrcorner\, d(X \lrcorner\, \rho) + \sum_{i<j} (-1)^{i+j-1} X^{(i,j)} \lrcorner\, \rho \\ + (-1)^{s-1}(s-1) X_1 \lrcorner\, \dots \lrcorner\, X_s \lrcorner\, d\rho.\end{aligned}$$

PROOF. For $s = 1$, the assertion is a tautology. For $s = 2$, the result follows from the formula $i_{[X,Y]} = [L_X, i_Y]$ (see, for example, [**43**]), where L_X is the Lie derivative and $i_X(\rho) = X \lrcorner\, \rho$. In fact,

$$\begin{aligned}[X,Y] \lrcorner\, \rho &= X(Y \lrcorner\, \rho) - Y \lrcorner\, X(\rho) \\ &= X \lrcorner\, d(Y \lrcorner\, \rho) + d(X \lrcorner\, Y \lrcorner\, \rho) - Y \lrcorner\, X \lrcorner\, d\rho - Y \lrcorner\, d(X \lrcorner\, \rho),\end{aligned}$$

or $d(X \lrcorner\, Y \lrcorner\, \rho) = Y \lrcorner\, X \lrcorner\, d\rho + Y \lrcorner\, d(X \lrcorner\, \rho) - X \lrcorner\, d(Y \lrcorner\, \rho + [X,Y] \lrcorner\, \rho$. The rest of the proof is an obvious induction based on the formula above. □

Since the Spencer Diff-complexes are exact for any geometric $C^\infty(M)$-module (see [**71**]), one has

COROLLARY. *The complex* $\mathcal{S}_k$

$$0 \leftarrow \Lambda^n \xleftarrow{\mu} \mathrm{Diff}_k^+ \Lambda^n \xleftarrow{S} \mathrm{Diff}_{k-1}^+ \Lambda^{n-1} \xleftarrow{S} \cdots \xleftarrow{S} \mathrm{Diff}_{k-n}^+ A \leftarrow 0$$

is exact for all $k = 1, \dots, \infty$.

Applying now the functor $\mathrm{Hom}^\bullet(P, \bullet)$ to the complex $\mathcal{S}_k$ and taking into account the isomorphism $\mathrm{Hom}^\bullet(P, \mathrm{Diff}_k^+ Q) = \mathrm{Diff}_k(P, Q)$, we get a new complex $\mathcal{S}_k P$ of the form

$$0 \to \mathrm{Diff}_{k-n}(P, A) \xrightarrow{S} \cdots \xrightarrow{S} \mathrm{Diff}_{k-1}(P, \Lambda^{n-1}) \xrightarrow{S} \mathrm{Diff}_k(P, \Lambda^n) \xrightarrow{\mu} \hat{P} \to 0.$$

For simplicity, we still write μ and $\mathcal{S}$ instead of $\mathrm{Hom}^\bullet(P, \mu)$ and $\mathrm{Hom}^\bullet(P, \mathcal{S})$, respectively. If P is a projective module, then the complex $\mathcal{S}_k P$ is obviously exact.

Finally, by applying the functor $\mathrm{Diff}_l(Q, \bullet)$, $l = 0, \dots, \infty$, to the complex $\mathcal{S}_k P$, we get the complex $\mathcal{S}_{k,l} PQ$:

$$\begin{aligned} 0 \to \mathrm{Diff}_l(Q, \mathrm{Diff}_{k-n}(P, A)) \xrightarrow{\bar{S}} \cdots \xrightarrow{\bar{S}} \mathrm{Diff}_l(Q, \mathrm{Diff}_k(P, \Lambda^n)) \\ \xrightarrow{\bar{\mu}} \mathrm{Diff}_l(Q, \hat{P}) \to 0, \end{aligned}$$

where $\bar{\mu} = \mathrm{Diff}_l(Q, \mu)$, $\bar{S} = \mathrm{Diff}_l(Q, \mathcal{S})$. As it will be proved below, this complex is exact if $l = \infty$ and P and Q are projective modules.

1.3.3. In this subsection, we describe the adjoint complexes of the de Rham complex and of the complexes $\mathcal{S}_k P$. Let $\cdots \to P_i \xrightarrow{\Delta_i} P_{i+1} \to \cdots$ be a complex of projective finite type A-modules. Then $\hat{\hat{P}}_i = P_i$, and one can consider Δ_i^* to act from $\hat{P}_{i+1}$ to $\hat{P}_i$. Since $\Delta_{i-1}^* \circ \Delta_i^* = (\Delta_i \circ \Delta_{i-1})^* = 0$, we get the complex $\cdots \leftarrow \hat{P}_i \xleftarrow{\Delta_i^*} \hat{P}_{i+1} \leftarrow \cdots$ which is called the *adjoint complex* to the initial one.

Below we identify Λ^{n-s} with $\hat{\Lambda}^s$ via the correspondence

$$\Lambda^{n-s} \ni \theta \mapsto h_\theta \in \mathrm{Hom}(\Lambda^s, \Lambda^n), \quad h_\theta(\rho) = \theta \wedge \rho, \ \rho \in \Lambda^s.$$

The following statement is a direct consequence of the definition and the above identification.

PROPOSITION. *The following statements are valid.*

1. *If* n *is even, then the de Rham complex is skew-adjoint.*
2. *For* n *odd, the de Rham complex is adjoint to the complex* $\{\Lambda^s, (-1)^s d_s\}$.
3. *The complex* SP_k *is adjoint to the "twisted" Spencer jet-complex* $\{\mathcal{J}^{k-s}(P) \otimes \Lambda^s, (-1)^{s+1} \Sigma_{s,k-s}\}$.

1.3.4. The Green formula. For any $k = 1, \dots, \infty$ consider the anti-commutative square (see Theorem 1.3.2)

$$\begin{array}{ccc} \mathrm{Diff}_k \Lambda^n & \xleftarrow{S} & D^\bullet(\mathrm{Diff}_{k-1}^+ \Lambda^n) \\ {\scriptstyle *}\downarrow & & \downarrow{\scriptstyle \psi} \\ \mathrm{Diff}_k^+ \Lambda^n & \xleftarrow{S} & \mathrm{Diff}_{k-1}^+ \Lambda^{n-1} \end{array}$$

Note that $\mathrm{im}\, S = \widetilde{\mathrm{Diff}_k \Lambda^n} = \{\Delta \in \mathrm{Diff}_k \Lambda^n \mid \Delta(1) = 0\}$. Since $\mathrm{Diff}_k \Lambda^n$ is a projective module, there exists a homomorphism $\lambda \colon \widetilde{\mathrm{Diff}_k \Lambda^n} \to D^\bullet(\mathrm{Diff}_k^+ \Lambda^n)$ such that $S \circ \lambda = 1$. Applying the equality $* \circ S + \mathcal{S} \circ \psi = 0$ to $\lambda(\Box)$, $\Box \in \widetilde{\mathrm{Diff}_k \Lambda^n}$, one gets $\Box^* + d \circ \tilde{\mathcal{K}}_\lambda(\Box) = 0$, where $\tilde{\mathcal{K}}_\lambda = \psi \circ \lambda$. In particular,

$$\Box^*(1) + d(\tilde{\mathcal{K}}_\lambda(\Box)(1)) = 0.$$

Note now that $\tilde{\mathcal{K}}_\lambda(\Box)(1) = (\mathcal{I}^+_{k-1} \circ \tilde{\mathcal{K}}_\lambda)(\Box)$. If $\Box = \nu(\Delta)$ with $\nu(\Delta) = \Delta - \Delta(1)$, $\Delta \in \mathrm{Diff}_k \Lambda^n$, then $\tilde{\mathcal{K}}_\lambda(\Box)(1) = \mathcal{K}_\lambda(\Delta)$, where $\mathcal{K}_\lambda = \mathcal{I}^+_{k-1} \circ \tilde{\mathcal{K}}_\lambda \circ \nu$. In addition, $\Box^*(1) = \Delta^*(1) - \Delta(1)$. So, the last equality can be rewritten in the form

$$\Delta(1) - \Delta^*(1) = d\mathcal{K}_\lambda(\Delta).$$

This is the *Green formula for the operator* $\Delta \in \mathrm{Diff}_k \Lambda^n$. Note that since $\tilde{\mathcal{K}}_\lambda$ is a homomorphism, $\mathcal{K}_\lambda \colon \mathrm{Diff}_k \Lambda^n \to \Lambda^{n-1}$ is an operator of order $k-1$.

If $\Delta \in \mathrm{Diff}_k(P, \hat{Q})$, then for any $p \in P$, $q \in Q$ one has $\Delta(p,q) \in \mathrm{Diff}_k \Lambda^n$ and $\Delta(p,q)(1) = \langle \Delta(p), q \rangle$, where $\langle \bullet, \bullet \rangle$ denotes the natural pairing $\hat{Q} \otimes Q \to \Lambda^n$. Hence, the Green formula for the operator $\Delta(p,q)$ can be written down as

$$\langle \Delta(p), q \rangle - \langle p, \Delta^*(q) \rangle = d\mathcal{K}_\lambda(\Delta(p,q)),$$

and this is the *general Green formula*.

1.3.5. Consider the expression $a(p,q) = \mathcal{K}_\lambda(\Delta(p,q))$. It is a bidifferential operator with respect to p and q. More precisely, let us put $b_q(p) = a(p,q) = c_p(q)$. Then $c_p \in \mathrm{Diff}_{k-1}(Q, \Lambda^{n-1})$. In fact, c_p is the composition $\mathcal{I}^+_{k-1} \circ \tilde{\mathcal{K}}_\lambda \circ \nu \circ h$, where $h(q) = q^* \circ \Delta \circ p$ and h, ν, $\tilde{\mathcal{K}}_\lambda$ are homomorphisms, while $\mathcal{I}^+_{k-1}$ is an operator of order $k-1$. If the splitting λ is homogeneous (see below), then it can be proved that $b_q \in \mathrm{Diff}_{k-1} \Lambda^n$.

We say that a splitting $\lambda \colon \widetilde{\mathrm{Diff}_k \Lambda^n} \to D^\bullet(\mathrm{Diff}^+_{k-1} \Lambda^n)$ is *homogeneous* if

$$\lambda(\widetilde{\mathrm{Diff}_i \Lambda^n}) \subset D^\bullet(\mathrm{Diff}^+_{i-1} \Lambda^n), \quad i = 1, \dots, k.$$

Due to the projectivity of the modules $\mathrm{Diff}_s \Lambda^n / \mathrm{Diff}_{s-1} \Lambda^n$, homogeneous splittings exist and are easily constructed by induction on k.

If the splitting λ is not homogeneous, one can only state that b_q belongs to $\mathrm{Diff}_{2k-2}(P, \Lambda^{n-1})$.

Remark. If $k = 1$ or $\dim M = 1$, the splitting λ is uniquely determined, since in these cases $S_{1,k-1}$ is an isomorphism.

1.3.6. Coordinates. To express the Green formula in coordinates, one needs to keep in mind Subsection 1.2.4 and to describe the operator $\mathcal{K}_\lambda$ only. In the notation of Subsection 1.2.4, one of possible splittings λ can be described as

$$\lambda(\Delta)_\omega = \sum_\sigma a_\sigma \frac{\partial^{|\bar\sigma|}}{\partial x_{\bar\sigma}} \otimes \frac{\partial}{\partial x_{i(\sigma)}},$$

where $\Delta_\omega = \sum_{|\sigma|>0} a_\sigma \partial^{|\sigma|}/\partial x_\sigma$, $\sigma = (i_1, \dots, i_s)$, $\bar\sigma = (i_1, \dots, i_{s-1})$, $i(\sigma) = i_s$, and the indices i_j are in ascending order. In other words,

$$\lambda(\Delta)(f) = \sum_\sigma \left(a_\sigma \frac{\partial^{|\bar\sigma|}}{\partial x_{\bar\sigma}} \circ \frac{\partial f}{\partial x_{i(\sigma)}} \right) \omega.$$

Since the Spencer complex is exact and $\widetilde{\mathrm{Diff}_k \Lambda^n}$ is a projective module, for any two splittings λ and $\lambda' \colon \widetilde{\mathrm{Diff}_k \Lambda^n} \to D^\bullet(\mathrm{Diff}^+_{k-1} \Lambda^n)$ there exists a homomorphism $h \colon \widetilde{\mathrm{Diff}_k \Lambda^n} \to D^\bullet_2(\mathrm{Diff}^+_{k-2} \Lambda^n)$ such that $\lambda' - \lambda = S_{2,k-2} \circ h$ and vice versa. Hence, any splitting λ can be obtained from the above one by adding a summand $S_{2,k-2}\circ h$ with an arbitrary h.

By setting $\omega_i = (\partial/\partial x_i) \lrcorner\, \omega = (-1)^i dx_1 \wedge \cdots \wedge \widehat{dx_i} \wedge \cdots \wedge dx_n$, one gets directly from the definition that

$$\mathscr{K}_\lambda(\Delta) = \sum_{|\sigma|>0} (-1)^{|\bar{\sigma}|} \frac{\partial^{|\bar{\sigma}|} a_\sigma}{\partial x_{\bar{\sigma}}} \omega_{i(\sigma)}.$$

From the above it follows that $\mathscr{K}_\lambda(\Delta) = \mathscr{K}_{\lambda'}(\Delta) + dh'(\Delta)$, where $h' = -\varPi^+_{k-2} \circ \psi \circ h \circ \nu\colon \mathrm{Diff}_k \Lambda^n \to \Lambda^{n-2}$, with h being an arbitrarily chosen homomorphism.

1.3.7. Transformations. Using the notation of Subsection 1.2.5, let us set

$$F(\mathscr{K}_\lambda) = F_\lambda \circ \mathscr{K}_\lambda \circ F^{-1}_{\mathrm{Diff}}, \qquad F(\mu) = F_\Lambda \circ \mu \circ F^{-1}_{\mathrm{Diff}},$$
$$X(\mathscr{K}_\lambda) = X_\Lambda \circ \mathscr{K}_\lambda - \mathscr{K}_\lambda \circ X_{\mathrm{Diff}}, \qquad X(\mu) = X_\Lambda \circ \mu - \mu \circ X_{\mathrm{Diff}}.$$

In particular,

$$X(\mathscr{K}_\lambda)(\Delta) = X(\mathscr{K}_\lambda(\Delta)) - \mathscr{K}_\lambda([X, \Delta]), \tag{1.14}$$
$$X(\mu)(\Delta) = X(\mu(\Delta)) - \mu([X, \Delta]). \tag{1.15}$$

Note that $F(\mu) = \mu$ and $X(\mu) = 0$. In fact, due to (1.8),

$$F(\mu)(\Delta) = F(\mu(F^{-1}(\Delta))) = F(F^{-1}(\Delta)^*(1))$$
$$= F(F^{-1}(\Delta^*))(F(1)) = \Delta^*(1) = \mu(\Delta)$$

and, by (1.9),

$$X(\mu)(\Delta) = X(\mu(\Delta)) - \mu([X, \Delta]) = X(\Delta^*(1)) - [X, \Delta]^*(1) = 0.$$

As a consequence of the above, we get

$$F(D_i(\mu)) = D_i(\mu), \; X(D_i(\mu)) = 0,$$

where $D_i(\mu)\colon D_i(\mathrm{Diff}^+_k \Lambda^n) \to D_i(\Lambda^n) = \Lambda^{n-i}$. Indeed, $[F, D_i] = [X, D_i] = 0$ by functorial reasons.

In the sequel, we shall need the following

PROPOSITION. $X(\mathscr{K}_\lambda) = d \circ v$, *where* $v \in \mathrm{Diff}(\mathrm{Diff}_k(\Lambda^n, \Lambda^{n-2}))$.

PROOF. In the notation of Subsection 1.3.4, one has $\mathscr{K}_\lambda = \varPi^+ \circ \psi \circ \lambda \circ \nu$. Setting $\lambda' = \lambda \circ \nu$ and $\bar{\psi} = \varPi^+ \circ \psi$, we get $X(\mathscr{K}_\lambda) = X(\bar{\psi}) \circ \lambda' + \bar{\psi} \circ X(\lambda')$. To compute $X(\bar{\psi})$, note that $\bar{\psi} = \Pi \circ D(\mu)$ with $D(\mu)\colon D^\bullet(\mathrm{Diff}^+_{k-1}\Lambda^n) \to D(\Lambda^n)$. Indeed, it is straightforward to check that both operators $\varPi^+ \circ \psi$ and $\Pi \circ D(\mu)$ take $\Delta \otimes X \in D^\bullet(\mathrm{Diff}^+_{k-1}\Lambda^n)$, where $\Delta \in \mathrm{Diff}_{k-1}\Lambda^{n-1}$ and $X \in D(A)$, to $X \lrcorner\, \Delta^*(1)$. Thus, since $X(\Pi) = 0$ and $X(D(\mu)) = 0$, one has

$$X(\bar{\psi}) = X(\Pi \circ D(\mu)) = X(\Pi) \circ D(\mu) + \Pi \circ X(D(\mu)) = 0$$

by functorial reasons. Hence, $X(\mathscr{K}_\lambda) = \bar{\psi} \circ X(\lambda')$.

Now we show that $\operatorname{im} X(\lambda') \subset S_{2,k-2}$. Since the Spencer complex is exact, it suffices to show that $S_{1,k-1} \circ X(\lambda') = 0$. But $S_{1,k-1} \circ \lambda' = \nu$ and both $S_{1,k-1}$ and ν are natural operators. Therefore, $X(S_{1,k-1}) = 0$, $X(\nu) = 0$ and we have

$$0 = X(\nu) = X(S_{1,k-1} \circ \lambda') = X(S_{1,k-1}) \circ \lambda' + S_{1,k-1} \circ X(\lambda')$$
$$= S_{1,k-1} \circ X(\lambda').$$

The exactness of the Spencer complex shows that

$$D^\bullet(\mathrm{Diff}^+_{k-1}\Lambda^n) = \operatorname{im} S_{2,k-2} \oplus \operatorname{im}\lambda.$$

Since $\operatorname{im}\lambda = \widetilde{\mathrm{Diff}}_k\Lambda^n$ is a projective module, this implies that $\operatorname{im} S_{2,k-2}$ is also projective. Therefore, there exists a splitting

$$\alpha\colon \operatorname{im} S_{2,k} \to D^\bullet(\mathrm{Diff}^+_{k-1}\Lambda^n), \quad S_{2,k-2}\circ\alpha = \mathrm{id}\,.$$

According to Theorem 1.3.2, $d\circ\bar\psi + \bar\psi\circ S_{2,k-2} = 0$. Hence, if $v = -\bar\psi\circ\alpha\circ X(\lambda')$, then $d\circ v = -d\circ\bar\psi\circ\alpha\circ X(\lambda') = \bar\psi\circ S_{2,k-2}\circ\alpha\circ X(\lambda') = \bar\psi\circ X(\lambda')$. □

1.4. Quadratic Lagrangians and the Euler operator

Below we shall consider linear Lagrangian formalism which results in linear Euler–Lagrange equations. The complex $\mathcal{S}_kPQ = \mathcal{S}_{k,\infty}PQ$ (see Subsection 1.3.2) plays the basic role here.

1.4.1. The $\mathcal{S}_kPQ$ complex. We shall prove acyclicity of this complex. The following elementary result will be needed.

LEMMA. *Let*

$$0 \leftarrow K_1 \xleftarrow{\partial_1} K_2 \xleftarrow{\partial_2} \cdots \xleftarrow{\partial_{s-1}} K_s \xleftarrow{\partial_s} \cdots$$

be an acyclic complex, where K_i are projective A-modules and ∂_i are A-homomorphisms. Then the modules $\operatorname{im}\partial_i$ *are projective as well.*

PROOF. If $\operatorname{im}\partial_i$ is a projective module, then there exists a splitting $\alpha\colon \operatorname{im}\partial_i \to K_{i+1}$ and $K_{i+1} = \alpha(\operatorname{im}\partial_i)\oplus\operatorname{im}\partial_{i+1}$. Hence, $\operatorname{im}\partial_{i+1}$ is projective. But $\operatorname{im}\partial_1 = K_1$; therefore, the lemma can be proved by induction. □

Consider the complex $\mathcal{S}^+_kP$ obtained from the complex $\mathcal{S}_k$ by applying the functor $\mathrm{Hom}(P,\bullet)$ to it:

$$0 \leftarrow \hat P \xleftarrow{\mu^+} \mathrm{Diff}^+_k(P,\Lambda^n) \xleftarrow{\mathcal{S}^+_{k,0}} \mathrm{Diff}^+_{k-1}(P,\Lambda^{n-1}) \xleftarrow{\mathcal{S}^+_{k-1,1}} \cdots .$$

If P is a projective module, then all the terms of this complex are projective as well. In addition, μ^+ and $\mathcal{S}^+_{k-l,l}$ are homomorphisms. Therefore, one has

COROLLARY. *If P is a projective module, then $\operatorname{im}\mathcal{S}^+_{k-l,l}$ is projective as well.*

Let $\varepsilon\colon \mathcal{S}_kP \to \mathcal{S}^+_kP$ be the chain mapping which is defined as the identity mapping of the corresponding sets. Let $\alpha_i\colon \operatorname{im}\mathcal{S}^+_{k-i,i} \to \mathrm{Diff}^+_{k-i+1}(P,\Lambda^{n-i+1})$ be a splitting of the projection $\mathcal{S}^+_{k-i,i}\colon \mathrm{Diff}^+_{k-i+1}(P,\Lambda^{n-i+1}) \to \operatorname{im}\mathcal{S}^+_{k-i,i}$, which exists because of the projectivity of the module $\operatorname{im}\mathcal{S}^+_{k-i,i}$. Since both operators $\varepsilon_i = \varepsilon|_{\mathrm{Diff}_{k-i}(P,\Lambda^{n-i})}$ and ε_i^{-1} are of order $\le k-i$ (see Subsection 1.1.3), the mapping $\beta_i = \varepsilon_{i+1}^{-1}\circ\alpha_i\circ(\varepsilon_i|_{\operatorname{im}\mathcal{S}_{k-i,i}})$ is a differential operator of order $\le 2(k-i)$ such that $\mathcal{S}_{k-i,i}\circ\beta_i = \mathrm{id}$. If $\square\in\mathrm{Diff}(Q,\mathrm{Diff}_{k-1}(P,\Lambda^{n-1}))$ is such that $\bar{\mathcal{S}}(\square) = 0$, then the acyclicity of the complex $\mathcal{S}_kP$ implies the inclusion $\operatorname{im}\square\subset\operatorname{im}\mathcal{S}_{k-i,k}$. Therefore, $\square = \bar{\mathcal{S}}(\beta_i\circ\square)$, where $\beta_i\circ\square\in\mathrm{Diff}(Q,\mathrm{Diff}_{k-i-1}(P,\Lambda^{n-i-1}))$. Thus we have proved

THEOREM. *The complex $\mathcal{S}_kPQ$ is acyclic in the case where P is a projective module.*

1.4.2. The involution w. One has a natural identification

$$w = w_{r,s}\colon \operatorname{Diff}_r(P, \operatorname{Diff}_s(P, Q)) \to \operatorname{Diff}_s(P, \operatorname{Diff}_r(P, Q)).$$

Namely, if $\Delta \in \operatorname{Diff}_r(P, \operatorname{Diff}_s(P, Q))$, then $\nabla = w(\Delta)$ is defined by the formula $\nabla(p)(p') = \Delta(p')(p)$. Obviously, $w^2 = \mathrm{id}$. Denote by $\mathcal{S}PP$ the direct limit of the complexes $\mathcal{S}_k PP$ as $k \to \infty$. Then w induces an involution in $\mathcal{S}PP$ also denoted by w. This involution acts on elements $\Delta \in \operatorname{Diff}_r(P, \operatorname{Diff}_s(P, Q))$ as described above. We also put $w(\square) = \square^*$ for $\square \in \operatorname{Diff}(P, \hat{P})$. The following fact is obvious.

PROPOSITION. *The involution w is an automorphism of the complex $\mathcal{S}PP$.*

Consider the complexes $\mathcal{S}_{\mathrm{sym}} P = \ker(1 - w)$ and $\mathcal{S}_{\mathrm{alt}} P = \ker(1 + w)$. From Theorem 1.4.1 one has the following

COROLLARY. *The complexes $\mathcal{S}_{\mathrm{sym}} P$ and $\mathcal{S}_{\mathrm{alt}} P$ are acyclic.*

Let P and Q be A-modules, and $(p_1, \dots, p_s) \mapsto \Delta(p_1, \dots, p_s) \in Q$, $p_1, \dots, p_s \in P$, an $\mathbb{F}$-multilinear function. We say that the function Δ is a *multidifferential operator* acting from P to Q if

1. for any fixed $p_1, \dots, p_{s-1} \in P$ and for any i, $1 \le i \le s$, the correspondence $p \mapsto \Delta(p_1, \dots, p_{i-1}, p, p_{i+1}, \dots, p_{s-1})$ is a differential operator, and
2. the order of this operator is bounded from above when $p_1, \dots, p_{s-1}$ change.

The least upper bound of these orders is called the order with respect to the i-th argument. The module of multidifferential operators of order $\le k_i$ with respect to the i-th argument is identified with the module

$$\operatorname{Diff}_{(k_1, \dots, k_s)}(P, Q) = \operatorname{Diff}_{k_1}(P, \operatorname{Diff}_{k_2}(P, \dots \operatorname{Diff}_{k_s}(P, Q) \dots))$$

if to each operator $\Delta \in \operatorname{Diff}_{(k_1, \dots, k_s)}(P, Q)$ one puts into correspondence the operator $\nabla\colon (p_1, \dots, p_s) \mapsto (\dots (\Delta(p_1)(p_2)) \dots (p_s))$.

One can consider all s-differential Q-valued operators of order $\le k$, both symmetric and skew-symmetric. The corresponding A-modules are denoted by $\operatorname{Diff}_{k,s}(P, Q)$, $\operatorname{Diff}^{\mathrm{sym}}_{k,s}(P, Q)$, and $\operatorname{Diff}^{\mathrm{alt}}_{k,s}(P, Q)$, respectively. In particular, the module of i-chains of the complex $\mathcal{S}_{\mathrm{sym}} P$ (resp., $\mathcal{S}_{\mathrm{alt}} P$) is identified with the module $\operatorname{Diff}^{\mathrm{sym}}_{(2)}(P; \Lambda^{n-i})$ (resp., with $\operatorname{Diff}^{\mathrm{sym}}_{(2)}(P; \Lambda^{n-i})$), $i \ge 0$, while that of (-1)-chains is identified with the module $\operatorname{Diff}^{\mathrm{sym}}(P, \hat{P})$ (resp., $\operatorname{Diff}^{\mathrm{alt}}(P, \hat{P})$) of selfadjoint (resp., skew-adjoint) operators acting from P to $\hat{P}$. Symbols $\operatorname{Diff}_{(s)}$, $\operatorname{Diff}^{\mathrm{sym}}_{(s)}$, and $\operatorname{Diff}^{\mathrm{alt}}_{(s)}$ denote the "direct limits" of the symbols $\operatorname{Diff}_{k,s}$, $\operatorname{Diff}^{\mathrm{sym}}_{k,s}$, and $\operatorname{Diff}^{\mathrm{alt}}_{k,s}$, respectively, as $k \to \infty$. Note that a symmetric multidifferential operator Δ is uniquely determined by its values on the diagonal, i.e., by a nonlinear differential operator of the form $p \mapsto \Delta(p, \dots, p)$. For $s = 2$ such operators will be called *quadratic*. Thus, the chains of the complex $\mathcal{S}_{\mathrm{sym}} P$, except for (-1)-dimensional ones, can be identified with quadratic operators from P to Λ^i, $0 \le i \le n$.

1.4.3. In this subsection we describe linear Lagrangian formalism.

DEFINITION. An operator $L \in \operatorname{Diff}^{\mathrm{sym}}_{(2)}(P, \Lambda^n)$ is called a *quadratic Lagrangian density* of order $\le k$ on P. The coset $L \bmod \bar{\mathcal{S}}(\operatorname{Diff}^{\mathrm{sym}}_{(2)}(P, \Lambda^{n-1}))$ of the operator L in $\operatorname{Diff}^{\mathrm{sym}}_{(2)}(P, \Lambda^n)$ is called the *quadratic Lagrangian* corresponding to the density L.

Denote by $\mathcal{J}(P)$ the set of all quadratic Lagrangians on P. Then one can see that, due to acyclicity of the complex $\mathcal{S}_{\mathrm{sym}} P$, the operator $\bar{\mu}$ is an A-module isomorphism of $\mathcal{J}(P)$ onto a submodule in $\mathrm{Diff}(P, \hat{P})$. This submodule consists of all self-adjoint operators and is denoted by $\mathrm{Diff}^{\mathrm{sym}}(P, \hat{P})$. Let us denote this isomorphism by $\mathcal{E}$ and call it the *Euler operator*. In other words, one has

$$\mathcal{J}(P) = \frac{\mathrm{Diff}_{*,2}(P, \Lambda^n)}{\bar{S}(\mathrm{Diff}_{*,2}(P, \Lambda^n))} \xrightarrow{\mathcal{E}} \mathrm{Diff}^{\mathrm{sym}}(P, \hat{P}).$$

DEFINITION. The equation $\mathcal{E}(\mathcal{L}) = 0$ is called the *Euler–Lagrange equation* corresponding to the Lagrangian $\mathcal{L} \in \mathcal{J}(P)$.

If L is the density of the Lagrangian $\mathcal{L}$, then obviously $\mathcal{E}(\mathcal{L}) = \bar{\mu}(L)$. It is convenient to put $\mathcal{E}_L = \bar{\mu}(L)$ so that $\mathcal{E}_L = \mathcal{E}(\mathcal{L})$.

Coincidence of the quadratic densities introduced above with the "usual" ones can be seen, for example, from their coordinate representation. Motivations for the above definition of Lagrangians will become clear later, when the nonlinear Lagrangian formalism is considered. Finally, to establish coincidence of the Euler–Lagrange equations with the "usual" ones, it suffices to write them down in coordinates as well. Instead of doing this, we shall show directly that the solutions of these equations are extremals of the corresponding variational problem.

So, let $L \in \mathrm{Diff}^{\mathrm{sym}}_{k,2}(P, \Lambda^n)$ and $L(p)(q) = L(p, q) = L_p(q)$, $p, q \in P$. Consider a variation of the form $L(p, p)$ (i.e., the density value at $p \in P$) corresponding to the variation $p \mapsto p + \varepsilon h$, $\varepsilon \in \mathbb{F}$, $h \in P$, of the argument. Then

$$L(p + \varepsilon h, p + \varepsilon h) = L(p, p) + 2\varepsilon L(p, h) + \varepsilon^2 L(h, h).$$

Extremality of p means that $\int_V L(p, h) = 0$ for all $h \in P$, $h|_{\partial V} = 0$, $V \subset M$ being the domain where the variational problem is considered. Since $L_p(h, 1)(1) = L_p(h) = L(p, h)$, the Green formula for the operator $L_p(h, 1) \in \mathrm{Diff}_k \Lambda^n$ can be written in the form

$$L(p, h) - \langle L_p^*(1), h \rangle = d\mathcal{K}_\lambda(L_p(h, 1)). \tag{1.16}$$

Therefore,

$$\int_V L(p, h) = \int_V (\langle L_p^*(1), h \rangle + d\mathcal{K}(L_p(h, 1))).$$

Let W be the support of h. Then the support of $\mathcal{K}_\lambda(L_p(h, 1))$ lies in W because the correspondence $h \mapsto \mathcal{K}_\lambda(L_p(h, 1))$ is a differential operator (see Subsection 1.3.5). If $\partial V \cap W = 0$, then it results from the Stokes formula that $\int_V L(p, h) = \int_V \langle L_p^*(1), h \rangle$. Note that the expression $\langle L_p^*(1), h \rangle$ is A-linear with respect to h. Hence, according to the Du Bois-Reymond lemma, from the equality $\int_V \langle L_p^*(1), h \rangle = 0$ valid for all h with compact support in V, it follows that $L_p^*(1) = 0$. It remains to note that $L_p^*(1) = \bar{\mu}(L)(p) = \mathcal{E}_L(p)$ (for the definition of μ see the end of Subsection 1.3.2).

In these terms, formula (1.16) can be rewritten as

$$L(p, h) - \langle \mathcal{E}_L(p), h \rangle = d\mathcal{K}_\lambda(L_p(h, 1)).$$

The last expression will be called the *Green $\mathcal{L}$-formula.* Note that $\mathcal{E}_L$, being the composition $P \xrightarrow{L} \mathrm{Diff}_k(P, \Lambda^n) \xrightarrow{\mu} \hat{P}$ of the k-th order operators, is an operator of order $\leq 2k$.

1.5. Conservation laws in the linear theory

Using the techniques of Spencer sequences again, we shall consider the theory of linear conservation laws and later establish its relations with the linear Noether theorem.

1.5.1. Let $\Delta \in \mathrm{Diff}_s(P, Q)$ and $E = \{\Delta(u) = 0\}$ be the corresponding equation.

DEFINITION (naive). An operator $\nabla \in \mathrm{Diff}(P, \Lambda^{n-1})$ is called the *density of a linear conservation law*, or *c-density*, if the form $\nabla(p) \in \Lambda^{n-1}$ is closed for any solution $p \in P|_{\mathcal{U}}$ of the equation E over an open set $\mathcal{U} \subset M$.

The integral of the form $\nabla(p)$ along an $(n-1)$-dimensional submanifold $N \subset M$ is a "conserved quantity". This means that it remains constant when we replace N by a homological submanifold. In the classical setting, the role of N is played by submanifolds of the form $t = \text{const}$ in $\mathbb{R}^{n+1} = \{(x,t)\}$, and in this case $\int_{t=t_0} \nabla(p)$ does not depend on time t_0. This justifies the terminology.

Suppose now that the equation E is such that any formal solution of E of order k at any point $x \in M$ can be extended to a real solution in a neighborhood $\mathcal{U} \ni x$. Such equations will be called *FP-equations*. Note that practically all interesting equations are of this type.

PROPOSITION. *Let E be an FP-equation and $\Box \in \mathrm{Diff}(P, P')$ be an operator such that $\Box(p) = 0$ for all p satisfying $\Delta(p) = 0$. Then there exists an operator $\Box' \in \mathrm{Diff}(Q, P')$ such that $\Box = \Box' \circ \Delta$.*

PROOF. Let $E_k \subset \mathcal{J}^{k+s}(P)$ be the submodule generated by formal solutions of the equation E of order $k+s$. In other words, $\theta \in E_k$ if and only if $\theta(x)$ is a formal solution of order $k+s$ of E at $x \in M$ for any x. Since E is an FP-equation, this submodule coincides with the submodule in $\mathcal{J}^{k+s}(P)$ generated by the elements $j_{k+s}(p)$, where p is a local solution. If $\Box \in \mathrm{Diff}_{k+s}(P, P')$ and $\varphi_\Box \colon \mathcal{J}^{k+s}(P) \to P'$ is the corresponding homomorphism, then $\ker \phi_\Box \supset E_k$. Since $E_k = \ker \varphi_{\Delta^{(k)}}$, where $\Delta^{(k)} = j_k \circ \Delta \colon P \to \mathcal{J}^k(Q)$, the homomorphism $\varphi_\Box$ can be represented as the composition $\mathcal{J}^{k+s}(P) \xrightarrow{\varphi_{\Delta^{(k)}}} \mathcal{J}^k(Q) \xrightarrow{\varphi} P'$. It remains to set $\Box' = \varphi \circ j_k$. $\Box$

COROLLARY. *Let E be an FP-equation and ∇ be a linear c-density for this equation. Then $d \circ \nabla = \Box \circ \Delta$, where $\Box \in \mathrm{Diff}(Q, \Lambda^n)$, and vice versa.*

1.5.2. Corollary 1.5.1 allows one to interpret c-densities of FP-equations in the following homological sense. Consider the complex $\widetilde{S_k P}$ (including the case $k = \infty$)

$$0 \to \mathrm{Diff}_{k-n}(P, A) \xrightarrow{\mathcal{S}} \cdots \xrightarrow{\mathcal{S}} \mathrm{Diff}_{k-1}(P, \Lambda^{n-1}) \xrightarrow{\mathcal{S}} \mathrm{Diff}_k(P, \Lambda^n) \to 0$$

which is obtained from $\mathcal{S}_k P$ by skipping the fragment $\cdots \to \hat{P} \to 0$. Since $\mathcal{S}_k P$ is acyclic, the homology of this complex is trivial in all dimensions different from n, and its n-th homology coincides with $\hat{P}$. Any operator $\Delta \in \mathrm{Diff}_s(P, Q)$ generates a chain mapping $\mathcal{S}_\Delta^k : \widetilde{\mathcal{S}_k Q} \to \widetilde{\mathcal{S}_{k+s} P}$. Namely, if $\Box \in \mathrm{Diff}_{k-i}(Q, \Lambda^{n-i})$, we set $\mathcal{S}_\Delta^k(\Box) = \Box \circ \Delta \in \mathrm{Diff}_{k+s-i}(P, \Lambda^{n-i})$. Evidently, one can consider the complex $\widetilde{\mathcal{S}P} = \widetilde{\mathcal{S}_\infty P}$ to be filtered by its subcomplexes $\widetilde{\mathcal{S}_k P}, k \geq 0$. The mapping $\mathcal{S}_\Delta = \mathcal{S}_\Delta^\infty$ increases the filtration by s.

Consider the complex $\operatorname{coker} \mathcal{S}_\Delta$ and denote by $\alpha = \alpha_\Delta$ the natural projection $\widetilde{\mathcal{S}P} \to \operatorname{coker} \mathcal{S}_\Delta$. Then, obviously, Corollary 1.5.1 shows the equivalence of the following two statements:

1. ∇ is a c-density of the FP-equation $\Delta = 0$, and
2. $\alpha(\Delta)$ is a cocycle in the complex $\operatorname{coker} \mathcal{S}_\Delta$.

Definition 1.5.1 is not constructive since it requires to know the solutions of $\Delta(f) = 0$. This is rather unsatisfactory. In particular, for this reason, in general it is impossible to describe the set of c-densities for a concrete equation. That is why we shall use below the following definition motivated by the previous arguments.

DEFINITION. An operator $\nabla \in \operatorname{Diff}(P, \Lambda^n)$ is called a *c-density* for the equation $\Delta = 0$ if $\alpha_\Delta(\nabla)$ is a cocycle in the complex $\operatorname{coker} \mathcal{S}_\Delta$.

REMARK. One can see that Definition 1.5.1 becomes Definition 1.5.2 if the notion of a solution is extended to include formal solutions as well.

1.5.3. If $p \in \ker \Delta$ and ∇, ∇' are c-densities such that $\nabla' = \nabla + \square \circ \Delta$, then $\nabla(p) = \nabla'(p)$. Moreover, if $\nabla' = \nabla + d \circ \square$, $\square \in \operatorname{Diff}(P, \Lambda^{n-2})$, then, due to the Stokes formula, $\int_N \nabla(p) = \int_N \nabla(p')$ for any $N^{n-1} \subset M$ and $p \in \ker \Delta$. Hence the integral $\int_N \nabla(p)$ depends only on the homology class of the cycle $\alpha(\nabla)$ in the complex $\operatorname{coker} \mathcal{S}_\Delta$. This motivates the following

DEFINITION. A *linear conservation law* for the equation $\Delta = 0$ is a class of the $(n-1)$-dimensional homology of the complex $\operatorname{coker} \mathcal{S}_\Delta$.

Let us compute the homology of the complex $\operatorname{coker} \mathcal{S}_\Delta$ under the assumption that $\ker \mathcal{S}_\Delta = 0$. Recall that $H^i(\widetilde{\mathcal{S}P}) = H^i(\widetilde{\mathcal{S}Q}) = 0$ for $i \neq n$. Hence, the nontrivial part of the exact homology sequence corresponding to the short sequence of complexes $0 \to \widetilde{\mathcal{S}Q} \xrightarrow{\mathcal{S}_\Delta} \widetilde{\mathcal{S}P} \xrightarrow{\alpha} \operatorname{coker} \mathcal{S}_\Delta \to 0$ is

$$0 \to H^{n-1}(\operatorname{coker} \mathcal{S}_\Delta) \to H^n(\widetilde{\mathcal{S}Q}) \to H^n(\widetilde{\mathcal{S}P}) \to H^n(\operatorname{coker} \mathcal{S}_\Delta) \to 0.$$

As was noted before, $H^n(\widetilde{\mathcal{S}Q}) = \hat{Q}$, $H^n(\widetilde{\mathcal{S}P}) = \hat{P}$. Under these identifications, the mapping $H^n(\widetilde{\mathcal{S}Q}) \to H^n(\widetilde{\mathcal{S}P})$ corresponds to $\hat{Q} \xrightarrow{\Delta^*} \hat{P}$. This follows from commutativity of the diagram

$$\begin{array}{ccc} \operatorname{Diff}(Q, \Lambda^n) & \xrightarrow{\mathcal{S}_\Delta} & \operatorname{Diff}(P, \Lambda^n) \\ {\scriptstyle\mu}\downarrow & & \downarrow{\scriptstyle\mu} \\ \hat{Q} & \xrightarrow{\quad\Delta^*\quad} & \hat{P} \end{array} \tag{1.17}$$

which can be checked directly. Therefore,

$$H^n(\operatorname{coker} \mathcal{S}_\Delta) = \operatorname{coker} \Delta^*, \quad H^{n-1}(\operatorname{coker} \mathcal{S}_\Delta) = \ker \Delta^*.$$

We shall now clarify the meaning of the condition $\ker \mathcal{S}_\Delta = 0$. To do this, consider the graded complexes $\widetilde{\mathcal{S}P}_{\mathrm{gr}}$ and $\widetilde{\mathcal{S}Q}_{\mathrm{gr}}$ associated with the filtered complexes $\widetilde{\mathcal{S}Q}$ and $\widetilde{\mathcal{S}Q}$. Consider also the mapping $\mathcal{S}_\Delta^{\mathrm{gr}} \colon \widetilde{\mathcal{S}Q}_{\mathrm{gr}} \to \widetilde{\mathcal{S}Q}_{\mathrm{gr}}$ generated by $\mathcal{S}_\Delta$. If $\ker \mathcal{S}_\Delta^{\mathrm{gr}} = 0$, then, obviously, $\ker \mathcal{S}_\Delta = 0$ as well.

By definition, one has

$$\mathcal{S}_\Delta^{\mathrm{gr}}(\operatorname{smbl}_k \square) = \operatorname{smbl}_{k+s}(\square \circ \Delta) = \operatorname{smbl}_k \square \circ \operatorname{smbl}_s \Delta \in \operatorname{Hom}_A(P, \operatorname{smbl}_{k+s} \Lambda^{n-i}),$$

where $\square \in \operatorname{Diff}_k(Q, \Lambda^{n-i})$.

Let $P = \Gamma(\xi)$, $Q = \Gamma(\eta)$. According to Subsection 1.1.9 we have the following identifications:

$$\operatorname{smbl}(P, \Lambda^j) = \operatorname{Hom}^{\text{poly}}(\tau^*(\xi), \tau^*(\tau_j)), \quad \operatorname{smbl}(Q, \Lambda^j) = \operatorname{Hom}^{\text{poly}}(\tau^*(\eta), \tau^*(\tau_j)),$$

where τ_j stands for the j-th exterior power of the cotangent bundle τ. Interpreting $\operatorname{smbl}\Delta$ as a vector bundle homomorphism $s_\Delta \colon \tau^*(\xi) \to \tau^*(\eta)$ (Subsection 1.1.8), we can interpret $\mathcal{S}_\Delta^{\text{gr}}$ as a mapping

$$\mathcal{S}_\Delta^{\text{Hom}} \colon \operatorname{Hom}^{\text{poly}}(\tau^*(\eta), \tau^*(\tau_j)) \to \operatorname{Hom}^{\text{poly}}(\tau^*(\xi), \tau^*(\tau_j)), \quad h \mapsto h \circ s_\Delta,$$

where $h \in \operatorname{Hom}^{\text{poly}}(\tau^*(\eta), \tau^*(\tau_j))$. Evaluation of this mapping at a point $\theta \in T^*M$ looks as

$$s_{\Box,\theta} \mapsto s_{\Box,\theta} \circ s_{\Delta,\theta},$$

if h corresponds to $\operatorname{smbl}\Box$, $\Box \in \operatorname{Diff}(Q, \Lambda^j)$. It is easy to see that this mapping has a nontrivial kernel if and only if $s_{\Delta,\theta}$ has a nontrivial cokernel, i.e., if and only if θ is a characteristic vector for Δ. In turn, this shows that $\ker \mathcal{S}_\Delta^{\text{Hom}} = 0$ if $\operatorname{char}\Delta$ is nowhere dense in T^*M. Hence, we get

THEOREM. *If the set* $\operatorname{char}\Delta$ *is nowhere dense in* $T^*(M)$, *then* $\ker \mathcal{S}_\Delta = 0$ *and hence the group* $\mathcal{Z}(\Delta)$ *of linear conservation laws for the equation* $\Delta = 0$ *is isomorphic to* $\ker \Delta^*$.

REMARK. The hypothesis of the theorem is "almost equivalent" to the fact that the equation $\Delta = 0$ is not overdetermined.

1.5.4. The Green formula for the operator $\Delta \in \operatorname{Diff}(P, Q)$ allows one to describe the isomorphism $\ker \Delta^* = \mathcal{Z}(\Delta)$ explicitly. Indeed, let $\Delta^*(\hat{q}) = 0$, $\hat{q} \in \hat{Q}$. Then the Green formula for $p \in P$, $\hat{q} \in \hat{Q}$ acquires the form $\langle \Delta(p), \hat{q} \rangle = d\mathcal{K}(\Delta(p, \hat{q}))$. Hence, if $\Delta(p) = 0$, then $d\mathcal{K}_\lambda(\Delta(p, \hat{q})) = 0$, i.e., the operator $p \mapsto \mathcal{K}_\lambda(\Delta(p, \hat{q}))$ is a c-density. Thus we obtain a mapping $\zeta \colon \ker \Delta^* \to \mathcal{Z}(\Delta)$, where $\zeta(\hat{q})$ is the conservation law corresponding to the c-density $p \mapsto \mathcal{K}_\lambda(\Delta(p, \hat{q}))$. Let us show that ζ coincides with the isomorphism of Theorem 1.5.3.

Let ∇ be a c-density and $[\nabla]$ be the corresponding conservation law. Then the element $\hat{q} \in \ker \Delta^*$ associated with ∇ by the construction from Subsection 1.5.3 is obtained as follows: $\hat{q} = \mu(\Box)$, where $\Box \in \operatorname{Diff}(Q, \Lambda^n)$ is an operator satisfying the relation $d \circ \nabla = \Box \circ \Delta$. If one understands $\hat{q}$ as an operator acting from Q to Λ^n, the equality $\langle \Delta(p), \hat{q} \rangle = d\mathcal{K}_\lambda(\Delta(p, \hat{q}))$ can be rewritten in the operator form: $\hat{q} \circ \Delta = d \circ \nabla$, where $\nabla(p) = \mathcal{K}_\lambda(\Delta(p, \hat{q}))$. Hence, one can set $\Box = \hat{q}$ and note that $\mu(\hat{q}) = \hat{q}$.

Let us now skip the condition $\ker \mathcal{S}_\Delta = 0$. Consider the exact homology sequences corresponding to the exact sequences of complexes

$$0 \to \ker \mathcal{S}_\Delta \to \widetilde{SQ} \to \operatorname{im} \mathcal{S}_\Delta \to 0 \quad \text{and} \quad 0 \to \operatorname{im} \mathcal{S}_\Delta \to \widetilde{SP} \to \operatorname{coker} \mathcal{S}_\Delta \to 0.$$

Since both complexes, $\widetilde{SP}$ and $\widetilde{SQ}$, are acyclic in dimensions $i \neq n$ and $H^n(\widetilde{SP}) = \hat{P}$, $H^n(\widetilde{SQ}) = \hat{Q}$ (see Subsection 1.5.2), one obtains

$$0 \to H^{n-1}(\operatorname{im} \mathcal{S}_\Delta) \xrightarrow{i_0} H^n(\ker \mathcal{S}_\Delta) \xrightarrow{i_1} \hat{Q} \xrightarrow{i_2} H^n(\operatorname{im} \mathcal{S}_\Delta) \to 0$$

and

$$0 \to H^{n-1}(\operatorname{coker} \mathcal{S}_\Delta) \xrightarrow{j_1} H^n(\operatorname{im} \mathcal{S}_\Delta) \xrightarrow{j_2} \hat{P} \to H^n(\operatorname{coker} \mathcal{S}_\Delta) \to 0.$$

Note that $j_2 \circ i_2 = \Delta^*$. Hence, i_2 induces an isomorphism of $\ker\Delta^*/\operatorname{im} i_1$ to $H^{n-1}(\operatorname{coker}\mathcal{S}_\Delta) = \mathcal{Z}(\Delta)$ whereas the composition

$$\ker\Delta^* \to \ker\Delta^*/\operatorname{im} i_1 \to H^{n-1}(\operatorname{coker}\mathcal{S}_\Delta) = \mathcal{Z}(\Delta)$$

coincides with ζ. Thus, we have proved

PROPOSITION. *The sequence*

$$0 \to H^{n-1}(\operatorname{im}\mathcal{S}_\Delta) \xrightarrow{i_0} H^n(\ker\mathcal{S}_\Delta) \xrightarrow{i_1} \ker\Delta^* \xrightarrow{\zeta} \mathcal{Z}(\Delta) \to 0$$

is exact.

1.5.5. To deal with the Noether theorem, we need another representation of the mapping ζ for Euler–Lagrange equations. Let us choose an operator $L \in \operatorname{Diff}_{k,2}^{\mathrm{sym}}(P;\Lambda^n)$ and set $\Delta = \mathcal{E}_L$. Consider the difference of the Green $\mathcal{L}$-formulas (see Subsection 1.4.3) for the pairs (p,q) and (q,p). Then, taking into account the identity $L(p,q) = L(q,p)$, we get

$$\langle\Delta(q),p\rangle - \langle\Delta(p),q\rangle = d\mathcal{K}_\lambda\big(L_p(q,1) - L_q(p,1)\big). \tag{1.18}$$

Therefore, the form $c(p,q) = c_{L,2}(p,q) = \mathcal{K}_\lambda\big(L_p(q,1)-L_q(p,1)\big)$ is closed if $\Delta(p) = \Delta(q) = 0$. In other words, the operator $\nabla_p\colon P \to \Lambda^{n-1}$, $\nabla_p(q) = c(p,q)$, is a linear c-density for the equation $\Delta = 0$. If $\Delta(p) = 0$, then, according to formula (1.18), $(p^*\circ\Delta)(q) = d\nabla(q)$, i.e., $d\circ\nabla_p = p^*\circ\Delta$. This relation shows that the conservation law corresponding to p by Theorem 1.5.3 is represented by the c-density ∇_p and thus coincides with $\zeta(p)$ as it was shown in Subsection 1.5.4.

1.5.6. The developed "linear algebra" allows one to construct easily nonlinear conservation laws for linear equations.

DEFINITION (naive). A nonlinear operator $\nabla\colon P \to \Lambda^{n-1}$ is called a (*nonlinear*) *c-density* for the equation $\Delta = 0$ if $d\nabla(p) = 0$ for all local solutions of this equation.

It is natural to say that two c-densities ∇_1, ∇_2 are *equivalent* if there exists an operator $\square\colon P \to \Lambda^{n-2}$ such that $\nabla_2 = \nabla_1 + d\circ\square$ on the equation $\Delta = 0$. Such an equivalence class of c-densities is called a *conservation law* for the equation $\Delta = 0$.

REMARK (1). A nonnaive version of these definitions will be given when constructing nonlinear Lagrangian formalism.

Let $\Delta_i\colon P_i \to Q_i$ be two differential operators. A nonlinear differential operator $\square\colon P_1 \to P_2$ is called a *morphism* of the equation $\Delta_1 = 0$ to the equation $\Delta_2 = 0$ if $\square$ takes formal solutions of the first equation to those of the second. In particular, $\square(\ker\Delta_1) \subset \ker\Delta_2$. In the case $P_1 = P_2$, we speak about an *endomorphism.* The following result immediately follows from Subsections 1.3.4, 1.4.3 and the basic definitions.

THEOREM. 1. *Let $\Delta \in \operatorname{Diff}(P,Q)$ and $\square_1$ be an endomorphism of the equation $\Delta = 0$. Let $\square_2$ be a morphism of $\Delta = 0$ to $\Delta^* = 0$. Then the operator $p \mapsto \mathcal{K}_\lambda\big(\Delta(\square_1(p),\square_2(p))\big)$ is a c-density of the equation $\Delta = 0$ the equivalence class of which is independent of λ.*

2. *If $\Delta = \mathcal{E}_L$ and $\square_i$ are endomorphisms of the equation $\Delta = 0$, then the operator $p \mapsto c_{L,\lambda}\big(\square_1(p),\square_2(p)\big)$ is a c-density of the equation $\Delta = 0$ the equivalence class of which is independent of λ.*[1]

[1]See Subsection 1.5.5 for the definition of $c_{L,\lambda}$.

3. *The c-densities described in* (1), (2) *are equivalent when* $\Delta = \mathcal{E}_L$.

REMARK (2). If multivalued conservation laws are admitted, then suitable Bäcklund transformations can be taken for $\Box_i$, $i = 1, 2$.

REMARK (3). If $[\Delta, \Box] = 0$, then $\Box$ is an endomorphism of the equation $\Delta = 0$ and hence, in the case where $\Delta = 0$ is the Euler–Lagrange equation, it determines a c-density $p \mapsto \mathcal{K}_\lambda(\Delta(\Box(p), p))$.

Note that any infinitesimal symmetry of a Lagrangian is a symmetry, i.e., an automorphism of the corresponding equation $\mathcal{E}_L = 0$ as well. Hence, the theorem above gives a more general (and, it seems, a more transparent) way of constructing conservation laws than the classical Noether theorem. Relations of these two theorems will be clarified below.

Note in conclusion that we have obtained a way to construct conservation laws from symmetries for arbitrary linear equations and not only for the Euler–Lagrange equations.

1.6. Automorphisms and the linear Noether theorem

In this section, we consider the linear Noether theorem in the algebraic setting, to compare it further with Theorem 1.5.6. To do this, one needs an algebraic description of the transformation laws for the object treated in the theorem. This is the topic of the first half of this section.

1.6.1. Let $F\colon A_1 \to A_2$ and $F_P\colon P_1 \to P_2$ be the same as in Subsection 1.2.5. Consider an operator $L \in \operatorname{Diff}_{k,2}(P_1, \Lambda^n(A_1))$ and define $F_P(L)$ in such a way that the diagram

$$\begin{array}{ccc} P_1 & \xrightarrow{\ L\ } & \operatorname{Diff}_k(P_1, \Lambda^n(A_1)) \\ {\scriptstyle F_P}\downarrow & & \downarrow{\scriptstyle F_{\mathrm{Diff}}} \\ P_2 & \xrightarrow{F_P(L)} & \operatorname{Diff}_k(P_2, \Lambda^n(A_2)) \end{array}$$

is commutative. Here

$$F_{\mathrm{Diff}}(\Delta) = F_{\Lambda^n} \circ \Delta \circ F_P^{-1}, \quad F_P(L) = F_{\mathrm{Diff}} \circ L \circ F_P^{-1},$$

$\Delta \in \operatorname{Diff}_k(P_1, \Lambda^n(A_1))$.

To obtain the infinitesimal version of these definitions, one should set $X_P(L) = X_{\mathrm{Diff}} \circ L - L \circ X_P$ for $(X_P, X) \in \operatorname{Der} P$, where $X_{\mathrm{Diff}}(\Delta) = X_{\Lambda^n} \circ \Delta - \Delta \circ X_P$. In particular, if L is a Lagrangian density (i.e., $L \in \operatorname{Diff}^{\mathrm{sym}}_{k,2}(P_1, \Lambda^n(A_1))$), then $F_P(L) \in \operatorname{Diff}^{\mathrm{sym}}_{k,2}(P_2, \Lambda^n(A_2))$ is called a *transformation of the density* L, whereas $X_P(L) \in \operatorname{Diff}^{\mathrm{sym}}_{k,2}(P, \Lambda^n)$ is called its *variation* under the infinitesimal automorphism $X_P \in \operatorname{Der} P$. Since, for obvious reasons, $F_P \circ \bar{S} = \bar{S} \circ F_P$ and $X_P \circ \bar{S} = \bar{S} \circ X_P$ and also $F(\mu) = \mu$, $X(\mu) = 0$ (see Subsection 1.3.7), one sees that

$$F_P\colon \mathcal{S}_{\mathrm{sym}} P_1 \to \mathcal{S}_{\mathrm{sym}} P_2, \quad X_P\colon \mathcal{S}_{\mathrm{sym}} P \to \mathcal{S}_{\mathrm{sym}} P$$

are chain mappings, which generate the mappings

$$\mathcal{J}(P_1) \to \mathcal{J}(P_2), \quad \mathcal{J}(P) \to \mathcal{J}(P),$$

respectively. We still denote them by F_p and X_P.

For brevity, set $\tilde{L} = X_P(L)$. Recall that in the notation of Subsection 1.4.3 we have $L_p = L(p) \in \mathrm{Diff}(P, \Lambda^n)$:

$$\tilde{L}_p = X_P(L)(p) = X_{\mathrm{Diff}}(L_p) - L(X_P(p)) = X_\Lambda \circ L_p - L_p \circ X_P - L_{X_P(p)}. \tag{1.19}$$

Since $\tilde{L}_p(q) = \tilde{L}(p,q)$ and $L_p(q) = L(p,q)$, by applying the last operator relation to $q \in P$, one finds

$$\tilde{L}(p,q) = X(L(p,q)) - L(p, X_P(q)) - L(X_P(p), q), \quad p, q \in P. \tag{1.20}$$

On the other hand, formula (1.19) implies

$$\tilde{L}_p(q,1) = \tilde{L}_p \circ q = X \circ L_p(q,1) - L_p \circ X_P \circ q - L_{X_P(p)}(q,1).$$

Since $X_P \circ q = q \circ X + X_P(q)$, the last expression can be rewritten as

$$X \circ L_p(q,1) - L_p(q,1) \circ X - L_p(X_P(q), 1) - L_{X_P(p)}(q,1).$$

Therefore,

$$\tilde{L}_p(q,1) = [X, L_p(q,1)] - L_p(X_P(q), 1) - L_{X_P(p)}(q,1). \tag{1.21}$$

Formulas (1.20) and (1.21) will be used in the proof of the Noether theorem. Let us transform the terms in (1.20) using the $\mathcal{L}$-Green formula:

$$\begin{aligned} X\big(L(p,q)\big) &= d\big(X \lrcorner\, L(p,q)\big), \\ L\big(p, X_P(q)\big) &= d\mathcal{K}_\lambda\big(p, X_P(q)\big) + \langle \mathcal{E}_L(p), X_P(q)\rangle, \\ L\big(X_P(p), q\big) &= L\big(q, X_P(p)\big) = d\mathcal{K}_\lambda\big(q, X_P(p)\big) + \langle \mathcal{E}_L(q), X_P(p)\rangle, \end{aligned}$$

where $\mathcal{K}_\lambda(p,q) = \mathcal{K}_\lambda(L_p(q,1))$. Hence, we obtain the following formula for the first variation in the linear theory:

$$\begin{aligned} \tilde{L}(p,q) = d\big(X \lrcorner\, L(p,q) - \mathcal{K}_\lambda(p, X_P(q)) - \mathcal{K}_\lambda(q, X_P(p))\big) \\ - \langle \mathcal{E}_L(p), X_P(q)\rangle - \langle \mathcal{E}_L(q), X_P(p)\rangle. \end{aligned} \tag{1.22}$$

1.6.2. The linear Noether theorem. In the classical calculus of variations, a Lagrangian $\mathcal{L} = \int L$ is called invariant with respect to some transformation if the action $\int_V L$ does not change for any compact domain V. According to the Du Bois-Reymond lemma this is equivalent to the invariance of the Lagrangian density L with respect to the transformation in question. For this reason, F_P (resp., X_P) will be called a *symmetry* (resp., an *infinitesimal symmetry*) of the Lagrangian density L if $F_P(L) = L$ (resp., $X_P(L) = 0$). Since F_P (when $P_1 = P_2$) and X_P are endomorphisms of the complex $\mathcal{S}_{\mathrm{sym}}P$, one has $F_P(\mathcal{E}_L) = \mathcal{E}_L$ and $X_P(\mathcal{E}_L) = 0$ when F_P and X_P are symmetries of the density L. In particular, $X_P(p) \in \ker \mathcal{E}_L$ for $p \in \ker \mathcal{E}_L$. This remark together with formula (1.22) gives the following result.

THEOREM (The Noether theorem). *If X_P is a symmetry of a Lagrangian density L, then the $(n-1)$-form*

$$n_\lambda(p,q) = X \lrcorner\, L(p,q) - \mathcal{K}_\lambda\big(p, X_P(q)\big) - \mathcal{K}_\lambda\big(q, X_P(p)\big)$$

is closed for $p, q \in \ker \mathcal{E}_L$ and its cohomology class does not depend on the choice of λ.

The form $n_\lambda(p) = n_\lambda(p,p)$ is the *classical Noether current*.

REMARK. Both the classical proof of the Noether theorem and the one given above show that the assumptions of this theorem can be weakened. Namely, it would be sufficient to assume that the action does not vary on the extremals, i.e., $\int_V \tilde{L}(p,q) = 0$ or, which is the same, $\int_V \tilde{L}(p,p) = 0$ for $p, q \in \ker \mathcal{E}_L$ and for any domain of integration V. However, usually this weakening is only imaginary. Namely, put $\Delta = \mathcal{E}_L$, $L \in \mathrm{Diff}^{\mathrm{sym}}_{k,2}(P, \Lambda^n)$. Then $\ker \varphi_\Delta \subset \mathcal{J}^{2k}(P)$, where $\varphi_\Delta \colon \mathcal{J}^{2k}(P) \to \hat{P}$ is the homomorphism corresponding to Δ. If $\nu_{2k,k}(\ker \varphi_\Delta) = \mathcal{J}^k(P)$, where $\nu_{a,b} \colon \mathcal{J}^a(P) \to \mathcal{J}^b(P)$, $a > b$, is the natural projection, then it can be easily seen that the weakened hypothesis implies the density invariance. The condition $\nu_{2k,k}(\ker \varphi_\Delta) = \mathcal{J}^k(P)$ essentially means that the equation $\mathcal{E}_L = 0$ does not degenerate to an equation of order $\leq k$.

1.6.3. The Noether theorem can be really strengthened by enlarging the class of operators X_P. Indeed, suppose that $X_P(L) = \bar{\mathcal{S}}(L')$, $L' \in \mathrm{Diff}^{\mathrm{sym}}_{k,2}(P, \Lambda^{n-1})$. Then $\tilde{L}(p,q) = (\bar{\mathcal{S}}(L')(p))(q) = (d \circ L'(p))(q) = dL'(p,q)$, where $L'(p,q) = L'(p)(q)$, $p, q \in P$, while from the first variation formula it follows that the $(n-1)$-form

$$\bar{n}_\lambda(p,q) \overset{\mathrm{def}}{=} n_\lambda(p,q) - L'(p,q), \quad p, q \in \ker \mathcal{E}_L,$$

is closed. Just as it was done before, one can show that its cohomology class is independent of λ. Since the complex $\mathcal{S}_{\mathrm{sym}} P$ is exact, this class also does not depend on the choice of L' such that $X_P(L) = \bar{\mathcal{S}}(L')$. Again, due to the exactness of $\mathcal{S}_{\mathrm{sym}} P$, this condition is equivalent to $\bar{\mu}(X_P(L)) = 0$. But $\bar{\mu}(X_P(L)) = X_P(\bar{\mu}(L)) = X_P(\mathcal{E}_L)$. Hence, X_P is an infinitesimal symmetry of the operator $\mathcal{E}_L$. The converse is obviously also valid. Therefore, we have proved

THEOREM. *Suppose that X_P is a symmetry of the operator $\mathcal{E}_L$, i.e., $[X_P, \mathcal{E}_L] = 0$, and P is a projective module. Then the $(n-1)$-form $\bar{n}_\lambda(p,q)$ is closed for $p, q \in \ker \mathcal{E}_L$ and its cohomology class depends neither on the splitting λ nor on the solution L' of the equation $X_P(L) = \bar{\mathcal{S}}(L')$. In particular, the mapping $p \mapsto \bar{n}_\lambda(p,p)$ is a c-density for the equation $\mathcal{E}_L = 0$ and the corresponding conservation law does not depend on λ and L'.*

REMARK (1). Note that $\bar{n}_\lambda$ can be efficiently computed from X_P. To show this, consider the homomorphism

$$\lambda_1 \colon \ker \mu = \operatorname{im} \mathcal{S} \to \mathrm{Diff}^+_{k-1} \Lambda^{n-1}, \quad \lambda_1 = -\psi \circ \lambda \circ *,$$

where the image $\operatorname{im} \mathcal{S}$ is understood as a submodule in $\mathrm{Diff}^+_k \Lambda^n$. Then Theorem 1.3.2 shows that $\mathcal{S} \circ \lambda_1 = \mathrm{id}_{\operatorname{im} \mathcal{S}}$, and therefore, that the module $\mathrm{Hom}_A(P, \operatorname{im} \mathcal{S})$ coincides with the image of the homomorphism $\mathcal{S}_{k,0} = \mathcal{S} \colon \mathrm{Diff}^+_{k-1}(P, \Lambda^{n-1}) \to \mathrm{Diff}^+_k(P, \Lambda^n)$. Hence, the mapping

$$\lambda_2 = \mathrm{Hom}(P, \lambda) \colon \mathrm{Hom}(P, \operatorname{im} \mathcal{S}) \to \mathrm{Diff}^+_{k-1}(P, \Lambda^{n-1})$$

is the right inverse to $\mathcal{S}_{k,0}$, while the mapping

$$\bar{\lambda} = \mathrm{Diff}(P, \lambda_2) \colon \operatorname{im} \bar{\mathcal{S}} \to \mathrm{Diff}(P, \mathrm{Diff}_{k-1}(P, \Lambda^{n-1}))$$

is the right inverse to $\bar{\mathcal{S}}$. Therefore, one can set $L' = \bar{\lambda}(X_P(L))$.

REMARK (2). A weaker requirement that could be imposed on X_P, namely $\tilde{L}(p,q) = dL'(p,q)$ on extremals, gives nothing new in essence, since under the conditions of the remark in Subsection 1.6.2, it turns out to be equivalent to the condition $X_P(L) = \bar{\mathcal{S}}(L')$.

1.6.4. Now we show that the linear Noether theorem is a rather special case of Theorem 1.5.6.

PROPOSITION. *The closed forms* $n_\lambda(p,q)$ *and* $c_\lambda(X_P(p),q)$, *where* $X_P \in \operatorname{Der} P$ *is a symmetry of the Lagrangian density* L, *are cohomologous to each other for all* $p, q \in \ker \mathcal{E}_L$. *In particular, the conservation law corresponding to the Noetherian c-density* $n_\lambda(p)$ *coincides with the conservation law corresponding to the c-density* $c_\lambda(X_P(p),p)$.

PROOF. Let

$$\begin{aligned} w(p,q) &= n_\lambda(p,q) - c_\lambda(X_P(p),q) \\ &= X \lrcorner L(p,q) - \mathcal{K}_\lambda(p, X_P(q)) - \mathcal{K}_\lambda(X_P(p),q). \end{aligned}$$

If $\mathcal{E}_L(p) = 0$, then according to the $\mathcal{L}$-Green formula one has

$$X \lrcorner L(p,q) = X \lrcorner d\mathcal{K}_\lambda(p,q) = X\big(\mathcal{K}_\lambda(p,q)\big) - d\big(X \lrcorner \mathcal{K}_\lambda(p,q)\big).$$

On the other hand, due to (1.14),

$$X\big(\mathcal{K}_\lambda(p,q)\big) = X\big(\mathcal{K}_\lambda(L_p(q,1))\big) = X(\mathcal{K}_\lambda)(L_p(q,1)) + \mathcal{K}_\lambda\big([X, L_p(q,1)]\big).$$

Using Proposition 1.3.7, one finds that

$$X\big(\mathcal{K}_\lambda(p,q)\big) = dv(p,q) - \mathcal{K}_\lambda\big([X, L_p(q,1)]\big),$$

where $v(p,q) = v(L_p(q,1))$ (see Subsection 1.3.7). Hence,

$$\begin{aligned} w(p,q) &= d\big(v(p,q) - X \lrcorner \mathcal{K}_\lambda(p,q)\big) + \mathcal{K}_\lambda([X, L_p(q,1)]) \\ &\quad + \mathcal{K}_\lambda(L_p(X_P(q),1)) - \mathcal{K}_\lambda(L_{X_P(p)}(q,1)) \\ &= du(p,q) + \mathcal{K}_\lambda\big([X, L_p(q,1)] - L_p(X_P(q,1)) - L_{X_P(p)}(q,1)\big), \end{aligned}$$

where $u(p,q) = v(p,q) - X \lrcorner \mathcal{K}_\lambda(P,q)$. According to (1.21), the argument of the function $\mathcal{K}_\lambda$ in the last equality coincides with $\tilde{L}_p(q,1)$ and consequently vanishes, since X_P is a symmetry of L. ☐

Note the following equality obtained during the proof:

$$n_\lambda(p,q) - c_\lambda(X_P(p),q) = du(p,q) + \mathcal{K}_\lambda(\tilde{L}_p(q,1)), \quad p \in \ker \mathcal{E}_L. \tag{1.23}$$

From the last formula we get

COROLLARY. *If a symmetry* X_P *of the equation* $\mathcal{E}_L = 0$ *is such that the form* $n_\lambda(p,q)$ *is closed for all* $p, q \in \ker \mathcal{E}_L$ *(i.e., the Noetherian current is "conserved"), then, under the assumption of Remark* 1.6.2, X_P *is actually a symmetry of the density* L.

PROOF. Since X_P is a symmetry of the equation L, then $X_P(p) \in \ker \mathcal{E}_L$ for $p \in \ker \mathcal{E}_L$. Therefore, $dc_\lambda(X_P(p),q) = 0$ if $p, q \in \ker \mathcal{E}_L$. Hence, from (1.23) it follows that

$$\begin{aligned} dn_\lambda(p,q) = d\mathcal{K}_\lambda\big(\tilde{L}_p(q,1)\big) &= \tilde{L}(p,q) - \langle \mathcal{E}_{\tilde{L}(p)}, q\rangle \\ &= \tilde{L}(p,q) - \langle X_P(\mathcal{E}_L)(p), q\rangle = \tilde{L}(p,q) \end{aligned}$$

in this case. Consequently, $\tilde{L}(p,q) = 0$ for $p, q \in \ker \mathcal{E}_L$. Therefore, under the assumptions of Remark 1.6.2, we have $\tilde{L} = X_P(L) = 0$. ☐

Thus, in general, a Noether current ceases to be a conserved quantity under a natural extension of the class of infinitesimal transformations.

1.7. Concluding remarks

The functors D_i, $\mathrm{Diff}_k^{(+)}$ and their combinations considered in this chapter are examples of *differential calculus functors*. Moreover, if we want to understand the words *differential calculus* in a rigorous mathematical way, then we have no alternative other than the *calculus of differential functors over commutative algebras* (in a most general sense). The final compilation of the list of such functors and main operations on them, started in [**136**], is still to be finished, though many facts are already known (see [**124, 133, 71, 120, 138, 63, 141**]). Systematic use of algebraic language in various situations where methods of differential calculus are applied, with rare exceptions (see [**61, 22, 21**]), was mainly the approach of the Moscow school (see, for example, [**73**] and references therein). Applications of the introduced differential calculus go far beyond the bounds of this book. They spread from fundamental problems of algebraic and differential geometry to subtle problems of field theory, not to mention the theory of differential equations itself. For example, the language of Spencer cohomology was proven to be quite useful in the theory of singularities of algebraic manifolds (see [**18**]).

It should be emphasized that in the contemporary literature the words "differential calculus" are often used in a much narrower sense than in this book, and their meaning varies from one paper to another. As a rule, they are understood as very particular differential operators or analogs of these operators in a nonstandard context. For example, in "noncommutative geometry" [**32**], they refer to noncommutative generalizations of the de Rham complex.

Naturality of the algebraic approach to Lagrangian formalism and related matters is confirmed by the fact that it directly generalizes to "super" and "quantum" cases, as it was demonstrated by Verbovetsky [**117, 116**] (see also [**145**] and [**72**]). The reader will find an elementary introduction to differential calculus over commutative algebras together with detailed motivations in [**95**].

CHAPTER 2

Geometry of Finite-Order Contact Structures and the Classical Theory of Symmetries of Partial Differential Equations

The first idea on how to draw a geometric portrait of a general k-th order system of partial differential equations is to interpret it as a submanifold in an appropriate k-th order jet space (manifold). In order to treat solutions of the system geometrically, the jet space is supplied with an additional structure called the *k-th order contact structure*, or the *Cartan distribution*. According to the philosophy of Klein's *Erlangen Program* [**57**], the theory of k-th order partial differential equations is governed by the automorphism group of the k-th order contact structure. This group, which is, in addition, the basis of the classical theory of symmetries of partial differential equations, is studied in this chapter. The result obtained seems to be very disappointing: this group does not depend on k. This means that in the framework of the proposed approach there is no difference between equations of the first order and those of higher orders. In particular, classical symmetries of a higher-order system of partial differential equations remain to be of the first order only. Thus higher order symmetries of such a system discovered in the theory of integrable systems are not covered by the classical approach. Therefore, we are forced to look for a more satisfactory geometric setting for partial differential equations. This is done in the next chapter by passing to infinite jet spaces. Thus, the results obtained in this chapter motivate and prepare this passage, which seems to be of a certain psychological difficulty.

2.1. Necessary facts from the geometry of jet spaces

In this section, we present basic facts concerning the geometry of finite-order jet spaces and nonlinear differential equations. Proofs of these basic results can be found in [**71**]. In what follows, we work in the category of smooth ($= C^\infty$) manifolds, and all the constructions and objects related to smooth manifolds are supposed to be smooth as well.

2.1.1. Jet spaces. Let E be a smooth manifold of dimension $m + n$. Let $L \subset E$ be an n-dimensional submanifold passing through a point $x \in E$. The class of n-dimensional submanifolds tangent to L with order k at the point x is called the *k-jet* of L at x and is denoted by $[L]_x^k$. Let

$$J_x^k(E,n) = \{[L]_x^k \mid L \ni x,\ \dim L = n\}, \quad J^k(E,n) = \bigcap_{x \in N} J_x^k(E,n).$$

The sets $J^k(E,n)$ are naturally endowed with the structure of a smooth manifold and are called the *sets* (or the *manifolds*) *of jets*. Local charts of a special type will be described below. Note that $J^0(E,n) = E$.

For $k \geq l$, the projection $\pi_{k,l}\colon J^k(E,n) \to J^l(E,n)$ is defined by $\pi_{k,l}([L]^k_x) = [L]^l_x$. If $L^n \subset E$, then one has a mapping

$$j_k(L)\colon L \to J^k(E,n), \quad j_k(L)(x) = [L]^k_x.$$

Obviously, $\pi_{k,l} \circ j_k(L) = j_l(L)$. Set $L^{(k)} = \operatorname{im} j_k(L) \subset J^k(E,n)$.

Note that all the above (except for the definition of the smooth structure) also makes sense if $k = \infty$. In this case we shall write $j(L)$ instead of $j_\infty(L)$.

Let $\pi\colon E = E_\pi \to M$ be a smooth submersion, $\dim M = n$, $\dim E = n + m$. Then the k-jets of the images of local sections of this submersion form an open set $J^k(\pi)$ in $J^k(E,n)$. If π is a fiber bundle, then $J^k(\pi)$ is called the *manifold of k-jets* or the *k-th jet bundle* of π. Let us also denote by $\Gamma(\pi)$ (resp., $\Gamma_{\mathrm{loc}}(\pi)$) the set of all (resp., local) sections of the submersion π. Denote by $\mathcal{U}_s$ the domain of a submersion $s \in \Gamma_{\mathrm{loc}}(\pi)$ and set

$$j_k(s) = j_k(L) \circ s\colon \mathcal{U}_s \to E, \quad L = s(\mathcal{U}_s),$$
$$\pi_k = \pi \circ \pi_{k,0}, \quad \mathcal{F}_k(\pi) = C^\infty(J^k(\pi)), \quad \mathcal{F}_{-1} = C^\infty(M).$$

We shall also write $[s]^k_y$, $y \in M$, instead of $[s(\mathcal{U}_s)]^k_{s(y)}$ and $M^{(k)}_s$ instead of $\operatorname{im} j_k(s)$.

If $A = C^\infty(M)$ and π is a smooth vector bundle over M, then $\Gamma(\pi_k) = \mathcal{J}^k(\Gamma(\pi))$, where $\mathcal{J}^k$ is the representing object for the functor Diff_k in the category of geometric modules over A.

Let $\mathcal{U} = V^n \times W^m$, where V^n and W^m are open domains in $\mathbb{R}^n$ and $\mathbb{R}^m$, $x = (x_1, \dots, x_n)$ and $u = (u_1, \dots, u_m)$ be Cartesian coordinates in V^n and W^m respectively. Then (x, u) forms a coordinate system in $\mathcal{U}$. Let also $\alpha\colon \mathcal{U} \to V^n$ be the projection onto the first factor. Then a coordinate system (x_j, p^i_σ), $1 \leq j \leq n$, $1 \leq i \leq m$, $|\sigma| \leq k$, arises in $J^k(\alpha)$, where σ stands for an unordered multiindex $(i_1, \dots, i_\nu)$, $1 \leq i_\mu \leq n$, and $|\sigma| = \nu$. The functions p^i_σ are uniquely determined by the following property:

$$j_k(\varphi)^*(p^i_\sigma) = \frac{\partial^{|\sigma|}\varphi_i}{\partial x_{i_1} \cdots \partial x_{i_\nu}}, \quad f \in \Gamma(\alpha),$$

where $\varphi(x) = (\varphi_1(x), \dots, \varphi_m(x))$. We shall often write u_i instead of $p^i_\emptyset$.

With any diffeomorphism $f\colon \mathcal{U} \to \mathcal{U}' = \operatorname{im} f \subset N$ one can naturally associate diffeomorphisms $f_{(k)}\colon J^k(\alpha) \to J^k(E,n)$. Namely, let us set

$$f_{(k)}([s]^k_y) = [L]^k_{f(s(y))}, \quad s \in \Gamma_{\mathrm{loc}}(\alpha), \ L = f(s(\mathcal{U}_s)).$$

Obviously, the sets $\operatorname{im} f_{(k)}$ are open in $J^k(E,n)$ and $f_{(k)}$ carries over the above described coordinates (x_j, p^i_σ) to these sets. The domain $\operatorname{im} f_{(k)} \subset J^k(E,n)$ equipped with these coordinates is called a *special chart*, while the coordinates (x_j, p^i_σ) are called *special coordinates*.

The set $J^\infty(E,n)$ is obviously identified with the inverse limit of the chain of mappings

$$E = J^0(E,n) \xleftarrow{\pi_{1,0}} \cdots \xleftarrow{\pi_{k,k-1}} J^k(E,n) \xleftarrow{\pi_{k+1,k}} \cdots.$$

Let $\mathcal{F}_k(E,n) = C^\infty(J^k(E,n))$. The direct limit of the sequence of mappings

$$C^\infty(N) = \mathcal{F}_0(E,n) \xrightarrow{\pi^*_{1,0}} \cdots \xrightarrow{\pi^*_{k,k-1}} \mathcal{F}_k(E,n) \xrightarrow{\pi^*_{k+1,1}} \cdots$$

is denoted by $\mathcal{F}(E,n)$. We identify the algebra $\mathcal{F}_k(E,n)$ with its image in $\mathcal{F}(E,n)$ under $\pi^*_{\infty,k}$. Since the homomorphisms $\pi^*_{\infty,k}$ are injective, we can assume that the algebra $\mathcal{F}(E,n)$ is filtered by its subalgebras

$$\mathcal{F}(E,n) \supset \cdots \supset \mathcal{F}_k(E,n) \supset \mathcal{F}_{k-1}(E,n) \supset \cdots .$$

In a similar way, we define $\mathcal{F}(\pi)$ to be the direct limit of algebras $\mathcal{F}_k(\pi)$ with respect to the sequence of homomorphisms $\pi^*_{k,k-1}$ and consider $\mathcal{F}(\pi)$ to be filtered by $\mathcal{F}_k(\pi)$.

2.1.2. Nonlinear differential operators. Let $\xi\colon E_\xi \to M$ be a bundle, and $F\colon M_1 \to M$ a smooth mapping. Denote by $F^*(\xi)$ the pullback of ξ by F. If $s\colon M \to M_1$ is such that $F \circ s = \mathrm{id}_M$ and $\varphi \in \Gamma_{\mathrm{loc}}(F^*(\xi))$, then we set $s^*(\varphi) = \bar{F} \circ \varphi \circ s \in \Gamma_{\mathrm{loc}}(\xi)$, where $\bar{F}\colon F^*(\xi) \to \xi$ is the canonical morphism.

If $\eta\colon E_\eta \to E$ is a bundle, then we shall denote by $\eta_{(k)}$ the bundle $\pi^*_{k,0}(\eta)$ over the space $J^k(E,n)$. Fix $\varphi \in \Gamma(\eta_{(k)})$. Generalizing slightly the standard terminology, we say that the operation

$$L \mapsto j_k(L)^*(\varphi) \in \Gamma(\eta|_L), \quad L \in N, \ \dim L = n,$$

is a (*nonlinear*) *differential operator* of order $\leq k$ acting on n-dimensional submanifolds of E.

If $\pi\colon E_\pi \to M$ is a bundle, $E = E_\pi$, and $\eta = \pi^*(\xi)$, where $\xi\colon E_\xi \to M$ is another bundle, then for $s \in \Gamma_{\mathrm{loc}}(\pi)$ one has $j_k(s)^*(\eta_{(k)}) = \xi$. In particular, if $L = s(\mathcal{U}_s)$, then $\eta|_L = \xi$. Obviously, the operation

$$\Delta_\varphi\colon s \mapsto j_k(s)^*(\varphi) \in \Gamma_{\mathrm{loc}}(\xi), \quad s \in \Gamma_{\mathrm{loc}}(\pi),$$

is a (*nonlinear*) *differential operator* of order $\leq k$ in the usual sense, taking $\Gamma_{\mathrm{loc}}(\pi)$ to $\Gamma_{\mathrm{loc}}(\xi)$. The correspondence $\varphi \mapsto \Delta_\varphi$ is a bijection between $\Gamma(\eta_{(k)}) = \Gamma(\pi^*_k(\xi))$ and the set of all nonlinear differential operators acting on $\Gamma_{\mathrm{loc}}(\pi)$ with values in $\Gamma_{\mathrm{loc}}(\xi)$. This set is denoted by $\mathrm{dif}_k(\pi,\xi)$. A section $\varphi \in \Gamma(\pi^*_k(\xi))$ corresponding to $\Delta \in \mathrm{dif}_k(\pi,\xi)$ under this bijection is denoted by φ_Δ. In the sequel, we also use the following notation: $\mathcal{F}_k(E,\eta) = \Gamma(\eta_{(k)})$, $\mathcal{F}_k(\pi,\xi) = \Gamma(\pi^*_k(\xi))$. The mappings $\pi_{k,l}$ induce embeddings $\mathcal{F}_l(E,\eta) \subset \mathcal{F}_k(E,\eta)$ and $\mathcal{F}_l(\pi,\xi) \subset \mathcal{F}_k(\pi,\xi)$. The embedding $\mathrm{dif}_l(\pi,\xi) \subset \mathrm{dif}_k(\pi,\xi)$ corresponds to the latter via the above bijection. We also set $\mathcal{F}(E,\eta) = \operatorname{inj\,lim}_{k\to\infty} \mathcal{F}_k(E,\eta)$ and $\mathcal{F}(\pi,\xi) = \operatorname{inj\,lim}_{k\to\infty} \mathcal{F}_k(\pi,\xi)$.

Note that $\mathcal{F}_k(\pi, \mathbb{1}_M) = \mathcal{F}_k(\pi)$, i.e., the algebra of smooth functions on $J^k(\pi)$ is identified with the algebra of scalar differential operators on $\Gamma_{\mathrm{loc}}(\pi)$. We define multiplication in the latter by the formula $(\Delta_1 \cdot \Delta_2)(s) = \Delta_1(s) \cdot \Delta_2(s)$, where $\Delta_i \in \mathrm{dif}_k(\pi, \mathbb{1}_M)$, $s \in \Gamma_{\mathrm{loc}}(\pi)$. In a similar way, if η (resp., ξ) is a vector bundle, then $\mathcal{F}_k(E,\eta)$ (resp., $\mathcal{F}_k(\pi,\xi)$) is an $\mathcal{F}_k(E,n)$- (resp., $\mathcal{F}_k(\pi)$-) module. Moreover, $\mathcal{F}(E,\eta)$ (resp., $\mathcal{F}(\pi,\xi)$) is a filtered $\mathcal{F}(E,n)$- (resp., $\mathcal{F}(\pi)$-) module with $\mathcal{F}_k(N,\eta)$ (resp., $\mathcal{F}_k(\pi,\xi)$) being the k-th filtration.

To any $\Delta \in \mathrm{dif}_k(\xi,\eta)$, where ξ and η are vector bundles over M, one can associate the operator

$$\hat{\Delta}\colon \mathcal{F}(\pi,\xi) \to \mathcal{F}(\pi,\eta), \quad \hat{\Delta}(\psi) = \varphi_{\Delta\circ\Delta_\psi}.$$

If $\Delta \in \mathrm{Diff}_k(\Gamma(\xi),\Gamma(\pi))$, then $\hat{\Delta} \in \mathrm{Diff}_k(\mathcal{F}(\pi,\xi),\mathcal{F}(\pi,\eta))$. In addition, one has $\hat{\Delta}(\mathcal{F}_l(\pi,\xi)) \subset \mathcal{F}_{k+l}(\pi,\eta)$ and

$$(2.1) \quad \widehat{\Delta_1 \circ \Delta_2} = \hat{\Delta}_1 \circ \hat{\Delta}_2, \quad \widehat{\mu\Delta_1 + \lambda\Delta_2} = \mu\hat{\Delta}_1 + \lambda\hat{\Delta}_2, \quad \Delta \circ j(s)^* = j(s)^* \circ \hat{\Delta},$$

where $\mu, \lambda \in C^\infty(M)$. Note that the last property can be used as the definition of $\hat{\Delta}$.

Let a special chart (see above) $f\colon \mathcal{U} \to E = E_\pi$ be such that f is a morphism of the bundle α to the bundle π. Then in the corresponding special coordinates on $\operatorname{im} f_{(\infty)}$ one has

$$D_i = \widehat{\frac{\partial}{\partial x_i}} = \frac{\partial}{\partial x_i} + \sum_\sigma p^\alpha_{\sigma i} \frac{\partial}{\partial p^\alpha_\sigma},$$

where $\sigma i = (i, i_1, \dots, i_\nu)$ for $\sigma = (i_1, \dots, i_\nu)$. The operator D_i is called the *total derivative* with respect to x_i.

The composition of the mappings

$$\mathcal{F}(\pi, \xi) \xrightarrow{\hat{j}_k} \mathcal{F}(\pi, \xi_k) \xrightarrow{i} \mathcal{J}^k(\mathcal{F}(\pi, \xi)),$$

where i is the natural embedding, will be denoted by ρ_k. Obviously, one has $\rho_k(\mathcal{F}_l(\pi, \xi)) \subset \mathcal{J}^k(\mathcal{F}_{k+l}(\pi, \xi))$ and

$$j(s)^*(\rho_k(\varphi)) = j_k(\Delta(s)) \tag{2.2}$$

for any $s \in \mathcal{F}(\pi, \xi)$, where $\Delta = \Delta_\varphi$.

For a linear fiber bundle π, it is useful to introduce the elements $\rho_k(\pi) = \rho_k(\varphi_{\mathrm{id}})$, where $\mathrm{id}\colon \Gamma(\pi) \to \Gamma(\pi)$ is the identity mapping. Then, as it follows from (2.1), for any $\theta \in \Gamma(\pi_k)$ we have

$$\theta^*(\rho_k(\pi)) = \theta. \tag{2.3}$$

Finally, let us denote by U_k the composition

$$\mathcal{F}_l(\pi, \xi) \xrightarrow{\rho_k} \mathcal{J}^k(\mathcal{F}_{k+l}(\pi, \xi)) \xrightarrow{\Sigma} \mathcal{J}^{k-1}(\mathcal{F}_{k+l}(\pi, \xi)) \otimes \Lambda^1(J^{k+l}(\pi)),$$

where Σ is the Spencer operator (see Subsection 1.1.10).

It can be easily seen that for $p \geq l$ the diagram

$$\begin{array}{ccc}
\mathcal{F}_l(\pi, \xi) & \xrightarrow{U_k} & \mathcal{J}^{k-1}(\mathcal{F}_{k+l}(\pi, \xi)) \otimes \Lambda^1(J^{k+l}(\pi)) \\
\Big\downarrow{\scriptstyle \pi^*_{p,l}} & & \Big\downarrow{\scriptstyle \pi^*_{k+p,k+l}} \\
\mathcal{F}_p(\pi, \xi) & \xrightarrow{U_k} & \mathcal{J}^{k-1}(\mathcal{F}_{k+p}(\pi, \xi)) \otimes \Lambda^1(J^{k+p}(\pi))
\end{array}$$

is commutative. This allows one, by passing to the limit as $l \to \infty$, to define the operator

$$U_k\colon \mathcal{F}(\pi, \xi) \to \mathcal{J}^{k-1}(\mathcal{F}(\pi, \xi)) \otimes \Lambda^1(J^\infty(\pi)).$$

It follows from the definitions of ρ_k and U_k that they are differential operators of order $\leq k$. Exactness of the jet-Spencer complex (Subsection 1.1.10) implies

PROPOSITION. *The following statements are valid*:

1. *For any* $s \in \Gamma_{\mathrm{loc}}(\pi)$, *one has* $j(s)^*(U_k(\varphi)) = 0$.
2. *If* $\theta \in \Gamma_{\mathrm{loc}}(\pi_k)$, $k > 0$, *and* $\theta^* U_k(\pi) = 0$, *then* θ *is of the form* $\theta = j_k(s)$ *for some* $s \in \Gamma_{\mathrm{loc}}(\pi)$.

This proposition shows that for a linear bundle π and for any $k > 0$ the element $U_k(\pi)$ distinguishes the sections $j_k(s)$ of the bundle $\pi_k\colon J^k(\pi) \to M$ from all others. Nonlinear bundles π will be discussed below.

EXAMPLE. Let π be a one-dimensional bundle. Then

$$U_1(\pi) \in \mathcal{J}^0(\mathcal{F}_1(\pi,\pi)) \otimes \Lambda^1(J^1(\pi)) = \Lambda^1(J^1(\pi)).$$

The expression for $U_1(\pi)$ in terms of special coordinates is $U_1(\pi) = du - \sum p_i\, dx_i$, where $u = p_\emptyset$. Hence, the form $U_1(\pi)$ determines the classical contact structure on $J^1(\pi)$ (see, for example, [**85**]). Thus one can consider the jet-valued differential form $U_k(\pi)$ as representing *generalized contact structures.*

To finish this subsection, note that to any section $\varphi \in \mathcal{F}_k(\pi,\xi)$ (or, which is the same, to an operator $\Delta \in \operatorname{dif}_k(\pi,\xi)$) there corresponds a fiber bundle morphism

$$\underline{\varphi}\colon J^k(\pi) \to J^0(\xi) = E_\xi, \quad \underline{\varphi}([s]_x^k) = \Delta_\varphi(s)(x).$$

If $\varphi^{(l)} = \widehat{j}_l(\varphi)$, then for any $l \geq t$ one has the commutative diagram

$$\begin{array}{ccc} J^{k+l}(\pi) & \xrightarrow{\underline{\varphi}^{(l)}} & J^l(\xi) \\ {\scriptstyle \pi_{k+l,k+t}}\big\downarrow & & \big\downarrow{\scriptstyle \pi_{l,t}} \\ J^{k+t}(\pi) & \xrightarrow{\underline{\varphi}^{(t)}} & J^t(\xi) \end{array}$$

This allows one to pass to the limit as $l \to \infty$ and to obtain a fiber bundle morphism $\underline{\varphi}^\infty$ such that the following diagram is commutative:

$$\begin{array}{ccc} J^\infty(\pi) & \xrightarrow{\underline{\varphi}^\infty} & J^\infty(\xi) \\ & {\scriptstyle \pi_\infty}\searrow \quad \swarrow{\scriptstyle \xi_\infty} & \\ & M & \end{array}$$

Obviously,

$$\underline{\varphi}^{(l)} \circ j_{k+l}(s) = j_l(\Delta_\varphi(s)), \quad 0 \leq s \leq \infty.$$

In particular, $\underline{\varphi} \circ j_k(s) = \Delta_\varphi(s)$.

2.1.3. Nonlinear differential equations. A submanifold (with singularities) $\mathcal{Y} \subset J^k(E,n)$ (resp., $\mathcal{Y} \subset J^k(\pi)$) is called a *system of nonlinear differential equations* of order $\leq k$ imposed on n-dimensional submanifolds of the manifold E (resp., on sections of the bundle π). Below we call them simply "equations" and not "systems of differential equations". A submanifold $L \subset E$ (resp., $s \in \Gamma_{\mathrm{loc}}(\pi)$) is called an (*ordinary*) *solution* of the equation $\mathcal{Y}$ if $L^{(k)} \subset \mathcal{Y}$ (resp., $M_s^{(k)} \subset \mathcal{Y}$).

Rewriting these definitions in local coordinates, one can see that they coincide with the usual ones. On the other hand, this emphasizes the fundamental role of the manifolds $J^k(E,n)$ and $J^k(\pi)$ in the theory of differential equations—they play the role of "universal receptacles" of all possible equations of order $\leq k$ imposed on n-dimensional submanifolds of E (resp., on sections of π). Roughly speaking, our approach to the theory of differential equations consists of analysis of "universal operations" (functors) on "universal" objects $J^k(E,n)$ and $J^k(\pi)$ and subsequent restriction of these operations onto particular submanifolds (i.e., equations) of $J^k(E,n)$ and $J^k(\pi)$.

Any equation $\mathcal{Y}$ can be represented in the form $\varphi = 0$, where $\varphi \in \mathcal{F}_k(N,\eta)$ or $\varphi \in \mathcal{F}_k(\pi,\xi)$ for appropriate vector bundles η or ξ. In the sequel, we shall assume that the graph of the section φ intersects the graph of the zero section transversally.

Define the *l-th prolongation* $\mathcal{Y}^{(l)}$ of the equation $\mathcal{Y} \subset J^k(E,n)$ (or $\mathcal{Y} \subset J^k(\pi)$) as follows: $[L]_x^{k+l} \in \mathcal{Y}^{(s)}$ if and only if $L^{(k)}$ is tangent to $\mathcal{Y}$ at the point $j_k(L)(x)$

with order l (resp., $[s]_x^{k+l} \in \mathcal{Y}^{(s)}$ if and only if $M_s^{(k)}$ is tangent to $\mathcal{Y}$ at the point $j_k(s)(x)$ with order l).

Obviously, $\pi_{k+l,k+t}(\mathcal{Y}^{(l)}) \subset \mathcal{Y}^{(t)}$, $l \geq t$, and the set of solutions for $\mathcal{Y}^{(l)}$ coincides with that of $\mathcal{Y}$.

If $\mathcal{Y} = \{\varphi = 0\} \subset J^k(\pi)$ and π is a linear bundle, then $\mathcal{Y}^{(l)} = \{\rho_l(\varphi) = 0\}$. If in a special chart (x^j, p^i_σ) the equation $\mathcal{Y}$ is described by $\varphi_i = 0$, $i = 1, \dots, p$, then in the corresponding special chart in $J^k(\pi)$ the equation $\mathcal{Y}^{(l)}$ is described by

$$D_\sigma(\varphi_i) = 0, \quad i = 1, \dots, p, \ |\sigma| \leq l,$$

where $D_\sigma = D_{i_1} \circ \dots \circ D_{i_\nu}$ for $\sigma = (i_1, \dots, i_\nu)$.

An equation $\mathcal{Y}$ is called *formally integrable* if the mapping $\pi_{\infty,k}|_{\mathcal{Y}_\infty} : \mathcal{Y}_\infty \to \mathcal{Y}$ is a surjection. It is called *strongly* formally integrable if for all $l \geq 0$ the mappings $\pi_{k+l,k+l-1}|_{\mathcal{Y}^{(l)}} : \mathcal{Y}^{(l)} \to \mathcal{Y}^{(l-1)}$ are fiber bundles. A detailed exposition of the modern theory of formal integrability can be found in [**45, 103**] and in the book [**98**].

Finally, equation $\mathcal{Y}$ is said to be *s-solvable* if $\mathcal{Y}^{(s)} \to \mathcal{Y}$ is a surjection.

2.1.4. The Cartan distribution. Jet manifolds themselves are not sufficient to develop geometric theory of partial differential equations in a self-contained way. In fact, while k-th order partial differential equations are interpreted as submanifolds of $J^k(E,n)$ or $J^k(\pi)$, a special class of n-dimensional submanifolds, namely, those of the form $L^{(k)}$ or M_s^k, is to be distinguished to define what are solutions of these equations. To do this in an intrinsic way, jet manifolds are to be supplied with an appropriate geometric structure. This structure is called the *Cartan distribution* or, alternatively, the *k-th order contact structure* and is defined as follows.

An *R-plane* at the point $\theta \in J^k(E,n)$ is a subspace in $T_\theta(J^k(E,n))$ of the form $T_\theta(L^{(k)})$ for some $L \subset E$. Consider the linear span $\mathcal{C}_\theta \subset T_\theta(J^k(E,n))$ of all R-planes at θ. The distribution $\theta \mapsto \mathcal{C}_\theta$ is called the *Cartan distribution* on $J^k(E,n)$.

Changing $J^k(E,n)$ to $J^k(\pi)$ and $L^{(k)}$ to $M_s^{(k)}$ in what was said above, one gets the Cartan distribution on $J^k(\pi)$. If $J^k(\pi)$ is understood as an open domain in $J^k(E,n)$, then, obviously, the Cartan distribution on $J^k(\pi)$ is the restriction of that on $J^k(E,n)$ to $J^k(\pi)$.

A submanifold $W \subset J^k(E,n)$ is said to be *integral* for the Cartan distribution if $T_\theta W \subset \mathcal{C}_\theta$ for any $\theta \in W$. Such a manifold is called *locally maximal* if no open subset of W can be embedded into another integral manifold of greater dimension.

Obviously, submanifolds of the form $L^{(k)}$ (resp., $M_s^{(k)}$) are integral manifolds of the Cartan distribution. Conversely, the following basic result holds.

THEOREM. *Let $V \subset J^k(E,n)$ (resp., $V \subset J^k(\pi)$) be an n-dimensional locally maximal integral manifold of the Cartan distribution. If $mk > 1$ (resp., $k \dim \pi > 1$), then locally (resp., locally and possibly after elimination of a submanifold of a lower dimension) V is of the form $L^{(k)}$ (resp., $M_s^{(k)}$).*

The case $mk > 1$ (resp., $k \dim \pi > 1$) is called *exceptional*. The Cartan distribution on $J^1(E,n)$ when $m = 1$ (resp., on $J^1(\pi)$ with $\dim \pi = 1$) determines a contact structure leading to the classical theory of scalar first-order partial differential equations (see, e.g., [**85, 71**]).

The above theorem shows that the Cartan distribution is precisely the structure on $J^k(E,n)$ or $J^k(\pi)$ we need.

Let $\theta = [L]_x^k \in J^k(E,n)$, $k > 0$. Then the R-plane

$$L_\theta = T_{\bar\theta}(L^{(k-1)}), \quad \bar\theta = \pi_{k,k-1}(\theta),$$

is well defined by the point θ. Moreover, $\theta_1 = \theta_2$ if $L_{\theta_1} = L_{\theta_2}$.

PROPOSITION (1). *For any $k \geq 1$ and $\theta \in J^k(E,n)$ one has*

$$\mathcal{C}_\theta = (d\pi_{k,k-1}|_\theta)^{-1}(L_\theta).$$

In other words, the projection $\pi_{k,k-1}$ maps the Cartan plane $\mathcal{C}_\theta$ onto the R-plane L_θ.

It is useful to describe the Cartan distribution on $J^k(\pi)$ in a dual way. Let $\varphi \in \mathcal{F}_{k-l}(\pi)$. Then $U_l(\varphi) \in \mathcal{J}^{l-1}(\mathcal{F}_k(\pi)) \otimes \Lambda^1(J^k(\pi))$. It is natural to interpret the elements $U_l(\varphi)$ as jet-valued differential one-forms so that the contraction operation is defined:

$$\xi \mapsto \xi \lrcorner U_l(\varphi) \in \mathcal{J}^{l-1}(\mathcal{F}_k(\pi))|_\theta, \quad \xi \in T_\theta(J^k(\pi)).$$

Set

$$\mathcal{C}_\theta^l = \{\xi \in T_\theta(J^k(\pi)) \mid \xi \lrcorner U_l(\varphi) = 0, \ \varphi \in \mathcal{F}_{k-l}(\pi)\}.$$

For a linear bundle π we also set

$$\mathcal{C}_\theta^\pi = \{\xi \in T_\theta(J^k(\pi)) \mid \xi \lrcorner U_k(\pi) = 0\}.$$

PROPOSITION (2). *For any l, $1 \leq l \leq k$, one has $\mathcal{C}_\theta^l = \mathcal{C}_\theta$. If π is a linear bundle, then $\mathcal{C}_\theta^\pi = \mathcal{C}_\theta$.*

In a particular case $l = 1$, this proposition shows that the system of forms $U_1(\varphi)$, $\varphi \in \mathcal{F}_{k-1}(\pi)$, generates the annihilator $\operatorname{Ann}\mathcal{C}$ of the Cartan distribution on $J^k(\pi)$. Since the operator U_1 is a derivation, i.e.,

$$U_1(\varphi\psi) = \psi U_1(\varphi) + \varphi U_1(\psi),$$

in a special coordinate system $\operatorname{Ann}\mathcal{C}$ is generated by the forms

$$U_1(p_\sigma^\alpha) = dp_\sigma^\alpha - \sum p_{\sigma i}^\alpha \, dx_i, \quad \alpha = 1, \dots, m, \ |\sigma| < k.$$

As above, let us take $\theta \in J^k(\pi)$, $\bar\theta = \pi_{k,k-1}(\theta)$, $x = \pi_k(\theta)$, $k > 0$. Since $d\pi_{k-1}|_{\bar\theta} \colon T_x M$ is an isomorphism, one has the direct sum decomposition

$$T_{\bar\theta}(J^{k-1}(\pi)) = L_\theta \oplus T_{\bar\theta}(\pi_{k-1}^{-1}(x)).$$

The corresponding projection

$$p_\theta \colon T_{\bar\theta}(J^{k-1}(\pi)) \to T_{\bar\theta}(\pi_{k-1}^{-1}(x))$$

allows one to define the linear mapping

$$U_k^\theta = p_\theta \circ (d\pi_{k,k-1}|_\theta) \colon T_\theta(J^k(\pi)) \to T_{\bar\theta}(\pi_{k-1}^{-1}(x)).$$

If π is a linear bundle, the fiber $\pi_{k-1}^{-1}(x)$ of the linear bundle π_{k-1} can be identified with $T_\theta(\pi_{k-1}^{-1}(x))$.

PROPOSITION (3). *If π is a linear bundle, then, under the above-made identification, one has $U_k^\theta(\xi) = \xi \lrcorner U_k(\pi)$ for any $\xi \in T_\theta(J^k(\pi))$.*

Let $\mathcal{C}_k(E) = \bigcup_{\theta \in J^k(E,n)} \mathcal{C}_\theta$. Then $\mathcal{C}_k(E) \subset T(J^k(E,n))$ is a linear subbundle of the tangent bundle $T(J^k(E,n))$:

$$\mathcal{C}_k \colon \mathcal{C}_k(E) \to J^k(E,n).$$

Consider the quotient bundle $l_k \colon T(J^k(E,n))/\mathcal{C}_k(N) \to J^k(E,n)$. From Propositions 1 and 3 it follows that the operation $X \mapsto U_k^\bullet(X) \in \Gamma(l_k)$, where $U_k^\bullet(X) = X \bmod \mathcal{C}_k(E)$, is a generalization of the jet-valued one-form $U_k(\pi)$ to the case of manifolds $J^k(E,n)$.

2.1.5. Two types of contact geometry. The classical theory of scalar first-order differential equations is geometrically formulated as the contact geometry on $J^1(E,n)$, $m=1$ (or $J^1(\pi)$, $\dim \pi = 1$). On the other hand, when π is a linear bundle, the contact structure on $J^1(\pi)$ is determined by zeros of the form $U_1(\pi)$. As we have seen above, higher analogs of $U_1(\pi)$ are jet-valued forms $U_k(\pi)$. Therefore, higher contact geometry should be defined as the geometry of "zeros" for the form $U_k(\pi)$ on $J^k(\pi)$. However, what are "zeros" for $U_k(\pi)$ can be understood in two different ways. First, they can be interpreted as manifolds $V \subset J^k(\pi)$ such that $U_k(\pi)|_V = 0$. Geometry arising from this viewpoint will be called the *U-geometry*. Second, $U_k(\pi)$ is a jet-valued one-form and its "zeros" can be understood as vector fields X such that $X \lrcorner U_k(\pi) = 0$. From Proposition 2.1.2 it follows that these fields lie in the Cartan distribution and vice versa. Hence, the Cartan distribution determines the second kind of higher contact geometry.

Both types of higher contact geometry will be studied below. But even now one can see that the first geometry is more "rigid" than the second. Nevertheless, the latter plays a more important role in the theory of differential equations.

2.1.6. Structure of integral manifolds in U-geometry. The bundle π is assumed to be linear in this subsection.

A submanifold $L \subset J^k(\pi)$ is called a *U-manifold* if

(1) $U_k(\pi)|_L = 0$, and
(2) L is locally maximal in the class of manifolds satisfying condition (1).

We describe the structure of U-manifolds below. To do this, we shall need the following construction.

Let $\alpha \colon V \hookrightarrow M$ be an l-dimensional submanifold, $J_V^k = \pi_k^{-1}(V)$, $\pi_V = \pi|_{\pi^{-1}(V)}$. Consider the natural embedding $\alpha_k \colon J_V^k(\pi) \to J^k(\pi)$ and the natural projection $\beta_k \colon J_V^k(\pi) \to J^k(\pi_V)$ defined by $[s]_x^k \mapsto [s|_V]_x^k$, $s \in \Gamma_{\mathrm{loc}}(\pi)$, $x \in V$.

PROPOSITION. $\alpha_k^*(U_k(\pi)) = \beta_k^*(U_k(\pi_V))$.

Let $s \in \Gamma_{\mathrm{loc}}(\pi_V)$ and $W = \beta_k^{-1}(V_s^{(k)})$. Recall that $V_s^{(k)} \subset J^k(\pi_V)$ is the image of $j_k(s)$ (see Subsection 2.1.1). From the last proposition and Proposition 2.1.2 (1) it follows that $U_k(\pi)|_W = 0$. It turns out that the converse is also "almost" true.

THEOREM. *Locally, possibly after elimination of a lower dimensional submanifold, any U-manifold is an open domain $O \subset \beta_k^{-1}(V_s^{(k)})$, where $V = \pi_k(O)$ and $s \in \Gamma(\pi_V)$.*

If $\dim V = l$, then

$$\dim O = l + m\left[\binom{n+k}{n} - \binom{l+k}{l}\right] \stackrel{\text{def}}{=} \lambda(n,m,k,l).$$

The function $\lambda(n, m, k, l)$ strictly decreases with respect to l in all cases except for $m = k = 1$ (when $\lambda(n, 1, 1, l) = n$). In this exceptional case, we obtain the usual contact geometry on $J^1(\pi)$, $\dim \pi = 1$. In all other cases, monotonicity of λ with respect to $l = \dim V$ shows that the number λ in the theorem does not depend on the domain O in the U-manifold under consideration. This number is called the *type* of this manifold.

Thus, when $mk > 1$, U-manifolds of the type 0 coincide with the fibers of the projection $\pi_{k,0}$ or with open domains of these fibers. U-manifolds of the type n are locally, possibly after elimination of a submanifold of lower dimension, of the form $M_s^{(k)}$.

In the sequel, a U-manifold of the form $\beta_k^{-1}(V_s^{(k)})$, where $s \in \Gamma_{\mathrm{loc}}(\pi_V)$, is denoted by $\bar{V}_s^{(k)}$.

2.1.7. Structure of integral manifolds of the Cartan distribution. Recall that an integral manifold of the Cartan distribution is locally maximal if it does not contain open subsets which can be embedded into an integral manifold of greater dimension.

Consider the following construction. Let an integral manifold L of the Cartan distribution on $J^{k-1}(E, n)$ (resp., on $J^{k-1}(\pi)$), $k > 0$, be such that $\pi_{k-1,0}|_L$ is an immersion (resp., $\pi_k|_L$ is an immersion). Set

$$W(L) = \{\theta \in J^k(E, n) \text{ (or } J^k(\pi)) \mid \bar{\theta} = \pi_{k,k-1}(\theta) \in L,\ L_\theta \supset T_{\bar{\theta}}(L)\}.$$

It is easy to see that

$$\dim W(L) = m\binom{p + k - 1}{k} + n - p,$$

where $n - p = \dim L$. The number $\dim L$ is called the *type* of $W(L)$. One can check directly that the inequality $\dim L < \dim L'$ implies $\dim W(L) > \dim W(L')$, except for the cases

(1) $n = m = 1$, and
(2) $k = m = 1$.

The situation when $nm > 1$ or $km > 1$ will be called *nonexceptional.*

THEOREM. *In the nonexceptional situation, every locally maximal integral manifold of the Cartan distribution on $J^k(E, n)$ or on $J^k(\pi)$, possibly after elimination of a lower dimensional submanifold, is locally an open domain in a manifold of the form $W(L)$.*

From the above it follows that in the nonexceptional case the dimension of L does not depend on the choice of a domain O in the integral manifold Z under consideration. Hence, we can introduce the notion of *type* of Z by putting $\operatorname{type} Z = \dim L$.

Note that in the nonexceptional case locally maximal integral manifolds of the type 0 are the fibers of the projection $\pi_{k,k-1}$ or open domains of these fibers. Integral manifolds of the maximal type ($= n$) are locally, possibly after elimination of a lower dimensional submanifold, of the form $L^{(k)}$ or $M_s^{(k)}$. Such integral manifolds will be called *R-manifolds*. These R-manifolds play an important role in the theory of partial differential equations. In particular, they allow one to generalize in a natural way the concept of a solution for an arbitrary, generally nonlinear, system of PDE. This generalization for linear equations is essentially equivalent to those constructed on the basis of functional analysis (distribution solutions, etc.). Further details in this direction can be found in [**71, 82, 86, 83, 128, 123, 124**].

The exceptional case $m = k = 1$ has been studied extensively: this is the standard contact geometry. In the case $m = n = 1$, locally maximal integral manifolds are curves consisting of pieces of the types 0 and 1.

2.2. The structure of U-transformations

In this section, we describe the structure of automorphisms in U-geometry. These automorphisms are called *U-transformations*. The results obtained allow one to clarify, in the spirit of Klein's *Erlangen Program* [**57**], the general structure of this kind of contact geometry.

Recall (see Subsection 2.1.5) that U-geometry arises as one of alternative generalizations of the classical contact geometry, which describes first-order scalar equations, to the case of arbitrary order equations. In fact, as it follows from the results of this section, this kind of geometry plays a minor role in the theory of differential equations. For this reason, we restrict ourselves to the case of linear bundles just not to complicate the exposition with additional constructions, which will be of no use in the subsequent text (see [**71**] for "nonlinear" details). In addition, we describe the structure of transformations preserving the elements $U_s(\pi)$ and $\rho_s(\pi)$ in the case of "functorial" bundles.

Throughout this section the basic bundle π is assumed to be linear.

2.2.1. We start with the following

DEFINITION. A diffeomorphism $\varphi\colon V_1 \to V_2$ of the domains $V_1, V_2 \subset J^k(\pi)$ is called a *U-transformation* if $U_k(\pi)|_{\varphi(L)} = 0$ whenever $U_k(\pi)|_L = 0$, $L \subset V_1$.

One could define U-transformations using U-manifolds only. But as the following proposition shows, we shall obtain nothing new.

PROPOSITION. *Let $\varphi\colon V_1 \to V_2$ be a diffeomorphism taking U-submanifolds of V_1 to U-submanifolds of V_2. Then φ is a U-transformation.*

PROOF. Let $L \subset V_1$ and $U_k(\pi)|_L = 0$. Choose a domain $O \subset V_1$ such that the rank of $\pi_k|_O$ is constant and $V = \pi(O)$ is a submanifold in M. Then O lies in the manifold $\bar{V}_s^{(k)}$ (see Subsection 2.1.6) and thus $U_k(\pi)|_{\varphi(O)} = 0$. Further, since the manifold L (possibly, except for a nowhere dense set) can be covered by domains of such a type, one has $U_k(\pi)|_{\varphi(L)} = 0$. □

REMARK. Obviously, the converse is also valid: any U-transformation takes U-manifolds to U-manifolds.

The structure of U-transformations in the exceptional case is well known. As it follows from the definition, they coincide with contact transformations of the contact structure in $J^1(\pi)$ determined by the form $U_1(\pi) \in \Lambda^1(J^1(\pi))$. See [**85, 71**] for the details. For this reason, we consider below the nonexceptional case only.

2.2.2. As follows from Subsection 2.1.6, in the class of submanifolds L such that $U_k(\pi)|_L = 0$, the submanifolds of the highest possible dimension are exactly U-manifolds of the type 0. But as was noted in the same subsection, manifolds of the type 0 are open subsets in the fibers of the projection $\pi_{k,0}\colon J^k(\pi) \to J^0(\pi)$. Hence, U-transformations take a connected component of the fiber $\pi_{k,0}^{-1}(x) \cap V_1$,

$x \in E_\pi = J^0(\pi)$, to a connected component of another fiber $\pi_{k,0}^{-1}(y) \cap V_2$, $y \in E_\pi$. Before formulating the main result of this subsection, we give the following

DEFINITION. Let $g\colon M_1 \to M_2$ be a mapping between two manifolds. A domain $W \subset M_1$ is called *g-connected* if all sets of the form $g^{-1}(x) \cap W$, $x \in M_2$, are connected.

PROPOSITION. *Any U-transformation of a $\pi_{s,0}$-connected domain $V_1 \subset J^s(\pi)$ takes the fibers of the projection $\pi_{s,0}|_{V_1}$ to the fibers of $\pi_{s,0}|_{V_2}$.*

2.2.3. Thus any U-transformation $\varphi\colon V_1 \to V_2$ of a $\pi_{s,0}$-connected domain in $J^s(\pi)$ is a fiberwise mapping and consequently generates a diffeomorphism $\tilde{\varphi}\colon \tilde{V}_1 \to \tilde{V}_2$, where $\tilde{V}_i = \pi_{s,0}(V_i)$, $i = 1, 2$, are domains in $J^0(\pi)$:

$$\tilde{\varphi}(x_0) = \pi_{s,0}(\varphi(x_s)), \quad \pi_{s,0}(x_s) = x_0.$$

Note that the diffeomorphism $\tilde{\varphi}$ preserves connected components of the fibers of $\pi|_{\tilde{V}_1}$. Indeed, otherwise there would exist a curve $C_0 \subset \tilde{V}_1$ diffeomorphically projected by π onto a curve $C \subset M$ and such that $\pi(\tilde{\varphi}(C_0))$ is a point, say $x \in M$. Then C_0 can be presented as $C_0 = \sigma(C)$ for some section $\sigma \in \Gamma_{\mathrm{loc}}(\pi_C)$. Obviously, C_0 can be chosen in such a way that the intersection $\tilde{C}_\sigma^{(k)} = \bar{C}_\sigma^{(k)} \cap V_1$ is nonempty (see Subsection 2.1.6).

Furthermore,

$$\pi_k(\varphi(\tilde{C}_\sigma^{(k)})) = (\pi \circ \pi_{k,0} \circ \varphi)(\tilde{C}_\sigma^{(k)}) = (\pi \circ \tilde{\varphi} \circ \pi_{k,0})(\tilde{C}_\sigma^{(k)}) = \pi(\varphi(C_0)) = x.$$

But on the other hand, from Theorem 2.1.6 it follows that any connected submanifold $L \subset \pi_k^{-1}(x)$ satisfying $U_k(\pi)|_L = 0$ lies in one of the fibers $\pi_{k,0}^{-1}(y)$, where $\pi(y) = x$.

Therefore, $\varphi(\tilde{C}_\sigma^{(k)})$ lies in a fiber of the projection $\pi_{k,0}$, whereas $\tilde{C}_\sigma^{(k)}$ does not. This contradicts the assumption that φ is $\pi_{k,0}$-fiberwise. Hence, we have proved the following

PROPOSITION. *Let $\tilde{\varphi}\colon \tilde{V}_1 \to \tilde{V}_2$ be the diffeomorphism generated by a transformation $\varphi\colon V_1 \to V_2$ and $\tilde{V}_1$ be a π-connected domain. Then $\tilde{\varphi}$ generates a diffeomorphism $\varphi'\colon V_1' \to V_2'$ such that $\pi \circ \tilde{\varphi} = \varphi' \circ \pi$, where $V_i' = \pi(\tilde{V}_i)$.*

2.2.4. Now let $\varphi\colon W_1 \to W_2$ be a diffeomorphism of the domains $W_i \subset J^0(\pi)$, $i = 1, 2$, preserving connected components of fibers of the projection $\pi|_{W_1}$. Then for any $k > 0$ the diffeomorphism φ generates a diffeomorphism $\varphi_k\colon \pi_{k,0}^{-1}(W_1) \to \pi_{k,0}^{-1}(W_2)$ by

$$\varphi_k(\theta) = [\varphi \circ f]_y^k, \quad \theta = [f]_x^k,$$

where $x \in M$, $f(x) \in W_1$, and $y = (\pi \circ \varphi \circ f)(x) \in \pi(W_2)$. Note that from the definition of φ_k one gets

$$\varphi_l \circ \pi_{k,l} = \pi_{k,l} \circ \varphi_k, \quad k \geq l. \tag{2.4}$$

PROPOSITION. *For any $k \geq 0$, the mapping φ_k is a U-diffeomorphism.*

PROOF. According to Proposition 2.2.1, it is sufficient to prove that φ_k sends U-manifolds to U-manifolds. But this is obvious for manifolds of the form $V_s^{(k)}$ (see Theorem 2.1.6) and consequently for all U-manifolds. □

2.2.5. Now we shall prove the following

THEOREM. *Any U-transformation $\varphi\colon V_1 \to V_2$ of a $\pi_{k,0}$-connected domain $V_1 \subset J^k(\pi)$ in the nonexceptional case $k \cdot \dim \pi > 1$ is of the form g_k for some diffeomorphism $g\colon \tilde{V}_1 \to \tilde{V}_2$ preserving connected components of the fibers of $\pi|_{\tilde{V}_1}$, where $\tilde{V}_i = \pi_{k,0}(V_i)$.*

PROOF. First we show that any U-transformation $F\colon V_1 \to V_2$ such that $\tilde{F} =$ id, is the identity mapping. Toward this end, consider how F acts on submanifolds of the form $L = M_s^{(k)}$. According to Proposition 2.1.2, $U_k(\pi)|_L = 0$, and therefore, $U_k(\pi)|_{F(L)} = 0$. Moreover, since F is $\pi_{k,0}$-fiberwise (Proposition 2.2.1) and $\tilde{F} =$ id, the mapping

$$\pi_{k,0}|_{F(L)}\colon L \to J^0(\pi)$$

is an embedding such that $\pi_{k,0}(F(L)) = \pi_{k,0}(L)$. Moreover, $\pi_k|_{F(L)}$ is a diffeomorphism because $\pi_k = \pi \circ \pi_{k,0}$ and $\pi_{k,0}(L) = s(\mathcal{U}_s)$. Thus, by Theorem 2.1.6, $F(L) = M_{s'}^{(k)}$ for a suitable section $s' \in \Gamma_{\mathrm{loc}}(\pi)$ and $s'(\mathcal{U}_{s'}) = \pi_{k,0}(F(L)) = \pi_{k,0}(L) = s(\mathcal{U}_s)$. This shows that $s' = s$ and $F|_L = \mathrm{id}_L$. Obviously, any point of V_1 belongs to a submanifold of the form $M_s^{(k)}$. Hence, F is the identity mapping.

Consider now the transformation $F = \varphi \circ \tilde{\varphi}_k^{-1}$. By Proposition 2.2.4 (1), it is a U-transformation of the domain V_1. In addition, $\tilde{F} =$ id. Hence, $F =$ id and $g = \tilde{\varphi}$. □

REMARK. The assumption of $\pi_{k,0}$-connectivity in the theorem is essential. This can be illustrated by the following example. Let V and V' be two disjoint $\pi_{k,0}$-connected domains and $\pi_{k,0}(V) = \pi_{k,0}(V') = W$. If g and g' are two different diffeomorphisms of W preserving fibers of π, then the diffeomorphism φ of the domain $V \cup V'$ satisfying $\varphi|_V = g_k$ and $\varphi|_{V'} = g'_k$ is a U-transformation which, in general, does not preserve fibers of $\pi_{s,0}$.

2.2.6. Now we assume that a bundle π is of the form $\Phi(M)$, where Φ is a contravariant functor acting on the category of smooth manifolds. In this case, any diffeomorphism $\varphi\colon J^k(\pi) \to J^k(\pi)$ acts naturally on both elements $\rho_k(\pi)$ and $U_k(\pi)$. This allows one to define $\rho_k(\pi)$- and $U_k(\pi)$-*transformations* as those which preserve $\rho_k(\pi)$ or $U_k(\pi)$, respectively. In other words, φ is a ρ_k- (resp., $U_k(\pi)$-) transformation if $\varphi^*(\rho_k(\pi)) = \rho_k(\pi)$ (resp., $\varphi^*(U_k(\pi)) = U_k(\pi)$), where φ^* stands for the aforementioned action defined as follows.

Let $F\colon M_1 \to M_2$ be a mapping connecting M_1 and M_2, and let $\xi\colon E_\xi \to M_2$ be a vector bundle. Recall that the vector bundle

$$F^*(\xi)\colon F^*(\xi) \to M_1$$

induced from ξ by means of F is defined as

$$F^*(E_\xi) = \{(a, x) \in E_\xi \times M_1 \mid \xi(a) = F(x)\}, \quad \xi\colon (a, x) \to x.$$

Suppose now that the basic bundle π is *functorial*, i.e.,

$$\pi = \Phi_M\colon \Phi(M) \to M$$

with Φ being a contravariant functor on the category of smooth manifolds (for instance, $\Phi = T^*$, J^k, etc.). Functoriality means that any mapping $F\colon M_1 \to M_2$ is lifted up to a fiber bundle morphism

$$\Phi(F)\colon \Phi(M_2) \to \Phi(M_1)$$

which covers F and generates the mapping

$$F^*\colon \Gamma(\Phi_{M_2}) \to \Gamma(\Phi_{M_1}), \quad \gamma \mapsto \Phi(F)\circ\gamma\circ F,\ \gamma \in \Gamma(\Phi_{M_2}). \tag{2.5}$$

In this case, there exists a natural homomorphism of vector bundles over M_2:

$$i_\Phi\colon F^*(\Phi(M_2)) \to \Phi(M_1),$$

where $F^*(\Phi(M_2))$ stands for $F^*(E_\xi)$ with $\xi = \Phi_{M_2}$ defined by

$$i_\Phi((a,x)) = \Phi(F)(a), \quad a \in \Phi(M_2),\ x \in M_1.$$

The mapping i_Φ is injective if F is a submersion. The same is valid for the corresponding mapping of sections

$$\Gamma(i_\Phi)\colon \Gamma(F^*(\Phi_{M_2})) \to \Gamma(\Phi_{M_1}).$$

Specialize now the above constructions to the case $M_1 = J^k(\pi)$, $M_2 = M$, and $F = \pi_k$ with $\pi = \Phi_M$. Then, as it is easy to see,

$$\rho_k(\pi) \in \Gamma(\pi_k^*(\Psi_M)) \ \text{ with } \ \Psi = J^k \circ \Phi$$

and the inclusion $\Gamma(i_\Psi)$ allows us to interpret $\rho_k(\pi)$ as an element of $\Gamma(\Psi_{J^k(\pi)})$. The action φ^* on $\rho_k(\pi)$ we are interested in is defined according to formula (2.5) specialized to the case $M_1 = M_2 = J^k(\pi)$, $\Phi = \Psi$, and $F = \varphi$. Similarly, one gets the action φ^* on $U_k(\pi)$ considering the functor $\Psi = (J^{k-1}\circ\Phi)\otimes T^*$.

Observe that according to Proposition 2.1.2, $U_k(\pi)$-transformations are U-transformations as well. Also, the naturality of the jet-Spencer operator $\Sigma_{0,k}$ (see Subsection 1.1.10) implies that $\rho_k(\pi)$-transformations are $U_k(\pi)$-transformations. To make the relations between $\rho_k(\pi)$- and $U_k(\pi)$-transformations precise, we need the following

PROPOSITION. *Let $\Delta \in \operatorname{dif}_l(\pi,\xi)$ be an operator such that*

$$\rho_k(\varphi_\Delta) = j_k(\psi), \quad \psi \in \mathcal{F}_{k+l}(\pi,\xi),\ k>0, \tag{2.6}$$

or, which is the same, $U_k(\varphi_\Delta) = 0$. Then Δ is a constant operator, i.e., $\Delta(f)$ does not depend on $f \in \Gamma(\pi)$.

PROOF. Consider the mapping

$$F = \underline{\varphi}^{(k)}\colon J^{k+l}(\pi) \to J^k(\xi), \quad \varphi = \varphi_\Delta$$

(see Subsection 2.1.2). Let $L = u(\mathcal{U}_u)$ for $u \in \Gamma_{\mathrm{loc}}(\pi_{k+l})$ with $\mathcal{U}_u$ being the domain of u. Evidently, F is a morphism of fiber bundles over M. Thus, $F(L) = v(\mathcal{U}_u)$ for a suitable section $v \in \Gamma_{\mathrm{loc}}(\xi_k)$, i.e., $v = F\circ u$. The assumption (2.6) implies that $v = j_k(w)$ for a section $w \in \Gamma_{\mathrm{loc}}(\xi)$. This is a consequence of the relation

$$\rho(\varphi_\Delta) = (\underline{\varphi}_\Delta^{(k)})^*(\rho_k(\xi)) = F^*(\rho_k(\xi)),$$

which results directly from the definitions. Indeed, taking into account the characteristic property (2.3) of $\rho(\xi)$, one has

$$v = v^*(\rho_k(\xi)) = u^*(F^*(\rho_k(\xi))) = u^*(\rho_k(\varphi_\Delta)) = u^*(j_k(\psi)) = j_k(u^*(\psi)) = j_k(w),$$

where $w = u^*(\psi)$. Hence, for any $u \in \Gamma_{\mathrm{loc}}(\pi_{k+l})$, the section $F\circ u \in \Gamma_{\mathrm{loc}}(\xi_k)$ is of the form $j_k(w)$. Now we are going to show that this is possible for constant operators only.

To do this, note that Δ is a constant operator if and only if

$$F\big(J^{(k+l)}(\pi)\big) = M_s^{(k)}$$

for some $s \in \Gamma(\xi)$. Hence in the cases where Δ is not constant, the rank of F is greater than $n = \dim M$ at least at some points.

Let $a \in J^{k+l}(\pi)$ be one of these points and $a \in \tilde{L} = \tilde{u}(M)$, $\tilde{u} \in \Gamma_{\mathrm{loc}}(\pi_{k+l})$. Since $\tilde{v} = F \circ \tilde{u}$ is a section of ξ_k, the rank of $F|_{\tilde{L}}$ is equal to n and $F|_{\tilde{L}} : \tilde{L} \to J^k(\xi)$ is an embedding. Moreover, there exists an $(n+1)$-dimensional submanifold $W \subset J^{k+l}(\pi)$ such that

(1) $F|_W : W \to J^l(\xi)$ is an embedding,
(2) $W \supset L \ni a$ with $L = u(\mathcal{U}_u)$ and u being a restriction of $\tilde{u}$ to a smaller domain,
(3) $\pi_{k+l}|_W : W \to \mathcal{U}_u$ is a fibration.

Indeed, there obviously exists a submanifold $\tilde{W} \in J^{k+l}(\pi)$ such that $a \in \tilde{W}$ and the rank of $F|_{\tilde{W}}$ equals $n+1$. Then a suitable neighborhood W of a in $\tilde{W}$ satisfies (2) and (3).

Since F is a fiberwise mapping, $F(W)$ is also a fibration over $\mathcal{U}_u$. One of the two cases may occur:

(i) Everywhere in $V = F(W)$ the rank of $\xi_{k,0}|_V$ is equal to n.
(ii) There are points in V at which the rank of $\xi_{k,0}|_V$ is equal to $n+1$.

In case (i), $\xi_{k,0}(V)$ is an n-dimensional submanifold in E_π, which obviously coincides with the graph of the section $w = \xi_{k,0} \circ v$ of the bundle ξ, $v = F \circ u$. Thus, for any section u_1 of the fibration $\pi_{k+l}|_W$ (see (3) above) one has $\xi_{k,0} \circ F \circ u_1 = w$. By deforming u in W, one easily finds such a u_1 which is different from u. Then $v_1 = F \circ u_1$ differs from $v = F \circ u$ due to (1), while $\xi_{k,0} \circ v_1 = w$. On the other hand, $v_1 = j_k(w_1)$ for $w_1 = \xi_{k,0} \circ v_1$, as was shown earlier. Thus, $w_1 = w$ and $v_1 = j_k(w) = v$ and we get a contradiction.

In case (ii), by passing to a neighborhood of a point at which the rank of $\xi_{k,0}|_V$ is equal to $n+1$, one can assume that $\xi_{k,0}|_V : V \to E_\xi$ is an embedding. Consider submanifolds $V_0 = \xi_{k,0}(V) \subset E_\xi = J^0(\xi)$ and $V_1 = \xi_{k,1}(V) \subset J^1(\xi)$. Then $\xi_{1,0}|_{V_1} : V_1 \to V_0$ is a diffeomorphism. This allows us to construct the following n-dimensional distribution along V_0:

$$V_0 \ni \theta_0 \mapsto L_{\theta_1} \quad \text{with} \quad \theta_1 \in V_1,\ \xi_{1,0}(\theta_1) = \theta_0$$

(see Subsection 2.1.4). We emphasize that this distribution is not generally tangent to V_0. But if $M_\gamma^{(1)} \subset V_1$, $\gamma \in \Gamma_{\mathrm{loc}}(\xi)$, then the manifold $\gamma(\mathcal{U}_\gamma) = M_\gamma^{(0)} = \xi(M_\gamma^{(1)}) \subset V_0$ is obviously integral for this distribution.

It is easy to find a section $\gamma \in \Gamma_{\mathrm{loc}}(\xi)$ such that the manifold $\gamma(\mathcal{U}_\gamma) = M_\gamma^{(0)} \subset V_0$ is not integral. Since $\xi_{k,0}|_V : V \to V_0$ is a diffeomorphism, there exists a section $\bar{\gamma} \in \Gamma_{\mathrm{loc}}(\xi_k)$ such that $\bar{L} = \bar{\gamma}(\mathcal{U}_{\bar{\gamma}}) \subset V$ and $\xi_{k,0}(\bar{L}) = M_\gamma^{(0)}$. Then $\bar{L}$ is not of the form $M_\gamma^{(k)}$. Indeed, assuming the contrary, one sees that $\xi_{k,1}(\bar{L}) = M_\gamma^{(1)}$, which implies that $M_\gamma^{(0)}$ is integral for the distribution constructed above.

The fact that $\bar{L} \neq M_\gamma^{(k)}$ is equivalent to the fact that $\bar{\gamma}$ is not of the form $j_k(w)$ since otherwise $w = \xi_{k,0} \circ \bar{\gamma} = \gamma$.

Finally, it remains to note that $\bar{\gamma} = F \circ u$ for a section $u \in \Gamma_{\mathrm{loc}}(\pi_{k+l})$ such that $u(\mathcal{U}_u) \subset W$ because $F|_W : W \to V$ is a diffeomorphism. Thus, $F \circ u$ is not of the form $j_k(w)$ in contradiction with the property of F established earlier. □

COROLLARY. *An operator $\Delta \in \mathrm{dif}_l(\pi, \xi)$ is not constant if and only if one of the following equivalent conditions holds*:

1. *The rank of* $\underline{\varphi}_{\Delta}^{(k)}$ *is not equal to* n *for some* $k > 0$.
2. *There exists a* k*-jet* $\theta \in \mathcal{J}^k(\Gamma(\pi))$ *such that* $S(\underline{\varphi}_{\Delta}^{(n)*}(\theta)) = 0$, *where* S *is the Spencer operator.*

REMARK. The proposition and its corollary above are obviously valid for any $\pi_{k+l,0}$-connected domain $V \subset J^{k+l}(\pi)$ provided that $\rho_k(\varphi_{\Delta}) = j_k(\psi)$ in this domain.

2.2.7. Let us now give the following

DEFINITION. A diffeomorphism $\varphi\colon V_1 \to V_2$, where V_1 and V_2 are domains in $J^k(\pi)$, is called a *translation* by $f \in \Gamma(\pi|_{\mathcal{O}})$ if $\mathcal{O} = \pi_k(V_i)$ and $\varphi(a) = a + [f]_y^k$, $y = \pi_k(a)$.

Consider now a $U_k(\pi)$-transformation $\varphi\colon V_1 \to V_2$, where the domain V_1 is assumed to be π_k-connected. Since the Spencer sequence is exact, one has

$$\varphi^*(\rho_k(\pi)) = \rho_k(\pi) + j_k(\psi), \quad \psi \in \mathcal{F}(\pi,\pi).$$

If $k \cdot \dim \pi > 1$, then φ, being a U-transformation, preserves connected components of the fibers of π_k (Theorem 2.2.5). Hence both $\varphi^*(\rho_k(\pi))$ and $\rho_k(\pi)$ are horizontal jets, i.e., belong to the image of the mapping $i\colon \mathcal{F}(\pi,\xi_k) \to \mathcal{J}^k(\pi,\xi)$ (see Subsection 2.1.2). But

$$j_k(s)^*(\varphi^*(\rho_k(\pi))) = j_k(s)^*(\rho_k(\pi) + j_k(\psi)) = j_k(s + \Delta_\psi(s)).$$

Hence, according to the characteristic property of elements $\rho_k(\lambda)$ (see Subsection 2.1.2), $\varphi^*(\rho_k(\pi)) = \rho_k(\varphi_\square)$, where $\square = \mathrm{id}_{\Gamma(\pi)} + \Delta_\psi$, i.e., $j_k(\psi) = \rho_k(\varphi_\square) - \rho_k(\pi) = \rho_k(\Delta_\psi)$. Then from Proposition 2.2.6 it follows that Δ_ψ is a constant operator.

Let $f \in \Gamma(\pi)$ be the image of Δ_ψ. Then obviously $\psi = \pi_k^*(f)$, and if β is a translation by $-f$, then $\beta^*\bigl(\rho_k(\pi) + j_s(\psi)\bigr) = \rho_k(\pi)$. Thus we have proved the following

THEOREM. *Let* $s \cdot \dim \pi > 1$. *Any* $U_k(\pi)$*-transformation of a* π_k*-connected domain in* $J^k(\pi)$ *can be represented as a composition of a* $\rho_k(\pi)$*-transformation and a translation.*

In the case $s = \dim \pi = 1$, the theorem does not hold. Indeed, let $\pi = \mathbb{1}_M$. Then a $U_1(\pi)$-transformation F of the manifold $J^1(\pi)$ must preserve the fibers of the natural projection $\tau\colon J^1(\pi) \to T^*(M)$ and consequently must induce a canonical (= symplectic) transformation $F'\colon T^*(M) \to T^*(M)$ possessing a generating function. Conversely, any canonical transformation F' possessing a generating function induces a $U_1(\pi)$-transformation F which is defined uniquely up to a translation by $f = \text{const}$ (see [**85**, Theorem 1.3.3] for the proof). Note also that in this case the $\rho_1(\pi)$-transformations coincide with transformations that are natural liftings of diffeomorphisms of M.

2.2.8. Now we consider $\rho_k(\pi)$-transformations in more details.

THEOREM. *Any* $\rho_k(\pi)$*-transformation of a* π_k*-connected domain* $V \subset J^k(\pi)$ *is generated by a functorial transformation of the bundle* $\pi|_{\mathcal{U}} = \Phi(\mathcal{U})$ *corresponding to a diffeomorphism of the domain* $\pi_k(V) = \mathcal{U}$.

PROOF. Let $\varphi\colon V_1 \to V_2$ be a $\rho_k(\pi)$-diffeomorphism. According to Proposition 2.2.3, φ is a π_k-fiberwise diffeomorphism. Hence it generates a diffeomorphism

$\varphi'\colon V_1' \to V_2'$ such that $\varphi' \circ \pi_k = \pi_k \circ \varphi$. Due to the functorial nature of π, the diffeomorphism φ' generates the mapping $\bar{\varphi} = \Phi(\varphi')\colon \pi|_{V_1} \to \pi|_{V_2}$, and the diagram

$$\begin{array}{ccc} \pi^{-1}(V_1') & \xrightarrow{\bar{\varphi}} & \pi^{-1}(V_2') \\ {\scriptstyle\pi}\big\downarrow & & \big\downarrow{\scriptstyle\pi} \\ V_1' & \xrightarrow{\varphi'} & V_2' \end{array}$$

is commutative.

From the characteristic property of the elements $\rho_k(\pi)$ it immediately follows that $(\bar{\varphi}_k)^*(\rho_k(\pi)) = \rho_k(\pi)$, i.e., $\bar{\varphi}_k$ is a $\rho_k(\pi)$-transformation.

Consider now the $\rho_k(\pi)$-transformation $\psi = \bar{\varphi}_k^{-1} \circ \varphi$. For this transformation one has $\psi' = \mathrm{id}$. If ψ were not the identity mapping, there would exist a local section $f \in \Gamma_{\mathrm{loc}}(\pi_k)$ such that $f \neq g$, $g = \psi \circ f$. This obviously contradicts the fact that ψ is a $\rho_k(\pi)$-transformation. Therefore, $\psi = \mathrm{id}$. □

2.3. Infinitesimal automorphisms of the Cartan distribution

Actually, this section should have been placed in Chapter 3. We put it here for methodological reasons mostly, because the study of the Cartan distribution automorphisms on $J^k(\pi)$, $k \leq \infty$, becomes clearer when one uses techniques and results related to $J^\infty(\pi)$.

The description of infinitesimal automorphisms of the Cartan distribution on $J^\infty(\pi)$ (called $\mathcal{C}$*-fields* below) plays an important role throughout the exposition.

In this section the bundle π is assumed to be linear unless stated otherwise.

2.3.1. In this subsection, we assume that the Cauchy problem for the equation

$$\frac{\partial u}{\partial t} = \nabla(u), \quad u \in \Gamma(\pi),\ \nabla \in \mathrm{dif}_s(\pi,\pi), \tag{2.7}$$

is solvable (not uniquely, of course) for any initial data. Let $u(t) \in \Gamma(\pi)$ be a solution of this equation, $t \in \mathcal{U} \subset \mathbb{R}$, and $\Delta \in \mathrm{dif}_k(\pi,\xi)$, where ξ is another bundle over M (the base of the bundle π). Define the mapping

$$\ell_\Delta^s\colon \mathcal{F}_s(\pi,\pi) \to \mathcal{F}_{k+s}(\pi,\xi)$$

by setting

$$j_{k+s}(u(0))^*(\ell_\Delta^s(\varphi_\nabla)) = \left[\frac{d}{dt}\Delta(u(t))\right]_{t=0}. \tag{2.8}$$

The idea of this definition is quite transparent: it presents the derivative of the function $\varphi = \varphi_\nabla$ along the "curve" $u(t) \in \Gamma(\pi)$ at the "point" $u(0) \in \Gamma(\pi)$. On the other hand, it is not very convenient for at least two reasons. First, it is based on the assumption about solvability of equation (2.7), which can hardly be checked for a generic ∇. Second, its correctness, i.e., independence of the choice of the "curve" $u(t)$, is to be verified. Let us ignore, however, these shortcomings. Then it is easy to see that the operators ℓ_Δ^s are linear and, in addition,

$$\ell_\Delta^p \circ \pi_{p,s}^* = \pi_{p+k,s+k}^* \circ \ell_\Delta^s, \quad p \geq s.$$

Therefore, the set of operators ℓ_Δ^s, $s \geq 0$, determines the operator

$$\ell_\Delta\colon \mathcal{F}(\pi,\pi) \to \mathcal{F}(\pi,\xi)$$

which raises filtration of the filtered module $\mathcal{F}(\pi,\pi)$ by k. In what follows, we call ℓ_Δ the *operator of universal linearization* (for Δ). We also write ℓ_φ instead of ℓ_{Δ_φ}.

2.3.2. Now we give the rigorous definition of the operators ℓ^s_Δ. Let $\square_t \in \mathrm{dif}_s(\pi,\pi)$ be a smooth (with respect to t) family of operators, $t \in \mathcal{U} \subset \mathbb{R}$, and $\nabla = d\square_t/dt|_{t=0}$.

DEFINITION. Let $\Delta \in \mathrm{dif}_k(\pi,\xi)$. For any $s \geq 0$ and $\nabla \in \mathrm{dif}_s(\pi,\pi)$ set $\ell^s_\Delta(\varphi_\nabla) = \varphi_{\Delta(\nabla)}$, where $\Delta(\nabla) = (d/dt)(\Delta \circ \nabla_t)|_{t=0}$.

Obviously, $\Delta(\nabla)$ depends on Δ and ∇ only, i.e., it is well defined. Let us compare it with definition (2.8). To do this, consider the family $\mathrm{id}_\pi + t\nabla$, where $\mathrm{id}_\pi \colon \Gamma(\pi) \to \Gamma(\pi)$ is the identity mapping. If $u(t)$ is a solution of (2.7) and $u(0) = u$, then $du(t)/dt|_{t=0} = (d/dt)[u + t\nabla(u)]_{t=0}$. Consequently,

$$j_{k+s}(u)^*(\varphi_{\Delta(\nabla)}) = \frac{d}{dt}\Big(\Delta(u + t\nabla(u))\Big)\Big|_{t=0} = \frac{d}{dt}\Delta(u(t))\Big|_{t=0}.$$

This means that definition (2.8) coincides with Definition 2.3.2 when the former makes sense. Therefore, we take Definition 2.3.2 for the definition of the *universal linearization operator.*

Note that (see Subsection 2.1.2)

$$\ell_\Delta(\varphi_\nabla) = \frac{d}{dt}(\varphi_{\Delta\circ\square_t})\Big|_{t=0} = \frac{d}{dt}\hat{\Delta}(\varphi_{\square_t})\Big|_{t=0}.$$

Hence, when Δ is a linear operator, one has $(d/dt) \circ \hat{\Delta} = \hat{\Delta} \circ (d/dt)$ and

$$\frac{d}{dt}\hat{\Delta}(\varphi_{\square_t})\Big|_{t=0} = \hat{\Delta}\left(\frac{d}{dt}\varphi_{\square_t}\Big|_{t=0}\right) = \hat{\Delta}(\varphi_\nabla).$$

This means that for $\Delta \in \mathrm{Diff}(\Gamma(\pi),\Gamma(\xi))$ one has $\ell_\Delta(\varphi) = \hat{\Delta}(\varphi)$, or

$$\ell_\Delta = \hat{\Delta}, \quad \Delta \in \mathrm{Diff}(\pi,\xi). \tag{2.9}$$

2.3.3. Let us give another

DEFINITION. Let $u \in \Gamma(\pi)$. Then the operator

$$\ell(\Delta, u) = j_k(u)^* \circ \ell^0_\Delta \circ \pi^* = j(u)^* \circ \ell_\Delta \circ \pi^*_\infty \colon \Gamma(\pi) \to \Gamma(\xi)$$

is called the *linearization of the operator* Δ *on the section* u.

Obviously, $\ell(\Delta,u) \in \mathrm{Diff}_k(\Gamma(\pi),\Gamma(\xi))$. From equality (2.9) it follows that $\ell(\Delta,u) = \Delta$ for any $u \in \Gamma(\pi)$ if $\Delta \in \mathrm{Diff}_k(\Gamma(\pi),\Gamma(\xi))$.

An important property of the operator ℓ_Δ is described by the following

PROPOSITION. *If* $j(u)^*(\varphi) = j(u)^*(\varphi')$, *then* $j(u)^*\ell_\Delta(\varphi) = j(u)^*\ell_\Delta(\varphi')$ *as well.*

PROOF. Indeed,

$$j(u)^*\ell_\Delta(\varphi) = j(u)^*\varphi_{\Delta(\Delta_\varphi)} = \Big(\Delta(\Delta_\varphi)\Big)(u) = \frac{d}{dt}\Delta\Big(u + \Delta_\varphi(u)\Big)\Big|_{t=0}.$$

But $\Delta_\varphi(u) = j(u)^*(\varphi) = j(u)^*(\varphi') = \Delta_{\varphi'}(u)$. □

This property can be interpreted as follows: the operators ℓ_Δ admit restrictions to submanifolds of the form $M^{(\infty)}_u \subset J^\infty(\pi)$.

2.3.4. Now we study how ℓ_Δ depends on the operator Δ. To emphasize the fact that Δ acts on sections of the bundle ξ we shall write ℓ^ξ_Δ instead of ℓ_Δ. Suppose that Δ is a k-th order operator. Further, let δ be an r-th order operator taking $\Gamma(\pi)$ to $C^\infty(M)$. Consider the operator $\delta\cdot\Delta$ of order $\le \max(k,r)$ defined by $(\delta\cdot\Delta)(u)=\delta(u)\cdot\Delta(u)$. Then, according to Definition 2.3.2,

$$\begin{aligned} j(u)^*\left(\ell^\xi_{\delta\cdot\Delta}(\varphi_\Delta)\right) &= \frac{d}{dt}\left((\delta\cdot\Delta)(\square_t(u))\right)\Big|_{t=0} \\ &= \frac{d}{dt}\left(\delta(\square_t(u))\right)\Big|_{t=0}\cdot\Delta(u)+\delta(u)\cdot\frac{d}{dt}\left(\Delta(\square_t(u))\right)\Big|_{t=0} \\ &= j(u)^*\left(\ell^{\mathbb{1}_M}_\delta(\varphi_\nabla)\cdot\varphi_\Delta+\varphi_\delta\cdot\ell^\xi_\Delta(\varphi_\nabla)\right). \end{aligned}$$

In other words,

$$\ell_{f\varphi}=f\ell^{\mathbb{1}_M}_\varphi+\varphi\ell^\xi_f,\quad f\in\mathcal{F}(\pi),\ \varphi\in\mathcal{F}(\pi,\xi), \tag{2.10}$$

where $(\varphi\ell^\xi_f)(\psi)=\ell^\xi_f(\psi)\cdot\varphi$.

For any $\psi\in\operatorname{dif}(\pi,\pi)$ define the mapping $\partial_\psi=\partial^\xi_\psi\colon\mathcal{F}(\pi,\xi)\to\mathcal{F}(\pi,\xi)$ by setting $\partial^\xi_\psi(\varphi)=\ell^\xi_\varphi(\psi)$. From (2.10) we obtain

PROPOSITION. *Let $\psi\in\mathcal{F}(\pi,\pi)$, $\varphi\in\mathcal{F}(\pi,\xi)$, and $f\in\mathcal{F}(\pi,\mathbb{1}_M)$. Then*

$$\partial^\xi_\psi(f\varphi)=f\partial^\xi_\psi(\varphi)+\partial^{\mathbb{1}_M}_\psi(f)\cdot\varphi. \tag{2.11}$$

This result, together with (2.8), motivates the following

DEFINITION. The operator $\partial^\xi_\psi\colon\mathcal{F}(\pi,\xi)\to\mathcal{F}(\pi,\xi)$ is called the *evolutionary derivation* corresponding to $\psi\in\mathcal{F}(\pi,\pi)$.

In particular, if $\xi=\mathbb{1}_M$, then $\partial^{\mathbb{1}_M}_\psi$ is a derivation of the algebra $\mathcal{F}(\pi)$ which raises filtration by at most k, $\psi\in\mathcal{F}_l(\pi,\pi)$.

2.3.5. Now we derive some elementary properties of evolutionary derivations. Consider the embedding $\pi^*_k\colon\Gamma(\xi)\to\mathcal{F}_k(\pi,\xi)$. Then for any $v\in\Gamma(\xi)$ the operator $\Delta_{\pi^*_k(v)}$ is constant. In fact, if $u\in\Gamma_{\mathrm{loc}}(\pi)$, then

$$\Delta_{\pi^*_k(v)}(u)=j_k(u)^*(\pi^*_k(v))=(\pi_k\circ j_k(u))^*(v)=\mathrm{id}_{\Gamma(\xi)}(v)=v.$$

Thus, for $\psi=\varphi_\nabla$, according to Definition 2.3.2, one has

$$j(u)^*\left(\partial_\psi(\pi^*_k(v))\right)=\frac{d}{dt}\left(\Delta_{\pi^*_k(v)}(\square(u))\right)\Big|_{t=0}=\frac{dv}{dt}\Big|_{t=0}=0,$$

i.e., $\partial_\psi\circ\pi^*_k=0$. Thus ∂_ψ is a π_k-vertical derivation.

Now let $\pi=\xi$ and $\Delta=\mathrm{id}\in\operatorname{dif}(\pi,\pi)$. Then, since $\ell_{\mathrm{id}}=\mathrm{id}$, the equality

$$\partial_\psi(\varphi_{\mathrm{id}})=\psi \tag{2.12}$$

holds. Therefore, any evolutionary derivation is uniquely determined by its value at the element $\varphi_{\mathrm{id}}=\rho_0(\pi)$.

Let $\delta\in\operatorname{Diff}(\Gamma(\xi),\Gamma(\xi))$. Obviously, for all $\varphi_{\nabla_t}\in\mathcal{F}(\pi,\pi)$ and $\varphi_\Delta\in\mathcal{F}(\pi,\xi)$ the equality

$$\delta\circ\frac{d}{dt}\left(\Delta\circ\square_t\right)\Big|_{t=0}=\frac{d}{dt}\left(\delta\circ\Delta\circ\square_t\right)\Big|_{t=0}$$

holds. Then from Definition 2.3.2 and from the definition of the "cap operation" $\nabla \mapsto \hat{\nabla}$ one gets

$$\hat{\delta} \circ \partial^{\xi}_{\psi} = \partial^{\xi}_{\psi} \circ \hat{\delta}. \tag{2.13}$$

PROPOSITION. *Let* $\partial^{\xi}\colon \mathcal{F}(\pi,\xi) \to \mathcal{F}(\pi,\xi)$ *be a set of* $\mathbb{R}$*-linear continuous mappings satisfying relations* (2.11) *and* (2.13). *Then it is of the form* ∂^{ξ}_{ψ}, *where the section* ψ *is defined by the identity* $\partial^{\pi}(\varphi_{\mathrm{id}}) = \psi$.

PROOF. Consider the mapping $\partial \mapsto \psi_{\partial} = \partial^{\pi}(\varphi_{\mathrm{id}}) \in \mathcal{F}(\pi,\pi)$ which acts from the set of families $\{\partial^{\xi}\} = \partial$ satisfying the conditions of the proposition into $\mathcal{F}(\pi,\pi)$. This mapping is obviously $\mathbb{R}$-linear. To prove the proposition, it is sufficient to prove that this mapping is injective. Let $\psi_{\partial} = 0$ and $\delta \in \operatorname{Diff}(\Gamma(\pi),\Gamma(\xi))$. Then $\partial^{\xi}(\hat{\delta}(\varphi_{\delta})) = \partial^{\xi}(\varphi_{\delta\circ\mathrm{id}}) = \partial(\hat{\delta}(\varphi_{\mathrm{id}}))$. But due to (2.13), $\partial^{\xi}(\hat{\delta}(\varphi_{\mathrm{id}})) = \hat{\delta}(\partial^{\xi}(\varphi_{\mathrm{id}})) = 0$. Hence, $\partial^{\xi}(\varphi_{\delta}) = 0$ for any linear DO δ. Therefore, from (2.11) it follows that if $\delta_1,\dots,\delta_s \in \operatorname{Diff}(\Gamma(\pi),C^{\infty}(M))$, then $\partial^{\xi}(\varphi_{\delta_1}\cdots\varphi_{\delta_s}) = 0$. But the elements $\sum \varphi_{\delta_1}\cdots\varphi_{\delta_s}$ constitute an everywhere dense set in the algebra $\mathcal{F}(\pi)$. This shows that $\partial^{\xi} = 0$. Using (2.11) again, one obtains that ∂^{ξ} is an endomorphism of the $\mathcal{F}(\pi)$-module $\mathcal{F}(\pi,\xi)$. Since the elements $\sum f_i\varphi_{\delta_i}$, $f_i \in \mathcal{F}(\pi)$, $\delta_i \in \operatorname{Diff}(\Gamma(\pi),\Gamma(\xi))$, form an everywhere dense set in $\mathcal{F}(\pi,\xi)$ and $\partial^{\xi}(\varphi_{\delta}) = 0$ for linear δ, this proves the triviality of ∂^{ξ} for all bundles ξ. □

From the last proposition it follows that the commutator $[\partial_{\varphi},\partial_{\psi}]$ of two evolutionary derivations is an evolutionary derivation as well. Thus the set of all evolutionary derivations is a Lie algebra over $\mathbb{R}$.

2.3.6. Evolutionary derivations introduced above allow one to define a Lie $\mathbb{R}$-algebra structure in the $C^{\infty}(M)$-module $\mathcal{F}(\pi,\pi)$.

DEFINITION. Let $\varphi,\psi \in \mathcal{F}(\pi,\pi)$ be two sections. The section

$$\{\varphi,\psi\} = \partial^{\pi}_{\varphi}(\psi) - \partial^{\pi}_{\psi}(\varphi)$$

is called the *higher Jacobi bracket* of φ and ψ.

PROPOSITION. *The operation* $\{\bullet,\bullet\}$ *defines a Lie* $\mathbb{R}$*-algebra structure on* $\mathcal{F}(\pi,\pi)$.

PROOF. Skew-symmetry and $\mathbb{R}$-linearity of the higher Jacobi bracket are obvious. Let us prove the Jacobi identity, i.e., the equality

$$\{\{\varphi,\psi\},\chi\} + \{\{\psi,\chi\},\varphi\} + \{\{\chi,\varphi\},\psi\} = 0$$

for any $\varphi,\psi,\chi \in \mathcal{F}(\pi,\pi)$. We have

$$[\partial_{\varphi},\partial_{\psi}](\varphi_{\mathrm{id}}) = \partial_{\varphi}(\partial_{\psi}(\varphi_{\mathrm{id}})) - \partial_{\psi}(\partial_{\varphi}(\varphi_{\mathrm{id}})) = \partial_{\varphi}(\psi) - \partial_{\psi}(\varphi) = \{\varphi,\psi\}.$$

Therefore,

$$[\partial_{\varphi},\partial_{\psi}] = \partial_{\{\varphi,\psi\}}. \tag{2.14}$$

But

$$[[\partial_{\varphi},\partial_{\psi}],\partial_{\chi}] + [[\partial_{\psi},\partial_{\chi}],\partial_{\varphi}] + [[\partial_{\chi},\partial_{\varphi}],\partial_{\psi}] = 0,$$

which means, due to (2.14), that

$$\partial_{\{\{\varphi,\psi\},\chi\}} + \partial_{\{\{\psi,\chi\},\varphi\}} + \partial_{\{\{\chi,\varphi\},\psi\}} = 0.$$

Applying the last equality to φ_{id} we get the desired result. □

Using the last definition, one can now introduce the operators $\bar{X}_\psi\colon \mathcal{F}(\pi,\pi) \to \mathcal{F}(\pi,\pi)$ by setting

$$\bar{X}_\psi(\varphi) = \{\psi,\varphi\} = (\partial_\psi - \ell_\psi)(\varphi), \quad \varphi,\psi \in \mathcal{F}(\pi,\pi). \tag{2.15}$$

For reasons which will become clear in the next chapter, the operator $\bar{X}_\psi$ can be called the *characteristic* of the equation $\psi = 0$. Obviously, $\bar{X}_\psi$ is an s-th order operator if $\psi \in \mathcal{F}_s(\pi,\pi)$, and $\bar{X}_\psi(\psi) = 0$.

2.3.7. Coordinates. Consider a neighborhood $\mathcal{U}$ in the manifold M with local coordinates $x = (x_1,\dots,x_n)$ in it. Let $e_1,\dots,e_m$ and $f_1,\dots,f_{m'}$ be local bases of sections of the bundles π and ξ, respectively, over $\mathcal{U}$. The corresponding special coordinates in $\mathcal{U}_k = \pi_k^{-1}(\mathcal{U})$ are (x_i, p_σ^j), $i = 1,\dots,n$, $j = 1,\dots,m$, $|\sigma| \le k$. The sections $\varphi_{\square_t}$ and φ_Δ (see Subsection 2.3.2) can be represented in the form

$$\varphi_{\square_t} = \sum_{\alpha=1}^{m} \mu_\alpha(x,\dots,p_\sigma^j,\dots,t)e_\alpha, \quad \varphi_\Delta = \sum_{\beta=1}^{m'} \nu_\beta(x,\dots,p_\sigma^j,\dots)f_\beta.$$

Therefore,

$$\varphi_{\Delta\circ\square_t} = \sum_{\beta=1}^{m'} \nu_\beta(x,\dots,D_\sigma(\mu_j),\dots)f_\beta,$$

where $D_\sigma = \widehat{\partial^{|\sigma|}/\partial x_\sigma}$. Taking into account that $\square_t|_{t=0} = \mathrm{id}$ and $(d\square_t/dt)|_{t=0} = \nabla$, one gets:

$$\left.\frac{d\varphi_{\Delta\circ\square_t}}{dt}\right|_{t=0} = \sum_{\beta,\sigma,j} \frac{\partial\nu_\beta}{\partial p_\sigma^j} D_\sigma(\mu_j) f_\beta,$$

where $\mu_j = \mu_j(t)|_{t=0}$. Thus

$$\ell_\Delta = \sum_{\beta,\sigma,j} \frac{\partial\nu_\beta}{\partial p_\sigma^j} f_\beta D_\sigma^{(j)}. \tag{2.16}$$

Here and later the symbol $D_\sigma^{(j)}$ means that the operator D_σ is applied to the j-th component of the section φ.

If the section $\psi \in \mathcal{F}(\pi,\pi)$ is locally of the form $\psi = \sum_j \mu_j e_j$, then

$$\partial_\psi = \sum_{\beta\sigma j} D_\sigma(\mu_j) f_\beta \frac{\partial^{(\beta)}}{\partial p_\sigma^j}. \tag{2.17}$$

From (2.17) one gets

$$\{\varphi,\psi\} = \sum_{i,j,\sigma} \left(D_\sigma(\mu_j)\frac{\partial\nu_i}{\partial p_\sigma^j} - D_\sigma(\nu_j)\frac{\partial\mu_i}{\partial p_\sigma^j} \right) e_i, \tag{2.18}$$

$$\bar{X}_\psi = \sum_{i,j,\sigma} \left(D_\sigma(\mu_j) e_i \frac{\partial^{(i)}}{\partial p_\sigma^j} - \frac{\partial\mu_i}{\partial p_\sigma^j} e_i D_\sigma^{(i)} \right). \tag{2.19}$$

Recall that $\mathcal{F}(\pi,\pi) = \mathcal{F}(\pi)$ for $\pi = \mathbb{1}_M$. If $\varphi,\psi \in \mathcal{F}_1(\pi)$, then from (2.18) it follows that $\{\varphi,\psi\}$ coincides with the standard contact bracket of φ and ψ on $J^1(\mathbb{1}_M)$ (see **[71, 85]**).

2.3.8. Now we give the main definition of this section, that of an infinitesimal automorphism of the Cartan distribution on $J^\infty(\pi)$, or a *$\mathcal{C}$-field.* Here the bundle π can be nonlinear.

Consider the sequence of monomorphisms

$$\Lambda^*(M) \xrightarrow{\pi^*} \Lambda^*(J^0(\pi)) \xrightarrow{\pi^*_{1,0}} \cdots \xrightarrow{\pi^*_{k,k-1}} \Lambda^*(J^k(\pi)) \xrightarrow{\pi^*_{k+1,k}} \cdots .$$

Denote by $\Lambda^* = \Lambda^*(\pi)$ the direct limit of this sequence and call it the *algebra of differential forms* on $J^\infty(\pi)$. Homogeneous components of this algebra are denoted by $\Lambda^i = \Lambda^i(\pi)$. Below we identify the module $\Lambda^*(J^k(\pi))$ (resp., $\Lambda^*(M)$) with $\pi^*_{\infty,k}(\Lambda^*(J^k(\pi)))$ (resp., with $\pi^*_\infty(\Lambda^*(M))$).

We use the operator $U_1\colon \mathcal{F}(\pi) \to \Lambda^1$ constructed in Subsection 2.1.2 to introduce the $\mathcal{F}(\pi)$-submodule $\mathcal{C}\Lambda^1$ in Λ^1 generated by $\operatorname{im} U_1$. We call it the *Cartan submodule*. Let $\mathcal{C}_k\Lambda^1 = \mathcal{C}\Lambda^1 \cap \Lambda^1(J^k(\pi))$. Then by Proposition 2 in Subsection 2.1.2 $\mathcal{C}_k\Lambda^1$ is the annihilator of the Cartan distribution on $J^k(\pi)$. For this reason, we consider $\mathcal{C}\Lambda^1$ to be the annihilator of the Cartan distribution on $J^\infty(\pi)$.

DEFINITION. A derivation $X \in D(\mathcal{F}(\pi))$ is called a *$\mathcal{C}$-field* if $X(\mathcal{C}\Lambda^1) \subset \mathcal{C}\Lambda^1$.

Denote by $D_\mathcal{C} = D_\mathcal{C}(\pi)$ the set of all $\mathcal{C}$-fields. Obviously, it is a Lie subalgebra in $D(\mathcal{F}(\pi))$.

LEMMA. *Any $\mathcal{C}$-field annihilating the subalgebra $\mathcal{F}_0(\pi) \subset \mathcal{F}(\pi)$ vanishes.*

PROOF. Let us prove by induction that X annihilates all the algebras $\mathcal{F}_k(\pi)$, $k \geq 0$. The first step coincides with the assumption of the lemma. Suppose that $X(\mathcal{F}_{k-1}(\pi)) = 0$. Since $U_1(f) = df - \hat{d}f$ for any $f \in \mathcal{F}(\pi)$, one has $X(U_1(f)) = dX(f) - X(\hat{d}(f))$. In particular, if $f \in \mathcal{F}_{k-1}(\pi)$, then due to the induction hypothesis $X(U_1(f)) = -X(\hat{d}f)$. But X is a $\mathcal{C}$-field, and therefore,

$$X(U_1(f)) = \sum_i g_i U_1(f_i), \quad f_i, g_i \in \mathcal{F}(\pi).$$

Consequently, $j(\psi)^*(X(U_1(f))) = 0$ for any section ψ of the bundle π. Thus $j(\psi)^*(X(\hat{d}(f))) = 0$. But $\hat{d}f$ is a horizontal form, i.e., can be presented as $\hat{d}f = \sum_j a_j dh_j$, where $h_j \in C^\infty(M)$, $a_j \in \mathcal{F}(\pi)$. Due to the conditions of the lemma, $X(h_j) = 0$ and hence $X(\hat{d}f) = \sum_j X(a_j)\, dh_j$, which means that $X(\hat{d}f)$ is a horizontal form as well. But for any horizontal form $\omega = \sum_j b_j\, dh_j$, the fulfillment of the equality $j(\psi)^*(\omega) = 0$ for all ψ implies that $\omega = 0$. Therefore, $X(\hat{d}f) = 0$ for every $f \in \mathcal{F}_{k-1}(\pi)$. In particular, $X(\hat{d}p^j_\sigma) = 0$, $|\sigma| < k$, where p^i_σ are special coordinates on $J^k(\pi)$. But $\hat{d}p^j_\sigma = \sum_i p^j_{\sigma i}\, dx_i$ and consequently $X(\hat{d}p^j_\sigma) = \sum_i X(p^j_{\sigma i})\, dx_i = 0$. Hence, $X(p^j_{\sigma i}) = 0$ for all i, j and σ, $|\sigma| < k$. Therefore, $X = 0$ on $\mathcal{F}_k(\pi)$. □

2.3.9. We show in this subsection that every evolutionary derivation is a $\mathcal{C}$-field.

PROPOSITION. *For each section $\psi \in \mathcal{F}(\pi, \pi)$, $k \geq 0$, one has*

$$\partial_\psi \circ \rho_k = \rho_k \circ \partial_\psi.$$

PROOF. Let $\nabla = \Delta_\psi$. Then from Definition 2.3.2 and that of ρ_k it follows:

$$\begin{aligned} j(u)^* \partial_\psi(\rho_k(\varphi_\Delta)) &= j(u)^*\big(\ell_{\rho_k(\varphi_\Delta)}(\psi)\big) = j_k(u)^*(\varphi_{(j_k\circ\Delta)(\nabla)}) = ((j_k\circ\Delta)(\nabla)) \\ &= \frac{d}{dt}((j_k\circ\Delta)(u+t\nabla(u)))\Big|_{t=0} = j_k\left(\frac{d}{dt}(\Delta(u+t\nabla(u)))\Big|_{t=0}\right) \\ &= \big(j_k\circ\Delta(\nabla)\big)(u) = j_k(u)^*(\varphi_{j_k\circ\Delta(\nabla)}) = j_k(u)^*\big(\hat{j}_k(\varphi_{\Delta(\nabla)})\big) \\ &= j_k(u)^*\big(\rho_k(\ell_\Delta(\psi))\big) = j_k(u)^*\rho_k(\partial_\psi(\varphi_\Delta)). \end{aligned}$$

However, $u \in \Gamma_{\mathrm{loc}}(\pi)$ is an arbitrary section, whereas the jets $\partial_\psi(\rho_k(\varphi_\Delta))$ and $\rho_k(\partial_\psi(\varphi_\Delta))$ are π_k-horizontal. Hence, $\partial_\psi\circ\rho_k = \rho_k\circ\partial_\psi$. □

COROLLARY. *For any section* $\psi\in\mathcal{F}(\pi,\pi)$, $k\geq 0$, *one has*

$$\partial_\psi\circ U_k = U_k\circ\partial_\psi.$$

PROOF. This follows from the proposition and the fact that $U_k = S\circ\rho_k$, where S is the Spencer operator. Indeed, Spencer operators, being natural, commute with Lie derivatives. □

If $k=1$, one has $\partial_\psi(U_1(f)) = U_1(\partial_\psi(f))$, which shows that ∂_ψ is a $\mathcal{C}$-field. Note also the following useful relation:

$$\hat{X}\circ\rho_k = \rho_k\circ\hat{X},\quad X\in D(M). \tag{2.20}$$

It is proved along the same lines as Proposition 2.3.9. From this relation (see the proof of Corollary 2.3.9) one has

$$\hat{X}\circ U_k = U_k\circ\hat{X}. \tag{2.21}$$

Hence, operators of the form $\hat{X}$ are also $\mathcal{C}$-fields. Moreover,

$$\hat{X} \lrcorner\, U_1(f) = 0 \tag{2.22}$$

for any $f\in\mathcal{F}(\pi)$. This results from the relations $U_1 = d-\hat{d}$ and $\hat{X}\lrcorner\,\hat{d}f = \hat{X}(f)$.

2.3.10. Now let $Y\colon C^\infty(M)\to\mathcal{F}_k(\pi)$ be a derivation of the algebra $C^\infty(M)$ with values in the $C^\infty(M)$-module $\mathcal{F}_k(\pi)$. Define the mapping $\overline{Y}\colon\mathcal{F}(\pi)\to\mathcal{F}(\pi)$ in the following way. For arbitrary $u\in\Gamma_{\mathrm{loc}}(\pi)$ and $\varphi\in\mathcal{F}(\pi)$ set

$$j(u)^*\overline{Y}(\varphi) = j(u)^*\big(Y(j(u)^*(\varphi))\big).$$

Obviously, $\overline{Y}$ is a derivation of the algebra $\mathcal{F}(\pi)$.

LEMMA. *The operator* $\overline{Y}$ *is a* $\mathcal{C}$*-field for any derivation* $Y\colon C^\infty(M)\to\mathcal{F}_k(\pi)$.

PROOF. Let $Y\in D(M)$, i.e., $k=1$. Then evidently $\overline{Y}=\hat{Y}$. Hence, as was shown in the previous subsection, $\overline{Y}$ is a $\mathcal{C}$-field. For $k>1$, any derivation $Y\colon C^\infty(M)\to\mathcal{F}_k(\pi)$, at least locally, can be represented as $Y=\sum_i\varphi_iY_i$, where $\varphi_i\in\mathcal{F}_k(\pi)$, $Y_i\in D(M)$. Hence $\overline{Y}=\sum_i\varphi_i\hat{Y}_i$. Note now the general relation

$$(aX)(\omega) = aX(\omega)+\omega(X)\,da,$$

where $a\in A$, $\omega\in\Lambda^1(A)$, $X\in D(A)$, valid for any commutative algebra A. Hence

$$\overline{Y}(U_1(f)) = \sum_i\big(\varphi_i\hat{Y}_i(U_1(f)) + (\hat{Y}_i\lrcorner\,U_1(f))\,d\varphi_i\big).$$

But as was proved in the previous subsection, $\hat{Y}_i \lrcorner U_1(f) = 0$ and $\hat{Y}_i \circ U_1 = U_1 \circ \hat{Y}_i$. Therefore, $\overline{Y}(U_1(f)) = \sum_i \varphi_i U_1(\hat{Y}_i(f)) \in \mathcal{C}\Lambda^1$. □

2.3.11. Let $X\colon \mathcal{F}(\pi) \to \mathcal{F}(\pi)$ be a $\mathcal{C}$-field. Consider the derivation $X_{-1} = X|_{\mathcal{F}_{-1}(\pi)}\colon C^\infty(M) \to \mathcal{F}_k(\pi)$. Then $X - \overline{Y}_{-1}\colon \mathcal{F}(\pi) \to \mathcal{F}(\pi)$ is a vertical derivation, i.e., annihilates the functions of the form $\pi_k^*(f)$, $f \in C^\infty(M)$.

LEMMA. *Any vertical derivation $Y\colon \mathcal{F}_0(\pi) \to \mathcal{F}_k(\pi)$ can be uniquely prolonged to an evolutionary derivation $\mathcal{Э}_\psi$ such that $Y = \mathcal{Э}_\psi|_{\mathcal{F}_0(\pi)}$.*

PROOF. Consider the mapping which takes any evolutionary derivation $\mathcal{Э}_\psi$, $\psi \in \mathcal{F}_k(\pi,\pi) \subset \mathcal{F}(\pi,\pi)$, to the vertical derivation $Y = \mathcal{Э}_\psi|_{\mathcal{F}_0(\pi)}\colon \mathcal{F}_0(\pi) \to \mathcal{F}_k(\pi)$. This mapping is obviously $\mathbb{R}$-linear and, as is seen from (2.12), injective. On the other hand, using (2.12) again, one can see that any derivation satisfying the conditions of the lemma can be prolonged to some evolutionary derivation. Indeed, the components of the section φ_{id} in a special coordinate system are u_i. Therefore, the prolongation needed can be defined by setting $\psi = (\psi_1, \dots, \psi_m)$, where $\psi_i = Y(u_i)$. But from this fact and from injectivity of the constructed mapping it follows that Y can be prolonged to an evolutionary derivation globally. □

2.3.12. Let X be a $\mathcal{C}$-field. Consider the field $X - \overline{X}_{-1}$ and its restriction to the algebra $\mathcal{F}_0(\pi)$. Then according to Lemma 2.3.11, this restriction can be prolonged to some evolutionary derivation $\mathcal{Э}_\psi$. Therefore, the restriction of $X - \overline{X}_{-1} - \mathcal{Э}_\psi$ to $\mathcal{F}_0(\pi)$ vanishes. This fact, together with the above proved lemmas, implies the following

THEOREM. *Any $\mathcal{C}$-field is of the form $X = \mathcal{Э}_\psi + \overline{Y}$ for some section $\psi \in \mathcal{F}(\pi,\pi)$ and a derivation $Y\colon C^\infty(M) \to \mathcal{F}_k(\pi)$.*

COROLLARY. *There exists a one-to-one correspondence between $\mathcal{C}$-fields and derivations of $\mathcal{F}_0(\pi)$ with values in $\mathcal{F}(\pi)$.*

PROOF. Indeed, a $\mathcal{C}$-field restricted to $\mathcal{F}_0(\pi)$ determines a derivation of this algebra with values in $\mathcal{F}(\pi)$. According to Lemma 2.3.8, this correspondence is injective. On the other hand, if $Y\colon \mathcal{F}_0(\pi) \to \mathcal{F}(\pi)$ is a derivation, then the mapping $Y - \overline{Y}_{-1}|_{\mathcal{F}_0(\pi)}$, where $Y_{-1} = Y|_{C^\infty(M)}$, can be prolonged to an evolutionary derivation $\mathcal{Э}_\psi$ (Lemma 2.3.11). Then $(\overline{Y}_{-1} + \mathcal{Э}_\psi)|_{\mathcal{F}_0(\pi)} = Y$. □

2.3.13. The following interpretation of Theorem 2.3.12 will be very important later. From formula (2.22) it follows that $\mathcal{C}$-fields of the form $\hat{Y}$ (and consequently of the form $\overline{Y}$) are tangent to manifolds $M_s^{(\infty)}$, $s \in \Gamma_{\mathrm{loc}}(\pi)$. Hence, a translation along trajectories of the field $\overline{Y}$, if it existed, would preserve manifolds $M_s^{(\infty)}$. On the other hand, a translation along trajectories of a $\mathcal{C}$-field X, if it existed, would send $M_s^{(\infty)}$ to $M_{s_t}^{(\infty)}$ generating a "flow" s_t on the "manifold" $\Gamma_{\mathrm{loc}}(\pi)$ (strictly speaking, this flow is only "virtual" since a generic $\mathcal{C}$-field does not have trajectories; see [**30**]). It can be easily seen that $(ds_t/dt)|_{t=0} = j(s)^*\psi$, and this is the meaning of the section ψ. Hence, a flow is trivial if and only if $X = \overline{Y}$. Therefore, it is reasonable to say that the $\mathcal{C}$-fields of the form $\overline{Y}$ are *trivial*. We denote the set of such fields by $\mathcal{C}D = \mathcal{C}D(\pi)$.

Note that $\mathcal{C}D$ is a Lie subalgebra in $D_{\mathcal{C}}$. Besides, since $[\mathcal{Э}_\psi, \hat{X}] = 0$ (see (2.13)) and any field $\overline{Y}$ can be represented as $\overline{Y} = \sum_i f_i \hat{X}_i$, one has $[\mathcal{Э}_\psi, \overline{Y}] \in \mathcal{C}D$.

Thus $\mathcal{C}D$ is an ideal of the Lie algebra $D_\mathcal{C}$ and the quotient algebra $\varkappa = \varkappa(\pi) = D_\mathcal{C}(\pi)/\mathcal{C}D(\pi)$ is isomorphic to the Lie algebra of all evolutionary derivations (or to the Lie algebra $\mathcal{F}(\pi,\pi)$ with respect to the higher Jacobi bracket $\{\bullet,\bullet\}$). Thus Theorem 2.3.12 establishes a canonical decomposition of any $\mathcal{C}$-field into "trivial" and "evolutionary" parts.

2.4. The structure of automorphisms of the Cartan distribution on the manifolds $J^k(E,m)$, $k<\infty$

In this section, we give a complete description of automorphisms for higher contact geometry of the second type, i.e., for the Cartan distribution on $J^k(E,n)$, $k<\infty$ (see Subsection 2.1.5). We consider here both finite and infinitesimal automorphisms constituting the basis of the theory of symmetries for differential equations.

Automorphisms (infinitesimal automorphisms) of the Cartan distribution on $J^k(E,n)$, $k<\infty$, are called here *Lie transformations* (resp., *Lie fields*).

2.4.1. Let us give the following

DEFINITION. A diffeomorphism $F\colon \mathcal{U}_1 \to \mathcal{U}_2$ of two domains $\mathcal{U}_1,\mathcal{U}_2 \subset J^k(E,n)$ is called a *Lie transformation* if $dF_\theta(\mathcal{C}_\theta) \subset \mathcal{C}_{F(\theta)}$ for every $\theta \in \mathcal{U}_1$.

Any Lie transformation $F\colon \mathcal{U}_1 \to \mathcal{U}_2$, where $\mathcal{U}_i \subset J^k(E,n)$, generates a Lie transformation

$$F^{(s)}\colon \pi_{k+s,k}^{-1}(\mathcal{U}_1) \to \pi_{k+s,k}^{-1}(\mathcal{U}_2), \quad s \geq 0.$$

Namely, any point $x_{k+s} \in J^{k+s}(E,n)$ can be understood as a class $\mathrm{cl}_s(x_{k+s})$ of R-manifolds passing through the point $x_k = \pi_{k+s,k}(x_{k+s})$ and tangent to each other with order s at this point (see Subsections 2.1.3 and 2.1.7). Lie transformations preserve classes of this type since under these transformations R-manifolds are transformed to R-manifolds. Thus we can define $F^{(s)}$ by

$$\mathrm{cl}_s\left(F^{(s)}(x_{k+s})\right) = F\left(\mathrm{cl}_s(x_{k+s})\right). \tag{2.23}$$

It should be noted though that this definition makes sense only for those points x_{k+s} for which $F(\mathrm{cl}_s(x_{k+s}))$ consists of R-manifolds having the form $L^{(k)}$, $L \subset N$, in the vicinity of the point $F(x_k)$. It is equivalent to the fact that their common tangent plane $dF(L_{x_{k+1}})$, where $x_{k+1} = \pi_{k+s,k+1}(x_{k+s})$, does not degenerate under the projection $d\pi_{k,0}$. The set of points x_{k+s} where $F^{(s)}$ is not defined is obviously a closed submanifold in $\pi_{k+s,k}^{-1}(\mathcal{U}_1)$. Nevertheless, for brevity shall say that $F^{(s)}$ is defined on $\pi_{k+s,k}^{-1}(\mathcal{U}_1)$. Note also that

$$\pi_{k+s,k+t} \circ F^{(s)} = F^{(t)} \circ \pi_{k+s,k+t}, \quad s \geq t \geq 0.$$

PROPOSITION. *The mapping $F^{(s)}$ is a Lie transformation for any $s>0$.*

PROOF. Note that for all $s,t \geq 0$ one has $F^{(s+t)} = (F^{(s)})^{(t)}$. Hence, it is sufficient to prove the proposition for $s=1$ only.

Let $x_{k+1} \in J^{k+1}(E,n)$ and $x_k = \pi_{k+1,k}(x_{k+1})$. Then by (2.23), we have $dF(L_{x_{k+1}}) = L_{F^{(1)}(x_{k+1})}$. Hence $dF^{(1)}(\mathcal{C}_{x_{k+1}}) = \mathcal{C}_{F^{(1)}(x_{k+1})}$ because of the equality $\mathcal{C}_{x_{k+1}} = (d\pi_{k+1,k})^{-1}(L_{x_{k+1}})$ (see Proposition 1 in Subsection 2.1.4). □

2.4.2. Let us introduce an infinitesimal analog of Lie transformations.

DEFINITION. A vector field on $J^k(E,n)$ is called a *Lie field* (or *infinitesimal Lie transformation*) if the (local) flow generated by this field consists of Lie transformations.

Immediately from the definition one has the following

PROPOSITION (1). *A field $X \in D(\mathcal{F}_k(\pi))$ is a Lie field if and only if $X(\mathcal{C}_k) \subset \mathcal{C}_k$.*

Any Lie field X on $J^k(E,n)$ can be lifted up to a field $X^{(s)}$ on $J^{k+s}(E,n)$ in the following way. Let $\{A_t\}$ be a one-parameter (local) group generated by X. Then one can consider a one-parameter group $\{A_t^{(s)}\}$ of Lie transformations lifted to $J^{k+s}(E,n)$. It generates the field $X^{(s)}$ which we call the *lifting* of X. From the above it follows that

$$X^{(s)} \circ \pi^*_{k+s,k+t} = \pi^*_{k+s,k+t} \circ X^{(t)}, \quad s \geq t \geq 0.$$

Hence, according to Proposition 2.4.1, the derivation $X^{(\infty)} \in D(\mathcal{F}(\pi))$ defined by $X^{(\infty)}|_{\mathcal{F}_s(\pi)} = X^{(s)}$ is a $\mathcal{C}$-field. Therefore, one has

PROPOSITION (2). *A $\mathcal{C}$-field Y is a Lie field (i.e., is of the form $X^{(\infty)}$ for some Lie field X on $J^k(E,n)$) if and only if it preserves filtration in $\mathcal{F}(\pi)$.*

Note the following obvious fact: the U-transformations (both finite and infinitesimal) are Lie transformations (finite and infinitesimal, respectively).

2.4.3. The description of Lie fields is based on the following

LEMMA. *Let X be a Lie field on the manifold $J^k(\pi)$. Then $X = 0$ if and only if $X \lrcorner U_1(\pi) = 0$.*

PROOF. Let us consider a $\mathcal{C}$-field $X^{(\infty)}$ corresponding to X. Then, obviously, $X^{(\infty)} \lrcorner U_1(\pi) = X \lrcorner U_1(\pi)$. Let us represent $X^{(\infty)}$ in the form $X^{(\infty)} = \mathcal{Э}_f + \overline{Y}$, where $Y\colon C^\infty(M) \to \mathcal{F}_k(\pi)$. Then one has

$$X^{(\infty)} \lrcorner U_1(\pi) = \mathcal{Э}_f \lrcorner U_1(\pi) = f.$$

Hence, $X \neq 0$ for $f \neq 0$.

Suppose now that $f = 0$. Then $X^{(\infty)} = \overline{Y}$. But nontrivial operators of the form $\overline{Y}$ increase filtration in $\mathcal{F}(\pi)$ by one at least. Therefore, according to Proposition 2.4.2, one can state that $\overline{Y} = 0$. □

PROPOSITION. *Let X be a $\mathcal{C}$-field on the space $J^k(\pi)$ and $f = X \lrcorner U_1(\pi)$. Then $f \in \mathcal{F}_1(\pi, \pi)$.*

PROOF. Let us use the formula $X \lrcorner U_k(\pi) = \rho_{k-1}(\pi)$ (see [**71**, p. 93]). If the filtration of the element f equals p, then the filtration of the jet $\rho_{k-1}(f)$ equals $p + k - 1$. But k is not less than the filtration of $X \lrcorner U_k(\pi)$ while the latter is $p + k - 1$. This shows that $p \leq 1$. □

Lemma and Proposition 2.4.3 show that any Lie field X is uniquely determined by the section $f = X \lrcorner U_1(\pi) \in \mathcal{F}_1(\pi,\pi)$. For this reason it will be called the *generating function* of X. It is natural to ask whether a given section $f \in \mathcal{F}_1(\pi,\pi)$ is a generating function of a Lie field X. A complete description of such sections will be obtained below. In this section, we restrict ourselves to the classical case of the one-dimensional vector bundle $\pi = \mathbb{1}_M$. Then the form $U_1(\pi)$ determines the standard contact structure on $J^1(\pi)$ (see [**85**]) and Lie transformations coincide with contact transformations. It is well known [**85**] that there exists a one-to-one correspondence between contact fields on $J^1(\mathbb{1}_M)$ and functions $f \in C^\infty(J^1(\mathbb{1}_M)) = \mathcal{F}_1(\mathbb{1}_M)$. This correspondence is given by $X \mapsto X \lrcorner U_1(\pi)$ and hence any function on $J^1(\mathbb{1}_M)$ is a generating function. Denote by Y_f the contact field such that $Y_f \lrcorner U_1(\pi) = f$. Then we get the following

THEOREM. *Let X be a Lie field on the space $J^k(\mathbb{1}_M)$ and $f = X \lrcorner U_1(\pi)$. Then $X = Y_f^{(k)}$.*

COROLLARY. *Let $\dim \pi = 1$. Then the flow generated by a Lie field on $J^k(\pi)$ consists of $\pi_{k,l}$-fiberwise transformations for all $k \geq l \geq 1$.*

PROOF. The result immediately follows from the theorem for $\pi = \mathbb{1}_M$. The case of an arbitrary one-dimensional bundle reduces to $\pi = \mathbb{1}_M$ because of local triviality of π. □

2.4.4. Let $X_f = Y_f^{(\infty)}$. Then according to Theorem 2.3.12, $X_f = \mathcal{Э}_f + Z_f$, where Z_f is a field of the form $\overline{W}$. To describe Z_f, we use the following property of the fields Y_f:

$$Y_f \lrcorner dU_1(\pi) = Y_1(f)U_1 - df;$$

see [**85**]. But $\overline{W}(\psi) = 2dU_1(\overline{W}, \mathcal{Э}_\psi)$ and $dU_1(\mathcal{Э}_f, \mathcal{Э}_\psi) = 0$ for any f, ψ (see [**71**, §3.4]). Therefore,

$$Z_f(\psi) = 2dU_1(\pi)(Z_f, \mathcal{Э}_\psi) = (Y_f \lrcorner dU_1(\pi))(\mathcal{Э}_\psi) = (Y_1(f) - \ell_f)(\psi)$$

for any $\psi \in \mathcal{F}(\pi)$. Hence, one has

$$Z_f = X_1(f) - \ell_f, \tag{2.24}$$

$$X_f = \mathcal{Э}_f - \ell_f + X_1(f) = \bar{X}_f + X_1(f). \tag{2.25}$$

Recall that the trajectories of the field $\bar{X}_f$ are characteristics for the first-order equation $f = 0$. According to (2.25), the operators $\bar{X}_f$, $f \in \mathcal{F}(\pi, \pi)$, generalize this concept for higher order equations.

2.4.5. Let us describe the structure of Lie transformations now.

THEOREM. *Let $F\colon \mathcal{U}_1 \to \mathcal{U}_2$ be a Lie transformation, $\mathcal{U}_i \subset J^k(E, n)$. Assume that $k > \varepsilon$, where $\varepsilon = 0$ for $m > 1$, and $\varepsilon = 1$ for $m = 1$, and that the domain $\mathcal{U}_1$ is $\pi_{k,l}$-connected for any $\varepsilon \leq l < k$. Then*

1. *The transformation F is $\pi_{k,\varepsilon}$-fiberwise and consequently generates a diffeomorphism $F_\varepsilon\colon \pi_{k,\varepsilon}(\mathcal{U}_1) \to \pi_{k,\varepsilon}(\mathcal{U}_2)$ such that $F_\varepsilon \circ \pi_{k,\varepsilon} = \pi_{k,\varepsilon} \circ F$.*
2. *$F_\varepsilon\colon \pi_{k,\varepsilon}(\mathcal{U}_1) \to \pi_{k,\varepsilon}(\mathcal{U}_2)$ is a Lie transformation.*
3. *$F = F_\varepsilon^{(k)}|_{\mathcal{U}_1}$.*

PROOF. (a) We start with the case $m > 1$ or $kn > 1$. This is the nonexceptional situation, and the fibers of the projection $\pi_{k,k-1}|_{\mathcal{U}_1}$ and only these fibers are integral manifolds of the Cartan distribution of maximum dimension (see Theorem 2.1.7). Therefore, connected components of these fibers are taken to each other by Lie transformations. Note that $\pi_{k,l}$-connectivity implies $\pi_{t,l}$-connectivity,

$k \geq t \geq \varepsilon$. The fibers of the projection $\pi_{k,k-1}|_{\mathcal{U}_1}$ are connected and therefore F is a $\pi_{k,k-1}$-fiberwise transformation. Let $F_{k-1}\colon \pi_{k,k-1}(\mathcal{U}_1) \to \pi_{k,k-1}(\mathcal{U}_2)$ be the transformation induced by F, i.e., $F_{k-1} \circ \pi_{k,k-1} = \pi_{k,k-1} \circ F_{k-1}$. Then F_{k-1} is a Lie transformation. In fact, according to Proposition 1 from Subsection 2.1.4,

$$\begin{aligned}(dF_{k-1})(L_{x_k}) &= (dF_{k-1})(d\pi_{k,k-1}(\mathcal{C}_{x_k})) \\ &= (d\pi_{k,k-1})(dF(\mathcal{C}_{x_k})) = (d\pi_{k,k-1})(\mathcal{C}_{F(x_k)}) = L_{F(x_k)}.\end{aligned}$$

On the other hand, if $W \subset \pi_{k,k-1}^{-1}(z)$ is an open set, then the set $L(W) = (\bigcup_{w\in W} L_w) \setminus \{0\}$ is open in $\mathcal{C}_z$ and the linear span $\mathcal{L}(W)$ of $L(W)$ coincides with $\mathcal{C}_z$. Hence, if $W_z = \pi_{k,k-1}^{-1}(z) \cap \mathcal{U}_1$, then

$$\begin{aligned}dF_{k-1}(L(W_z)) = dF_{k-1}\left(\bigcap_{w\in W_z} L_w\right) &= \bigcap_{w\in W_z} dF_{k-1}(L_w) \\ &= \bigcap_{w\in W_z} L_{F(w)} = L(F(W_z)) = L(W_{F_{k-1}(z)}).\end{aligned}$$

Thus, the differential dF_{k-1} maps the linear span of $L(W_z)$, i.e., $\mathcal{C}_z$, onto the linear span of $L(W_{F_{k-1}(z)})$, i.e., $\mathcal{C}_{F_{k-1}(z)}$. Therefore, F_{k-1} is a Lie transformation. Similar to the proof of Theorem 2.2.5, one can see that $F = F_{k-1}^{(1)}|_{\mathcal{U}_1}$.

Now, the mapping $F_{k-1}\colon \pi_{k,k-1}(\mathcal{U}_1) \to \pi_{k,k-1}(\mathcal{U}_2)$ satisfies the conditions of the theorem if $k-1 > \varepsilon$, and by repeating the above reasoning we can construct F_{k-2}, etc., up to F_ε.

(b) Consider the case $m = n = 1$. According to Subsection 2.1.7, in this case the integral manifolds of the Cartan distribution are one-dimensional. Nevertheless, the Lie transformations still remain $\pi_{k,k-1}$-fiberwise for $k > 1$. Due to its local nature, it suffices to prove this fact for $J^k(\pi) \subset J^k(E,n)$, $\dim \pi = 1$.

Let $x_k \in J^k(\pi)$, and let $L \ni x_k$ be an integral curve nondegenerately projected to $J^{k-1}(\pi)$ in the vicinity of x_k. Then there exists a Lie field X on $J^k(\pi)$ such that the corresponding flow preserves the point x_k but moves $T_{x_k}(L)$. Any field $\mathfrak{I}_f$, $f \in \mathcal{F}_0(\pi)$, can be taken for such a field, if $[f]_x^k = 0$ and $[f]_x^{k+1} \neq 0$, $x = \pi_k(x_k)$. Suppose now that F is not $\pi_{k,k-1}$-fiberwise. This means that there exists a point $x'_k \in \mathcal{U}_1$ such that the curve $L = F(\pi_{k,k-1}^{-1}(x'_{k-1}) \cap \mathcal{U}_1)$ has nondegenerate projection to $J^{k-1}(\pi)$ in the vicinity of $x_k = F(x'_k)$, where $x'_{k-1} = \pi_{k,k-1}(x'_k)$. The field $F(X) = F^* \circ X \circ F^{*-1}$ is obviously a Lie field. Translations along its trajectories preserve x'_k and move $T_{x'_k}(F^{-1}(L)) = T_{x'_k}(\pi_{k,k-1}^{-1}(x'_k))$. But this contradicts the fact that translations along Lie fields are $\pi_{k,k-1}$-fiberwise (see Corollary 2.4.3). The rest of the proof coincides with that of part (a). □

Thus we see that the Lie transformations (Lie fields) on $J^k(E,n)$, $k \geq 1$, are the liftings of Lie transformations (resp., Lie fields) from $J^1(E,n)$ when $m = 1$, while in the case $m > 1$ they are lifted from $E = J^0(E,n)$. The Cartan distribution on $J^1(E,n)$ determines a contact structure for $m = 1$. Hence in this case the Lie transformations (resp., Lie fields) on $J^1(E,n)$ coincide with contact transformations (resp., with contact fields). In addition, any diffeomorphism of a domain $\mathcal{U} \subset E$ is a Lie transformation (resp., a Lie field). This shows that the theory of Lie transformations reduces to the theory of contact transformations in $J^1(E,n)$ for $m = 1$ and to the theory of diffeomorphisms of E and their liftings to $J^k(E,n)$ for $m > 1$.

2.4.6. To conclude this section, we describe the sections f corresponding to the Lie fields on $J^1(\pi)$. It is sufficient to consider only the case $\dim \pi > 1$ (see Subsection 2.4.3).

DEFINITION. A section $f \in \mathcal{F}_1(\pi, \pi)$ is called a *generating function* if it corresponds to a Lie field $X^{(1)} \in D(J^1(\pi))$, i.e., if $f = X^{(1)} \lrcorner U_1(\pi)$.

The subsequent exposition becomes more transparent by using the notion of a *relative vector field*. Recall that an $\mathbb{R}$-linear mapping $Y\colon C^\infty(M_2) \to C^\infty(M_1)$ is called a *relative vector field along a mapping* $F\colon M_1 \to M_2$ of smooth manifolds if

$$Y(fg) = Y(f)F^*(g) + F^*(f)Y(g), \quad f, g \in C^\infty(M_2).$$

Given a point $x \in M_1$, let us define the vector $Y_x \in T_{F(x)}(M_2)$ by setting

$$Y_x(f) = Y(f)(x), \quad f \in C^\infty(M_2).$$

Thus, a relative vector field along F may be viewed as a "field" assigning to any point $x \in M_1$ a tangent vector to M_2 at the point $F(x)$. If $\alpha \in \Lambda^1(M_2)$, then the *insertion* (or *inner product*) $Y \lrcorner \alpha \in C^\infty(M_1)$ of Y into α is defined as

$$(Y \lrcorner \alpha)(x) = Y_x \lrcorner \alpha_{F(x)}, \quad x \in M_1.$$

"Absolute" vector fields on M are interpreted as relative ones along the identity mapping id_M. Moreover, if $X \in D(M_1)$, then $X \circ F^*$ is a relative vector field along F as well as $F^* \circ Z$ for $Z \in D(M_2)$.

Let $X \in D(E)$, $E = J^0(\pi)$, be a vector field whose lifting to $J^1(\pi)$ coincides with $X^{(1)}$. Then, by Proposition 3 in Subsection 2.1.4, $f(x_1) = \big(X^{(1)} \lrcorner U_1(\pi)\big)(x_1)$ coincides with the projection of the vector X_{x_0} to E_x along L_{x_1}, where $x_0 = \pi_{1,0}(x_1)$, $x = \pi_1(x_1)$. The property that distinguishes generating functions of Lie fields from other elements of $\mathcal{F}_1(\pi, \pi)$ comes from the fact that the form $U_1(\pi)$ is $\pi_{1,0}$-horizontal. Therefore, to describe it precisely, one must study the behavior of f along the fibers of the projection $\pi_{1,0}$. To do this, it is sufficient to compute the difference $f(\bar{x}_1) - f(x_1)$ for $\pi_{1,0}(\bar{x}_1) = \pi_{1,0}(x_1) = x_0$. The last implies that $\bar{x}_1 = x_1 + \theta$ with $\nu_{1,0}(\theta) = 0$.

If $\psi \in \Gamma(\pi)$ is such that $[\psi]_x^1 = \theta$, then

$$f(x_1 + \theta) - f(x_1) = -d\pi(X_{x_0}) \lrcorner \delta[\psi]_x^1,$$

where $\delta\colon \mathcal{J}^1(\Gamma(\pi)) \to \Gamma(\pi) \otimes \Lambda^1(M)$ is the δ-Spencer operator (see, for instance, [**71**, Proposition 4.2.5]). In particular, if $\theta = \omega \otimes e$, $\omega \in T_x^*(M)$, $e \in E$, then $\delta(\omega \otimes e) = \omega \otimes e$ and consequently

$$f(x_1 + \theta) - f(x_1) = -\langle d\pi(X_{x_0}), \omega\rangle e. \tag{2.26}$$

So, a generating section f must satisfy (2.26) at any point of the manifold $J^1(\pi)$.

Consider the relative vector field $Y = X \circ \pi^*\colon C^\infty(M) \to C^\infty(E)$. Then $\omega(Y_x) = Y_x \lrcorner \omega = d\pi(X_{x_0}) \lrcorner \omega = \langle d\pi(X_{x_0}), \omega\rangle$ and formula (2.26) can be rewritten in the form

$$f(x_1 + \omega \otimes e) = f(x_1) - \omega(Y_x)e. \tag{2.27}$$

To prove the converse, assume that $f \in \mathcal{F}(\pi, \pi)$ satisfies (2.27) with a relative vector field Y along π. We show that in this case there exists a field $X \in D(E)$ such that $f = X^{(1)} \lrcorner U_1(\pi)$. Set

$$X_{x_0} = f(x_1) + dh(Y_x), \tag{2.28}$$

where $h \in \Gamma_{\mathrm{loc}}(\pi)$, $[h]_x^1 = x_1$, while $f(x_1)$, by linearity of E_x, is understood as a vector in $T_{x_0}(E_x)$.

Let us prove that (2.27) implies that the vector X_{x_0} depends on the point $x_0 \in E$ only and not on $x_1 \in J^1(\pi)$. To do this, compare the right-hand sides of (2.28) at the points x_1 and $x_1 + \omega \otimes e$. A section $\tilde{h} \in \Gamma_{\mathrm{loc}}(\pi)$ such that $[\tilde{h}]_x^1 = x_1 + \omega \otimes e$ can be chosen in the form $\tilde{h} = h + gs$ with $g \in C^\infty(M)$, $s \in \Gamma_{\mathrm{loc}}(\pi)$, such that $g(x) = 0$, $dg_x = \omega$, and $s(x) = e$. It is straightforward to verify that $d\tilde{h}_x = dh_x + dg_x \otimes e$, where $dg_x \otimes e$ is understood to be a linear mapping from T_xM to E_x taking $\xi \in T_xM$ to $(\xi \lrcorner dg_x)e$. Here dh_x (resp., $d\tilde{h}_x$) stands for the differential of the mapping $h\colon \mathcal{U}_h \to E$ (resp., $\tilde{h}\colon \mathcal{U}_{\tilde{h}} \to E$) at $x \in M$. Then

$$
\begin{aligned}
&f(x_1 + \omega \otimes e) + d(h + gs)(Y_x) \\
&\qquad = f(x_1) - (Y_x \lrcorner \omega)e + dh(Y_x) + (Y_x \lrcorner dg_x)e = f(x_1) + dh(Y_x).
\end{aligned}
$$

Thus in the case of $\dim \pi > 1$ we have proved the following

THEOREM. *A section $f \in \mathcal{F}_1(\pi, \pi)$ is a generating function of a Lie field if and only if*

$$
(\theta + \alpha \otimes s)^*(f) = \theta^*(f) - \theta_0^*(Y \lrcorner \alpha) \tag{2.29}
$$

for arbitrary $\theta \in \Gamma(\pi_1)$, $\theta_0 = \nu_{1,0}(\theta) \in \Gamma(\pi)$, $\alpha \in \Lambda^1(M)$, $s \in \Gamma(\pi)$, and some relative vector field Y along π. The field $X^{(1)} = X_f^{(1)}$ corresponding to f satisfies the relation

$$
\theta_0^*(Y \lrcorner \alpha) = \theta^*\bigl(X_f^{(1)} \lrcorner \pi_1^*(\alpha)\bigr).
$$

DEFINITION. A vector field on $J^k(\pi)$ generating the flow consisting of U-transformations is called a *U-field.*

Obviously, the U-fields are Lie fields as well.

COROLLARY. *A Lie field is a U-field if and only if the corresponding vector field Y, relative along π, is of the form $\pi^* \circ Z$ for some $Z \in D(M)$.*

PROOF. Let $X \in D(E)$ be a vector field whose lifting is the U-field in question. Then, according to Theorem 2.2.5, the flow generated by X is π-fiberwise. This is equivalent to the fact that X can be projected with respect to π. The latter means that $X \circ \pi^* = \pi^* \circ Z$ with $Z \in D(M)$ being the projection of X to M. But $Y = X \circ \pi^*$. □

2.4.7. Let us find a coordinate presentation for the Lie fields $X_f^{(1)} \in D(J^1(\pi))$.

Choose local coordinates $x_1, \dots, x_n$ in M and a basis $e_1, \dots, e_m$ of sections in $\Gamma(\pi)$ over a coordinate neighborhood $\mathcal{O}$ under consideration. Denote by $e_1^*, \dots, e_m^*$, $e_i^* = \pi_1^*(e_i)$, the basis in $\mathcal{F}_1(\pi, \pi)$ over $\pi_1^{-1}\mathcal{O}$.

Let

$$
X = \sum_{i=1}^n a_i \frac{\partial}{\partial x_i} + \sum_{j=1}^m b_j \frac{\partial}{\partial u_j}, \qquad a_i, b_j \in C^\infty(\mathcal{O}),
$$

be the field on E whose lifting is $X_f^{(1)}$. The field X is the projection of $X_f^{(1)}$ to E by $\pi_{1,0}$, so that $X_f^{(1)} \circ \pi_{1,0}^* = \pi_{1,0} \circ X$. But the form

$$
U_1(\pi) = \sum_{j=1}^m \left(du_j - \sum_{i=1}^n p_i^j \, dx_i \right) e_j^*
$$

is $\pi_{1,0}$-horizontal. Hence, the insertion of $X_f^{(1)}$ into $U_1(\pi)$ coincides with that of $X_f^{(1)} \circ \pi_{1,0}^*$ (the relative vector field along $\pi_{1,0}$ corresponding to $X_f^{(1)}$). Thus,

$$f = X_f^{(1)} \lrcorner U_1(\pi) = \left(X_f^{(1)} \circ \pi_{1,0}^*\right) \lrcorner U_1(\pi) = (\pi_{1,0}^* \circ X) \lrcorner U_1(\pi).$$

Noticing now that the coordinate expression of X is identical to that of $\pi_{1,0}^* \circ X$, one finds

$$f = \sum_{i=1}^{m} \Big(- \sum_{j=1}^{n} a_j p_i^j + b_j \Big) e_i^* = \sum_{i=1}^{m} f_i e_i^*.$$

To find the coordinate expression of the field $X^{(1)} = X_f^{(1)}$, we note that, according to Proposition 2.4.2 (2), $X^{(1)} = X^{(\infty)}|_{\mathcal{F}_1(\pi)}$. Therefore, the coefficients of the field $X_f^{(1)}$ can be found by applying $X^{(\infty)}$ to coordinate functions x_i, u_j, p_i^j in $\pi_{1,0}^{-1}(\mathcal{O})$. By Theorem 2.3.12, we have $X^{(\infty)} = \partial_f + \overline{Y}$ with $Y = X \circ \pi^*$ being the relative field discussed in the previous subsection. Obviously, $Y = \sum_{i=1}^n a_i \partial/\partial x_i$, so that $\overline{Y} = \sum_{i=1}^n a_i D_i$. Then a direct computation using (2.27) shows that

$$X_f^{(1)} = \sum_{i=1}^{n} a_i \frac{\partial}{\partial x_i} + \sum_{j=1}^{m} b_j \frac{\partial}{\partial u_j} + \sum_{i,j} \left(\frac{\partial f_j}{\partial x_i} + \sum_{l} \frac{\partial f_j}{\partial u_l} p_i^l \right) \frac{\partial}{\partial p_i^j}. \tag{2.30}$$

2.5. Classical theory of symmetries of partial differential equations

Similarity to the classical theory of scalar first-order partial differential equations, as well as a number of other reasons, makes it advisable to generalize the notion of a solution for a differential equation $\mathcal{Y} \subset J^k(E,n)$. Namely, it is natural to consider any R-manifold (see Subsection 2.1.7) lying in $\mathcal{Y}$ as a solution of this equation. For this reason, it is also natural to call an (*intrinsic*) *symmetry* of $\mathcal{Y}$ a diffeomorphism of $\mathcal{Y}$ into itself taking generalized solutions to solutions of the same type. It is easily seen that a diffeomorphism of $\mathcal{Y}$ is a symmetry if and only if it is an automorphism of the Cartan distribution restricted to $\mathcal{Y}$. Any Lie transformation $F\colon J^k(E,n) \to J^k(E,n)$ such that $F(\mathcal{Y}) = \mathcal{Y}$, is obviously an (*extrinsic*) *symmetry* of $\mathcal{Y}$.

The classical symmetry theory by Sophus Lie is, from the above viewpoint, the theory of extrinsic symmetries. The results obtained in the previous section show that the group of extrinsic symmetries for any equation $\mathcal{Y} \subset J^k(E,n)$ is a subgroup in the group of diffeomorphisms of the domain $\pi_{k,0}(\mathcal{Y}) \subset E$ if $m > 1$, and a subgroup in the group of all contact transformations of the domain $\pi_{k,1}(\mathcal{Y}) \subset J^1(E,n)$ if $m = 1$. If $\mathcal{Y} \subset J^k(\pi)$, then, in the nonexceptional case, the U-transformations preserving $\mathcal{Y}$ are those extrinsic symmetries that preserve the class of usual solutions of $\mathcal{Y}$. This explains the role of the theory of Lie transformations and U-transformations in the theory of symmetries of differential equations.

In this context, we have a natural problem of computing intrinsic symmetries, or more generally, of finding relations between intrinsic and extrinsic theories. Evidently, a proper viewpoint is the "intrinsic" one, while "extrinsic" approach is more convenient.

In this section, we address these questions. In particular, it is shown that for the majority of equations arising in practice the two points of view coincide. This means that for such equations the classical theory cannot be generalized by passing to the "intrinsic viewpoint".

2.5.1. Let us give formal definitions.

DEFINITION. Let $\mathcal{Y} \subset J^k(E,n)$ be a differential equation.

1. A Lie transformation $F\colon J^k(E,n) \to J^k(E,n)$ is called an *extrinsic symmetry* of the equation $\mathcal{Y}$ if $F(\mathcal{Y}) \subset \mathcal{Y}$.
2. A Lie field $X \in D(J^k(E,n))$ is called an *extrinsic infinitesimal symmetry* of the equation $\mathcal{Y}$ if it is tangent to $\mathcal{Y}$.

It is easily seen that these definitions possess local analogs.

Extrinsic symmetries obviously form a group of transformations which we denote by $\mathfrak{Lie}_0(\mathcal{Y})$. The corresponding Lie algebra consists of extrinsic infinitesimal symmetries and is denoted by $\mathfrak{lie}_0(\mathcal{Y})$.

Let $\mathcal{C}(\mathcal{Y})\colon x \mapsto \mathcal{C}_x(\mathcal{Y}) = \mathcal{C}_x \cap T_x(\mathcal{Y})$, $x \in \mathcal{Y}$, be the restriction of the Cartan distribution to $\mathcal{Y}$.

DEFINITION. A diffeomorphism of $\mathcal{Y}$ (a vector field on $\mathcal{Y}$) is called an *intrinsic (infinitesimal) symmetry* of the equation $\mathcal{Y}$ if it preserves the distribution $\mathcal{C}(\mathcal{Y})$.

As above, the set of intrinsic symmetries is a group denoted by $\mathfrak{Lie}(\mathcal{Y})$ with the Lie algebra $\mathfrak{lie}(\mathcal{Y})$ consisting of all infinitesimal intrinsic symmetries.

In the sequel, when it is clear from the context, infinitesimal symmetries will be called simply symmetries.

Obviously, the restriction of an extrinsic symmetry to $\mathcal{Y}$ determines an intrinsic symmetry and thus one has the homomorphisms

$$\mathfrak{Lie}_0(\mathcal{Y}) \to \mathfrak{Lie}(\mathcal{Y}), \qquad \mathfrak{lie}_0(\mathcal{Y}) \to \mathfrak{lie}(\mathcal{Y}).$$

An equation $\mathcal{Y}$ is said to be *common* if these mappings are epimorphisms. In other words, $\mathcal{Y}$ is a common equation if for any intrinsic symmetry F of this equation there exists an extrinsic symmetry (a Lie transformation) $\tilde{F}$ such that $\tilde{F}|_{\mathcal{Y}} = F$. A new question arises immediately: what is the ambiguity in choosing $\tilde{F}$? This problem is closely related to the following one: to what extent intrinsic geometry of the equation (i.e., the manifold $\mathcal{Y}$ together with the distribution $\mathcal{C}(\mathcal{Y})$) determines its extrinsic geometry (i.e., the ambient space $J^k(E,n)$ equipped with the Cartan distribution $\mathcal{C}$ and the embedding $i\colon \mathcal{Y} \hookrightarrow J^k(E,n)$ satisfying $i_*(\mathcal{C}_x(\mathcal{Y})) = T_x(\mathcal{Y}) \cap \mathcal{C}_x$, $x \in \mathcal{Y}$)? Equations whose intrinsic geometry determines extrinsic one in the above sense are called *rigid.* It is clear that any rigid equation is common. The converse is obviously not true. As an example, one can consider first-order equations with one independent variable.

Let us note that any intrinsic infinitesimal symmetry of a rigid equation can be uniquely extended to its extrinsic infinitesimal symmetry. This immediately follows from the fact that the flow generated by an intrinsic infinitesimal symmetry of a rigid equation extends uniquely to a flow of Lie transformations on the manifold $J^k(E,n)$.

2.5.2. Our goal is to establish sufficient conditions for the rigidness of an equation $\mathcal{Y}$. To do this, some notions introduced below will be useful.

Recall that when proving the theorem on the structure of Lie transformations in the nonexceptional case (Section 2.4), we used the fact that the fibers of the projection $\pi_{k,k-1}\colon J^k(E,n) \to J^{k-1}(E,n)$ and these fibers only are integral manifolds of the maximum dimension for the Cartan distribution on $J^k(E,n)$. This justifies the following

DEFINITION. An equation $\mathcal{Y} \subset J^k(E,n)$ is called $\mathcal{C}$-*general* if the following two conditions hold:

1. The set $\pi_{k,k-1}(\mathcal{Y})$ is everywhere dense in some domain $\mathcal{U}_\mathcal{Y} \subset J^{k-1}(E,n)$.
2. Fibers of the projection $\pi_\mathcal{Y} = \pi_{k,k-1}|_\mathcal{Y} \colon \mathcal{Y} \to \pi_{k,k-1}(\mathcal{Y})$ are connected, and they and they only are integral manifolds of the maximum dimension for the distribution $\mathcal{C}(\mathcal{Y})$.

Any intrinsic symmetry of a $\mathcal{C}$-general equation $\mathcal{Y} \subset J^k(E,n)$ determines a transformation of the domain $\mathcal{U}_\mathcal{Y} \subset J^{k-1}(E,n)$.

PROPOSITION. *If $\mathcal{Y} \subset J^k(E,n)$ is a $\mathcal{C}$-general equation and $F \in \mathfrak{Li}(\mathcal{Y})$, then there exists a diffeomorphism F' of the domain $\mathcal{U}_\mathcal{Y}$ such that the diagram*

$$\begin{array}{ccc} \mathcal{Y} & \xrightarrow{F} & \mathcal{Y} \\ {\scriptstyle \pi_\mathcal{Y}}\downarrow & & \downarrow{\scriptstyle \pi_\mathcal{Y}} \\ \mathcal{U}_\mathcal{Y} & \xrightarrow{F'} & \mathcal{U}_\mathcal{Y} \end{array}$$

is commutative.

PROOF. Let $x_{k-1} \in \pi_{k,k-1}(\mathcal{Y})$. Then, according to Definition 2.5.2, the set $\pi_\mathcal{Y}^{-1}(x_{k-1})$ is a maximum dimension integral manifold of the Cartan distribution $\mathcal{C}(\mathcal{Y})$ on $\mathcal{Y}$. If $F \in \mathfrak{Li}(\mathcal{Y})$, then F is an automorphism of the distribution $\mathcal{C}(\mathcal{Y})$. Therefore, $F(\pi_\mathcal{Y}^{-1}(x_{k-1}))$ is also an integral manifold of maximum dimension, i.e., $F(\pi_\mathcal{Y}^{-1}(x_{k-1})) = \pi_\mathcal{Y}^{-1}(x'_{k-1})$ for some point $x'_{k-1} \in \mathcal{U}_\mathcal{Y}$. Set $F'(x_{k-1}) = x'_{k-1}$. Obviously, $\pi_\mathcal{Y}(F(x_{k-1})) = x'_{k-1}$. □

2.5.3. Now we deduce the conditions under which the mapping F' constructed in the previous subsection is a Lie transformation of the domain $\mathcal{U}_\mathcal{Y}$. We need another definition.

DEFINITION. An equation $\mathcal{Y} \subset J^k(E,n)$ is said to be $\mathcal{C}$-*complete* if for any point $x_{k-1} \in \pi_{k,k-1}(\mathcal{Y})$ the linear span of the set

$$\bigcup_{x_k \in \pi_\mathcal{Y}^{-1}(x_{k-1})} L_{x_k} \subset T_{x_{k-1}}(J^k(E,n))$$

coincides with the Cartan plane $\mathcal{C}_{x_{k-1}}$.

Thus the equation is $\mathcal{C}$-complete if it completely determines Cartan planes at the points of $\pi_{k,k-1}(\mathcal{Y})$.

PROPOSITION. *If a $\mathcal{C}$-general equation is 1-solvable and $\mathcal{C}$-complete, then the mapping F' constructed in Proposition 2.5.2 is a Lie transformation. Moreover, the lifting of F' restricted to $\mathcal{Y}$ coincides with the symmetry F.*

PROOF. Since the equation $\mathcal{Y}$ is $\mathcal{C}$-complete, to prove the first statement it is sufficient to show that if $Y \in L_{x_k}$, $x_k \in \pi_\mathcal{Y}^{-1}(x_{k-1})$, $x_{k-1} \in \pi_{k,k-1}(\mathcal{Y})$, then $dF'(Y) \in \mathcal{C}_{F'(x_{k-1})}$.

Consider a point $x_{k+1} \in \mathcal{Y}^{(1)}$ satisfying $\pi_{k+1,k}(x_{k+1}) = x_k$. Then $L_{x_{k+1}} \subset T_{x_k}(\mathcal{Y})$ and $d\pi_{k,k-1}(L_{x_{k+1}}) = L_{x_k}$. Hence, there exists a vector $Y_1 \in L_{x_{k+1}}$ such that $d\pi_{\mathcal{Y}}(Y_1) = Y$. Then one has

$$dF'(Y) = dF'(d\pi_{\mathcal{Y}}(Y_1)) = d\pi_{\mathcal{Y}}(dF(Y_1)).$$

But F is an automorphism of $\mathcal{C}(\mathcal{Y})$ and therefore $dF(Y_1) \in \mathcal{C}_{F(x_k)}(\mathcal{Y}) \subset \mathcal{C}_{F(x_k)}$. On the other hand, one has $d\pi_{k,k-1}(\mathcal{C}_{F(x_k)}) = L_{F(x_k)}$ (see Proposition 1 in Subsection 2.1.4), hence $dF'(Y) \in L_{F(x_k)} \subset \mathcal{C}_{F'(x_{k-1})}$.

The fact that F coincides with the restriction of the Lie transformation F' to $\mathcal{Y}$ means that $dF'(L_{x_k}) = L_{F(x_k)}$, $x_k \in \mathcal{Y}$. But we have shown above that $dF'(L_{x_k}) \subset L_{F(x_k)}$. It remains to note that F' is a diffeomorphism and $\dim L_{x_k} = n = \dim L_{F(x_k)}$. □

Equations satisfying the assumptions of the above proposition are *normal.*

The condition of $\mathcal{C}$-completeness together with 1-solvability can be substituted by the following one: the span of the subspaces $d\pi_{k,k-1}(\mathcal{C}_{x_k}(\mathcal{Y})) \subset \mathcal{C}_{x_{k-1}}$, $x_k \in \pi_{\mathcal{Y}}^{-1}(x_{k-1})$, coincides with $\mathcal{C}_{x_{k-1}}$. Indeed, in this case $L_{x_{k+1}} \subset \mathcal{C}_{x_k}(\mathcal{Y}) \subset \mathcal{C}_{x_k}$, $x_{k+1} \in \mathcal{Y}^{(1)}$, and $d\pi_{k,k-1}(L_{x_{k+1}}) = L_{x_k} = d\pi_{k,k-1}(\mathcal{C}_{x_k})$. Therefore, $d\pi_{k,k-1}(\mathcal{C}_{x_k}(\mathcal{Y})) = L_{x_k}$.

This remark shows that the concept of normality is an intrinsic one, i.e., it can be formulated in terms of the manifold $\mathcal{Y}$ and of the distribution $\mathcal{C}(\mathcal{Y})$ over it. This is in accordance with Cartan's interpretation of differential equations as Pfaff systems corresponding to $\mathcal{Y}$. Indeed, in this case the integral manifolds of the dimension of the distribution $\mathcal{C}(\mathcal{Y})$ determine a bundle ν whose base B can be equipped with the distribution

$$B \ni b \mapsto \mathcal{C}_b = \{\text{linear span of subspaces } d\nu(\mathcal{C}_y(\mathcal{Y})),\ y \in \nu^{-1}(b)\}.$$

In the setting under consideration, one has $B = \mathcal{U}_{\mathcal{Y}}$ and the described distribution coincides with the Cartan distribution on $\mathcal{U}_{\mathcal{Y}}$.

Now, if one considers the manifold B^1 consisting of n-dimensional maximal integral subspaces of the distribution constructed on B, then $\mathcal{Y} \subset B^1$ and $B^1 = \pi_{k,k-1}^{-1}(\mathcal{U}_{\mathcal{Y}})$. This procedure shows that the normality condition allows one to reconstruct "the external environment" of the equation $\mathcal{Y}$, i.e., to reconstruct the embedding $\mathcal{Y} \subset J^k(E,n)$ using the distribution $\mathcal{C}(\mathcal{Y})$ on $\mathcal{Y}$ only. Thus we have proved the following

THEOREM. *Normal equations are rigid.*

2.5.4. We shall now formulate normality conditions for the equation $\mathcal{Y}$ in more efficient terms. We start with 1-solvability.

Let $\mathcal{Y} \subset J^k(\pi)$ be defined as $\varphi_\Delta = 0$ for an operator $\Delta \in \operatorname{dif}_k(\pi,\xi)$, where $\xi\colon E' \to M$ is a vector bundle. In other words,

$$\mathcal{Y} = \{x_k \in J^k(\pi) \mid \varphi_\Delta(x_k) = 0\}.$$

Since $\mathcal{Y}^{(1)} = (\varphi_{\Delta^{(1)}}^{-1} \circ j_1(0))(M)$, where 0 is the zero section of the bundle ξ, the commutativity of the diagram

$$\begin{array}{ccccc}
J^{k+1}(\pi) & \xrightarrow{\pi_{k+1,k}} & J^k(\pi) & \xrightarrow{\pi_k} & M \\
\downarrow{\scriptstyle \varphi_{\Delta^{(1)}}} & & \downarrow{\scriptstyle \varphi_\Delta} & & \downarrow{\scriptstyle \xi} \\
J^1(\xi) & \xrightarrow{\pi_{1,0}} & J^0(\pi) & \xrightarrow{\text{id}} & E'
\end{array}$$

implies that the set $\mathcal{Y}^{(1)} \cap \pi_{k+1,k}^{-1}(\theta)$ is not empty, provided the mapping

$$\varphi_{\Delta^{(1)}} : \pi_{k+1,k}^{-1}(\theta) \to \pi_{1,0}^{-1}(\varphi_{\Delta}(\theta))$$

is surjective. But $\varphi_{\Delta^{(1)}}|_{\pi_{k+1,k}^{-1}(\theta)}$ is the first prolongation of the symbol of the operator Δ at the point θ. This symbol is a family of affine mappings with coefficient depending on the point θ. If $\mathcal{Y}$ is a determined equation, i.e., if $\dim \pi = \dim \xi$, then the dimension of the projection $J^{k+1}(\pi) \to J^k(\pi)$ fiber is much greater than that of $J^1(\xi) \to J^0(\xi)$. Hence, in "general position" the first prolongation of the operator symbol is surjective. For example, if $\dim \pi = \dim \xi = 1$ and $\mathcal{Y} = \{f = 0\}$, then surjectiveness takes place when $(\partial f/\partial p_\sigma)(\theta) \neq 0$ for at least one σ, $|\sigma| = k$.

These considerations show that 1-solvability is a rather weak condition, always valid for equations one usually meets in practice.

2.5.5. The conditions of $\mathcal{C}$-generality are also related to simple dimension considerations.

PROPOSITION. *Let $\mathcal{Y} \subset J^k(\pi)$, $\dim \pi = m$, $\dim M = n$, and let the fibers of the projection $\pi_{\mathcal{Y}}$ be connected. Then from the inequality*

$$\operatorname{codim} \mathcal{Y} \leq m \frac{(n+k-2)!}{(k-1)!(n-1)!} - 2$$

it follows that $\mathcal{Y}$ is a $\mathcal{C}$-general equation.

PROOF. Indeed,

$$\lambda = m \frac{(n+k-2)!}{(k-1)!(n-1)!} - 1$$

is the difference between the dimensions of maximal integral manifolds of types 0 and 1 (see Subsection 2.1.7). Therefore, when one restricts the Cartan distribution to $\mathcal{Y}$, the dimension of integral manifolds of maximum dimension will change by a number less than λ. This guarantees that the fibers of $\pi_{k,k-1}|_{\mathcal{Y}}$ and only they are integral submanifolds of $\mathcal{C}(\mathcal{Y})$ of maximum dimension. □

If $\mathcal{Y}$ is a determined equation, then the assumptions of the proposition hold in all cases except

(1) $k = 1$ or $n = 1$ and arbitrary m;
(2) $m = 1$, $k = 2$.

2.5.6. We now deduce a criterion of $\mathcal{C}$-completeness. Let Y be a vector in $d\pi_{k,k-1}(\mathcal{C}_{x_k}(\mathcal{Y})) \subset \mathcal{C}_{x_{k-1}}$, $x_k \in \mathcal{Y}$. Then for any other vector $Y' \in \mathcal{C}_{x_{k-1}}$ such that $d\pi_{k-1}(Y) = d\pi_{k-1}(Y')$, one has $Y - Y' = d\pi_{k-1}(Y) \lrcorner \delta\psi$ (see [**71**, Proposition 4.2.5]), where δ is the differential of the δ-Spencer complex and $\psi \in S^k T_x^*(M) \otimes E_x$, $x = \pi_k(x_k)$. From the same proposition it follows that Y' lies in the span of subspaces $\{L_{y^\alpha}\}$, where $y^\alpha \in \pi_{k,k-1}^{-1}(x_k)$, if and only if ψ is a linear combination of the elements $y^\alpha - x_k \in S^k T_x^*(M) \otimes E_x$, where the point x_k is such that $Y \in L_{x_k}$. Since $\mathcal{C}$-completeness of $\mathcal{Y}$ means that the span of subspaces L_{y^α}, $y^\alpha \in \mathcal{Y} \cap \pi_{k,k-1}^{-1}(x_{k-1})$, $x_{k-1} \in \pi_{k,k-1}(\mathcal{Y})$, coincides with $\mathcal{C}_{x_{k-1}}$, we get the following

PROPOSITION. *Let $A(\mathcal{Y}, x_{k-1})$ be the affine span of the set $\mathcal{Y} \cap \pi_{k,k-1}^{-1}(x_{k-1})$ in the space $\pi_{k,k-1}^{-1}(x_{k-1})$, and $l(\mathcal{Y}, x_{k-1}) \subset S^k T_x^*(M) \otimes E_x$ the corresponding linear* *space. Then the equation $\mathcal{Y}$ is $\mathcal{C}$-complete at the point x_{k-1} if and only if for any $X \in T_x(M)$, $X \neq 0$, one has*

$$X \rfloor \delta(l(\mathcal{Y}, x_0)) = S^k T_x^*(M) \otimes E_x.$$

2.5.7. The obtained results can be easily generalized. Let, as before, $\mathcal{Y} \subset J^k(E,n)$ be a k-th order equation. Assume that $\mathcal{Y}_1 = \pi_{k,k-1}(\mathcal{Y}) \subset J^{k-1}(E,n)$, possibly after elimination of a measure zero set, is a normal equation in $J^{k-1}(E,n)$. If the equation $\mathcal{Y}$ is 1-solvable, then the fibers of the projection $\pi_{\mathcal{Y}} \colon \mathcal{Y} \to \mathcal{Y}_1$ and only they are integral manifolds of the maximum dimension for the distribution $\mathcal{C}(\mathcal{Y})$, and for any point $x_{x-1} \in \mathcal{Y}_1$ the plane $\mathcal{C}_{x_{k-1}}$ lies in the linear span of the subspaces L_{x_k}, $x_k \in \mathcal{Y} \cap \pi_{k,k-1}^{-1}(x_{k-1})$. Then repeating word for word the arguments of Subsections 2.5.2–2.5.3, one can deduce that $\mathcal{Y}$ is a rigid equation. For the same reasons we have the following

THEOREM. *Let $\mathcal{Y} \subset J^k(E,n)$ be a differential equation such that the sets $\mathcal{Y}_i = \pi_{k,k-i}(\mathcal{Y}) \subset J^{k-i}(E,n)$, $i = 0, \dots, s$, $s \leq k-1$, possibly after elimination of a measure zero set, are submanifolds in $J^{k-i}(E,n)$. Further, let*

- *$\mathcal{Y}$ be a 1-solvable equation,*
- *for each $i = 0, 1, \dots, s-1$, the fibers of the projection $\pi_{k-i,k-i-1}|_{\mathcal{Y}_i} \colon \mathcal{Y}_i \to \mathcal{Y}_{i+1}$ and only these fibers are maximal integral manifolds of the distribution $\mathcal{C}(\mathcal{Y}_i)$, and*
- *for any point $x_{k-i-1} \in \mathcal{Y}_{i+1}$ the plane $\mathcal{C}_{x_{k-i-1}}(\mathcal{Y}_{i+1})$ lie in the linear span of the subspaces $L_{x_{k-i}}$, $x_{k-i} \in \mathcal{Y}_i \cap \pi_{k-i,k-i-1}^{-1}(x_{k-i-1})$.*

Then normality of $\mathcal{Y}_s$ implies rigidity of $\mathcal{Y}$.

PROOF. As was said above, the proof repeats the arguments of Subsections 2.5.2–2.5.3. We must only note that 1-solvability of $\mathcal{Y}$ implies that of $\mathcal{Y}_i$. □

2.5.8. Consider in more details the most important case of nonrigid equations when $k = m = 1$. To avoid standard topological considerations, we assume that $\mathcal{Y} \subset J^1(\mathbb{1}_M)$.

The contact structure on $J^1(\mathbb{1}_M)$ is determined by an arbitrary form $\omega_\lambda = \lambda U_1(\mathbb{1}_M)$, where the function λ vanishes nowhere. In the sequel, we assume that there exists a function λ such that $d\omega_\lambda|_{\mathcal{Y}}$ is a nondegenerate form. This is always true locally. Below we set $\omega = \omega_\lambda$ and $\tilde{\omega} = \omega_\lambda|_{\mathcal{Y}}$.

Under these assumptions, it is not difficult to describe the structure of intrinsic infinitesimal symmetries of the equation $\mathcal{Y}$. Evidently, $X \in \mathfrak{li}(\mathcal{Y})$ if and only if $X(\tilde{\omega}) = \mu\tilde{\omega}$ for some $\mu \in C^\infty(\mathcal{Y})$. From the equality $X(\tilde{\omega}) = X \rfloor d\tilde{\omega} + d(X \rfloor \tilde{\omega})$ it follows that

$$X \in \mathfrak{li}(\mathcal{Y}) \text{ if and only if } \mu\tilde{\omega} = X \rfloor d\tilde{\omega} + d\tilde{f},$$

where $\tilde{f} = X \rfloor \tilde{\omega}$.

Define the mapping $\Gamma \colon D(\mathcal{Y}) \to \Lambda^1(\mathcal{Y})$ by setting $\Gamma(Y) = -Y \rfloor d\tilde{\omega}$. Since $d\tilde{\omega}$ is a nondegenerate form, Γ is an isomorphism of $C^\infty(\mathcal{Y})$-modules. Applying Γ^{-1} to the above equality, we obtain that $X \in \mathfrak{li}(\mathcal{Y})$ if and only if

$$\mu Y_{\tilde{\omega}} = Y_{d\tilde{f}} - X,$$

where $Y_\theta = \Gamma^{-1}(\theta)$, $\theta \in \Lambda^1(\mathcal{Y})$.

Note now that $Y_\theta \rfloor \theta = Y_\theta \rfloor (-Y_\theta \rfloor d\tilde{\omega}) = 0$. Therefore,

$$(\mu Y_{\tilde{\omega}})(\tilde{\omega}) = (\mu Y_{\tilde{\omega}}) \rfloor d\tilde{\omega} + d(\mu Y_{\tilde{\omega}} \rfloor Y_{\tilde{\omega}}) = -\mu\tilde{\omega},$$

i.e., $\mu Y_{\tilde{\omega}} \in \mathfrak{li}(\mathcal{Y})$ (it can be easily seen that the fields of the form $Y_{\tilde{\omega}}$ are directed along characteristics of the equation $\mathcal{Y}$). Hence, $X \in \mathfrak{li}(\mathcal{Y})$ if and only if $Y_{d\tilde{f}} \in \mathfrak{li}(\mathcal{Y})$. But $Y_{d\tilde{f}} \lrcorner\, \tilde{\omega} = X \lrcorner\, \tilde{\omega} = \tilde{f}$, since $Y_{\tilde{\omega}} \lrcorner\, \tilde{\omega}$. Therefore,

$$Y_{d\tilde{f}}(\tilde{\omega}) = Y_{d\tilde{f}} \lrcorner\, d\tilde{\omega} + d(Y_{d\tilde{f}} \lrcorner\, \tilde{\omega}) = -d\tilde{f} + d\tilde{f} = 0.$$

This shows that $X \in \mathfrak{li}(\mathcal{Y})$ if and only if

$$X = Y_{d\tilde{f}} - \mu Y_{\tilde{\omega}} \quad \text{and} \quad Y_{d\tilde{f}} \lrcorner\, \tilde{\omega} = \tilde{f}.$$

The last condition can be rewritten in a more convenient form:

$$Y_{d\tilde{f}} \lrcorner\, \tilde{\omega} = -Y_{d\tilde{f}} \lrcorner\, (Y_{\tilde{\omega}} \lrcorner\, d\tilde{\omega}) = -Y_{\tilde{\omega}} \lrcorner\, d\tilde{f} = -Y_{\tilde{\omega}}(\tilde{f}).$$

Summing up, one has the following

THEOREM. *A field $X \in D(\mathcal{Y})$ is an intrinsic symmetry of the equation $\mathcal{Y}$ if and only if it is of the form $X = \mu Y_{\tilde{\omega}} + Y_{d\tilde{f}}$, where $\mu \in C^\infty(\mathcal{Y})$ and $Y_{\tilde{\omega}}(\tilde{f}) = -\tilde{f}$. One also has $X(\tilde{\omega}) = \mu\tilde{\omega}$.*

Note that Y_{dh}, $h \in C^\infty(\mathcal{Y})$, is a Hamiltonian vector field with the Hamiltonian h with respect to the symplectic structure on $\mathcal{Y}$ defined by the form $d\tilde{\omega}$. Hence, a Hamiltonian field on $\mathcal{Y}$ is an intrinsic symmetry of the equation if and only if it corresponds to a Hamiltonian h such that $Y_{\tilde{\omega}}(h) = -h$.

2.5.9. Now we describe the algebra $\mathfrak{li}_0(\mathcal{Y})$. To do this, recall that the Lie fields on $J^1(\mathbb{1}_M)$ coincide with the contact fields with respect to the Cartan distribution on $J^1(\mathbb{1}_M)$. These contact fields are described as follows (see, for instance, [**85**]). Let the field X_1 be a (unique) solution of the system $X_1 \lrcorner\, \omega = 1$, $X_1 \lrcorner\, d\omega = 0$. Then any contact field is uniquely determined by its generating function $f = X \lrcorner\, \omega$ satisfying the equation

$$X \lrcorner\, d\omega + df = X_1(f)\omega.$$

It should be also noted that any function f is a generating function for some vector field denoted by X_f.

The following result (in slightly different terms) was proved in [**85**].

THEOREM. *A field X_f lies in $\mathfrak{li}_0(\mathcal{Y})$ if and only if $Y_{\tilde{\omega}}(\tilde{f}) = \tilde{f}$, where $\tilde{f} = f|_{\mathcal{Y}}$.*

Comparing the last result with Theorem 2.5.8, we obtain

COROLLARY. *Let $Y = \mu Y_{\tilde{\omega}} + Y_{d\tilde{f}} \in \mathfrak{li}(\mathcal{Y})$, $X_f \in \mathfrak{li}_0(\mathcal{Y})$. Then $X_f|_{\mathcal{Y}} = Y$ if and only if $f|_{\mathcal{Y}} = \tilde{f}$ and $X_1(f)|_{\mathcal{Y}} = \mu$.*

Thus, under the assumptions made above, the mapping $\mathfrak{li}_0(\mathcal{Y}) \to \mathfrak{li}(\mathcal{Y})$ is an epimorphism, i.e., the equation $\mathcal{Y}$ is common but not rigid.

Note in conclusion that "ordinary differential equations" (the case $k = 1$), as a rule, are not common (and not rigid).

2.5.10. The obtained results allow us to get an overview of the general classification problem for partial differential equations. It is natural to call two equations *equivalent* if there exists a diffeomorphism $f\colon \mathcal{Y} \to \mathcal{Y}'$ transforming $\mathcal{C}(\mathcal{Y})$ into $\mathcal{C}(\mathcal{Y}')$. In a similar manner, one can define local equivalence of two equations in the vicinity of given points. Such a formulation is a natural generalization to arbitrary equations of the classification problem posed by Sophus Lie for first-order equations in one dependent variable and for equations of any order in one independent variable (i.e., for "ordinary differential equations"). In these cases the solution is well known: any two equations with the same n, m, and k are locally equivalent at nonsingular points. But results of this kind are exceptions rather than a rule. The reasons are as follows.

If two equations $\mathcal{Y}$ and $\mathcal{Y}'$ are rigid, then a transformation f realizing their equivalence (even locally) preserves the fibers of the projection $\pi_{\mathcal{Y}}\colon \mathcal{Y} \to \pi_{k,k-1}(\mathcal{Y})$, $\mathcal{Y} \subset J^k(E,n)$. But the fibers of the projection $\pi_{k,k-1}$, in the cases $k > 1$, $m > 1$ and $k > 2$, $m > 0$, possess a natural affine structure. Projections $\pi_{1,0}$, $m > 1$, and $\pi_{2,1}$, $m = 1$, possess a natural projective structure. Hence, classification problem posed above includes affine (projective) classification of smooth submanifolds in an affine (projective) space.

The last problem, in principle, admits a solution in terms of scalar differential invariants (see [**2**, Chapter 7] and [**131**]). According to the general theory, the corresponding equivalence classes are labeled by a number of functions (invariants), which are subject to some differential constraints. The number of independent invariants increases very rapidly with m, n, and k. Moreover, their analytic expressions are so complicated that it becomes hardly possible to manage them in practice, even for small values of m, n, and k.

These arguments show that the complexity of the classification problem depends on the rigidity of the class of equations in question. Classification results depending on a finite-dimensional parameter may occur only for "very" nonrigid equations. In its turn, "very nonrigid" means, as a rule, "sufficiently overdetermined". Some infrequent situations when this is not the case, are described in Subsection 2.5.5.

It is worth mentioning also that the method used above to get the rigidity results (Theorems 2.4.5 and 2.5.7) can be also applied to the classification problem in cases of particular interest.

2.6. Concluding remarks

Due to works by C. Ehresmann, jet manifolds became part of mathematical fundamentals. After that, it seemed that the geometric theory of scalar first-order differential equations constructed by Sophus Lie could be easily extended to higher-order equations in a straightforward way. The result was described in this chapter. The main conclusion is that contact geometry of finite order $k > 1$ does not allow us to obtain a satisfactory geometric theory for differential equations of higher orders, and this drastically differs from the case $k = 1$. This fact makes it inevitable to pass to infinite-order jets, which will be done in subsequent chapters.

In this context, it is useful to mention that many attempts to construct "higher" analogs of differential calculus based on finite-order jets have failed. An instructive example is the problem of constructing Hamiltonian formalism for variational problems where functionals depend on higher-order derivatives (see, for example, [**34**]). On the other hand, some aspects of the finite jet geometry, interpreted in the context of infinite jets, may happen to be useful. For example, this approach helped the author to find the basics of the theory of multivalued solutions for general (nonlinear) partial differential equations (see [**123**, **82**]).

The facts exposed in this chapter and related to the structure of higher contact transformations were mainly known to Lie [**81**] and Bäcklund [**11**]. What is new is the description of the structure of integral manifold of the Cartan distribution and the "rigidness theory" as its consequence (Theorem 2.5.7 on p. 91). This approach was first described in a short note [**128**]. Later, its details and generalizations appeared in [**142**] and [**71**].

CHAPTER 3

Geometry of Infinitely Prolonged Differential Equations and Higher Symmetries

The passage to infinite jets and infinitely prolonged differential equations accomplished in this chapter is prompted not only by the results of the preceding one, but by a number of different arguments not to be neglected. One of them is the following natural parallel to algebraic geometry.

When constructing a general image of a system of algebraic equations, one has to deal with all *algebraic consequences* of it, which form an ideal J of the corresponding polynomial algebra A. Then the required geometric image is, as is well known, the spectrum of the quotient algebra A/J. This scheme, repeated literally in the context of systems of differential equations, was presented in Chapter 2 according to a common point of view on the geometry of partial differential equations. However, this approach completely ignores the fact that in the case of differential equations one must take into account all *differential consequences* of the original system. It becomes clear from this simple observation what are incurable shortcomings of the standard approach and how it should be modified: instead of the ideal composed of all algebraic consequences, one has to consider the ideal composed of all differential consequences. The implementation of this idea leads directly to infinite jets and infinitely prolonged equations.

Infinite jet spaces, or, more generally, infinitely prolonged differential equations are a kind of generally infinite-dimensional manifolds. Thus, the paradigm of contemporary mathematics forces one to supply these objects with something like a "good" norm, topology, etc. It would be however suicidal for our purposes to follow these temptations. At this point, it is worth recalling that numerous attempts to develop a reasonably rich calculus over function spaces (Hilbert, Banach, Sobolev, etc.) failed invariably. What really prevents us from making this error is the purely algebraic nature of calculus as was discussed in Chapter 1. As we show in this chapter, it suffices to pass to a filtered version of algebraically understood calculus and obtain all that is really needed.

The "normless" and "topologyless" approach developed below, starting from this chapter, may create a feeling that this is a *formal* theory. A very rough and informal explanation why it is not so is that Secondary Calculus is no more formal than elementary particles which are described very naturally in its terms. See the collection [**50**] for first ideas about these matters.

3.1. Geometry of infinitely prolonged equations, and related differential calculus

In this section, we present basic facts about infinitely prolonged equations and construct an adequate differential calculus over such objects.

3.1.1. Below we assume that $\mathcal{Y} \subset J^k(E,n)$ or $\mathcal{Y} \subset J^k(\pi)$. Let us set $\mathcal{Y}_s = \pi_{\infty,s}(\mathcal{Y}^{(\infty)})$, $s \le \infty$ (in particular, $\mathcal{Y}_\infty = \mathcal{Y}^{(\infty)}$), and for simplicity take $\mathcal{Y}_s$ to be a closed submanifold (possibly, with singularities) in $J^s(E,n)$ or $J^s(\pi)$. If $\mathcal{Y} \subset J^k(\pi)$, we set in addition $\mathcal{Y}_{-1} = M = \pi_\infty(\mathcal{Y}^{(\infty)})$. We also use the notation $\mathcal{F}_s(\mathcal{Y},\xi) = \mathcal{F}_s(E,\xi)|_{\mathcal{Y}_s}$ (or $\mathcal{F}_s(\pi,\xi)|_{\mathcal{Y}_s}$). Obviously, $\pi_{s,s-1}(\mathcal{Y}_s) = \mathcal{Y}_{s-1}$ and $\mathcal{Y}_\infty$ is the inverse limit of the chain of mappings

$$\cdots \leftarrow \mathcal{Y}_{s-1} \xleftarrow{\pi_{s,s-1}} \mathcal{Y}_s \leftarrow \cdots .$$

The direct limit of the chain of mappings

$$\cdots \to \mathcal{F}_{s-1}(\mathcal{Y},\xi) \xrightarrow{\pi^*_{s,s-1}} \mathcal{F}_s(\mathcal{Y},\xi) \to \cdots$$

is denoted by $\mathcal{F}(\mathcal{Y},\xi)$. We identify $\mathcal{F}_s(\mathcal{Y},\xi)$ with its image in $\mathcal{F}(\mathcal{Y},\xi)$. In a similar way, we set $\mathcal{F}_s(\mathcal{Y}) = \mathcal{F}_s(E,n)|_{\mathcal{Y}_s}$, $\mathcal{F}(\mathcal{Y}) = \operatorname{proj\,lim}_{s\to\infty} \mathcal{F}_s(\mathcal{Y})$ and consider the algebra $\mathcal{F}(\mathcal{Y})$ to be filtered by the subalgebras $\mathcal{F}_s(\mathcal{Y})$. Let us also set $\mathcal{F}_{-1}(\mathcal{Y}) = C^\infty(M)$ when $\mathcal{Y} \subset J^k(\pi)$.

All the constructions below make sense provided the equation $\mathcal{Y}$ under consideration is *quasi-integrable.* This means that for any point $\theta \in \mathcal{Y}_\infty$ there exist n-dimensional submanifolds $V \subset \mathcal{Y}_\infty$ and $L \subset E$ (resp., $s \in \Gamma_{\mathrm{loc}}(\pi)$) such that $L^{(\infty)}$ (resp., $M_s^{(\infty)}$) is tangent to $\pi_{\infty,s}(V)$ at the point $\pi_{\infty,s}(\theta)$ for all $s < \infty$. Formally integrable equations are obviously quasi-integrable. The manifold V in the definition of quasi-integrability will be called *θ-integral.*

In what follows, $\mathcal{Y}_\infty$ is equipped with the Zariski topology determined by the algebra $\mathcal{F}(\mathcal{Y})$, i.e., $\mathcal{Y}_\infty$ is considered to be a subset in $\operatorname{Spec}\mathcal{F}(\mathcal{Y})$.

3.1.2. Let us define the *Cartan distribution* on $\mathcal{Y}_\infty$. As usual, a tangent vector at the point $\theta \in \mathcal{Y}_\infty$ is understood as an $\mathbb{R}$-linear operator $\xi\colon \mathcal{F}(\mathcal{Y}) \to \mathbb{R}$ satisfying the identity

$$\xi(fg) = f(\theta)\xi(g) + \xi(f)g(\theta), \quad f, g \in \mathcal{F}(\mathcal{Y}).$$

Obviously, the operator $\xi_i = \xi|_{\mathcal{F}_i(\mathcal{Y})}$ determines a tangent vector to the manifold $\mathcal{Y}_i$ and $d\pi_{i,i-1}(\xi_i) = \xi_{i-1}$. Hence, ξ can be regarded geometrically as the inverse limit of the system $\xi_i \in T_{\theta_i}(\mathcal{Y}_i)$, $d\pi_{i,i-1}(\xi_i) = \xi_{i-1}$, where $\theta_i = \pi_{\infty,i}(\theta)$. The linear space formed by all vectors tangent to $\mathcal{Y}_\infty$ at the point θ will be denoted by $T_\theta(\mathcal{Y})$. From the above-said it follows that $T_\theta(\mathcal{Y})$ is the direct limit of the following sequence of linear spaces:

$$\cdots \xleftarrow{d\pi_{i,i-1}} T_{\theta_i}(\mathcal{Y}_i) \xleftarrow{d\pi_{i+1,i}} T_{\theta_{i+1}}(\mathcal{Y}_{i+1}) \xleftarrow{d\pi_{i+2,i+1}} \cdots .$$

DEFINITION. The direct limit of the sequence

$$\cdots \xleftarrow{d\pi_{i,i-1}} \mathcal{C}_{\theta_i}(\mathcal{Y}_i) \xleftarrow{d\pi_{i+1,i}} \mathcal{C}_{\theta_{i+1}}(\mathcal{Y}_{i+1}) \xleftarrow{d\pi_{i+2,i+1}} \cdots$$

of Cartan planes is called the *Cartan plane* $\mathcal{C}_\theta = \mathcal{C}_\theta(\mathcal{Y}_\infty)$ on $\mathcal{Y}_\infty$ at the point $\theta \in \mathcal{Y}_\infty$.

Obviously, $\mathcal{C}_\theta(\mathcal{Y}_\infty)$ is a linear subspace of $T_\theta(\mathcal{Y}_\infty)$.

THEOREM. *If $\mathcal{Y}$ is a quasi-integrable equation, then $\mathcal{C}_\theta(\mathcal{Y}_\infty) = T_\theta(V)$ for some θ-integral manifold V. In particular, all θ-integral manifolds are mutually tangent, $\dim \mathcal{C}_\theta(\mathcal{Y}_\infty) = n$, while the mappings $d\pi_{\infty,s}|_{\mathcal{C}_\theta(\mathcal{Y}_\infty)}$ are always monomorphic.*

PROOF. From Proposition 2.1.4 (1) it follows that

$$d\pi_{\infty,s}(\mathcal{C}_\theta(\mathcal{Y}_\infty)) = d\pi_{s+1,s}\big(d\pi_{\infty,s+1}(\mathcal{C}_\theta(\mathcal{Y}_\infty))\big) = d\pi_{s+1,s}(\mathcal{C}_{\theta_{s+1}}) = L_{\theta_{s+1}}.$$

On the other hand, if the manifold V is tangent to $L^{(\infty)}$ at the point θ, then one has $d\pi_{\infty,s}(T_\theta(V)) = d\pi_{\infty,s}(T_\theta(L^{(\infty)})) = L_{\theta_{s+1}}$, i.e., $L_{\theta_{s+1}} \subset d\pi_{\infty,s}(\mathcal{Y}_\infty) \cap \mathcal{C}_{\theta_s}(\mathcal{Y}_s)$. Therefore, the direct limit of isomorphisms

$$\cdots \leftarrow L_{\theta_i} \xleftarrow{d\pi_{i+1,i}} L_{\theta_{i+1}} \leftarrow \cdots$$

coincides with $T_\theta(V)$ and lies in $\mathcal{C}_\theta(\mathcal{Y}_\infty)$. But from the embedding $d\pi_{\infty,i}(\mathcal{C}_\theta(\mathcal{Y}_\infty)) \subset L_{\theta_{i+1}}$ it results that $T_\theta(V)$ lies in $\mathcal{C}_\theta(\mathcal{Y}_\infty)$. □

COROLLARY. *If $\theta = [L]_x^\infty$ (resp., $\theta = [s]_y^\infty$, $s \in \Gamma_{\mathrm{loc}}(\pi)$), then $\mathcal{C}_\theta(\mathcal{Y}_\infty) = T_\theta(L^{(\infty)})$ (resp., $\mathcal{C}_\theta(\mathcal{Y}_\infty) = T_\theta(M_s^{(\infty)})$).*

Let $\mathcal{Y} \subset J^k(\pi)$, so that $\mathcal{Y}_\infty \subset J^\infty(\pi)$. Then from the last result it follows that $d\pi_\infty$ maps $\mathcal{C}_\theta(\mathcal{Y}_\infty)$ onto $T_x(M)$, $x = \pi_\infty(\theta)$, isomorphically. In other words, in this case $\theta \mapsto \mathcal{C}_\theta$ is a family of horizontal subspaces with respect to the projection $\pi_\infty \colon \mathcal{Y}_\infty \to M$, hence it determines a connection in this bundle (when it is really a bundle). It will be shown below that this connection is integrable.

Note that when $L \subset E$ (resp., $s \in \Gamma_{\mathrm{loc}}(\pi)$) is a solution of the equation $\mathcal{Y}$, then $L^{(\infty)}$ (resp., $M_s^{(\infty)}$) is an integral manifold of the Cartan distribution on $\mathcal{Y}_\infty$. Conversely, if V is an integral manifold, then $V_k = \pi_{\infty,k}(V) \subset \mathcal{Y}_k \subset \mathcal{Y}^{(k)}$ is an integral manifold of the Cartan distribution on $J^k(E,n)$ (resp., on $J^k(\pi)$) and therefore is a solution of $\mathcal{Y}$. Since the type of V_k equals n, V_k is (locally) of the form $L^{(k)}$, i.e., $V = L^{(\infty)}$ (resp., $V = M_s^{(\infty)}$ for some $s \in \Gamma_{\mathrm{loc}}(\pi)$).

The above argiments show, in particular, that n-dimensional integral manifolds of the Cartan distribution on $J^\infty(E,n)$ (resp., on $J^\infty(\pi)$) are locally of the form $L^{(\infty)}$ (resp., $M_s^{(\infty)}$). Since the Cartan distribution on $J^\infty(E,n)$ (or on $J^\infty(\pi)$) is n-dimensional, these manifolds and only they are locally maximal (see Subsection 2.1.7). As in Subsection 2.1.7, we call them *R-manifolds.*

In the sequel, we shall also need the dual definition of the Cartan distribution. This definition will be given below.

3.1.3. We start with a general algebraic scheme of the theory of filtered differential operators (DO). Let A be a K-algebra filtered by K-algebras A_i, $A_i \subset A_{i+1}$, $\bigcup_i A_i = A$. The structure of a filtered A-module in the A-module Q is determined by fixing a system of A_i-submodules $Q_i \subset Q$ satisfying $Q_i \subset Q_{i+1}$, $\bigcup_i Q_i = Q$. We use the notation $Q = \{Q_i\}$.

DEFINITION (1). Let $\Delta \in \mathrm{Diff}(P,Q)$, and let $P = \{P_i\}$, $Q = \{Q_i\}$ be filtered A-modules. Then Δ is called a *filtered operator* if there exists $s \in \mathbb{Z}$ such that $\Delta(P_i) \subset Q_{i+s}$ for all i.

REMARK. If Δ is a filtered DO and $s \geq 0$, then $\Delta|_{P_i} \colon P_i \to Q_{i+s}$ is a DO over the embedding $A_i \hookrightarrow A_{i+s}$.

In particular, filtered DO of order zero are called *filtered homomorphisms* in the sequel.

DEFINITION (2). A filtered A-module $Q = \{Q_i\}$ is said to be *geometric* if Q_i is a geometric A_i-module for any i.

The category whose objects are filtered geometric A-modules and morphisms are filtered homomorphisms of these modules is called the *FG-category* over A. Differential calculus in the FG-category over A is constructed along the same lines as in Section 1.1. More precisely, instead of "simple" basic functors Diff_s and D one should consider the functors Diff_s^Φ and D^Φ "consisting" of filtered DO and filtered derivations, respectively. Based on these functors, one can generate more "complicated" ones, such as $D_2^\Phi = D^\Phi$ $(D^\Phi \subset (\mathrm{Diff}_s^\Phi)^+)$. It can be checked directly that all constructions of Section 1.1 remain valid in this situation. Let us emphasize that the geometric nature of the modules under consideration did not affect the construction of the general theory so far. As in Chapter 1, this is important when dealing with representing objects only, which are described in the next subsection.

3.1.4. Let $\Delta \in \mathrm{Diff}_k^\Phi(P,Q)$ and $\Delta_i = \Delta|_{P_i} : P_i \to Q_{i+s}$, $s \geq 0$. Then Δ_i is a k-th order DO acting from the A_i-module P_i to A_{i+s}-module Q_{i+s} over the embedding $A_i \hookrightarrow A_{i+s}$. In other words, $\Delta_i \in \mathrm{Diff}_k(P_i, Q_{i+s})$, where Q_{i+s} is understood as an A_i-module due to the embedding $A_i \hookrightarrow A_{i+s}$. Moreover, if P_i and Q_{i+s} are geometric modules, then one has the diagram

$$\begin{array}{ccc} P_i & \xrightarrow{\Delta_i} & Q_{i+s} \\ & {\scriptstyle j_k}\searrow \quad \nearrow{\scriptstyle f^{\Delta_i}} & \\ & [\mathcal{J}^k_{A_i}(P_i)]_g & \end{array}$$

which is commutative. Here f^{Δ_i} is a homomorphism of A_i-modules, while S_g denotes geometrization of the module S. Moreover, the diagram

$$\begin{array}{ccccc} P_i & \xrightarrow{j_k} & [\mathcal{J}^k(P_i)]_g & \xrightarrow{f^{\Delta_i}} & Q_{i+s} \\ \downarrow & & \downarrow{\scriptstyle \alpha_{i,l}} & & \downarrow \\ P_l & \xrightarrow{j_k} & [\mathcal{J}^k(P_l)]_g & \xrightarrow{f^{\Delta_l}} & Q_{i+l} \end{array}$$

is also commutative, where $i \leq l$ and $\alpha_{i,l}$ is a natural mapping of jets corresponding to the extension $P_i \subset P_l$ of modules over the extensions $A_i \subset A_l$ of algebras. Denote by $[\mathcal{J}^k(P)]_{g\Phi}$ the direct limit of the sequence

$$\cdots \to [\mathcal{J}^k(P_{i-1})]_g \xrightarrow{\alpha_{i-1,i}} [\mathcal{J}^k(P_i)]_g \xrightarrow{\alpha_{i,i+1}} [\mathcal{J}^k(P_{i+1})]_g \to \cdots$$

and consider the module $[\mathcal{J}^k(P)]_{g\Phi}$ to be filtered by the images of natural mappings $[\mathcal{J}^k(P_i)]_g \to [\mathcal{J}^k(P)]_{g\Phi}$. Then we have the following result.

PROPOSITION. *The module* $[\mathcal{J}^k(P)]_{g\Phi}$ *is the representing object for the functor* $\mathrm{Diff}_k(P, \bullet)$ *in the FG-category.*

PROOF. Let us denote the direct limits of the mappings $j_k : P_i \to [\mathcal{J}^k(P_i)]_g$ and $f^{\Delta_i} : [\mathcal{J}^k(P_i)]_g \to Q_{i+s}$ by j_k and f^Δ, respectively. Then we have the following commutative diagram:

$$\begin{array}{ccc} P & \xrightarrow{\Delta} & Q \\ & {\scriptstyle j_k}\searrow \quad \nearrow{\scriptstyle f^{\Delta}} & \\ & [\mathcal{J}^k(P)]_{g\Phi} & \end{array}$$

where $f^{\Delta} \in \mathrm{Hom}_A^{\Phi}(\mathcal{J}^k(P), Q)$, i.e., f^{Δ} is a homomorphism of filtered modules. Obviously, the correspondence $\Delta \mapsto f^{\Delta}$ determines an isomorphism $\mathrm{Diff}_k^{\Phi}(P, Q) \to \mathrm{Hom}_A^{\Phi}([\mathcal{J}^k(P)]_{g\Phi}, Q)$. □

3.1.5. The proposition proved above is a particular case of the following general situation. Let $\mathfrak{F}$ be a functor of differential calculus (e.g., Diff_s, D_l, $D_l(\mathrm{Diff}_s^+)$, etc.) and Φ its representing object in the geometric category over A_i. Then one has natural mappings

$$\alpha_{i,j} \colon \Phi_i \to \Phi_j, \quad i \leq j,$$

with $\alpha_{j,k} \circ \alpha_{i,j} = \alpha_{i,k}$, $i \leq j \leq k$. If Φ is the direct limit of the sequence

$$\cdots \to \Phi_{i-1} \xrightarrow{\alpha_{i-1,i}} \Phi_i \xrightarrow{\alpha_{i,i+1}} \Phi_{i+1} \to \cdots$$

filtered by the images of canonical mappings $\Phi_i \to \Phi$, then we have the following

PROPOSITION. *Φ is the representing object for the functor $\mathfrak{F}$ in the FG-category, i.e., the functors $\mathfrak{F}$ and $\mathrm{Hom}_A^{\Phi}(\Phi, \bullet)$ are isomorphic.*

PROOF. The proof is a word-for-word repetition of that of the previous proposition. □

COROLLARY. *In the FG-category, the Spencer and the de Rham complexes are defined. They are the direct limits of the corresponding complexes in the categories of geometric modules over A_i.*

3.1.6. In the sequel, we shall work in the FG-category over the filtered algebra $\mathcal{F}(\mathcal{Y})$. Typical filtered modules which we shall deal with are $\mathcal{F}(\mathcal{Y}, \xi)$. To avoid cumbersome notation, the functors of differential calculus and their representing objects in the FG-category over $\mathcal{F}(\mathcal{Y}) = \{\mathcal{F}_s(\mathcal{Y})\}$ will be denoted without additional indices (e.g., Diff_s instead of Diff_s^{Φ} or $\mathcal{J}^s(P)$ instead of $[\mathcal{J}^s(P)]_{g\Phi}$). It will not cause ambiguity since from this moment all the DO over $\mathcal{F}(\mathcal{Y})$ will be assumed to be filtered and all $\mathcal{F}(\mathcal{Y})$-modules will be geometric. The above-said concerns also FG-categories over filtered algebras $\mathcal{F}(E, n)$ and $\mathcal{F}(\pi)$.

In [**71**, Chapter 1], it was shown that all usual operations concerning differential forms and vector fields (such as d or Lie derivative) are of a pure algebraic-categorical nature. Therefore, all these operations together with basic formulas relating them are carried over to the filtered theory. We use this fact without mentioning. We emphasize the role of the language developed in Section 1.1, which allows us to use the above remark without loss of generality and accuracy. The standard language of differential geometry becomes cumbersome and obscure when one works with "manifolds" $\mathcal{Y}_\infty$. In the situation with singularities it also loses its validity.

3.1.7. Let $\mathfrak{F}$ be a functor of differential calculus in the FG-category over $\mathcal{F}(\mathcal{Y}_\infty)$, and $\Phi = \Phi(\mathcal{Y}_\infty)$ its representing object. We set

$$\mathcal{C}\Phi = \mathcal{C}\Phi(\mathcal{Y}_\infty) = \{\varphi \in \Phi \mid [j(L)^*(\varphi)](x) = 0, \ \forall [L]_x^\infty \in \mathcal{Y}_\infty\} \tag{3.1}$$

for $\mathcal{Y} \subset J^k(E, n)$. Changing $L \subset E$ by $s \in \Gamma_{\mathrm{loc}}(\pi)$, we get the definition of $\mathcal{C}\Phi$ for $\mathcal{Y} \subset J^k(\pi)$.

Evidently, $\mathcal{C}\Phi$ is a submodule in Φ filtered by the submodules $\mathcal{C}\Phi_i = \mathcal{C}\Phi \cap \Phi_i$, where Φ_i is the representing object for $\mathfrak{F}$ in the geometric category over $C^\infty(\mathcal{Y}_i) = \mathcal{F}_i(\mathcal{Y})$. The module $\mathcal{C}\Phi$ is called the *Cartan submodule* in Φ.

Let $\Delta\colon \Phi_1 \to \Phi_2$ be a natural differential operator relating two representing objects in the theory under consideration. Then from (3.1) it follows that $\Delta(\mathcal{C}\Phi_1) \subset \mathcal{C}\Phi_2$, i.e., Cartan submodules are invariant with respect to natural operators.

Consider the quotient module $\bar{\Phi} = \Phi/\mathcal{C}\Phi$. From the above it follows that there exists the quotient operator

$$\bar{\Delta}\colon \bar{\Phi}_1 \to \bar{\Phi}_2, \quad \bar{\Delta}(\varphi \bmod \mathcal{C}\Phi_1) = \Delta(\varphi) \bmod \mathcal{C}\Phi_2, \ \varphi \in \Phi_1.$$

This fact will be used below in the cases where $\Delta = d$, $\Delta = j_k$, or Δ is the Spencer operator. In particular, an important role is played by the de Rham $\bar{d}$-complex $\dots \xrightarrow{\bar{d}} \bar{\Lambda}^i \xrightarrow{\bar{d}} \bar{\Lambda}^{i+1} \xrightarrow{\bar{d}} \dots$, where $\bar{\Lambda}^i = \bar{\Lambda}^i(\mathcal{Y}_\infty)$. This complex is also called the *horizontal de Rham complex* over $\mathcal{Y}_\infty$.

The operation $\mathcal{C}$ allows one to consider the following subtheory in differential calculus over $\mathcal{Y}_\infty$. For example, let $\mathfrak{F}$ be a unary functor of differential calculus, and Φ its representing object. Set $\mathcal{C}\mathfrak{F} = \operatorname{Ann}\mathcal{C}\Phi$. To be more precise, if one identifies $\mathfrak{F}(P)$ with $\operatorname{Hom}_{\mathcal{F}(\mathcal{Y})}(\Phi, P)$, then

$$\mathcal{C}\mathfrak{F}(P) = \{h \in \operatorname{Hom}_{\mathcal{F}(\mathcal{Y})}(\Phi, P) \mid h(\mathcal{C}\Phi) = 0\} = \operatorname{Hom}_{\mathcal{F}(\mathcal{Y})}(\bar{\Phi}, P).$$

The above subtheory studies the functors $\mathcal{C}\mathfrak{F}$ instead of $\mathfrak{F}$. In this case the representing object is $\bar{\Phi}$.

For instance, let $\mathfrak{F}\colon P \Rightarrow \operatorname{Diff}_s(Q, P)$, where Q is fixed and P is a "variable" $\mathcal{F}(\mathcal{Y})$-module. Then elements of $\mathcal{C}\mathfrak{F}(P) = \mathcal{C}\operatorname{Diff}_s(Q, P)$ are called *$\mathcal{C}$-differential operators*. These operators are uniquely determined by their characteristic property: they admit restrictions to submanifolds of the form $L^{(\infty)}$ (or $M_s^{(\infty)}$). The same property characterizes all other $\mathcal{C}$-objects (i.e., elements of modules $\mathcal{C}\mathfrak{F}(P)$). Indeed, the construction

$$(\Delta|_{L^{(\infty)}})(q|_{L^{(\infty)}}) = \Delta(q)|_{L^{(\infty)}}, \quad q \in Q, \ \Delta \in \mathcal{C}\operatorname{Diff}_s(Q, P),$$

is well defined. Conversely, any operator on $J^\infty(E, n)$ (or on $J^\infty(\pi)$) possessing this property is $\mathcal{C}$-differential. The operators of universal linearization (see Subsection 2.3.2) and derivations of the form $\overline{Y}$ (see Subsection 2.3.10) are $\mathcal{C}$-differential operators.

In a dual way, one can consider a "quotient theory" in differential calculus over $\mathcal{Y}_\infty$. In this theory, $\mathcal{C}\Phi$ is the representing object for the functor $\bar{\mathfrak{F}}$. Thus for any module P one has $\bar{\mathfrak{F}}(P) = \operatorname{Hom}_{\mathcal{F}(\mathcal{Y})}(\mathcal{C}\Phi, P)$. If Φ is a projective module, then obviously $\bar{\mathfrak{F}}(P) = \mathfrak{F}(P)/\mathcal{C}\mathfrak{F}(P)$. It will be shown below that if P is a projective module, then the modules $\bar{\Lambda}^i$ and $\bar{\mathcal{J}}^s(P)$ are projective as well. Therefore, for example, an element $\square \in \overline{\operatorname{Diff}}_s(Q, P)$ is a class of differential operators modulo $\mathcal{C}\operatorname{Diff}_s(Q, P)$.

3.2. Horizontalization operation, the structure of Cartan submodules and of $\mathcal{C}$-differential operators

In the case where $\mathcal{Y} \subset J^k(\pi)$, i.e., $\mathcal{Y}_\infty \subset J^\infty(\pi)$, a *horizontalization operation* $\rceil = \rceil_\Phi\colon \Phi \to \Phi_0$ arises on representing objects $\Phi = \Phi(\pi)$, where $\Phi_0 = \Phi_0(M) \otimes_{\mathcal{F}_{-1}(\pi)} \mathcal{F}(\pi)$ is the "horizontal part" of the object Φ and $\mathcal{F}_{-1}(\pi) = C^\infty(N)$. The operation $\rceil_\Phi$ can be efficiently described for all Φ we are interested in. This operation gives the representations

$$\mathcal{C}\Phi = \operatorname{im}(1 - \rceil_\Phi) = \ker \rceil_\Phi \quad \text{and} \quad \bar{\Phi} = \Phi_0 = \operatorname{im} \rceil_\Phi = \ker(1 - \rceil_\Phi).$$

These relations allow one to establish a number of useful facts concerning the modules $\mathcal{C}\Phi$ and to show, in particular, that $\mathcal{C}\Lambda^1$ is the annihilator of the Cartan distribution on $\mathcal{Y}_\infty$.

3.2.1. To define the horizontalization operation we need the notion of the *degree*, $\deg\mathfrak{F}$, *of a differential calculus functor* $\mathfrak{F}$. Putting aside the general case, we set

$$\deg \mathrm{Diff}_q = q, \quad \deg(D_p\mathrm{Diff}_q^+) = \deg(D_p^\bullet \mathrm{Diff}_q^+) = q+1,$$
$$\deg(\mathfrak{F}_1 \circ \cdots \circ \mathfrak{F}_k) = \max_i\{p_i\}, \quad \mathfrak{F}_i = \mathrm{Diff}_{p_i} \text{ or } \widetilde{\mathrm{Diff}}_{p_i}.$$

We shall use the following property of the degree (which can be taken for its definition in the category of smooth manifolds): for any pair of manifolds $M_1 \subset M$ the value of the module $\Phi(M_1)$ at a point $x \in M_1$ is uniquely determined by $[M_1]_x^k$ if $k \geq \deg\mathfrak{F}$ (here $\Phi(N)$ denotes the representing object of $\mathfrak{F}$ in the geometric category over N).

This is obviously valid in the particular cases considered above.

3.2.2. Now let $\pi\colon E_\pi \to M$ be a submersion, $\mathcal{Y}_\infty \subset J^\infty(\pi)$, $\mathfrak{F}$ a functor of differential calculus, and $\Phi = \Phi(\mathcal{Y}_\infty)$ the representing object in the FG-category over $\mathcal{F}(\mathcal{Y})$. Set

$$\Phi_\Gamma = \Phi_\Gamma(\mathcal{Y}_\infty) = \Phi(M) \otimes_{\mathcal{F}_{-1}(\pi)} \mathcal{F}(\mathcal{Y}),$$

where $\Phi(M)$ is the representing object for $\mathfrak{F}$ in the geometric category over M. Evidently, Φ_0 is a submodule in Φ generated by the image of the monomorphism $\pi_\infty^*\colon \Phi(M) \to \Phi(\mathcal{Y}_\infty)$. The elements of the module Φ_Γ are said to be *horizontal*. Let $\deg\mathfrak{F} < \infty$. Then the operation $\rceil = \rceil_\Phi\colon \Phi \to \Phi_\Gamma$ is defined by

$$\rceil(\varphi)(\theta) = (\pi_\infty|_{\mathcal{Y}_\infty})^*[j(f)^*(\varphi)](\theta), \quad \theta = [f]_x^\infty,\ x = \pi_\infty(\theta). \tag{3.2}$$

Due to the condition $\deg\mathfrak{F} < \infty$, this is a well-defined operation. Indeed, if $[f]_x^\infty = [g]_x^\infty$, $\varphi \in \Phi_l$, and $k \geq \max(l, \deg\mathfrak{F})$, then according to the property of the degree (see the previous subsection) one has

$$\big(j(f)^*(\varphi)\big)(\theta) = (\varphi|_{M_f^{(k)}})(\theta_k) = (\varphi|_{M_g^{(k)}})(\theta_k) = \big(j(g)^*(\varphi)\big)(\theta),$$

where $\theta_k = \pi_{\infty,k}(\theta)$.

Immediately from the definition one gets

PROPOSITION. *The horizontalization operation defined above possesses the following properties*:

1. $\rceil$ *is a homomorphism of* $\mathcal{F}(\mathcal{Y})$*-modules,*
2. $\rceil|_{\Phi_\Gamma} = \mathrm{id}$,
3. $\rceil^2 = \rceil$,
4. $\rceil$ *preserves natural multiplicative structures.*

Property (4) means in particular that

$$\rceil(\omega_1 \wedge \omega_2) = (\rceil\omega_1) \wedge (\rceil\omega_2), \quad \omega_i \in \Lambda^*.$$

Property (2) shows that $\rceil$ is an epimorphism, while (3) yields the direct sum decomposition

$$\Phi = \mathrm{im}\rceil_\Phi \oplus \mathrm{im}(\mathrm{id}_\Phi - \rceil_\Phi) = \Phi_\Gamma \oplus \mathrm{im}(\mathrm{id}_\Phi - \rceil_\Phi). \tag{3.3}$$

Note also that

$$\Phi_\Gamma = \operatorname{im}\rceil_\Phi = \ker(\operatorname{id}_\Phi - \rceil_\Phi), \quad \operatorname{im}(\operatorname{id}_\Phi - \rceil_\Phi) = \ker\rceil_\Phi. \tag{3.4}$$

THEOREM. *One has*

$$\mathcal{C}\Phi(\mathcal{Y}_\infty) = \ker\rceil_\Phi = \operatorname{im}(\operatorname{id}_\Phi - \rceil_\Phi)$$

for any quasi-integrable equation $\mathcal{Y} \subset J^k(\pi)$.

PROOF. The result is a consequence of the following three facts:

(a) $\operatorname{im}(\operatorname{id}_\Phi - \rceil_\Phi) \subset \mathcal{C}\Phi(\mathcal{Y}_\infty)$,
(b) direct sum decomposition (3.3), and
(c) $\Phi_\Gamma \cap \mathcal{C}\Phi = 0$.

To prove the last claim, we only need to note that if $\varphi \in \Phi_\Gamma$, then $\varphi(\theta) = (\varphi|_V)(\theta)$, where $V \subset \mathcal{Y}_\infty$ is the graph of a local section of the projection $\pi_\infty \colon \mathcal{Y}_\infty \to M$ passing through θ. Any θ-integral manifold can be taken for such a graph. □

Comparing this theorem with (3.3) we get

$$\Phi(\mathcal{Y}_\infty) = \Phi_\Gamma(\mathcal{Y}_\infty) \oplus \mathcal{C}\Phi(\mathcal{Y}_\infty). \tag{3.5}$$

Thus one has the decomposition

$$\varphi = \rceil\varphi + (\varphi - \rceil\varphi), \quad \rceil\varphi \in \Phi_\Gamma, \ \varphi - \rceil\varphi \in \mathcal{C}\Phi.$$

COROLLARY. *If* $\mathcal{Y} \subset J^k(\pi)$ *is a quasi-integrable equation, then*

$$\bar{\Phi}(\mathcal{Y}_\infty) = \Phi_\Gamma(\mathcal{Y}_\infty).$$

3.2.3. Let $\Delta \colon \Phi' \to \Phi''$ be a natural differential operator relating the representing objects of the functors $\mathfrak{F}'$ and $\mathfrak{F}''$, respectively (e.g., $\Delta = d$ or $\Delta = j_k$). If ξ' and ξ'' are fiber bundles over M such that $\Gamma(\xi') = \Phi'(M)$ and $\Gamma(\xi'') = \Phi''(M)$, then one has the following canonical isomorphisms on $J^\infty(\pi)$:

$$\mathcal{F}(\pi, \xi') = \Phi'_\Gamma(J^\infty(\pi)), \quad \mathcal{F}(\pi, \xi'') = \Phi''_\Gamma(J^\infty(\pi)).$$

Hence, any operator $\hat{\Delta} \colon \mathcal{F}(\pi, \xi') \to \mathcal{F}(\pi, \xi'')$ can be understood as an operator $\hat{\Delta} \colon \Phi'_\Gamma \to \Phi''_\Gamma$. On the other hand, the operator $\rceil \circ (\Delta|_{\Phi'_\Gamma}) \colon \Phi'_\Gamma \to \Phi''_\Gamma$ is defined (we use the same notation for natural operators acting over different algebras).

PROPOSITION. $\hat{\Delta} = \rceil \circ (\Delta|_{\Phi'_\Gamma})$.

PROOF. It suffices to check that the operator $\square = \rceil \circ (\Delta|_{\Phi'_\Gamma})$ satisfies the relation $\Delta \circ j(s)^* = j(s)^* \circ \square$ for any $s \in \Gamma_{\mathrm{loc}}(\pi)$ (see Subsection 2.1.2). Let $\varphi \in \Phi'_\Gamma$. Then according to the definition of $\rceil$ one has $j(s)^*(\rceil(\Delta(\varphi))) = j(s)^*(\Delta(\varphi))$. But $j(s)^*(\Delta(\varphi)) = \Delta(j(s)^*(\varphi))$ since Δ is a natural operator. □

Note that the identification of $\bar{\Phi}$ with Φ_Γ over $J^\infty(\pi)$ described in the previous subsection implies the identification of $\bar{\Delta} \colon \bar{\Phi}' \to \bar{\Phi}''$ with the operator $\rceil \circ (\Delta|_{\Phi'_\Gamma})$ and consequently with $\hat{\Delta} \colon \Phi'_\Gamma \to \Phi''_\Gamma$. Thus we obtain

COROLLARY. *Operators of the form* $\bar{\Delta}$ *are* $\mathcal{C}$*-differential.*

PROOF. Indeed, $\bar{\Delta}$ is a $\mathcal{C}$-differential operator on $J^\infty(\pi)$ or on $\mathcal{Y}_\infty \subset J^\infty(\pi)$, since $\hat{\Delta}$ is a $\mathcal{C}$-differential operator. But the property of an operator to be $\mathcal{C}$-differential is local. Hence, the result is valid in the general case. □

EXAMPLE. Let us describe the operator $\hat{d}$ in special coordinates. Due to multiplicative properties of the operation $\square \mapsto \hat{\square}$ (see Subsection 2.1.2), it suffices to compute $\hat{d}\colon \Lambda^0_\Gamma = \mathcal{F}(\pi) \to \Lambda^1_\Gamma$. But in a local coordinate system $x_1, \dots, x_n$ on M one has $d = \sum_i (\partial/\partial x_i) \otimes dx_i$. Therefore,

$$\hat{d} = \sum_i \widehat{\frac{\partial}{\partial x_i}} \otimes dx_i = \sum_i D_i \otimes dx_i,$$

and in special coordinates on $J^\infty(\pi)$ we obtain

$$\rceil(dx_i) = dx_i, \quad \rceil(dp^j_\sigma) = \hat{d}p^j_\sigma = \sum_i p^j_{\sigma i}\, dx_i.$$

These relations, together with the fourth statement of Proposition 3.2.2, allow one to compute easily $\rceil(\omega)$ for any $\omega \in \Lambda^*$ by using the identity $\rceil|_{\mathcal{F}(\pi)} = \mathrm{id}_{\mathcal{F}(\pi)}$.

3.2.4. The above definitions of the operations $\mathcal{C}$ and $\rceil$ are "extrinsic": they use the embedding $\mathcal{Y}_\infty \subset J^\infty(E,n)$ (or $\mathcal{Y}_\infty \subset J^\infty(\pi)$). For quasi-integrable equations, we shall give equivalent "intrinsic" definitions.

Let $\mathcal{Y}_\infty \subset J^\infty(E,n)$, and let Φ be the representing object of a functor $\mathfrak{F}$. Then, since $\Phi(\mathcal{Y}_k) = \Phi(J^k(E,n))|_{\mathcal{Y}_k}$, $k < \infty$, one has $\Phi(\mathcal{Y}_\infty) = \Phi(J^\infty(E,n))|_{\mathcal{Y}_\infty}$. From the last equality it follows that if $i\colon \mathcal{Y}'_\infty \hookrightarrow \mathcal{Y}''_\infty$ is an embedding, then $\Phi(\mathcal{Y}'_\infty) = i^*(\mathcal{Y}''_\infty)$. Naturally, similar facts are valid for $\mathcal{Y}_\infty \subset J^\infty(\pi)$. This remark, together with Theorem 3.1.2, implies

PROPOSITION. *Let $\mathfrak{F}$ be a functor of differential calculus and* $\deg \mathfrak{F} < \infty$. *Then for any quasi-integrable equation $\mathcal{Y}$,*

1. *The following description is valid*:

$$\mathcal{C}\Phi(\mathcal{Y}_\infty) = \{\varphi \in \Phi(\mathcal{Y}_\infty) \mid (\varphi|_W)(\theta) = 0,\ \forall \theta \in \mathcal{Y}_\infty,\ \forall\ \theta\text{-integral } W\}.$$

2. *If W is a θ-integral manifold, then $(\rceil\varphi)(\theta) = (\varphi|_W)(\theta)$.*

3.2.5. The operation $\rceil$ possesses the following important property: for any embedding $i\colon \mathcal{Y}'_\infty \hookrightarrow \mathcal{Y}''_\infty$, the diagram

$$\begin{array}{ccc} \Phi(\mathcal{Y}'_\infty) & \xleftarrow{i^*} & \Phi(\mathcal{Y}''_\infty) \\ \rceil\downarrow & & \downarrow\rceil \\ \Phi_\Gamma(\mathcal{Y}'_\infty) & \xleftarrow{i^*} & \Phi_\Gamma(\mathcal{Y}''_\infty) \end{array}$$

is commutative. This immediately follows from the definition. This fact, together with Theorem 3.2.2, proves the following

PROPOSITION. *If the equations $\mathcal{Y}'$, $\mathcal{Y}''$ are quasi-integrable and $\mathcal{Y}' \subset \mathcal{Y}''$, then* $\mathcal{C}\Phi(\mathcal{Y}'_\infty) = \mathcal{C}\Phi(\mathcal{Y}''_\infty)|_{\mathcal{Y}'_\infty}$.

As was noted in Subsection 2.3.8, the module $\mathcal{C}\Lambda^1(J^\infty(\pi))$ is the annihilator of the Cartan distribution on $J^\infty(\pi)$. Therefore, setting $\mathcal{Y}'' = J^\infty(\pi)$ and taking into account the last proposition, we see that $\mathcal{C}\Lambda^1(\mathcal{Y}_\infty)$ is the annihilator of the Cartan distribution on $\mathcal{Y}_\infty$.

COROLLARY. *If $\mathcal{Y}_\infty$ is a quasi-integrable equation, then $\mathcal{C}\Lambda^1(\mathcal{Y}_\infty)$ is the annihilator of the Cartan distribution on $\mathcal{Y}_\infty$.*

PROOF. The case $\mathcal{Y}_\infty \subset J^\infty(\pi)$ was already considered. The general case is implied by the local nature of the statement. $\square$

3.2.6. Let $P = \mathcal{F}(\pi, \xi)$ and $P_s = \mathcal{F}_s(\pi, \xi)$. Then from (3.5) it follows that there exists the direct sum decomposition on $J^\infty(\pi)$:

$$\mathcal{J}^k(P) = \mathcal{C}\mathcal{J}^k(P) \oplus \mathcal{J}^k(P)_\Gamma. \tag{3.6}$$

In turn, decomposition (3.6) implies, for any module Q, the decomposition

$$\mathrm{Diff}_k(P, Q) = \mathcal{C}\mathrm{Diff}_k(P, Q) \oplus \mathrm{Diff}_k^v(P, Q), \tag{3.7}$$

where under the identification of $\mathrm{Diff}_k(P, Q)$ with the module $\mathrm{Hom}_{\mathcal{F}(\pi)}(\mathcal{J}^k(P), Q)$ the module $\mathrm{Diff}_k^v(P, Q)$ consists of homomorphisms from $\mathcal{J}^k(P)$ to Q annihilating $\mathcal{J}^k(P)_\Gamma$. Our goal is to describe the decomposition (3.7) explicitly.

PROPOSITION. *For any P, the module $\mathcal{C}\mathcal{J}^k(P)$ (resp., $\mathcal{J}^k(P)_\Gamma$) is generated by the image of the operator $j_k - \rho_k \colon P \to \mathcal{J}^k(P)$ (resp., of $\rho_k \colon P \to \mathcal{J}^k(P)$).*

PROOF. Note first that $\rho_k = \rceil \circ j_k$. This immediately follows from the definition of the operator ρ_k (see Subsection 2.1.2). Further, since $\rceil(f) = f$ for any $f \in \mathcal{F}(\pi)$, we have $\rceil(f j_k(p)) = f\rho_k(p)$. Taking into account the fact that the module $\mathcal{J}^k(P)$ consists of the elements $\sum_i f_i j_k(p_i)$, we see that the decomposition

$$f j_k(p) = (f j_k(p) - \rceil(f j_k(p))) + \rceil(f j_k(p)) = f(j_k(p) - \rho_k(p)) + f\rho_k(p)$$

proves our result. □

For any $\Delta \in \mathrm{Diff}_k(P, Q)$, let us consider the operator $\overline{\Delta} \in \mathrm{Diff}_k(P, Q)$ defined by

$$\overline{\Delta}(p) = (\Delta, \rho_k(p)), \quad p \in P, \tag{3.8}$$

where $(\bullet, \bullet)$ denotes the natural pairing between the modules $\mathrm{Diff}_k(P, Q)$ and $\mathcal{J}^k(P)$. Since the isomorphism between $\mathrm{Hom}_{\mathcal{F}(\pi)}(\mathcal{J}^k(P), Q)$ and $\mathrm{Diff}_k(P, Q)$ is realized by the correspondence $h \mapsto h \circ j_k$, $h \in \mathrm{Hom}_{\mathcal{F}(\pi)}(\mathcal{J}^k(P), Q)$, the induced isomorphism between the module

$$\mathrm{Hom}_{\mathcal{F}(\pi)}(\overline{\mathcal{J}^k(P)}, Q) = \mathrm{Hom}_{\mathcal{F}(\pi)}(\mathcal{J}^k(P)_\Gamma, Q)$$

and $\mathcal{C}\mathrm{Diff}_k(P, Q)$ is the correspondence $h \mapsto h \circ \rho_k$. Hence, due to (3.7), if $\Delta = h \circ j_k$, then the projection of Δ to $\mathcal{C}\mathrm{Diff}_k(P, Q)$ equals $h \circ \rho_k = \overline{\Delta}$. In other words, the mapping $\Delta \mapsto \overline{\Delta}$ is the projection onto the first summand in (3.7). Thus we have proved the following

THEOREM. *The representation $\Delta = \overline{\Delta} + (\Delta - \overline{\Delta})$ gives the direct sum decomposition* (3.7). *In particular,*

$$\mathcal{C}\mathrm{Diff}(P, Q) = \{\overline{\Delta} \mid \Delta \in \mathrm{Diff}(P, Q)\}.$$

COROLLARY (1). *If $\Delta|_{P_{-1}} = \nabla|_{P_{-1}}$, then $\overline{\Delta} = \overline{\nabla}$, i.e., the operator $\overline{\Delta}$ is uniquely determined by $\Delta|_{P_{-1}}$.*

PROOF. It suffices to show that $(\square, \rho_k(p)) = 0$ if $\square|_{P_{-1}} = 0$. Indeed, $\rho_k(p)$ is a horizontal jet and hence is of the form $\rho_k(p) = \sum_i f_i j_k(p_i)$, $f_i \in \mathcal{F}(\pi)$, $p_i \in P_{-1}$. Therefore,

$$(\square, \rho_k(p)) = \sum_i f_i(\square, j_k(p_i)) = \sum_i f_i \square(p_i) = 0.$$

□

An operator $\square \in \operatorname{Diff}(P,Q)$ is called *vertical* if $\square|_{P_{-1}} = 0$. From the above result one gets

COROLLARY (2). *The $\mathcal{F}(\pi)$-module $\operatorname{Diff}_k^v(P,Q)$ (see (3.7)) consists of all vertical operators.*

3.2.7. Let $\square \in \operatorname{Diff}(P_{-1},Q)$, where Q is understood as an $\mathcal{F}_{-1}(\pi)$-module. Corollary 1 from the previous subsection shows that the following construction is well defined:

$$\overline{\square} = \overline{\Delta}, \quad \Delta \in \operatorname{Diff}(P,Q), \ \Delta|_{P_{-1}} = \square. \tag{3.9}$$

Note that (3.9) generalizes the definition of operators $\overline{Y}$ from Subsection 2.3.10.

PROPOSITION. *If $\Delta \in \operatorname{Diff}(\Gamma(\xi),\Gamma(\eta))$, then $\hat{\Delta} = \overline{\Delta}$.*

PROOF. First note that the operator $\hat{\Delta}$ is a $\mathcal{C}$-differential operator. Indeed, the relation $j(s)^* \circ \hat{\Delta} = \Delta \circ j(s)^*$, $s \in \Gamma_{\mathrm{loc}}(\pi)$, which can be regarded as a definition of $\hat{\Delta}$, means that $\hat{\Delta}$ admits restriction to $M_s^{(\infty)}$. Therefore, according to Theorem 3.2.6 and the first corollary from this theorem, it suffices to show that $\hat{\Delta}|_{P_{-1}} = \overline{\Delta}|_{P_{-1}}$. But $\hat{\Delta}|_{P_{-1}} = \Delta = \overline{\Delta}|_{P_{-1}}$. □

This proposition allows one to obtain a useful description of the operators $\overline{\Delta}$, i.e., of all $\mathcal{C}$-differential operators. To do this, note that any element $q \in Q$ is locally representable in the form $\sum_i g_i e_i$, where $e_i \in Q_{-1} = \Gamma(\pi)$ is a local basis of sections and $g_i \in \mathcal{F}(\pi)$. Therefore, any operator $\Delta \in \operatorname{Diff}(P_{-1},Q)$ is locally representable in the form $\Delta = \sum_i f_i \Delta_i$, $f_i \in \mathcal{F}(\pi)$, $\Delta_i \in \operatorname{Diff}(\Gamma(\xi),\Gamma(\eta))$. Hence, $\overline{\Delta} = \sum_i \overline{f_i \Delta_i} = \sum_i f_i \overline{\Delta}_i = \sum_i f_i \hat{\Delta}_i$.

COROLLARY. *The module $\mathcal{C}\operatorname{Diff}_*(P,Q)$ consists of all operators of the form $\sum_i f_i \hat{\Delta}_i$, where $f_i \in \mathcal{F}(\pi)$, $\Delta_i \in \mathcal{C}\operatorname{Diff}(\Gamma(\xi),\Gamma(\eta))$.*

PROOF. The local version of this statement has been already proved. The general case is obtained from it by the standard partition of unity techniques. □

3.2.8. Let $\mathcal{Y}'_\infty \subset \mathcal{Y}''_\infty \subset J^\infty(\pi)$ be quasi-integrable equations. Also, let $P = \mathcal{F}(\pi,\xi)$, $P' = P|_{\mathcal{Y}'_\infty} = \mathcal{F}(\mathcal{Y}',\xi)$, $P'' = P|_{\mathcal{Y}''_\infty} = \mathcal{F}(\mathcal{Y}'',\eta)$. Define the modules Q' and Q'' in a similar way. From Proposition 3.2.5 applied to $\Phi = \mathcal{J}^k(P)$ it follows that any operator $\Delta \in \mathcal{C}\operatorname{Diff}_k(P'',Q'')$ can be restricted to $\mathcal{Y}'_\infty$, i.e., the equality

$$(\Delta|_{\mathcal{Y}'_\infty})(p''|_{\mathcal{Y}'_\infty}) = \Delta(p'')|_{\mathcal{Y}'_\infty}, \quad p'' \in P'',$$

is well defined. Moreover, the mapping

$$\mathcal{C}\operatorname{Diff}(P'',Q'') \to \mathcal{C}\operatorname{Diff}(P',Q')$$

arising from this equality is an epimorphism. Hence, the operators ρ_k, $\overline{\Delta}$, $\hat{\Delta}$ admit restrictions from $J^\infty(\pi)$ to $\mathcal{Y}_\infty$ provided that $\mathcal{Y}_\infty$ is quasi-integrable. Moreover, all the results of the previous subsections remain valid if we take $\mathcal{Y}_{-1} = M$ (see Subsection 3.1.1).

In the sequel, the restrictions of operators ρ_k, $\overline{\Delta}$, $\hat{\Delta}$, etc., to $\mathcal{Y}_\infty$ are often denoted by the same symbols.

3.2.9. To conclude this section, we describe the Hamiltonian formalism in the "$\mathcal{C}$-theory".

Let $\mathcal{F} = \mathcal{F}(\mathcal{Y})$ and P, Q be $\mathcal{F}$-modules. Set

$$\mathcal{C}\,\mathrm{smbl}_k(P,Q) = \frac{\mathcal{C}\mathrm{Diff}_k(P,Q)}{\mathcal{C}\mathrm{Diff}_{k-1}(P,Q)}, \qquad \mathcal{C}\,\mathrm{smbl}_k\,P = \mathcal{C}\,\mathrm{smbl}_k(\mathcal{F},P),$$
$$\mathcal{C}\,\mathrm{smbl}(P,Q) = \sum_{k\geq 0}\mathcal{C}\,\mathrm{smbl}_k(P,Q), \qquad \mathcal{C}\,\mathrm{smbl}\,P = \sum_{k\geq 0}\mathcal{C}\,\mathrm{smbl}_k\,P.$$

Composition of $\mathcal{C}$-differential operators induces a multiplication in the module $\mathcal{C}\,\mathrm{smbl}(P,P)$ supplying it with an associative algebra structure. As usual, the algebra $\mathcal{C}\,\mathrm{smbl}\,\mathcal{F}$ is commutative, since for any $\Delta_1 \in \mathrm{Diff}_k\mathcal{F}$ and $\Delta_2 \in \mathrm{Diff}_l\mathcal{F}$ the commutator $[\Delta_1,\Delta_2]$ is of order $\leq k+l-1$. Moreover, the $\mathcal{F}$-module $\mathcal{C}\,\mathrm{smbl}(P,Q)$ acquires the structures of a left $\mathcal{C}\,\mathrm{smbl}(P,P)$-module and of a right $\mathcal{C}\,\mathrm{smbl}(Q,Q)$-module. In particular, $\mathcal{C}\,\mathrm{smbl}\,P$ is a $\mathcal{C}\,\mathrm{smbl}\,\mathcal{F}$-module.

If $\Delta \in \mathcal{C}\mathrm{Diff}_k(P,Q)$, then the image of the operator Δ under the natural projection $\mathcal{C}\mathrm{Diff}_k(P,Q) \to \mathcal{C}\,\mathrm{smbl}_k(P,Q)$ will be denoted by $\mathcal{C}\,\mathrm{smbl}_k\,\Delta$. Commutator of operators from $\mathcal{C}\mathrm{Diff}\,\mathcal{F}$ induces the *Poisson bracket* $\{\bullet,\bullet\}_{\mathcal{C}}$ in $\mathcal{C}\,\mathrm{smbl}\,\mathcal{F}$:

$$\{h_1,h_2\}_{\mathcal{C}} = \mathcal{C}\,\mathrm{smbl}_{i+j-1}[\Delta_1,\Delta_2],$$

where $\Delta_1 \in \mathcal{C}\mathrm{Diff}_i\mathcal{F}$, $\Delta_2 \in \mathcal{C}\mathrm{Diff}_j\mathcal{F}$, $h_1 = \mathcal{C}\,\mathrm{smbl}_i\,\Delta_1$, $h_2 = \mathcal{C}\,\mathrm{smbl}_j\,\Delta_2$. Obviously, $\mathcal{C}\,\mathrm{smbl}\,\mathcal{F}$ becomes a Lie algebra with respect to the bracket $\{\bullet,\bullet\}_{\mathcal{C}}$.

Let $\mathcal{Y}_\infty \subset J^\infty(\pi)$. In this case, one has a homomorphism of filtered algebras $\mathrm{Diff}\,C^\infty(M) \to \mathcal{C}\mathrm{Diff}\,\mathcal{F}$ which takes $\Delta \in \mathrm{Diff}\,C^\infty(M)$ to $\hat{\Delta} \in \mathrm{Diff}\,\mathcal{F}$. This homomorphism induces a monomorphism of Lie algebras $\mathrm{smbl}\,C^\infty(M) \to \mathcal{C}\,\mathrm{smbl}\,\mathcal{F}$. Therefore, Hamiltonian formalism on $T^*(M)$ is embedded into the $\mathcal{C}$-Hamiltonian formalism.

3.2.10. Let us use Corollary 3.2.7 to describe the algebra $\mathcal{C}\,\mathrm{smbl}\,\mathcal{F}$, $\mathcal{F} = \mathcal{F}(\mathcal{Y})$, and the bracket $\{\bullet,\bullet\}_{\mathcal{C}}$. From this proposition it follows that the algebra $\mathcal{C}\mathrm{Diff}\,\mathcal{F}$ is generated by $\mathcal{F} = \mathcal{C}\mathrm{Diff}_0\mathcal{F}$ and $\mathcal{C}D(\mathcal{F})$. Let $\mathcal{U} \subset M$ be a local chart in M. Then in special coordinates over the domain $(\pi_\infty|_{\mathcal{Y}_\infty})^{-1}(\mathcal{U})$ the module $\mathcal{C}D(\mathcal{F})$ is spanned by the operators $D_i = \widehat{\partial/\partial x_i}$. Therefore, the algebra $\mathcal{C}\,\mathrm{smbl}\,\mathcal{F}$ is generated by $\mathcal{F} = \mathrm{smbl}_0\,\mathcal{F}$ and by the elements $\Pi_i = \mathcal{C}\,\mathrm{smbl}_1\,D_i$ which are independent over $\mathcal{F}$. Hence, in the chart under consideration $\mathcal{C}\,\mathrm{smbl}\,\mathcal{F}$ is isomorphic to the polynomial algebra in the variables Π_i with coefficients in $\mathcal{F}$. The module $\mathcal{C}\,\mathrm{smbl}_k\,\mathcal{F}$ is identified with homogeneous polynomials of degree k.

Since $[D_i,D_j] = 0$ and $[D_i,f] = D_i(f)$, $f \in \mathcal{F}$, one has

$$\{\Pi_i,\Pi_j\}_{\mathcal{C}} = 0, \quad \{\Pi_i,f\}_{\mathcal{C}} = D_i(f).$$

Note also that from the commutator equality $[\Delta_1\circ\Delta_2,\Delta_3] = [\Delta_1,\Delta_3]\circ\Delta_2 + \Delta_1\circ[\Delta_2,\Delta_3]$ one obtains

$$\{h_1h_2,h_3\}_{\mathcal{C}} = h_1\{h_2,h_3\}_{\mathcal{C}} + h_2\{h_1,h_3\}_{\mathcal{C}}. \tag{3.10}$$

Hence, if we set $D_i(\Pi_j) = 0$, then

$$\{h,g\}_{\mathcal{C}} = \sum\left(\frac{\partial h}{\partial \Pi_i}D_i(g) - \frac{\partial g}{\partial \Pi_i}D_i(h)\right). \tag{3.11}$$

3.3. Higher infinitesimal symmetries of differential equations

Using the results of Sections 2.3 and 3.2, we establish here basic facts concerning higher infinitesimal symmetries of systems of nonlinear partial differential equations. Roughly speaking, any symmetry of such a type is uniquely determined by a "vector-valued function" ψ on $\mathcal{Y}_\infty$, while classical symmetries correspond to special "vector-valued functions" on $\mathcal{Y}_1$. This explains the use of the adjective "higher". We show that the Lie algebra $\operatorname{sym}\mathcal{Y}$ of all infinitesimal symmetries of the equation $\mathcal{Y} = \{\varphi = 0\}$ coincides with the kernel of the operator $\bar{\ell}_\varphi$, where $\bar{\ell}_\varphi$ is the restriction of the operator ℓ_φ to $\mathcal{Y}_\infty$. It gives the most efficient general method for computing symmetries for the equation $\mathcal{Y}$, even if we restrict ourselves to the classical case.

In this section, we suppose that the assumptions of Subsection 3.1.1 hold.

3.3.1. Let us start with

DEFINITION. A derivation $X \in D(\mathcal{F}(\mathcal{Y}))$ is called an *infinitesimal automorphism of the Cartan distribution* on $\mathcal{Y}_\infty$, or a *$\mathcal{C}$-field,* if $X(\mathcal{C}\Lambda^1(\mathcal{Y}_\infty)) \subset \mathcal{C}\Lambda^1(\mathcal{Y}_\infty)$. Equivalently, we have $[X, \mathcal{C}D(\mathcal{F}(\mathcal{Y}))] \subset \mathcal{C}D(\mathcal{F}(\mathcal{Y}))$.

Note that the set of vector fields lying in the Cartan distribution on $\mathcal{Y}_\infty$ coincides with the module $\mathcal{C}D(\mathcal{F}(\mathcal{Y}))$, whereas $\mathcal{C}\Lambda^1(\mathcal{Y}_\infty)$ (see Corollary 3.2.5) is the annihilator of the Cartan distribution. Hence, the two versions of the above definition are indeed equivalent.

Denote by $D_\mathcal{C}(\mathcal{Y}_\infty)$ the set of all $\mathcal{C}$-fields on $\mathcal{Y}_\infty$. We shall also write $\mathcal{C}D(\mathcal{Y}_\infty)$ instead of $\mathcal{C}D(\mathcal{F}(\mathcal{Y}))$. In the case $\mathcal{Y}_\infty = J^\infty(E,n)$ we set $D_\mathcal{C}(E) = D_\mathcal{C}(J^\infty(E,n))$ and $\mathcal{C}D(E) = \mathcal{C}D(J^\infty(E,n))$. Evidently, $D_\mathcal{C}(\mathcal{Y}_\infty)$ is a subalgebra in the Lie algebra $D(\mathcal{F}(\mathcal{Y}))$.

From the embedding $[\mathcal{C}D(\mathcal{Y}_\infty), \mathcal{C}D(\mathcal{Y}_\infty)] \subset \mathcal{C}D(\mathcal{Y}_\infty)$ it follows that $\mathcal{C}D(\mathcal{Y}_\infty) \subset D_\mathcal{C}(\mathcal{Y}_\infty)$. Moreover, $\mathcal{C}D(\mathcal{Y}_\infty)$ is obviously an ideal in the Lie algebra $D_\mathcal{C}(\mathcal{Y}_\infty)$. Let us set

$$\operatorname{sym}\mathcal{Y} = D_\mathcal{C}(\mathcal{Y}_\infty)/\mathcal{C}D(\mathcal{Y}_\infty).$$

DEFINITION. The elements of the algebra $\operatorname{sym}\mathcal{Y}$ are called *higher* (*intrinsic*) *infinitesimal symmetries* of the equation $\mathcal{Y}$.

Later we shall usually say "symmetry" instead of "higher infinitesimal symmetry".

An obvious interpretation of the concepts introduced is as follows (cf. Subsection 2.3.3). The flow generated by a $\mathcal{C}$-field X (if it existed) would send an n-dimensional integral manifold of the Cartan distribution on $\mathcal{Y}_\infty$ (i.e., a solution of the equation $\mathcal{Y}$) to another such manifold. If $X \in \mathcal{C}D(\mathcal{Y}_\infty)$, then X is tangent to any n-dimensional integral manifold, and the above transformation would slide any such manifold (a solution) along itself. In other words, the fields $X \in \mathcal{C}D(\mathcal{Y}_\infty)$ and only these fields determine trivial flows on "the manifold of solutions" of the equation $\mathcal{Y}$. Thus the Lie algebra $\operatorname{sym}\mathcal{Y}$ can be understood as the Lie algebra of vector fields on "the manifold of solutions".

3.3.2. First of all, let us describe the $\mathcal{C}$-fields on $J^\infty(E,n)$. Since $J^\infty(E,n)$ can be covered by special charts of the form $J^\infty(\pi)$, from Lemma 2.3.8 one obtains

LEMMA. *Let $X \in D_\mathcal{C}(E)$ be a $\mathcal{C}$-field. Then it is uniquely determined by its restriction $X|_{\mathcal{F}_0(E,n)}$ to $\mathcal{F}_0(E,n) = C^\infty(M)$.*

Corollary 2.3.12 shows that, inside a special chart, any derivation of the form $X\colon \mathcal{F}_0(E,n) \to \mathcal{F}(E,n)$ can be extended up to a $\mathcal{C}$-field in this chart. From the lemma above it follows that such an extension is independent of the chart choice. Therefore, the set of such extensions gives a well-defined $\mathcal{C}$-field on $J^\infty(E,n)$. Hence we have proved

PROPOSITION. *The operation $X \mapsto X|_{\mathcal{F}_0(E,n)}$ defines a one-to-one correspondence between $D_\mathcal{C}(E)$ and the $\mathcal{F}(E,n)$-module of $\mathcal{F}(E,n)$-valued derivations of the algebra $\mathcal{F}_0(E,n)$.*

The correspondence above allows us to provide $D_\mathcal{C}(E)$ with the structure of an $\mathcal{F}(E,n)$-module. Then $\mathcal{C}D(E)$ is a submodule in $D_\mathcal{C}(E)$. Hence the Lie algebra

$$\varkappa = \varkappa(E) = D_\mathcal{C}(E)/\mathcal{C}D(E)$$

carries the structure of an $\mathcal{F}(E,n)$-module.

Let us emphasize that if $X \in D_\mathcal{C}(E)$ and $f \in \mathcal{F}(E,n)$, then in general $fX \notin D_\mathcal{C}(E)$. As an example one can take $X = \mathcal{Э}_\varphi$ and $f \neq \text{const}$.

3.3.3. Let us describe now the structure of $\mathcal{C}$-fields on $\mathcal{Y}_\infty \subset J^\infty(\pi)$. First of all, consider the following generalization of Lemma 2.3.8.

LEMMA. *If $X \in D_\mathcal{C}(\mathcal{Y}_\infty)$ and $X|_{\mathcal{F}_0(\mathcal{Y})} = 0$, then $X = 0$.*

PROOF. The proof is identical to that of Lemma 2.3.8 with the only exception that the equality $X(\hat{d}f) = 0$ is deduced from the direct sum decomposition $\Lambda^1(\mathcal{Y}_\infty) = \mathcal{C}\Lambda^1(\mathcal{Y}_\infty) \oplus \Lambda^1_\Gamma(\mathcal{Y}_\infty)$. □

COROLLARY. *If $\mathcal{Y}_\infty \subset J^\infty(E,n)$, $X \in D_\mathcal{C}(\mathcal{Y}_\infty)$, and $X|_{\mathcal{F}_0(\mathcal{Y})} = 0$, then $X = 0$.*

PROOF. According to the lemma, the statement is valid inside any chart in $\mathcal{Y}_\infty$, i.e., in the domains of the form $\operatorname{im} f_{(\infty)} \cap \mathcal{Y}_\infty$ (Subsection 2.1.2). Due to its local nature, it is valid globally as well. □

Let us come back to the case $\mathcal{Y}_\infty \subset J^\infty(\pi)$ again and consider the direct sum decomposition induced by the decomposition in Theorem 3.2.6:

$$D_\mathcal{C}(\mathcal{Y}_\infty) = \mathcal{C}D(\mathcal{Y}_\infty) \oplus D^v_\mathcal{C}(\mathcal{Y}_\infty), \tag{3.12}$$

where $D^v_\mathcal{C}(\mathcal{Y}_\infty) = \{X \in D_\mathcal{C}(\mathcal{Y}_\infty) \mid X_{\mathcal{F}_{-1}(\mathcal{Y})} = 0\}$. Obviously, $\operatorname{sym}\mathcal{Y} = D^v_\mathcal{C}(\mathcal{Y}_\infty)$. The structure of the $\mathcal{F}(\mathcal{Y})$-module $\mathcal{C}D(\mathcal{Y}_\infty)$ is described by the same theorem:

$$\mathcal{C}D(\mathcal{Y}_\infty) = \{\hat{\Delta} \mid \Delta \in D(\mathcal{F}(\mathcal{Y}))\},$$

or equivalently, $\mathcal{C}D(\mathcal{Y}_\infty)$ consists of the fields X represented in the form $X = \sum_i f_i \hat{X}_i$, $f_i \in \mathcal{F}(\mathcal{Y})$, $X_i \in D(M)$. Thus it only remains to describe explicitly the vector fields $X \in D^v_\mathcal{C}(\mathcal{Y}_\infty)$.

Note the following important property of the fields $X \in D^v_\mathcal{C}(\mathcal{Y}_\infty)$.

PROPOSITION. *If $\Delta \in \operatorname{Diff} C^\infty(M)$, then $[X, \hat{\Delta}] = 0$.*

PROOF. Since each $\Delta \in \operatorname{Diff} C^\infty(M)$ is representable as the sum of monomials $\Delta = Y_1 \circ \cdots \circ Y_s$, $Y_i \in D(M)$, it suffices to show that $[X, \hat{Y}] = 0$, $Y \in D(M)$. On one hand, since X is a $\mathcal{C}$-field, one has $[X, \mathcal{C}D(\mathcal{Y}_\infty)] \subset \mathcal{C}D(\mathcal{Y}_\infty)$, i.e., $[X, \hat{Y}] \in \mathcal{C}D(\mathcal{Y}_\infty)$. On the other hand, $[X, \hat{Y}]$ is a vertical field. Indeed, identifying $C^\infty(M)$ and $\mathcal{F}_{-1}(\mathcal{Y})$, one has $\hat{Y}|_{\mathcal{F}_{-1}(\mathcal{Y})} = Y$, so that

$$X\big(\hat{Y}(\mathcal{F}_{-1}(\mathcal{Y}))\big) = X\big(Y(\mathcal{F}_{-1}(\mathcal{Y}))\big) \subset X(\mathcal{F}_{-1}(\mathcal{Y})) = 0$$

and $\hat{Y}\big(X(\mathcal{F}_{-1}(\mathcal{Y}))\big) = 0$, since X is a vertical field. Therefore, $[X, \hat{Y}] = 0$ due to the direct sum decomposition (3.12). □

3.3.4. Here we assume that $\mathcal{Y}_0 = J^0(\pi)$. If this is not the case, one can consider the mapping $\pi' = \pi|_{\mathcal{Y}_0}$ and then interpret $\mathcal{Y}_\infty \subset J^\infty(\pi')$ as a subset of $J^\infty(\pi)$. Thus, there is no loss of generality in the assumption that π' is a submersion. This assumption is equivalent to $i^*\colon \mathcal{F}_0(\pi) \to \mathcal{F}_0(\mathcal{Y})$ being an isomorphism, where $i\colon \mathcal{Y}_\infty \hookrightarrow J^\infty(\pi)$ is the canonical embedding.

Further, let $X_0'\colon \mathcal{F}_0(\pi) = \mathcal{F}_0(\mathcal{Y}) \to \mathcal{F}(\pi)$ be a derivation of the algebra $\mathcal{F}_0(\pi)$ such that

$$i^* \circ X_0' = X|_{\mathcal{F}_0(\pi)} \circ i^*, \quad X \in D^v_{\mathcal{C}}(\mathcal{Y}). \tag{3.13}$$

Such an extension of the field $X|_{\mathcal{F}_0(\pi)}$ obviously exists. Consider a $\mathcal{C}$-field $X' \in D_{\mathcal{C}}(\pi)$ uniquely determined by the equality $X_0' = X'|_{\mathcal{F}_0(\pi)}$ (see Corollary 2.3.12). Then we claim that

$$i^* \circ X' = X \circ i^*. \tag{3.14}$$

Indeed, let $\varphi \in \mathcal{F}_0(\pi)$ and $\Delta \in \operatorname{Diff} C^\infty(M)$. Then using Proposition 3.3.3 and the fact that $i^* \circ \hat{\Delta} = \hat{\Delta} \circ i^*$ (see Subsection 3.2.8), one has

$$i^*\big(X'(\hat{\Delta}(\varphi))\big) = i^*\big(\hat{\Delta}(X'(\varphi))\big) = \hat{\Delta}\big(i^*(X'(\varphi))\big).$$

But according to (3.13), $i^* \circ X' = X \circ i^*$ on $\mathcal{F}_0(\pi)$. Hence $i^*(X'(\varphi)) = X(i^*(\varphi))$ and consequently

$$i^*\big(X'(\hat{\Delta}(\varphi))\big) = \hat{\Delta}\big(X(i^*(\varphi))\big) = X\big(\hat{\Delta}(i^*(\varphi))\big) = X\big(i^*(\hat{\Delta}(\varphi))\big).$$

Therefore, (3.14) is valid for functions $\psi = \hat{\Delta}(\varphi)$, where $\varphi \in \mathcal{F}(\pi)$. But any $\psi \in \mathcal{F}_0(\pi)$ is a smooth function of $\hat{\Delta}(\varphi)$, $\varphi \in \mathcal{F}_0(\pi)$ (this follows, for example, from the fact that for the coordinate functions one has $p^i_\sigma = D_\sigma(u_i) = (\widehat{\partial^{|\sigma|}/\partial x_\sigma})(u_i)$). This proves (3.14) in full generality.

Thus we have proved the following

THEOREM. *Let $\mathcal{Y}$ be an equation such that the mapping $\pi|_{\mathcal{Y}_0}$ is a submersion, $\mathcal{Y}_0 = \pi_{\infty,0}(\mathcal{Y}_\infty)$. Then for any $X \in D^v_{\mathcal{C}}(\mathcal{Y}_\infty)$ there exists a vertical $\mathcal{C}$-field X' on $J^\infty(\pi)$ such that $i^* \circ X' = X \circ i^*$.*

Since $D^v_{\mathcal{C}}(\mathcal{Y}_\infty) = \operatorname{sym} \mathcal{Y}$, the theorem can be reformulated in the following way: any intrinsic symmetry of the equation $\mathcal{Y}_\infty$ is the restriction of an extrinsic symmetry. To be more precise, note that according to Subsection 3.2.8 any field $Y \in \mathcal{C}D(J^\infty(\pi))$ admits the restriction to $\mathcal{Y}_\infty$, i.e., is tangent to $\mathcal{Y}_\infty$. Therefore, in this case the coset $\{X + \mathcal{C}D(J^\infty(\pi))\} \in \varkappa(\pi)$ (see Subsection 2.3.13) consists of the fields which are simultaneously tangent to $\mathcal{Y}_\infty$. Hence, it makes sense to say that an element $\chi \in \varkappa(\pi)$ is *tangent* to $\mathcal{Y}_\infty$.

DEFINITION. An element $\chi \in \varkappa(\pi)$ tangent to $\mathcal{Y}_\infty$ is called a *higher extrinsic infinitesimal symmetry* of the equation $\mathcal{Y}$.

To conclude this subsection, note a particular case of decomposition (3.7):

$$D_{\mathcal{C}}(\pi) = \mathcal{C}D(\pi) \oplus D^v_{\mathcal{C}}(\pi), \tag{3.15}$$

where

$$\mathcal{C}D(\pi) = \mathcal{C}D(J^\infty(\pi)), \quad D^v_{\mathcal{C}}(\pi) = D^v_{\mathcal{C}}(J^\infty(\pi)) = \varkappa(\pi).$$

If $\chi \in \varkappa(\pi)$, then by $\mathfrak{Э}_\chi \in D^v_{\mathcal{C}}(\pi)$ we denote the $\mathcal{C}$-field corresponding to χ under the above-described isomorphism $\varkappa(\pi) = D^v_{\mathcal{C}}(\pi)$. The same notation is used for $\mathcal{Y}_\infty \subset J^\infty(\pi)$.

3.3.5. Let us specify the results of the previous subsection in the case when π is a linear bundle. In this case $\varkappa(\pi) = D^v_{\mathcal{C}}(\pi)$ can be identified with the set of all evolutionary derivations by the correspondence $\chi \mapsto \mathfrak{Э}_f$, where $f = X \lrcorner U_1(\pi)$ and $\chi = \{X + \mathcal{C}D(\pi)\}$ (see Subsection 2.3.11). If $X \in D^v_{\mathcal{C}}(\mathcal{Y}_\infty)$ and $\bar{f} = X \lrcorner i^*U_1(\pi) \in \mathcal{F}(\mathcal{Y}, \pi)$, then choosing $f \in \mathcal{F}(\pi, \pi)$ so that $i^*(f) = \bar{f}$ we get

$$X \circ i^* = i^* \circ \mathfrak{Э}_f. \tag{3.16}$$

In fact, as is seen from the proof of Theorem 3.3.4, it suffices to check this equality for the elements of the subalgebra $\mathcal{F}_0(\pi)$. If $x_1, \dots, x_n$, $u_1, \dots, u_m$ are standard coordinates in $J^0(\pi)$, then the equality $\bar{f} = i^*(f)$ shows that

$$X(\bar{u}_j) = \bar{f}_j = i^*(f_j) = i^*(\mathfrak{Э}_f(u_j)), \tag{3.17}$$

where $\bar{f} = \sum_j \bar{f}_j e_j$, $f = \sum_j f_j e_j$, and $e_1, \dots, e_m$ is a local basis of sections in π that was used to construct the coordinate system in question. But (3.17) is equality (3.16) restricted to $J^0(\pi)$. This obviously implies the validity of (3.17) for all functions on $J^0(\pi)$.

Let an evolutionary field $\mathfrak{Э}_f$ be tangent to $\mathcal{Y}_\infty$. Algebraically this fact means that $\mathfrak{Э}_f(I(\mathcal{Y}_\infty)) \subset I(\mathcal{Y}_\infty)$, where

$$I(\mathcal{Y}_\infty) = \{\varphi \in \mathcal{F}(\pi) \mid \varphi|_{\mathcal{Y}_\infty} = 0\}.$$

If the system $\{F_i = 0,\ i = 1, \dots, s\}$, $F_i \in \mathcal{F}_k(\pi)$, determines the manifold $\mathcal{Y} \subset J^k(\pi)$, then from Subsection 2.1.3 it follows that the ideal $I(\mathcal{Y}_\infty)$ is spanned by the functions of the form $h\hat{\Delta}(F_i)$. Hence the equality

$$\mathfrak{Э}_f(h\hat{\Delta}(F_i)) = \mathfrak{Э}_f(h)\hat{\Delta}(F_i) + h\hat{\Delta}(\mathfrak{Э}_f(F_i))$$

shows that $\mathfrak{Э}_f(I(\mathcal{Y}_\infty)) \subset I(\mathcal{Y}_\infty)$ if and only if $\mathfrak{Э}_f(F_i) \in I(\mathcal{Y}_\infty)$.

In general, for topological reasons a submanifold cannot be always described by a system $F_i = 0$. Nevertheless, it can always be represented in the form $\varphi = 0$, where $\varphi \in \mathcal{F}_k(\pi, \xi)$ for a suitable bundle ξ. Similar arguments show that in this case $\mathfrak{Э}_f$ is tangent to $\mathcal{Y}_\infty$ if and only if $\mathfrak{Э}_f(\varphi) = 0$ on $\mathcal{Y}_\infty$. Thus, taking into account that $\mathfrak{Э}_f(\varphi) = \ell_\varphi(f)$, one obtains

THEOREM. *A field $\mathfrak{Э}_f$ is an extrinsic symmetry of the equation $\mathcal{Y}_\infty$, where $\mathcal{Y} = \{\varphi = 0\} \subset J^\infty(\pi)$, if and only if $i^*(\ell_\varphi(f)) = 0$. In particular, if the submanifold $\mathcal{Y} \subset J^k(\pi)$ is determined by a system of equations $F_i = 0$, $i = 1, \dots, s$, then $\mathfrak{Э}_f$ is an extrinsic symmetry for $\mathcal{Y}$ if and only if $i^*(\ell_{F_i}(f)) = 0$, $i = 1, \dots, s$.*

Let $\bar{\ell}_\varphi$ be the restriction of the operator ℓ_φ to $\mathcal{Y}_\infty$. It exists according to Subsection 3.2.8 since ℓ_φ is a $\mathcal{C}$-differential operator (see Subsection 2.3.2). Taking into account the identity $i^* \circ \ell_\varphi = \bar{\ell}_\varphi$, the isomorphism $\operatorname{sym} \mathcal{Y} = D^v_{\mathcal{C}}(\mathcal{Y}_\infty)$, and Theorems 3.3.4 and 3.3.5, we obtain

COROLLARY. *Under the assumptions of Theorem* 3.3.5,

$$\operatorname{sym} \mathcal{Y} = \ker \bar{\ell}_\varphi. \tag{3.18}$$

We emphasize that in contrast with the "absolute" case $\mathcal{Y}_\infty = J^\infty(\pi)$, the Lie algebra $\operatorname{sym}\mathcal{Y}$ is not an $\mathcal{F}(\mathcal{Y})$-module. Note also that (3.18) reduces the computation of higher symmetries for the equation $\mathcal{Y}$ to the solution of the differential equation $\bar{\ell}_\varphi(f) = 0$ on $\mathcal{Y}_\infty$.

REMARK. Let $\mathcal{Y} = \{\varphi = 0\} \subset J^\infty(E,n)$. Then, using direct computations, one can show that the operator $\bar{\ell}_\varphi$ constructed in a special coordinate system $\operatorname{im}\alpha_{(\infty)} = J^\infty(\alpha) \subset J^\infty(E,n)$ does not depend on the choice of this system. Therefore, there exists a $\mathcal{C}$-differential operator $\square$ on $\mathcal{Y}_\infty$ such that $\square = \bar{\ell}_\varphi$ in any special coordinate system. Thus, the above arguments show that $\operatorname{sym}\mathcal{Y} = \ker\square$.

3.3.6. As should be expected, the classical theory of infinitesimal symmetries for differential equations considered in Section 2.4 is included in the theory of higher symmetries constructed above.

Indeed, if X is a classical symmetry, then the Lie field $X^{(l)}$ is tangent to the prolongation $\mathcal{Y}^{(l-k)}$ for any $l \geq k$. Hence the lifting $X^{(\infty)}$ of the Lie field X to $J^\infty(E,n)$ (or to $J^\infty(\pi)$) is tangent to $\mathcal{Y}_\infty$, i.e., is an extrinsic infinitesimal symmetry. Thus, the operation $X \mapsto X^{(\infty)}$ generates a mapping $\mathrm{cl}_0\colon \mathfrak{li}_0(\mathcal{Y}) \to \operatorname{sym}\mathcal{Y}$. Since $[X,Y]^{(\infty)} = [X^{(\infty)}, Y^{(\infty)}]$, this mapping is a Lie algebra homomorphism. In the case where $\mathcal{Y} \subset J^\infty(E,n)$ (or $J^k(\pi)$) and $\mathcal{Y}_k$ is everywhere dense in $\mathcal{Y}$, the homomorphism cl_0 is monomorphic.

Using similar arguments, one can construct a homomorphism of the Lie algebra $\mathfrak{li}(\mathcal{Y})$ of intrinsic symmetries of $\mathcal{Y}$ into the Lie algebra $\operatorname{sym}\mathcal{Y}$. To do this, it suffices to define prolongations $\mathcal{Y}^{(s)}$ of the equation $\mathcal{Y}$ in an intrinsic way. It can be done as follows.

We say that an n-dimensional manifold $W \subset \mathcal{Y}$ is *l-integral* at a point $\theta \in \mathcal{Y}$ if for any form $\omega \in \mathcal{C}\Lambda^1(\mathcal{Y})$ we have $j_l(\omega|_W)|_\theta = 0$. Then each point $\bar{\theta} \in \mathcal{Y}^{(s-k)}$ is uniquely determined by the class $\{W\}_\theta^{s-k}$ of $(s-k)$-integral manifolds (W being one of them) tangent to each other with order $s-k$ at the point θ. Indeed, let $\bar{\theta} \in \mathcal{Y}^{(s-k)}$ and $\bar{\theta} = [L]_x^s$, where $x = \pi_s(\bar{\theta}) = \pi_k(\theta)$. Then the submanifold $L^{(k)} \subset J^k(E,n)$ is tangent to the equation $\mathcal{Y}$ with order $s-k$. If $W \subset \mathcal{Y}$ is a submanifold tangent to $L^{(k)}$ at the point θ with order $s-k$, then it is an $(s-k)$-integral manifold. This immediately follows from the fact that the degree of the functor D equals 1 and from the equality $\mathcal{C}\Lambda^1|_{L^{(k)}} = 0$. Conversely, if $W \subset \mathcal{Y}$ is an $(s-k)$-integral manifold at the point θ, then there exists a manifold $L \subset E$ such that $L^{(k)}$ is tangent to W with order $s-k$. Thus the correspondence $\{W\}_\theta^{s-k} \mapsto [L]_x^s$ is a bijection of the set $\bigcup_\theta \{W\}_\theta^{s-k}$ onto $\mathcal{Y}^{(s-k)}$, and we can identify these two sets.

Now let $X \in \mathfrak{li}(\mathcal{Y})$ be a classical intrinsic infinitesimal symmetry of the equation $\mathcal{Y}$, and $A_t\colon \mathcal{Y} \to \mathcal{Y}$ the corresponding (local) flow. Since A_t are transformations preserving the Cartan distribution, they also preserve the class of l-integral manifolds. Therefore, the following "intrinsic definition" of the lifting $A_t^{(s)}$ of the transformation A_t is well posed:

$$A_t^{(s)}(\bar{\theta}) = \{A_t(W)\}_{A_t(\theta)}^s$$

if $\bar{\theta} = \{W\}_\theta^s$. One obviously has $\pi_{l,s} \circ A_t^{(l)} = A_t^{(s)} \circ \pi_{l,s}$.

The lifting $X^{(s)}$ of a field $X \in \mathfrak{li}(\mathcal{Y})$ to $\mathcal{Y}^{(s)}$ is defined as

$$X^{(s)} = \left(\frac{d}{dt}\right)\left(A_t^{(s)}\right)^*\bigg|_{t=0}.$$

Then the field $X^{(s)}$ preserves the Cartan distribution on $\mathcal{Y}^{(s)}$ and $\pi_{l,s}^* \circ X^{(s)} = X^{(l)} \circ \pi_{l,s}^*$. Thus the system of fields $X^{(s)}$ generates a $\mathcal{C}$-field $X^{(\infty)}$ on $\mathcal{Y}_\infty$ and, as before, the correspondence $X \mapsto X^{(\infty)}$ generates a Lie algebra homomorphism $\mathrm{cl}\colon \mathfrak{li}(\mathcal{Y}) \to \mathrm{sym}\,\mathcal{Y}$ which is injective if $\mathcal{Y}_k$ is everywhere dense in $\mathcal{Y}$.

Immediately from the definition of the homomorphisms cl_0 and cl one has the following commutative diagram:

$$\begin{array}{ccc} \mathfrak{li}_0(\mathcal{Y}) & \longrightarrow & \mathfrak{li}(\mathcal{Y}) \\ & {\scriptstyle \mathrm{cl}}\searrow \quad \swarrow{\scriptstyle \mathrm{cl}_0} & \\ & \mathrm{sym}\,\mathcal{Y} & \end{array}$$

REMARK. The Lie fields lying in the kernels of the homomorphisms cl_0 and cl are useless for studying the "manifold of solutions" of the equation in question, since these fields generate trivial "vector fields" on this "manifold" (see the end of Subsection 3.3.1).

3.3.7. According to the definition, $D_{\mathcal{C}}(\mathcal{Y}_\infty)$ is the normalizer of the subalgebra $\mathcal{C}D(\mathcal{Y}_\infty)$ in the algebra $D(\mathcal{F}(\mathcal{Y}))$. This notion can be naturally generalized: one can consider the normalizer $N(\mathcal{Y})$ of the subalgebra $\mathcal{C}\mathrm{Diff}\,\mathcal{F}(\mathcal{Y})$ in the Lie algebra $\mathrm{Diff}\,\mathcal{F}(\mathcal{Y})$ (with respect to the commutator operation $[\Delta_1, \Delta_2] = \Delta_1 \circ \Delta_2 - \Delta_2 \circ \Delta_1$). Since, obviously, $\mathcal{C}\mathrm{Diff}\,\mathcal{F}(\mathcal{Y}) \subset N(\mathcal{Y})$, it is interesting to consider the quotient algebra

$$\bar{N}(\mathcal{Y}) = N(\mathcal{Y})/\mathcal{C}\mathrm{Diff}\,\mathcal{F}(\mathcal{Y}).$$

Since $D(\mathcal{F}(\mathcal{Y})) \cap N(\mathcal{Y}) = D_{\mathcal{C}}(\mathcal{Y})$, one has a canonically determined Lie algebra homomorphism

$$\gamma\colon \mathrm{sym}\,\mathcal{Y} \to \bar{N}(\mathcal{Y}).$$

In this subsection we show that under assumptions of a rather general nature γ is an isomorphism. It suffices to prove this fact inside every affine chart, so we restrict ourselves to the case $\mathcal{Y}_\infty \subset J^\infty(\pi)$.

The direct sum decomposition of Theorem 3.2.6 gives in this case the decomposition

$$N(\mathcal{Y}) = N^v(\mathcal{Y}) \oplus \mathcal{C}\mathrm{Diff}\,\mathcal{F}(\mathcal{Y}),$$

where $N^v(\mathcal{Y}) = N(\mathcal{Y}) \cap \mathrm{Diff}^v\,\mathcal{F}(\mathcal{Y})$. Thus the problem reduces to the description of $N^v(\mathcal{Y})$.

Keeping this goal in mind, consider the following algebraic situation. Let $A_1 \subset A$ be a subalgebra of the basic algebra A and $A_1 \ni 1$. Denote by $\mathrm{Diff}'(A_1, A)$ the set of all differential operators acting from the algebra A_1 into the A_1-module A. Also let $i\colon \mathrm{Diff}\,A \to \mathrm{Diff}'(A_1, A)$ denote the restriction operation: $i(\Delta) = \Delta|_{A_1}$, $\Delta \in \mathrm{Diff}\,A$. Finally, for any submodule $I \subset \mathrm{Diff}\,A$ we set

$$N(I) = \{\Delta \in \mathrm{Diff}\,A \mid [\Delta, I] \subset I\}.$$

LEMMA. *Let a submodule $I \subset \mathrm{Diff}\,A$ be such that $I \cap \mathrm{Diff}_0 A = A$ and $(\ker i) \cap I = 0$. Let $\Delta \in N(I)$. Then from $i(\Delta) = 0$ it follows that $\Delta \in D(A)$.*

PROOF. Let a', $a'' \in A_1$. Then $\delta_{a'}(\Delta)(a'') = \Delta(a'a'') - a'\Delta(a'') = 0$, since $i(\Delta) = 0$. In other words, $i(\delta_{a'}(\Delta)) = 0$. But for any $a \in A$ one has $\delta_a(\Delta) =$

$[a, \Delta] \in I$, since $A \subset I$. Hence from $i(\delta_{a'}(\Delta)) = 0$ and $(\ker i) \cap I = 0$ it follows that $\delta_{a'}(\Delta) = 0$, i.e., Δ is an A_1-linear operator.

Further, let $a \in A$, $a' \in A_1$. Then

$$\big(\delta_a(\Delta) - \Delta(a)\big)(a') = \Delta(aa') - a\Delta(a') - a'\Delta(a).$$

By the A_1-linearity of the operator Δ and by hypothesis, $\Delta(a') = 0$ and hence $\Delta(a'a) = a'\Delta(a)$. Therefore, $(\delta_a(\Delta) - \Delta(a))(a') = 0$ for all $a' \in A_1$, i.e., $i(\delta_a(\Delta) - \Delta(a)) = 0$. But $\delta_a(\Delta) - \Delta(a) \in I$, since $\delta_a(\Delta) \in I$ and $\Delta(a) \in A \subset I$. Thus we have $\delta_a(\Delta) - \Delta(a) = 0$. But the equality $\delta_a(\Delta) = \Delta(a)$ for any $a \in A$ is equivalent to the fact that $\Delta \in D(A)$. □

THEOREM. $\bar{N}(\mathcal{Y}) = \operatorname{sym} \mathcal{Y}$.

PROOF. As was noted above, it suffices to consider the case $\mathcal{Y}_\infty \subset J^\infty(\pi)$. Then $\bar{N}(\mathcal{Y}) = N^v(\mathcal{Y})$. Set $A_1 = \mathcal{F}_{-1}(\mathcal{Y})$, $A = \mathcal{F}(\mathcal{Y})$, $I = \mathcal{C}\mathrm{Diff}\,\mathcal{F}(\mathcal{Y})$, $\Delta \in N^v(\mathcal{Y})$ and recall that according to Subsections 3.2.6–3.2.8 a $\mathcal{C}$-differential operator ∇ on $\mathcal{Y}_\infty$ vanishes if and only if $\nabla|_{\mathcal{F}_{-1}(\mathcal{Y})} = 0$. Thus we are in the setting of the previous lemma. Therefore, $\Delta \in D(\mathcal{F}(\mathcal{Y}))$, which means that $\Delta \in D^v_{\mathcal{C}}(\mathcal{Y}) = N^v(\mathcal{Y}) \cap D(\mathcal{F}(\mathcal{Y}))$. □

3.4. The structure of $\mathcal{C}$-transformations

By a $\mathcal{C}$-transformation we mean a diffeomorphism of a domain $\mathcal{U} \subset J^\infty(E, n)$ preserving the Cartan distribution. In the theory of infinite jet spaces this concept is similar to that of a Lie transformation. In Sections 2.3 and 3.3 we saw that the supply of $\mathcal{C}$-fields, i.e., of infinitesimal $\mathcal{C}$-transformations, is much greater than that of infinitesimal Lie transformations. That is why the theory of higher infinitesimal symmetries is much richer than the classical one. In this section, we study to what extent the class of finite $\mathcal{C}$-transformations is larger than the class of finite Lie transformations. Roughly speaking, the answer is as follows: the set of $\mathcal{C}$-transformations is not much larger than that of Lie transformations, and moreover, there are cases that the $\mathcal{C}$-transformations are exhausted by the Lie transformations. We emphasize that in the theory of differential equations the role of finite $\mathcal{C}$-transformations is more modest than that of infinitesimal ones.

This section begins with a number of auxiliary notions and constructions.

3.4.1. Consider a mapping preserving the Cartan distribution.

DEFINITION. A function on a domain $\mathcal{U} \subset J^\infty(E, n)$ (or $\subset J^\infty(\pi)$) is said to be *smooth*, if it can be represented in the form $\pi^*_{\infty,k}(f)$ for some $f \in \mathcal{F}(\pi)$.

Here the algebra of smooth functions on $\mathcal{U}$ is denoted by $\mathcal{F} = \mathcal{F}(\mathcal{U})$. Fix k and denote by $\mathcal{F}_k = \mathcal{F}_k(\mathcal{U})$ the subalgebra of functions locally representable as $\pi^*_{\infty,k}(f)$, where $f \in \mathcal{F}_k(E, n)$ (or $f \in \mathcal{F}_k(\pi)$). These subalgebras define a filtration in $\mathcal{F}$.

Let $\tilde{\mathcal{U}} \subset J^\infty(\tilde{E}, n)$, $\dim \tilde{N} = n + \tilde{m}$ (or $\tilde{\mathcal{U}} \subset J^\infty(\tilde{\pi})$, where $\tilde{\pi}\colon E_{\tilde{\pi}} \to \tilde{M}$ is a smooth bundle, $\dim \tilde{M} = n$), and let $F\colon \mathcal{U} \to \tilde{\mathcal{U}}$ be a smooth mapping preserving the Cartan distribution. The latter means that $F^*(\mathcal{F}(\tilde{\mathcal{U}})) \subset \mathcal{F}(\mathcal{U})$ and $F^*(\mathcal{C}\Lambda^1(\tilde{\mathcal{U}})) \subset \mathcal{C}\Lambda^1(\mathcal{U})$, where $\mathcal{C}\Lambda^1(\mathcal{U})$ is the $\mathcal{F}(\mathcal{U})$-module generated by $\mathcal{C}\Lambda^1|_{\mathcal{U}}$.

A mapping F is said to be *nondegenerate* if $\dim dF(\mathcal{C}_\theta) = n$ for all $\theta \in \mathcal{U}$. This definition is obviously equivalent to the fact that the image of any R-manifold in $\mathcal{U}$ is an R-manifold in $\tilde{\mathcal{U}}$.

PROPOSITION. *Let F be a nondegenerate mapping. Then the mapping F^*, and hence F itself, is uniquely determined by the restriction $F^*|_{\mathcal{F}_0(\tilde{\mathcal{U}})}$.*

PROOF. Let x_i, u_α, p_σ^α, $i = 1, \dots, n$, $\alpha = 1, \dots, m$, $0 \le |\sigma| < \infty$, be special local coordinates in $\mathcal{U}$, and let $\tilde{x}_j$, $\tilde{u}_\beta$, $\tilde{p}_\tau^\beta$, $j = 1, \dots, n$, $\beta = 1, \dots, \tilde{m}$, $0 \le |\tau| < \infty$, be special local coordinates in $\tilde{\mathcal{U}}$. Due to the local nature of the statement, it suffices to prove it for the coordinate neighborhoods under consideration. Moreover, it is sufficient to show that the functions $P_\tau^\beta = F^*(\tilde{p}_\tau^\beta)$ are uniquely determined by the functions $Q_j = F^*(\tilde{x}_j)$, $U_\beta = F^*(\tilde{u}_\beta)$. We prove it by induction on $|\tau|$ starting with $|\tau| = 0$.

Let $\bar{f} = F^*(f)$ and $\tilde{D}_j = \widehat{\partial/\partial\tilde{q}_j}$. Then

$$F^*(U_1(f)) = F^*(df - \hat{d}f) = d\bar{f} - \sum_{j=1}^{n} F^*(\tilde{D}_j(f))\, dQ_j$$

$$= U_1(\bar{f}) - \sum_{j=1}^{n} F^*(\tilde{D}_j(f)) U_1(Q_j) + \hat{d}\bar{f} - \sum_{j=1}^{n} F^*(\tilde{D}_j(f))\, \hat{d}Q_j.$$

Since $F^*(U_1(f)) \in \mathcal{C}\Lambda^1(\mathcal{U})$, the form $\hat{d}\bar{f} - \sum_j F^*(\tilde{D}_j(f))\, \hat{d}Q_j$ also lies in $\mathcal{C}\Lambda^1(\mathcal{U})$. But this is a horizontal form. Hence it vanishes, i.e.,

$$D_i(\bar{f}) = \sum_{j=1}^{n} F^*(\tilde{D}_j(f)) D_i(Q_j), \quad i = 1, \dots, n. \tag{3.19}$$

In particular, if $f = \tilde{p}_\tau^\beta$, then $F^*(\tilde{D}_j(f)) = F^*(\tilde{p}_{\tau j}^\beta) = P_{\tau j}^\beta$. Thus from (3.19) one obtains

$$D_i(P_\tau^\beta) = \sum_{j=1}^{n} P_{\tau j}^\beta D_i(Q_j). \tag{3.20}$$

Fixing β and τ and using Cramer's rule we get

$$P_{\tau j}^\beta = \frac{|D(Q_1), \dots, D(P_\tau^\beta), \dots, D(Q_n)|}{D(Q)} \tag{3.21}$$

if $D(Q) = \det \|D_i(Q_j)\| \neq 0$. The numerator in (3.21) is the determinant of the matrix whose j-th row is

$$D(Q_j) = (D_1(Q_j), \dots, D_n(Q_j)) \ \text{ if } \ j \neq i,$$

and

$$D(P_\tau^\beta) = (D_1(P_\tau^\beta), \dots, D_n(P_\tau^\beta)) \ \text{ if } \ i = j.$$

Thus, if we succeed to show that $D(Q) \neq 0$, then the desired statement will follow from (3.21).

Let $V \subset \mathcal{U}$ be an R-manifold. Then, since F is nondegenerate, $F(V) = \tilde{V}$ is an R-manifold in $\tilde{\mathcal{U}}$. Therefore, from Theorem 3.1.2 it follows that the functions $\tilde{x}_j|_{\tilde{V}}$ are local coordinates on $\tilde{V}$. Hence $Q_j' = Q_j|_V$ are local coordinates on V. On the other hand, the functions $x_i' = x_i|_V$ form another coordinate system on V, and $D(Q)|_V$ is obviously the Jacobian of $\|\partial Q_j'/\partial x_i'\|$. □

REMARK. The nondegeneracy condition for F is essential. But it can be weakened if one assumes that the set where $D(Q) = 0$, does not have interior points.

COROLLARY. *If F is nondegenerate, then there exists a number $r \in \mathbb{N}$ such that $F^*(\mathcal{F}_k(\tilde{\mathcal{U}})) \subset \mathcal{F}_{k+r}(\mathcal{U})$ for all $k \geq 0$. In other words, F^* agrees with filtration.*

PROOF. From (3.21) it follows that the filtration of $P^\beta_{\tau j}$ can exceed the filtration of P^β_τ by at most 1. Therefore, for r one can take the maximum of filtrations of the functions Q_j, U_β. □

3.4.2. In this subsection we prove the existence of degenerate mappings preserving the Cartan distribution. We do this by presenting a large class of examples. We accept the notation used in the proof of Proposition 3.4.1. For simplicity, we restrict ourselves to the case $m = \tilde{m} = 1$.

As we saw in the previous subsection, degeneracy of a mapping F is equivalent to the equality $D(Q) = 0$ or, equivalently, to linear dependence of the forms $\hat{d}Q_1, \ldots, \hat{d}Q_n$. On any R-manifold V, one has $\hat{d}Q_j|_V = dQ_j|_V$, so that the forms $dQ_1, \ldots, dQ_n$ are linearly dependent as well. Hence, there exists a functional dependence of the functions $Q_1|_V, \ldots, Q_n|_V$. Since this is true for all V, the system of functions $Q_1, \ldots, Q_n$ is functionally dependent. Furthermore, for simplicity we assume that there exists only one essential functional relation connecting $Q_1, \ldots, Q_n$, i.e., that the rank of the system of forms $\hat{d}Q_1, \ldots, \hat{d}Q_n$ equals $n - 1$. Thus we assume that $Q_n = g(Q_1, \ldots, Q_{n-1})$.

When proving Proposition 3.4.1, we have seen that the horizontal form $\omega_f = \hat{d}F^*(f) - \sum_j F^*(\tilde{D}_j(f))\,\hat{d}Q_j$ vanishes for all f. Repeating the above arguments and taking into account the assumption made, we find that the functions $F^*(f)$, $Q_1, \ldots, Q_{n-1}$ are functionally dependent. At generic points, one can consider $F^*(f)$ to be a function of $Q_1, \ldots, Q_{n-1}$. In particular, one can assume that $P_\tau = P^1_\tau$ is a function of $Q_1, \ldots, Q_{n-1}$. Note now that the equality $\omega_{P_\tau} = 0$ holds if and only if F preserves the Cartan distribution, and to define F completely one needs to define, in addition to $Q_1, \ldots, Q_n$, only the function P_τ. By fixing arbitrary functions $P_{(n,\ldots,n)} = \varphi_k(Q_1, \ldots, Q_{n-1})$, where n in the subscript is taken k times, $k \geq 0$, we satisfy these relations provided that

$$P_{\tau j} = \frac{\partial P_\tau}{\partial Q_j} - P_{\tau n}\frac{\partial g}{\partial Q_j}, \quad j < n.$$

These formulas allow one to express the functions P_τ by induction on $|\tau|$ in terms of $Q_1, \ldots, Q_{n-1}$ using g and $\varphi_0, \varphi_1, \ldots$. This describes the local structure of degenerate mappings in the case where the rank of $\hat{d}Q_1, \ldots, \hat{d}Q_n$ equals $n - 1$ and $m = \tilde{m} = 1$. In a similar way one can consider a general situation.

3.4.3. To study $\mathcal{C}$-transformations we shall need some facts about $\mathcal{C}_{(k)}$-fields, which are presented below.

Let $\mathcal{C}\Lambda^1 = \mathcal{C}\Lambda^1(J^\infty(\pi))$, $\mathcal{C}_k\Lambda^1 = \mathcal{C}_k\Lambda^1(J^\infty(\pi))$ (see Subsection 3.1.7). Let $\mathcal{C}_{(k)}\Lambda^1$ be the $\mathcal{F}(\pi)$-submodule in $\mathcal{C}\Lambda^1$ generated by $\mathcal{C}_k\Lambda^1$. Obviously, the forms $U_1(f)$, $f \in \mathcal{F}_{k-1}(\pi)$, can be taken for generators of $\mathcal{C}_{(k)}\Lambda^1$. Let us also set $\Lambda^s_k(J^\infty(\pi)) = \pi^*_{\infty,k}\big(\Lambda^s(J^k(\pi))\big)$.

DEFINITION. A field $X \in D(\mathcal{F}(\pi))$ is called a *$\mathcal{C}_{(k)}$-field* if $X(\mathcal{C}_{(k)}\Lambda^1) \subset \mathcal{C}_{(k)}\Lambda^1$.

Obviously, a Lie field is a $\mathcal{C}_{(k)}$-field. If a field $Z \in D(\mathcal{F}(\pi))$ satisfies $Z|_{\mathcal{F}_k(\pi)} = 0$ and $\omega \in \Lambda^*_k(J^\infty(\pi))$, then $Z(\omega) = 0$. Since $U_1(f) \in \Lambda^1_k(J^\infty(\pi))$ for $f \in \mathcal{F}_{k-1}(\pi)$, from this it follows also that Z is a $\mathcal{C}_{(k)}$-field as well. Such $\mathcal{C}_{(k)}$-fields are said to be *trivial.* In other words, the set of all trivial $\mathcal{C}_{(k)}$-fields coincides with the set of all $\pi_{\infty,k}$-vertical fields on $J^\infty(\pi)$.

LEMMA. *Any $\mathcal{C}_{(k)}$-field X is uniquely representable in the form $X = Y + Z$, where Y is a $\mathcal{C}$-field and Z is a trivial $\mathcal{C}_{(k)}$-field.*

PROOF. Let Y be a $\mathcal{C}$-field uniquely determined by the equality $Y|_{\mathcal{F}_0(\pi)} = X|_{\mathcal{F}_0(\pi)}$ (see Corollary 2.3.12). Repeating the proof of Lemma 2.3.8, we find that $(X - Y)|_{\mathcal{F}_k(\pi)} = 0$, i.e., $X = Y + Z$, where $Z = X - Y$ and $Z|_{\mathcal{F}_k(\pi)} = 0$. □

PROPOSITION. *A $\mathcal{C}_{(k)}$-field X is trivial if and only if $X \lrcorner \mathcal{C}_{(k)}\Lambda^1 = 0$.*

PROOF. Using the lemma, let us represent X in the form $Y + Z$. Since the forms lying in $\mathcal{C}_{(k)}\Lambda^1$ are $\pi_{\infty,k}$-horizontal, one has $X \lrcorner \omega = Y \lrcorner \omega$, $\omega \in \mathcal{C}_{(k)}\Lambda^1$. If $X \lrcorner \mathcal{C}_{(k)}\Lambda^1 = 0$, then, in particular, $0 = X \lrcorner U_1(\pi) = Y \lrcorner U_1(\pi)$, from which it follows that $Y = \widehat{V} \in \mathcal{C}D(\pi)$ (Theorem 2.3.12). We show that $V = 0$. Let $\widehat{V} = \sum_i h_i D_i$, $h_i \in \mathcal{F}(\pi)$, locally. Then

$$\begin{aligned}\widehat{V}(U_1(f)) &= \sum_i h_i D_i(U_1(f)) + \sum_i dh_i \wedge (D_i \lrcorner U_1(f)) \\ &= \sum_i h_i D_i(U_1(f)) = \sum_i h_i U_1(D_i(f)).\end{aligned}$$

If $f = p^\alpha_\sigma$, $|\sigma| < k$, then $\widehat{V}(U_1(p^\alpha_\sigma)) = \sum h_i U_1(p^\alpha_{\sigma i}) \in \mathcal{C}_{(k)}\Lambda^1$. On the other hand, if $|\sigma| = k - 1$, then $U_1(p^\alpha_{\sigma i}) \notin \mathcal{C}_{(k)}\Lambda^1$. In addition, the forms $U_1(p^\alpha_{\sigma 1}), \dots, U_1(p^\alpha_{\sigma n})$ are linearly independent, so that $h_i = 0$, i.e., $\widehat{V} = 0$. Thus we have proved sufficiency. The second part of the proposition is evident. □

Obviously, the results of this subsection are valid in any domain $\mathcal{U} \subset J^\infty(\pi)$.

3.4.4. In the sequel we shall need the following construction. Let

$$\chi = \{X + \mathcal{C}D(\pi)\} \in \varkappa(\pi), \quad X \in D_{\mathcal{C}}(\pi),$$

and $\omega \in \mathcal{C}\Lambda^1(J^\infty(\pi))$. Since $Y \lrcorner \omega = 0$ for each $Y \in \mathcal{C}D(\pi)$, the operation $\chi \lrcorner \omega \in \mathcal{F}(\pi)$,

$$\chi \lrcorner \omega = Y \lrcorner \omega,$$

is well defined. Consider the mapping $L_\omega \colon \varkappa(\pi) \to \mathcal{F}(\pi)$ defined by $L_\omega(\chi) = \chi \lrcorner \omega$.

LEMMA. *The mapping L_ω is a differential operator. Its order equals the number k such that $\omega \in \mathcal{C}_{(k+1)}\Lambda^1$ and $\omega \notin \mathcal{C}_{(k)}\Lambda^1$.*

PROOF. In special local coordinates on $J^\infty(\pi)$,

$$U_1(\varphi) = \sum_{\alpha,\sigma} \frac{\partial \varphi}{\partial p^\alpha_\sigma} U_1(p^\alpha_\sigma).$$

Since $\omega \in \mathcal{C}\Lambda^1$, one has $\omega = \sum a_i U_1(\varphi_i)$, and hence in local coordinates the form ω is uniquely representable as

$$\omega = \sum_{\alpha,\sigma} f^\alpha_\sigma U_1(p^\alpha_\sigma), \quad f^\alpha_\sigma \in \mathcal{F}(\pi).$$

In this case, $k = \max\{|\sigma| \mid f_\sigma^\alpha \neq 0\}$. Identifying $\varkappa(\pi)$ with $\mathcal{F}(\pi,\pi)$, we see that if $\chi = \vartheta_\varphi$, then

$$L_\omega(\chi) = L_\omega(\vartheta_\varphi) = \vartheta_\varphi \lrcorner \sum_{\alpha,\sigma} f_\sigma^\alpha U_1(p_\sigma^\alpha) = \sum_{\alpha,\sigma} f_\sigma^\alpha \vartheta_\varphi(p_\sigma^\alpha)$$
$$= \sum_{\alpha,\sigma} f_\sigma^\alpha D_\sigma(\varphi^\alpha) = \left(\sum_{\alpha,\sigma} f_\sigma^\alpha D_\sigma^{(\alpha)}\right)(\varphi),$$

where $D_\sigma^{(\alpha)}(\varphi) = D_\sigma(\varphi^\alpha)$ and $\varphi = \sum \varphi^\alpha e_\alpha$ (see Subsection 2.3.7). Therefore, $L_\omega = \sum_{\alpha,\sigma} f_\sigma^\alpha D_\sigma^{(\alpha)}$. □

PROPOSITION. *The following statements hold:*

1. $\mathcal{C}_{(k)}\Lambda^1$, $1 \leq k \leq \infty$, *is generated as an* $\mathcal{F}$-*module by the differential forms* $X_1(X_2(\dots(X_s(U_1(\varphi)))\dots))$, *where* $X_i \in \mathcal{CD}(\pi)$, $\varphi \in \mathcal{F}_0(\pi)$, *and* $s \leq k-1$.
2. *If* $X \in \mathcal{CD}(\pi)$ *and* $\omega \in \mathcal{C}\Lambda^1$, *then* $L_{X(\omega)} = X \circ L_\omega$.

PROOF. Since the first statement is local, it is sufficient to prove it inside a local coordinate system. But $\mathcal{C}_{(k)}\Lambda^1$ is obviously generated by the forms $U_1(p_\sigma^\alpha)$, $|\sigma| < k$ (see Subsection 3.4.1). Besides, according to Corollary 2.3.9, $D_i(U_1(p_\sigma^\alpha)) = U_1(D_i(p_\sigma^\alpha)) = U_1(p_{\sigma i}^\alpha)$, which gives the desired result immediately.

Furthermore, if $\omega = U_1(\psi)$, then $L_{X(\omega)}(\vartheta_\varphi) = \vartheta_\varphi \lrcorner X(\omega)$. But $X(\omega) = d(X \lrcorner \omega) + X \lrcorner d\omega = X \lrcorner d\omega$, since $X \in \mathcal{CD}(\pi)$. Consequently, $\vartheta_\varphi \lrcorner X(\omega) = dU_1(\psi)(X, \vartheta_\varphi)$. By the well-known formula

$$d\omega(X_1, X_2) = X_1(\omega(X_2)) - X_2(\omega(X_1)) - \omega([X_1, X_2]),$$

we have

$$dU_1(\psi)(X, \vartheta_\varphi) = X(U_1(\psi)(\vartheta_\varphi)) - \vartheta_\varphi(U_1(\psi)(X)) - \omega([X, \vartheta_\varphi]).$$

Since $X \in \mathcal{CD}(\pi)$, the commutator $[X, \vartheta_\varphi]$ also belongs to $\mathcal{CD}(\pi)$ (Subsection 3.3.1). Therefore, $U_1(\psi)(X) = U_1(\psi)([X, \vartheta_\varphi]) = 0$ and

$$dU_1(\psi)(X, \vartheta_\varphi) = X(U_1(\psi)(\vartheta_\varphi)) = X(\vartheta_\varphi(\psi)) = X(\ell_\psi(\varphi)) = (X \circ L_{U_1(\psi)})(\varphi).$$

This proves the second statement for the forms $U_1(\psi)$. In the general case, $\omega = \sum_i a_i U_1(\psi_i)$ and $X(\omega) = \sum_i \big(X(a_i)U_1(\psi_i) + a_i X(U_1(\psi_i))\big)$. Hence,

$$L_{X(\omega)}(\vartheta_\varphi) = \vartheta_\varphi \lrcorner X(\omega) = \sum_i \Big(X(a_i)(\vartheta_\varphi \lrcorner U_1(\psi_i)) + a_i(\vartheta_\varphi \lrcorner X(U_1(\psi_i)))\Big)$$
$$= \sum_i \Big(X(a_i)L_{U_1(\psi_i)}(\varphi) + a_i L_{X(U_1(\psi_i))}(\varphi)\Big)$$
$$= \left(\sum_i X(a_i)L_{U_1(\psi_i)} + a_i X \circ L_{U_1(\psi_i)}\right)(\varphi)$$
$$= \left(\sum_i X \circ a_i \circ L_{U_1(\psi_i)}\right)(\varphi) = (X \circ L_\omega)(\varphi).$$

□

3.4.5. We shall now proceed with the study of transformations preserving the Cartan distribution. Since the "manifolds" $J^\infty(E,n)$ (or $J^\infty(\pi)$) are infinite-dimensional, the notion of a transformation should be defined more precisely.

DEFINITION. Let $\mathcal{U}$ and $\tilde{\mathcal{U}}$ be two domains in $J^\infty(E,n)$ or $J^\infty(\pi)$.

1. A bijection $F\colon \mathcal{U} \to \tilde{\mathcal{U}}$ is called a *diffeomorphism* of the domain $\mathcal{U}$ onto $\tilde{\mathcal{U}}$ if $F^*\colon \mathcal{F}(\tilde{\mathcal{U}}) \to \mathcal{F}(\mathcal{U})$ is an isomorphism.
2. A diffeomorphism $F\colon \mathcal{U} \to \tilde{\mathcal{U}}$ is called a *$\mathcal{C}$-transformation* if $F^*(\mathcal{C}\Lambda^1(\tilde{\mathcal{U}})) \subset \mathcal{C}\Lambda^1(\mathcal{U})$.

Note that any $\mathcal{C}$-transformation is a nondegenerate mapping.

Let a diffeomorphism $F^{(k)}\colon V \to W$ preserving the Cartan distribution be given, where V is a domain in $J^k(E,n)$ and W is a domain in $J^{\tilde{k}}(E,\tilde{n})$. Taking into account the structure of integral manifolds of the Cartan distribution (see Subsection 2.1.7), one can deduce that $m = \tilde{m}$ and $k = \tilde{k}$, i.e., F is a Lie transformation. As we have seen in Subsection 2.4.1, $F^{(k)}$ induces a series of transformations $F^{(l)}\colon \pi_{l,s}^{-1}(V) \to \pi_{l,s}^{-1}(W)$, $l \geq k$, satisfying $\pi_{l,s} \circ F^{(l)} = F^{(s)} \circ \pi_{l,s}$, $l \geq s \geq k$. This defines a transformation $F^{(\infty)}\colon \pi_{\infty,k}^{-1}(V) \to \pi_{\infty,k}^{-1}(W)$. Obviously, $F^{(\infty)}$ is a $\mathcal{C}$-transformation. Such $\mathcal{C}$-transformations are still called *Lie transformations*.

LEMMA. *A $\mathcal{C}$-transformation $F\colon \mathcal{U} \to \tilde{\mathcal{U}}$ is a local Lie transformation if and only if there exists a number $k_0 \in \mathbb{N}$ such that $F^*(\mathcal{F}_k(\tilde{\mathcal{U}})) \subset \mathcal{F}_k(\mathcal{U})$ for all $k \geq k_0$, i.e., if and only if F^* preserves the filtration.*

PROOF. The fact that the Lie transformations preserve the filtration is obvious. Assume further that there exists an integer $k \geq 0$ such that $F^*(\mathcal{F}_k(\tilde{\mathcal{U}})) \subset \mathcal{F}_k(\mathcal{U})$. This means that for $\pi_{\infty,k}$-connected domains $\mathcal{U}$ and $\tilde{\mathcal{U}}$ the mapping F preserves the fibers of the projection $\pi_{\infty,k}$ and thus induces a mapping $F^{(k)}\colon \pi_{\infty,k}(\mathcal{U}) \to \pi_{\infty,k}(\tilde{\mathcal{U}})$ satisfying $\left(F^{(k)}\right)^* = F^*|_{\mathcal{F}_k(\tilde{\mathcal{U}})}$. The mapping $F^{(k)}$ takes the Cartan distribution on $\pi_{\infty,k}(\mathcal{U})$ to the Cartan distribution on $\pi_{\infty,k}(\tilde{\mathcal{U}})$ and, when $l \geq k$, $F^{(k)} \circ \pi_{l,k} = \pi_{l,k} \circ F^{(l)}$. Moreover, $F^{(l)}$ takes the fibers of the projection $\pi_{\infty,l}(\mathcal{U}) \to \pi_{\infty,k}(\mathcal{U})$ to those of $\pi_{\infty,l}(\tilde{\mathcal{U}}) \to \pi_{\infty,k}(\tilde{\mathcal{U}})$. Indeed, since F preserves the Cartan distribution, it is the lifting of the mapping $F^{(k)}$. In particular, $F^{(l)}$ is the $(l-k)$-th lifting of $F^{(k)}$. □

When the domains $\mathcal{U}$ and $\tilde{\mathcal{U}}$ are not $\pi_{\infty,k}$-connected, the above reasons obviously remain valid locally.

3.4.6. Here we describe the structure of $\mathcal{C}$-transformations $F\colon \mathcal{U} \to \tilde{\mathcal{U}}$ in the case when m or $\tilde{m}$ equals 1. We assume that $\tilde{m} = 1$. We do not lose generality, since the case $m = 1$ reduces to the latter by considering F^{-1}.

THEOREM. *Let $F\colon \mathcal{U} \to \tilde{\mathcal{U}}$ be a $\mathcal{C}$-transformation and $\tilde{m} = 1$. Then locally F is a Lie transformation.*

PROOF. According to Lemma 3.4.5, it is sufficient to show that $F^*(\mathcal{F}_1(\tilde{\mathcal{U}})) \subset \mathcal{F}_1(\mathcal{U})$.

Corollary 3.4.1 shows that for any $\mathcal{C}$-transformation F, the homomorphisms F^* and $(F^{-1})^*$ agree with filtration. Hence, the mapping F^* allows one to identify the FG-category over $\tilde{\mathcal{U}}$ with that over $\mathcal{U}$ (see Subsection 3.1.3). In particular, the correspondence $\Delta \mapsto \Delta^F = (F^{-1})^* \circ \Delta \circ F^*$ is a bijection of the set of all differential operators over $\mathcal{F}(\mathcal{U})$ onto the similar set over $\mathcal{F}(\tilde{\mathcal{U}})$. Since, in addition, F preserves the Cartan distribution, this correspondence establishes an isomorphism of the Lie algebra $D_{\mathcal{C}}(\mathcal{U})$ of all $\mathcal{C}$-fields over $\mathcal{U}$ onto $D_{\mathcal{C}}(\tilde{\mathcal{C}})$, and this isomorphism maps $\mathcal{C}D(\mathcal{U})$ onto $\mathcal{C}D(\tilde{\mathcal{U}})$. According to Proposition 3.4.4, this implies that the forms $X_1(\dots(X_s(\omega))\dots)$, where $\omega = F^*(U_1(\varphi))$, $\varphi \in \mathcal{F}_0(\tilde{\mathcal{U}})$, $X_i \in \mathcal{C}D(\mathcal{U})$, generate the $\mathcal{F}(\mathcal{U})$-module $\mathcal{C}\Lambda^1(\mathcal{U})$. Since $\tilde{m} = 1$, locally, inside the chart $J^\infty(\tilde{\pi}) \subset J^\infty(\tilde{E}, 1)$, one has $U_1(\varphi) = \lambda U_1(\tilde{\pi})$. Therefore, only one form $F^*(U_1(\tilde{\pi}))$ can be taken for the set of forms ω above. If

$$\Omega = \sum_i X_1^i(\dots(X_s^i(\omega))\dots), \quad X_1^i, \dots, X_s^i \in \mathcal{C}D(\mathcal{U}),$$

then, in accordance with the second statement of Proposition 3.4.4, one has $L_\Omega = \Delta \circ L_\omega$, where $\Delta = \sum_i X_1^i \circ \dots \circ X_s^i$. Since in special local coordinates a field $X \in \mathcal{C}D(\mathcal{U})$ is of the form $X = \sum_k a_k D_k$, the operator Δ can be represented as $\Delta = \sum_\tau g_\tau D_\tau$, $g_\tau \in \mathcal{F}(\mathcal{U})$. Therefore, the order of the operator

$$L_\Omega = \Delta \circ L_\omega = \left(\sum_\tau g_\tau D_\tau\right) \circ \left(\sum_\sigma f_\sigma D_\sigma\right),$$

where $L_\omega = \sum_\sigma f_\sigma D_\sigma$ (see Lemma 3.4.4), is not less than the order of the operator L_ω if $\Delta \neq 0$. Therefore, if $\deg L_\omega > 0$, then either $L_\Omega = 0$, or $\deg L_\Omega > 0$. In this case, independently of the special coordinate system, there are no forms $U_1(\varphi)$, $\varphi \in \mathcal{F}_0(\mathcal{U})$, among the forms obtained from ω by the above procedure, since due to Lemma 3.4.4, $\deg L_{U_1(\varphi)} = 0$. But this is impossible. Therefore, $\deg L_\omega = 0$ and hence $\omega = F^*(U_1(\tilde{\pi})) \in \mathcal{C}_{(1)}\Lambda^1(\mathcal{U})$ (Lemma 3.4.4). Thus we have proved that $F^*(\mathcal{C}_{(1)}\Lambda^1(\tilde{\mathcal{U}})) \subset \mathcal{C}_{(1)}\Lambda^1(\mathcal{U})$. The above arguments show that L_Ω is an operator of order 0 if and only if the order of Δ equals 0, i.e., if and only if $L_\Omega = \lambda L_\omega$, $\lambda \in \mathcal{F}(\mathcal{U})$. Hence $\Omega = \lambda\omega$ and therefore, due to Lemma 3.4.4, $\mathcal{C}_{(1)}\Lambda^1(\mathcal{U}) = \mathcal{F}(\mathcal{U})\omega$. Consequently, $F^*(\mathcal{C}_{(1)}\Lambda^1(\tilde{\mathcal{U}})) \supset \mathcal{C}_{(1)}\Lambda^1(\mathcal{U})$ and F^* induces an isomorphism of $\mathcal{C}_{(1)}\Lambda^1(\tilde{\mathcal{U}})$ onto $\mathcal{C}_{(1)}\Lambda^1(\mathcal{U})$. Finally, we can deduce that the mapping $\Delta \mapsto \Delta^F$ realizes a bijection of the set of $\mathcal{C}_{(1)}$-fields over $\tilde{\mathcal{U}}$ onto the set of $\mathcal{C}_{(1)}$-fields over $\mathcal{U}$ taking trivial $\mathcal{C}_{(1)}$-fields to trivial ones (Proposition 3.4.3). Since the fields of both types are $\pi_{\infty,1}$-vertical, one can see that F takes the fibers of $\pi_{\infty,1}$ into themselves, or, equivalently, $F^*(\mathcal{F}_1(\mathcal{U})) \subset \mathcal{F}_1(\mathcal{U})$. □

3.4.7. In this subsection we show why the theorem just proved is false when $m = \dim \pi > 1$. Assume for the time being that π and $\tilde{\pi}$ are smooth bundles over M and $\Delta\colon \Gamma(\pi) \to \Gamma(\tilde{\pi})$ is an operator of order k. Recall that for any $s \in \mathbb{N}$ the following diagram is commutative:

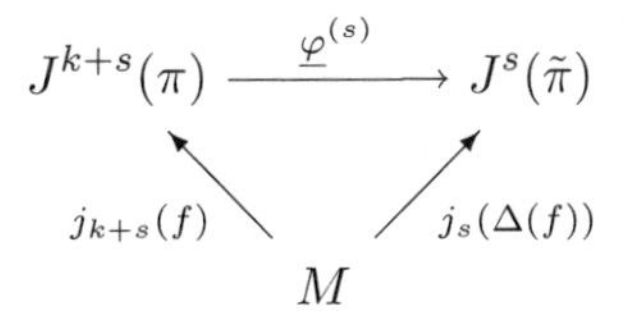

where $f \in \Gamma_{\mathrm{loc}}(\pi)$, $\varphi = \varphi_\Delta$ (see Subsection 2.1.2). Hence the mapping $\underline{\varphi}^{(s)}$ takes the Cartan distribution on $J^{k+s}(\pi)$ to the Cartan distribution on $J^s(\tilde{\pi})$. Since, in addition, $\underline{\varphi}^{(s)} \circ \pi_{k+t,k+s} = \tilde{\pi}_{t,s} \circ \underline{\varphi}^{(t)}$ for $t \geq s$, the set of mappings $\underline{\varphi}^{(s)}$ determines the mapping

$$\bar{\varphi}_\Delta = \underline{\varphi}^{(\infty)}\colon J^\infty(\pi) \to J^\infty(\tilde{\pi})$$

preserving the Cartan distribution and uniquely determined by the condition

$$\bar{\varphi}_\Delta \circ j(f) = j(\Delta(f)) \tag{3.22}$$

for any $f \in \Gamma_{\mathrm{loc}}(\pi)$.

An operator $\Delta \in \mathrm{Diff}(P, Q)$ is said to be *invertible* if there exists an operator $\nabla \in \mathrm{Diff}(Q, P)$ such that $\nabla \circ \Delta = \mathrm{id}_P$ and $\Delta \circ \nabla = \mathrm{id}_Q$. Obviously, if the operator ∇ exists, it is uniquely determined. We denote it by Δ^{-1}.

PROPOSITION. *A mapping $F = \bar{\varphi}_\Delta \colon J^\infty(\pi) \to J^\infty(\tilde{\pi})$, where Δ is an invertible operator, is a $\mathcal{C}$-transformation. If the order of Δ is positive, then F is not a Lie transformation.*

PROOF. Since for any $\square$ and Δ one has $\bar{\varphi}_{\square\circ\Delta} = \bar{\varphi}_\square \circ \bar{\varphi}_\Delta$ and $\bar{\varphi}_{\mathrm{id}} = \mathrm{id}$, the mapping F is invertible and $F^{-1} = \bar{\varphi}_{\Delta^{-1}}$. Maps of the form $\bar{\varphi}_\Delta$ preserve the Cartan distribution and consequently F is a $\mathcal{C}$-transformation. Note further that $F^*|_{\mathcal{F}_s(\tilde{\pi})} = (\varphi^{(s)})^*$. Hence $F^*(\mathcal{F}_0(\tilde{\pi})) \subset \mathcal{F}_k(\pi)$ and $F^*(\mathcal{F}_0(\tilde{\pi})) \not\subset \mathcal{F}_{k-1}(\pi)$, where k is the order of Δ. Besides, for the mappings of the form $\bar{\varphi}_\Delta$ one has

$$\bar{\varphi}_\Delta^* \circ \hat{\nabla} = \hat{\nabla} \circ \bar{\varphi}_\Delta^*, \quad \nabla \in \mathrm{Diff}\, C^\infty(M). \tag{3.23}$$

Indeed, for any $s \in \Gamma_{\mathrm{loc}}(\pi)$ and $f \in \mathcal{F}(\pi)$, using (3.22) and the definition of the operation $\nabla \mapsto \hat{\nabla}$ (see Subsection 2.1.2), one obtains

$$j(s)^*\big(\hat{\nabla}(\bar{\varphi}_\Delta^*(f))\big) = \nabla\big(j(s)^*(\bar{\varphi}_\Delta^*(f))\big) = \nabla\big(j(\Delta(s))^*(f)\big)$$

and

$$j(s)^*\big(\bar{\varphi}_\Delta^*(\hat{\nabla}(f))\big) = j(\Delta(s))^*(\hat{\nabla}(f)) = \nabla\big(j(\Delta(s))\big)^*(f),$$

which proves (3.23). In particular, if $F^*(f) \in F^*(\mathcal{F}_0(\tilde{\pi})) \setminus \mathcal{F}_{k-1}(\pi)$, then there exists a vector field $X \in D(M)$ such that $\hat{X}(F^*(f)) \notin \mathcal{F}_k(\pi)$ and $\hat{X}(F^*(f)) \in \mathcal{F}_{k+1}(\pi)$. Any field X which vanishes nowhere in a domain $X \subset M$ such that $F^*(f)|_{\pi_k^{-1}(V)} \notin \mathcal{F}_{k-1}(\pi_k^{-1}(V))$, satisfies this condition. Then according to (3.23), $F^*(\hat{X}(f)) \in \mathcal{F}_{k+1}(\pi)$ and $F^*(\hat{X}(f)) \notin \mathcal{F}_k(\pi)$, i.e., $F^*(\mathcal{F}_1(\tilde{\pi})) \subset \mathcal{F}_{k+1}(\pi)$ and $F^*(\mathcal{F}_1(\tilde{\pi})) \not\subset \mathcal{F}_k(\pi)$. Therefore, if $k > 0$, then F cannot be a Lie transformation, since for the Lie transformations one must have $F^*(\mathcal{F}_1(\tilde{\pi})) \subset \mathcal{F}_1(\pi)$. $\square$

If $P = \Gamma(\xi)$ and $\dim \xi > 1$, then there exists an invertible operator of an arbitrary order. A class of examples can be immediately obtained if one takes $P = \mathrm{Diff}_k R$, $Q = \mathrm{Diff}_k^+ R$, where $R = \Gamma(\eta)$, and $\Delta = i_+$, $\Delta^{-1} = i^+$ (see Subsection 1.1.3). Another class of examples for $M = \mathbb{R}^n$ is described below.

Let $\xi \colon \mathbb{R}^n \times \mathbb{R}^m \to \mathbb{R}^n$ be a trivial bundle over $\mathbb{R}^n$, and $x_1, \dots, x_n$, $u_1, \dots, u_m$, the corresponding Cartesian coordinates in $\mathbb{R}^n \times \mathbb{R}^m$. Then any section $s \in \Gamma(\xi)$ can be given by a system of functions $u_i = f(x_i)$, $i = 1, \dots, m$, and, respectively, any operator $\Delta \in \mathrm{Diff}(P, P)$ is given by an operator matrix $\|\Delta_{ij}\|$, $\Delta_{ij} \in \mathrm{Diff}\, C^\infty(\mathbb{R}^n)$. In this notation, any upper triangular operator, i.e., such that $\Delta_{ij} = 0$ for $i < j$, is invertible if $\Delta_{ii} = g_i(x) \in C^\infty(M)$, $i = 1, \dots, n$, with nowhere vanishing functions g_i. Thus, one gets examples of invertible operators of arbitrary order by choosing appropriate operators Δ_{ij} for $i > j$.

We emphasize that for $\dim \xi = 1$ the invertible operators acting on $P = \Gamma(\xi)$ are always of order 0, which explains the reasons for Theorem 3.4.6 to be true.

3.4.8. It is useful to note that, as a rule, a $\mathcal{C}$-field does not generate even a local flow, and this fact essentially distinguishes these fields from the Lie fields. Nevertheless, as follows from the above, there exist $\mathcal{C}$-fields which are not Lie fields but generate local flows. Such $\mathcal{C}$-fields will be called *integrable.* Examples of integrable $\mathcal{C}$-fields are easily constructed by defining one-parameter groups of invertible operators. Let $\nabla = \|\nabla_{ij}\|$, where $\nabla_{ij} = 0$ for $i \le j$, $\nabla_{ij} \in \mathrm{Diff}_k C^\infty(\mathbb{R}^n)$. Then the one-parameter group $t \mapsto \Delta_t = \exp(t\nabla)$ consists of invertible operators. An exhaustive description of integrable $\mathcal{C}$-fields is given in [**30**].

3.4.9. The category DE. The above results suggest the idea to place the theory of nonlinear differential equations in the category theory framework. A natural analogy with algebraic geometry, i.e., with categorically formulated theory of algebraic equations, suggests that the necessary and most important point in the construction of the category of differential equations (the category DE) is to define its morphisms as nonlinear differential operators. Proposition 3.4.1 shows that this will be achieved if for the objects of DE we take "manifolds" $J^\infty(E,n)$, $J^\infty(\pi)$, or more generally, $\mathcal{Y}_\infty$ equipped with the Cartan distribution, while the morphisms will be mappings preserving the Cartan distribution. Adopting this point of view for a start, we nevertheless shall keep in mind that since the "manifolds" $\mathcal{Y}_\infty$ are finite-dimensional, it is difficult and even inefficient to follow it literally.

Analysis of basic moments in this chapter shows that the following algebraic approach is useful.

A. An *object* of the category DE is a filtered F-algebra $A = \{A_i\}$ such that the differential calculus in the FG-category over A is equipped with a natural operation $\mathcal{C}\colon \Phi \to \Phi$ for any representing object Φ in a category such that $\mathcal{C}\Lambda^* = \mathcal{C}\Lambda^1 \wedge \Lambda^*$.

B. A *morphism* $F\colon A' \to A''$ in the category DE is an algebra homomorphism $f\colon A'' \to A'$ compatible with filtrations (i.e., such that $f(A''_i) \subset A'_{j(i)}$ for some $j(i)$) and transposable with the operation $\mathcal{C}$.

REMARK (1). Since we are interested not in the filtration of A itself, but in the FG-category generated by this filtration, it is natural to say that two filtrations are *equivalent* if they lead to the same FG-category over A. It also seems natural to fix a class of equivalent filtrations when defining objects of DE (cf. Theorem 3.4.6 and Subsections 3.4.7, 3.4.8).

REMARK (2). The condition $\mathcal{C}\Lambda^* = \mathcal{C}\Lambda^1 \wedge \Lambda^*$ (which reflects the fact that the objects under consideration are "infinitely prolonged") can be better formulated in the form of differential closure of the ideal $\mathcal{C}\tilde{\Lambda}^* \subset \tilde{\Lambda}^*$. Here $\tilde{\Lambda}^*$ denotes the algebra of higher differential forms in the FG-category over A corresponding to the sequence $\sigma = (\infty, \infty, \dots)$ (see [**133**]). Since in the quasi-integrable case this formulation is equivalent to the one given before, we shall use the latter, to avoid presenting the theory of higher differential forms.

A geometric image associated with an algebra $A = \{A_i\}$ is the inverse limit $\mathrm{Spec}_\infty A$ of the sequence of mappings $\dots \leftarrow \mathrm{Spec}\, A_i \leftarrow \mathrm{Spec}\, A_{i+1} \leftarrow \cdots$. Together with the corresponding Cartan distribution, the space $\mathrm{Spec}_\infty A$ can be regarded as geometric realization of an object of DE. It is clear that such spaces are rather inconvenient for direct use. In the theory of differential equations and for natural generalizations of this theory it is sufficient to consider only the algebras $A = \{A_i\}$ with $A_i = C^\infty(M_i)$ and M_i a smooth manifold (possibly with singularities). Then

$M_i \subset \operatorname{Spec} A_i$, while the "projections" $M_{i+1} \to M_i$ correspond to the embeddings $A_i \subset A_{i+1}$. If M_∞ is the inverse limit of the sequence $\cdots \leftarrow M_i \leftarrow M_{i+1} \leftarrow \cdots$, then $M_\infty \subset \operatorname{Spec}_\infty A$ and $\bar{M}_\infty = \operatorname{Spec}_\infty A$. Since M_∞ contains all "good" points of $\operatorname{Spec}_\infty A$, the geometric image of A we use in the sequel will be $\mathcal{O} = M_\infty$.

3.5. Examples of diffieties

These preliminary ideas will be developed in Chapter 5, where the fundamental concept of a *diffiety* is introduced. See also [**89**], where some nonstandard features of the category DE are discussed. Nevertheless, we consider here first examples of these objects and illustrate their usefulness.

In this section, the word diffiety refers to a triple $(\mathcal{O}, \mathcal{F}(\mathcal{O}), \mathcal{C}(\mathcal{O}))$, where $\mathcal{F}(\mathcal{O})$ is the function algebra on $\mathcal{O}$ and $\mathcal{C}(\mathcal{O})$ is a finite-dimensional distribution on $\mathcal{O}$. We also assume that locally this triple is of the form $(\mathcal{Y}_\infty, C^\infty(\mathcal{Y}_\infty), \mathcal{C}(\mathcal{Y}_\infty))$, where "locally" is used in the sense of the Zariski topology on $\mathcal{O}$ defined by the algebra $\mathcal{F}(\mathcal{O})$.

Let $\theta \in \mathcal{O}$ and μ_θ be the ideal of functions vanishing at θ. We define the tangent plane to $\mathcal{O}$ at the point θ as the $\mathbb{R}$-vector space of $\mathbb{R}$-linear derivations of $\mathcal{F}(\mathcal{O})$ with values in $R = \mathcal{F}(\mathcal{O})/\mu_\theta \mathcal{F}(\mathcal{O})$. The subspace $\mathcal{C}_\theta \subset T_\theta \mathcal{O}$ consists of the vectors of $\mathcal{C}(\mathcal{O})$ at θ. A morphism (called a smap; see Chapter 0) of one diffiety to another is a mapping $F\colon \mathcal{O}_1 \to \mathcal{O}_2$ such that

1. $F^*(\mathcal{F}(\mathcal{O}_2)) \subset \mathcal{F}(\mathcal{O}_1)$, and
2. $d_\theta F$ takes $\mathcal{C}_\theta(\mathcal{O}_1)$ to $\mathcal{C}_{F(\theta)}(\mathcal{O}_2)$ for any $\theta \in \mathcal{O}_1$.

If G is a "good" group of smaps acting on a diffiety $\mathcal{O}$, then the quotient $\mathcal{O}/G$ is also a diffiety. In particular, if (locally) $\mathcal{O} = \mathcal{Y}_\infty$, then $\mathcal{O}/G$ is (locally) of the form $\mathcal{Y}'_\infty$. In such a situation, the equation $\mathcal{Y}'$ is called the *quotient* of $\mathcal{Y}$ (by G).

EXAMPLE. Let $u_{xx} + u_{yy} = 0$ be the Laplace equation, and $G = \mathbb{R}^2$ the group of translations of the (x, y)-plane lifted naturally to $\mathcal{Y}_\infty$. Then the quotient of $\mathcal{Y}$ by G is again the Laplace equation; see [**19**]. It is instructive to note that the group of translations lifted to $J^2(\pi)$, $\pi\colon \mathbb{R}^2 \times \mathbb{R}^1 \to \mathbb{R}^2$, leaves $\mathcal{Y} \subset J^2(\pi)$ invariant, but the manifold $\mathcal{Y}/G$, being 2-dimensional, cannot be interpreted as a differential equation.

Other examples of quotient equations can be found in [**19**]. Below we show that factorization (i.e., the procedure of taking quotient objects) forms the basis of the theory of scalar differential invariants; see [**131**] and [**2**, Chapter 7]. For simplicity, we shall confine ourselves to the case of scalar differential invariants of geometric structures.

Consider the category of n-dimensional manifolds, with n fixed, whose morphisms are local diffeomorphisms. A *natural bundle* is a covariant functor on this category taking values in the category of smooth fiber bundles: $M \Rightarrow \{\pi\colon E \to M\}$ (see [**2, 59**]). If S is such a functor, then $S(M)$ is also called a natural bundle and S itself is called the *type* of this bundle.

A *geometric structure* on a manifold M will be understood as a section of a natural bundle. The *type* of a geometric structure is the type of the corresponding natural bundle.

Let $\pi_i\colon E_i \to M_i$, $i = 1, 2$, be natural bundles of the same type, and $\Phi\colon M_1 \to M_2$ a local diffeomorphism. Then Φ is canonically lifted to a local diffeomorphism $\hat{\Phi}\colon E_1 \to E_2$ such that $\pi_2 \circ \hat{\Phi} = \Phi \circ \pi_1$. In a similar way, any vector field X on M

is canonically lifted to a vector field $\hat{X}$ of the total space E of the natural bundle $\pi\colon E \to M$.

In turn, the diffeomorphism $\hat{\Phi}$ generates a $\mathcal{C}$-transformation $\hat{\Phi}^{(\infty)}\colon J^\infty(\pi) \to J^\infty(\pi)$. A function $I \in \mathcal{F}(\pi)$ is called a *scalar differential invariant* of geometric structures on M determined by the bundle π if $I \circ \hat{\Phi}^{(\infty)} = I$ for any local diffeomorphism $\Phi\colon M \to M$. If we confine ourselves to diffeomorphisms preserving orientation, this definition can be reformulated in infinitesimal terms. Namely, a function $I \in \mathcal{F}(\pi)$ is a scalar differential invariant if $\hat{X}^{(\infty)}(I) = 0$ for any vector field X on M, where $\hat{X}^{(\infty)}$ is the lifting of $\hat{X}$ to a $\mathcal{C}$-field on $J^\infty(\pi)$.

If I is a differential invariant and $\omega\colon M \to E$ is a particular geometric structure on M, then the function $I(\omega) = j_\infty(\omega)^*(I) \in C^\infty(M)$ is called the *value* of the invariant I on the structure ω. In local coordinates $x = (x_1, \dots, x_n)$ on M, an arbitrary structure ω is determined by its components $\omega^j(x)$. The corresponding local expression for $I(\omega)$ is of the form

$$I(\omega) = \psi(\omega^j, \omega^j_{(1)}, \omega^j_{(2)}, \dots), \tag{3.24}$$

where $\omega^j_{(k)}$ denotes the set of all derivations of order k of the function ω^j with respect to x.

The algebra of all scalar invariants of geometric structures determined by the bundle π will be denoted by $\mathcal{A}(\pi) \subset \mathcal{F}(\pi)$. If $\pi\colon E \to M$ and $\pi'\colon E' \to M'$ are natural bundles of the same type over the manifolds M and M', $\dim M = \dim M'$, then the algebras of invariants $\mathcal{A}(\pi)$ and $\mathcal{A}(\pi')$ are canonically isomorphic. In other words, the algebra of scalar invariants for geometric structures of a given type S is independent of a manifold where these structures are realized.

Let $x_1, \dots, x_n$ be independent scalar invariants of geometric structures of the type S on n-dimensional manifolds. We define new invariants p^j_σ in the following way. Let a structure ω on a manifold M be generic with respect to the invariants x_i, i.e., the functions $x_1(\omega), \dots, x_n(\omega)$ form a local coordinate system on M. We define the value of the invariant p^j_σ on ω by

$$p^j_\sigma(\omega) = \frac{\partial^{|\sigma|}\omega^j}{\partial x_\sigma(\omega)}(x_1(\omega), \dots, x_n(\omega)),$$

where ω^j are the components of ω in the coordinates $x_1(\omega), \dots, x_n(\omega)$. It is easy to see [**2**, Chapter 7] that any differential invariant is a function in x_i and p^j_σ. On the other hand, these invariants are related by the equations

$$\psi_i(\dots, p^j_\sigma, \dots) = x_i, \qquad i = 1, \dots, n, \tag{3.25}$$

where ψ_i is a local expression for the invariant x_i determined by the equation (3.24). Hence the algebra $\mathcal{A}$ of all scalar differential invariants coincides locally with the function algebra $\mathcal{F}(\mathcal{Y}_\infty)$, where the equation $\mathcal{Y}$ is given by (3.25).

The algebra of scalar differential invariants can be naturally identified with the algebra of smooth functions on the diffiety $J^\infty(\pi)/\operatorname{Diffeo} M$, where $\operatorname{Diffeo} M$ is the pseudogroup of all local diffeomorphisms of M acting on $J^\infty(\pi)$: $\Phi \mapsto \hat{\Phi}^\infty$, $\Phi \in \operatorname{Diffeo} M$. Then R-manifolds in $J^\infty(\pi)/\operatorname{Diffeo} M$ correspond to the classes of equivalent geometric structures on M. Note that the diffiety $J^\infty(\pi)/\operatorname{Diffeo} M$ is naturally stratified. Namely, a point $[\theta] \in J^\infty(\pi)/\operatorname{Diffeo} M$ belongs to the k-th stratum if $\dim(\mathcal{C}_\theta \cap K_\theta) = k$, where K_θ is the tangent space to the orbit of $\operatorname{Diffeo} M$ passing through θ. The 0-th stratum will be called *regular* while all others will be called *singular*. Below it will be shown that the regular stratum of $J^\infty(\pi)/\operatorname{Diffeo} M$

is a diffiety if the type S of geometric structures under consideration satisfies two conditions. The first is *homogeneity*, which means that the pseudogroup $\operatorname{Diffeo} M$ acts on $J^0(\pi)$ transitively. The second condition is that S admits $n = \dim M$ independent scalar invariants (we say that invariants are *independent*, if they are functionally independent as functions on $J^\infty(\pi)$). Below we consider geometric structures satisfying these conditions.

In fact, since the group $\operatorname{Diffeo} M$ acts transitively on the manifold M, there exists a natural diffeomorphism $J^\infty(\pi)/\operatorname{Diffeo} M = J^\infty_x(\pi)/\operatorname{Diffeo}_x M$, where, by definition, $J^\infty_x(\pi) = \pi_\infty^{-1}(x)$ and $\operatorname{Diffeo}_x M \subset \operatorname{Diffeo} M$ is the stabilizer of the point $x \in M$. The restriction $\hat{\Phi}^{(\infty)}_x \colon J^\infty_x(\pi) \to J^\infty_x(\pi)$ of the transformation $\hat{\Phi}^{(\infty)}$ to the fiber $J^\infty_x(\pi)$ is uniquely determined by the infinite jet of the diffeomorphism Φ at the point $x \in M$. This allows one to identify canonically $J^\infty(\pi)/\operatorname{Diffeo} M$ with $J^\infty_x(\pi)/G^\infty_x M$, where $G^\infty_x M = \{\, [\Phi]^\infty_x \mid \Phi \in \operatorname{Diffeo}_x M \,\}$.

The following example illustrates the above notions. A *web* on a 2-dimensional manifold M is formed by three line distributions on M which are mutually transversal. A web can also be understood as three regular families of curves on M such that the curves of one family intersect the curves of the other families transversally.

A web W can be given by three vector fields X_1, X_2, X_3 whose integral curves are the curves forming the corresponding three families. It is easy to see that locally one can find a local chart (x, y) on M such that these fields are

$$\frac{\partial}{\partial x}, \quad \frac{\partial}{\partial y}, \quad \frac{\partial}{\partial x} + u(x,y)\frac{\partial}{\partial y} \tag{3.26}$$

with a nowhere vanishing function $u(x, y)$. Obviously, such a chart is defined up to a transformation of the form

$$(x, y) \mapsto (f(x), g(y)). \tag{3.27}$$

The differential form

$$\omega = \omega(W) = \frac{\partial^2 \ln|u|}{\partial x \partial y}\, dx \wedge dy$$

is invariant with respect to transformations (3.27) and therefore is a differential invariant of the web. Generally, it is nonvanishing. Assuming this to be the case, vector fields (3.26) can be normalized uniquely to the fields Y_1, Y_2, Y_3 such that

$$Y_1 + Y_2 + Y_3 = 0, \quad \omega(Y_1, Y_2) + \omega(Y_2, Y_3) + \omega(Y_3, Y_1) = \pm 1.$$

Namely, one gets

$$Y_1 = F\frac{\partial}{\partial x}, \quad Y_2 = uF\frac{\partial}{\partial y}, \quad Y_3 = -F\left(\frac{\partial}{\partial x} + u\frac{\partial}{\partial y}\right)$$

with

$$F = \left| u \frac{\partial^2 \ln|u|}{\partial x \partial y} \right|^{-\frac{1}{2}}.$$

Basic scalar differential invariants I and J of W are defined by the relation

$$[Y_1, Y_2] = IY_1 - JY_2.$$

The invariants I and J can be taken for a natural local chart on M, and moreover, the local expressions for the vector fields Y_1 and Y_2 are as follows:

$$Y_1 = \zeta_1\frac{\partial}{\partial I} + (\zeta_1 + 1)\frac{\partial}{\partial J}, \quad Y_2 = \zeta_2\frac{\partial}{\partial I} + \zeta_3\frac{\partial}{\partial J}.$$

The functions ζ_1, ζ_2, ζ_3 appearing here are, obviously, other differential invariants of W. It can be shown [**144**] that they are subject to the following relations:

$$(3.28)\qquad \begin{aligned} \zeta_1\frac{\partial\zeta_2}{\partial I}+(\zeta_2+1)\frac{\partial\zeta_2}{\partial J}-\zeta_2\frac{\partial\zeta_1}{\partial I}-\zeta_3\frac{\partial\zeta_1}{\partial J}-J\zeta_1+I\zeta_2=0,\\ \zeta_1\frac{\partial\zeta_3}{\partial I}+(\zeta_2+1)\frac{\partial\zeta_3}{\partial J}-\zeta_2\frac{\partial\zeta_2}{\partial I}-\zeta_3\frac{\partial\zeta_2}{\partial J}-J(\zeta_2+1)+I\zeta_3=0.\end{aligned}$$

If the invariants I and J are dependent, then all differential invariants are functions of one of them, say I. Then, among the vector fields X_1, X_2, X_3 defining the web, there are two, say X_1 and X_2, such that the functions $X_1(I)$ and $X_2(I)$ vanish nowhere. Let us normalize them by $X_1(I)=X_2(I)=1$. Hence, $[X_1,X_2](I)=0$ and $(X_2-X_1)(I)=0$. This implies that $[X_1,X_2]=\mu(X_2-X_1)$ and also that $X_3=X_1+\alpha(X_2-X_1)$. The functions $\alpha=\alpha(I)$, $\mu=\mu(I)$ are differential invariants of this kind of webs. It is possible to prove that functions $\alpha=\alpha(I)$ and $\mu=\mu(I)$ determine the web W completely and that they satisfy the following relations:

$$(3.29)\qquad \frac{\lambda}{\beta}\frac{d}{dI}\left(\beta H(\beta)\right)^{\frac{1}{2}}=-I,$$

where $\beta=\alpha^{-1}-\alpha$, $d\lambda/dI=\mu$, and

$$H(\beta)=\left(\lambda\frac{d}{dI}\right)^2(\ln|\beta|).$$

Finally note that if all differential invariants of a web are constant, then it is (locally) a flat web, i.e., it is equivalent to a web on the plane composed of three families of parallel lines.

Now we are ready to describe the diffiety $\mathcal{O}=J^\infty(\pi)/\operatorname{Diffeo} M$, where π is the fiber bundle on M whose sections are webs on M. Namely, we have:

(i) the regular, i.e., 2-stratum of $\mathcal{O}$ is $\mathcal{Y}_\infty$, where $\mathcal{Y}$ is given by (3.28);
(ii) the 1-stratum of $\mathcal{O}$ is (locally) equivalent to $\mathcal{Y}'_\infty$, where $\mathcal{Y}'$ is given by (3.29);
(iii) the 0-stratum of $\mathcal{O}$ is just a point.

Such a situation is typical for other quotient diffieties.

3.6. Concluding remarks

We started this chapter with general arguments in favor of the use of infinite jets instead of finite ones in the construction of geometric theory of partial differential equations. Now we can be more specific by taking into account the results obtained in this chapter. One of them is very instructive: while finite-order contact structures on finite-order jet spaces are in a sense *completely nonintegrable*, the infinite-order contact structure is *completely integrable*, i.e., Frobenius. This fact has not only important theoretical consequences, but also quite important practical ones. For instance, even the problem of how to describe practically the defining equations for classical symmetries of PDE is more complicated if considered from the classical, i.e., finite-order jet, point of view. For example, in the book [**97**] dedicated to the classical theory, an algorithm together with a computer program is suggested to this end. But the infinite jet point of view leads directly to a simple solution of this problem as is done in Corollary 3.3.5 on p. 110 (equation (3.18)). The reader will find further examples of this kind in Chapters 4 and 5.

Nonclassical behavior of higher symmetries, as displayed in this chapter, is also worth a comment. Namely, as we have seen, the number of (finite) symmetries, i.e., of $\mathcal{C}$-transformations, does not increase much when passing to infinite jets—in sharp contrast with infinitesimal ones. In other words, it may seem that the classical relation between the Lie groups and Lie algebras no longer holds in the nonclassical context. One gets an explanation of this phenomenon in the framework of Secondary Calculus (Chapter 5), which reveals the fact that higher symmetries are, in fact, a kind of cohomology classes and as such do not possess trajectories, i.e., are not, generally, integrable (see Chetverikov's theorem in [**30**]). The quotient mapping $D_{\mathcal{C}}(\mathcal{Y}_\infty) \to \operatorname{sym} \mathcal{Y} = D_{\mathcal{C}}(\mathcal{Y}_\infty)/\mathcal{C}D(\mathcal{Y}_\infty)$ is exactly the mapping associating to a "cochain" the corresponding cohomology class.

Nonclassical period in the symmetry theory of partial differential equations started with the discovery of integrable systems in the famous series of works by Kruscal, Miura, et al. [**93, 94, 105, 37, 76, 38**]. It was understood that higher analogs of the Korteweg–de Vries equation and of other integrable systems are in fact some kind of symmetries of these equations. Mathematically rigorous fundamentals of the higher symmetry theory were developed by the author in [**137, 121**] (see also [**55, 124, 142**]). A detailed exposition of the mathematical background, apart from [**142**], which passed almost unnoticed, was given for the first time in [**143, 71**], while in [**53**] and [**96**] the subject was presented as seen by experts in mathematical physics. In the recent monograph [**19**] an attempt was made to put together mathematical rigor and computational aspects important for applications. In this book, the reader will find numerous examples and applications of higher symmetries along the lines developed in this chapter. For an alternative user-oriented approach see [**96**]. It is worth mentioning that in the literature higher infinitesimal symmetries are also called Lie–Bäcklund transformations [**54**], or generalized symmetries [**96**].

The theory presented in this chapter can in fact be developed in a purely algebraic context. This is now on the agenda, and all necessary data to begin with seem to be at our disposal [**65, 67**]. This possibility is assured first of all by differential calculus over arbitrary commutative algebras [**95**].

CHAPTER 4

$\mathcal{C}$-spectral Sequence and Some Applications

This chapter deals with the $\mathcal{C}$-spectral sequence, which, as will be shown in Chapter 5, is a central construction in Secondary Calculus. Originally, the $\mathcal{C}$-spectral sequence was discovered when the author was looking for conceptual foundations of the Lagrangian formalism in full generality. Afterwards, the $\mathcal{C}$-spectral sequence turned out to be crucial for capturing the first idea of Secondary Calculus. Moreover, subsequent more accurate analysis of the terms of this spectral sequence allowed one to express this idea in a rather precise form. It became clear later that associated with any functor of differential calculus as well as with its representing object is a pair of $\mathcal{C}$-spectral type sequences, one dual to another. The spectral sequence introduced and studied in this chapter corresponds in this sense to differential forms. $\mathcal{C}$-spectral sequences corresponding to some other functors are discussed in Chapter 5.

The original paper [**121**] appeared in 1978 (see also [**134**]). Four years later, T. Tsujishita [**112**] published an alternative interpretation of a local version of the $\mathcal{C}$-spectral sequence, which he called the *variational bicomplex*. Tsujishita's approach was based on the use of the Koszul complex techniques combined with a direct coordinate description of the $\mathcal{C}$-spectral sequence terms. This approach seemed to be simpler, and some expositions of similar nature were subsequently written on that basis. However, the use of coordinate considerations instead of the natural language of differential calculus in commutative algebras, makes it hardly possible to reveal the underlying basic structures of the theory and creates some additional difficulties in problems of global nature. It is unlikely that Secondary Calculus could have been discovered by starting from the "bicomplex" point of view.

In this chapter and later, only formally integrable systems are considered.

4.1. Spencer complexes and the Green formula in the $\mathcal{C}$-theory

To construct nonlinear Lagrangian formalism and nonlinear conservation laws, we shall need, just like in the linear situation, the theory of the adjoint operator in the framework of Spencer complexes. Below we show that in the $\mathcal{C}$-theory on $\mathcal{Y}_\infty$ it can be done by the same techniques as those used in Chapter 1. This becomes possible due to the functorial nature of the language used there.

4.1.1. Action of operators on forms in the $\mathcal{C}$-theory. Let $A = \mathcal{F}(E, n)$ or $A = \mathcal{F}(\pi)$. In this case $\bar{\Lambda}^n$ (see Subsection 3.1.7) is a one-dimensional locally free A-module, where "locally" means "inside any special chart". In fact, the module $\bar{\Lambda}^n$ on $J^\infty(\pi)$ can be identified with $\bar{\Lambda}^n_\Gamma = A \otimes_{C^\infty(M)} \Lambda^n(M)$ (see Corollary 3.2.2).

If $A = \mathcal{F}(\mathcal{Y})$, then the module $\bar{\Lambda}^n = \bar{\Lambda}^n(\mathcal{Y}_\infty)$ is also locally free since $\bar{\Lambda}^n$ is the restriction of $\bar{\Lambda}^n(J^\infty(E,n))$ (or of $\bar{\Lambda}^n(J^\infty(\pi))$) to $\mathcal{Y}_\infty \subset J^\infty(E,n)$ (or $\subset J^\infty(\pi)$). In particular, $\bar{\Lambda}^i = 0$ for $i > n$ and consequently $\bar{d}(\bar{\Lambda}^n) = 0$.

PROPOSITION. *Let A be equal to $\mathcal{F}(E,n)$, $\mathcal{F}(\pi)$, or $\mathcal{F}(\mathcal{Y})$. Then there exists a uniquely defined action $\Delta[\bar{\omega}]$ of operators $\Delta \in \mathcal{C}\mathrm{Diff}\, A$ on forms $\bar{\omega} \in \bar{\Lambda}^n$ satisfying axioms* 1 – 4 *of Subsection* 1.2.1.

PROOF. The arguments follow the proof of Proposition 1.2.1, since according to Corollary 3.2.7 the module $\mathcal{C}\mathrm{Diff}\, A$ is generated by $\mathcal{C}D(A)$ and A. The only point to be clarified is the notion of the Lie derivative $X(\bar{\omega}) \in \bar{\Lambda}^k$, $X \in \mathcal{C}D(A)$, $\bar{\omega} \in \bar{\Lambda}^k$, satisfying the relation $X(\bar{\omega}) = \bar{d}(X \lrcorner \bar{\omega}) + X \lrcorner \bar{d}\bar{\omega}$. To do this, note that the substitution $X \lrcorner \bar{\Lambda} = \overline{X \lrcorner \omega} \in \bar{\Lambda}^{k-1}$, $X \in \mathcal{C}D(A)$, $\omega \in \Lambda^k$, is well defined, where the bar denotes the coset mod $\mathcal{C}\bar{\Lambda}^k$. Indeed, since $\mathcal{C}\Lambda^* = \mathcal{C}\Lambda^1 \wedge \Lambda^*$, from $X \lrcorner \mathcal{C}\Lambda^1 = 0$ it follows that $X \lrcorner \mathcal{C}\Lambda^k \subset \mathcal{C}\Lambda^{k-1}$. But since $\mathcal{C}D(A) \subset D_{\mathcal{C}}(A)$, the definition $X(\bar{\omega}) = \overline{X(\omega)}$ is well posed and $\overline{d(X \lrcorner \omega)} + \overline{X \lrcorner d\omega} = \bar{d}(\overline{X \lrcorner \omega}) + X \lrcorner \overline{d(\omega)} = \bar{d}(X \lrcorner \bar{\omega}) + X \lrcorner \bar{d}\bar{\omega}$. □

4.1.2. The adjoint operator in the $\mathcal{C}$-theory. Below we assume that $A = \mathcal{F}(E,n)$, $\mathcal{F}(\pi)$, or $\mathcal{F}(\mathcal{Y})$. In any of these cases the theory of the adjoint operator is constructed in exactly the same way as in Section 1.2. To be more precise, let $\Delta \in \mathcal{C}\mathrm{Diff}\, \bar{\Lambda}^n$. Then the operator $\Delta^* \in \mathrm{Diff}\, \bar{\Lambda}^n$ is defined by the formula $\Delta^*(f) = \Delta_{\bar{\omega}}(f\bar{\omega})$, where $\bar{\omega} \in \bar{\Lambda}^n$ is a local horizontal "volume form", i.e., a local generator of the A-module $\bar{\Lambda}^n$, while $\Delta(f) = \Delta_{\bar{\omega}}(f)\bar{\omega}$. Correctness of this definition is checked in the same way as in Subsection 1.2.2. It only remains to verify that Δ^* is a $\mathcal{C}$-differential operator. It is obvious for $\Delta = f$, $f \in A$. If $\Delta = \square \circ X$, $\square \in \mathcal{C}\mathrm{Diff}\, \bar{\Lambda}^n$, $X \in \mathcal{C}D(A)$, then

$$\Delta^*(f) = (\square \circ X)_\omega[f\bar{\omega}] = (\square_{\bar{\omega}} \circ X)[f\bar{\omega}] = -X(\square_{\bar{\omega}}[f\bar{\omega}]) = -X(\square^*(f)),$$

i.e., $\Delta^* = -X_\Lambda \circ \square^*$, where $X_\Lambda(\bar{\omega}) = X(\bar{\omega})$. But the composition of $\mathcal{C}$-differential operators is a $\mathcal{C}$-differential operator as well. So, one can prove that Δ^* is a $\mathcal{C}$-differential operator by induction on $\deg \Delta$ using the representation $\Delta_{\bar{\omega}} = f + \sum X^i_{i_1} \circ \cdots \circ X^i_{s(i)}$, $f \in A$, $X^i_j \in \mathcal{C}D(A)$.

Let $\hat{Q} = \mathrm{Hom}_A(Q, \bar{\Lambda}^n)$. As in Section 1.2, for each $\Delta \in \mathcal{C}\mathrm{Diff}(P, \hat{Q})$ we define the adjoint operator $\Delta^* \in \mathcal{C}\mathrm{Diff}(Q, \hat{P})$ by setting $\Delta^*(q,p) = \Delta(p,q)^*$, $q \in Q$. Since the operators $\Delta^*(q,p)$ are $\mathcal{C}$-differential for all p, q, the operator Δ^* is $\mathcal{C}$-differential as well.

PROPOSITION. *Propositions* 1.2.2 *and* 1.2.3 *remain valid for $\mathcal{C}$-differential operators if one considers the module $\bar{\Lambda}^n$ instead of Λ^n and the functors $\mathcal{C}\,\mathrm{smbl}_s$ instead of* smbl_s.

PROOF. The proof of this statement repeats word for word the proofs of Propositions 1.2.2 and 1.2.3. □

Let A be $\mathcal{F}(E,n)$, $\mathcal{F}(\pi)$, or $\mathcal{F}(\mathcal{Y})$. Choose a special chart and take the form $\bar{\omega} = dx_1 \wedge \cdots \wedge dx_n$ as a generator in $\bar{\Lambda}^n$. Taking into account that any $\Delta \in \mathcal{C}\mathrm{Diff}\, A$ is representable in the form $\sum_\sigma a_\sigma D_\sigma$, $a_\sigma \in A$ (Corollary 3.2.7), one can find that

$$\Delta^*_{\bar{\omega}} = \sum_\sigma (-1)^{|\sigma|} D_\sigma \circ a_\sigma,$$

where $\Delta \in \mathcal{C}\mathrm{Diff}\bar{\Lambda}^n$, $\Delta_{\bar{\omega}} = \sum_\sigma a_\sigma D_\sigma$. If, as in Subsection 1.2.4, the operator $\Delta \in \mathcal{C}\mathrm{Diff}_s(P,Q)$ is described by a matrix $\|(\Delta_{ij})_{\bar{\omega}}\|$, $\Delta_{ij} \in \mathcal{C}\mathrm{Diff}\bar{\Lambda}^n$, then Δ^* is given by the matrix $\|c_{kl}\|$, where $c_{kl} = (\Delta_{lk})^*_{\bar{\omega}}$.

4.1.3. The Spencer $\mathcal{C}$Diff-complexes. Here we prove the exactness of the Spencer complexes in the $\mathcal{C}$-theory. To avoid cumbersome notations, we denote the composition of differential calculus functors $\mathfrak{F}_1 \circ \cdots \circ \mathfrak{F}_k$ in the $\mathcal{C}$-theory by $\mathcal{C}\mathfrak{F}_1 \circ \cdots \circ \mathfrak{F}_k$ instead of $\mathcal{C}\mathfrak{F}_1 \circ \cdots \circ \mathcal{C}\mathfrak{F}_k$, e.g., $\mathcal{C}D_i\mathrm{Diff}^+_k = \mathcal{C}D_i(\mathcal{C}\mathrm{Diff}^+_k)$, etc.

The proof of exactness is based on the following considerations. Let $P \Rightarrow S_kP$ (resp., $P \Rightarrow SP$) be the functor taking the module P to the corresponding Spencer complex S_kP (resp., SP) in the differential calculus in question, e.g., in the $\mathcal{C}$-theory. Let this functor be representable. Then its representing object is a complex consisting of representing objects for the functors $D^\bullet_i\mathrm{Diff}^+_j$. Since these objects are $\mathcal{J}^j(\Lambda^i)$, this complex, denoted by $(\mathcal{J}_k)$, is of the form

$$0 \to A \xrightarrow{j_k} \mathcal{J}^k \xrightarrow{S_{\mathcal{J}}} \mathcal{J}^{k-1}(\Lambda^1) \xrightarrow{S_{\mathcal{J}}} \cdots \xrightarrow{S_{\mathcal{J}}} \mathcal{J}^{k-i}(\Lambda^i) \xrightarrow{S_{\mathcal{J}}} \cdots .$$

From the explicit formula for the operators $S_{i,j}$ (see Subsection 1.3.1) it immediately follows that

$$S_{\mathcal{J}}(fj_s(\omega)) = fj_{s-1}(d\omega), \quad \omega \in \Lambda^i$$

(see also [**71**]). The following result is obvious.

LEMMA. *If the complex $(\mathcal{J}_k)$ is exact and the modules $\mathcal{J}^j(\Lambda^i)$ are projective, then the Spencer complexes S_kP and SP are also exact in the category under consideration.*

Let us apply this lemma to the $\mathcal{C}$-theory. The Spencer Diff-complex in this theory, denoted by $\mathcal{C}S_kP$ or $\mathcal{C}SP = \mathcal{C}S_\infty P$, looks as follows:

$$0 \leftarrow P \leftarrow \mathcal{C}\mathrm{Diff}^+_k P \xleftarrow{S_{1,k-1}} \mathcal{C}D^\bullet(\mathrm{Diff}^+_{k-1}P) \xleftarrow{S_{2,k-2}} \cdots$$
$$\xleftarrow{S_{i,k-i}} \mathcal{C}D^\bullet_i(\mathrm{Diff}^+_{k-i}P) \xleftarrow{S_{i+1,k-i-1}} \cdots .$$

According to Subsection 3.1.7, $\overline{\mathcal{J}^s(\Lambda^i)}$ is the representing object for the functor $\mathcal{C}D^\bullet_i(\mathrm{Diff}^+_s)$. Thus, it suffices to show that the complex $(\bar{\mathcal{J}}_k)$,

$$0 \to A \xrightarrow{\bar{\imath}_k} \bar{\mathcal{J}}^k \xrightarrow{\bar{S}_{\mathcal{J}}} \overline{\mathcal{J}^{k-1}(\Lambda^1)} \xrightarrow{\bar{S}_{\mathcal{J}}} \cdots \xrightarrow{\bar{S}_{\mathcal{J}}} \overline{\mathcal{J}^{k-i}(\Lambda^i)} \xrightarrow{\bar{S}_{\mathcal{J}}} \cdots ,$$

satisfies the assumptions of the lemma.

If A is $\mathcal{F}(\pi)$ or $\mathcal{F}(\mathcal{Y})$, where $\mathcal{Y}_\infty \subset J^\infty(\pi)$, then it follows from the isomorphism $\bar{\Phi} = \Phi_\Gamma = \Phi(M) \otimes_{C^\infty(M)} A$ (see Subsection 3.2.2) that the modules $\overline{\mathcal{J}^s(\Lambda^i)}$ are projective. In the case $A = \mathcal{F}(E,n)$ or $\mathcal{F}(\mathcal{Y})$, where $\mathcal{Y}_\infty \subset J^\infty(E,n)$, this isomorphism takes place inside any chart, and consequently $\overline{\mathcal{J}^s(\Lambda^i)}$ are also projective modules.

Finally, the complex $\bar{\mathcal{J}}_k$ is exact. In fact, the operators $S_{\mathcal{J}}$ are A-homomorphisms and $\bar{\mathcal{J}}_k = \mathcal{J}_kM \otimes_{C^\infty(M)} A$, where $\mathcal{J}_kM$ is the Spencer $\mathcal{J}_k$-complex in the category of geometric modules over $C^\infty(M)$. But $\mathcal{J}_kM$ is an exact complex (see [**71, 133**]). Thus we have proved

THEOREM. *The complexes $\bar{\mathcal{J}}_k$, as well as the complexes $\mathcal{C}S_kP$ and $\mathcal{C}SP$, over A are exact for $A = \mathcal{F}(\mathcal{Y})$ and for all $k \leq \infty$.*

4.1.4. A $\mathcal{C}$-analog of the theory of Section 1.3. All the results and constructions from Section 1.3, appropriately modified, remain valid in the $\mathcal{C}$-theory. It is demonstrated below with necessary comments.

Constructions and formulas of Subsection 1.3.1 can be repeated verbatim in the $\mathcal{C}$-theory if all operators are assumed to be $\mathcal{C}$-differential and each functor $\mathfrak{F}$ is replaced by $\mathcal{C}\mathfrak{F}$.

The Poincaré duality in the $\mathcal{C}$-theory (which is needed to realize constructions of Subsection 1.3.2) is the isomorphism $\Pi\colon \mathcal{C}D_i(\bar{\Lambda}^n) \to \bar{\Lambda}^{n-i}$ defined by

$$\Pi(X_1 \wedge \cdots \wedge X_i \otimes \bar{\omega}) = X_1 \lrcorner (\ldots (X_i \lrcorner \bar{\omega}) \ldots).$$

Here $X_i \in \mathcal{C}D(A)$ and $\bar{\omega} \in \bar{\Lambda}^n$ is a local volume form, whereas the operation $Y \lrcorner \bar{\rho}$, $Y \in \mathcal{C}D(A)$, $\bar{\rho} \in \bar{\Lambda}^k$, is understood in the same sense as in Subsection 4.1.1. The fact that Π is an isomorphism follows from the local representation $\bar{\Lambda}^i = \Lambda^i(M) \otimes_{C^\infty(M)} A$ valid inside any special chart.

The complex

$$0 \leftarrow \bar{\Lambda}^n \xleftarrow{-\mu} \mathcal{C}\mathrm{Diff}_k^+ \xleftarrow{\mathcal{S}} \mathcal{C}\mathrm{Diff}_{k-1}^+ \bar{\Lambda}^{n-1} \xleftarrow{\mathcal{S}} \cdots \xleftarrow{\mathcal{S}} \mathcal{C}\mathrm{Diff}_{k-n}^+ A \leftarrow 0 \tag{4.1}$$

is defined in the same way: $\mathcal{S}(\nabla) = \bar{d} \circ \nabla$, $\mu(\nabla) = \nabla^*(1)$. It remains to note that the operator $\bar{d} \circ \nabla$ is also $\mathcal{C}$-differential since $\bar{d}$ is a $\mathcal{C}$-differential operator (Corollary 3.2.3).

We can now construct analogs of the isomorphisms

$$\psi\colon \mathcal{C}D_i^\bullet(\mathrm{Diff}_s^+ \bar{\Lambda}^n) \to \mathcal{C}\mathrm{Diff}_s^+ \bar{\Lambda}^{n-i}$$

and prove analogs of Theorem 1.3.2 and of Corollary 1.3.2. The same is valid for the complexes $\mathcal{S}_k P$ and $\mathcal{S}_{k,l} PQ$. For the sake of briefness, we keep the same notations for their $\mathcal{C}$-analogs. Theorem 1.4.1 together with its proof is also carried over to the $\mathcal{C}$-theory.

Proposition 1.3.3 is valid for the de Rham $\bar{d}$-complex

$$\cdots \to \bar{\Lambda}^i \xrightarrow{\bar{d}} \bar{\Lambda}^{i+1} \to \cdots$$

and for the Spencer $\overline{\mathcal{J}_k P}$-complex.

Now the only thing we need to prove the $\mathcal{C}$-Green formula is the existence of a splitting $\lambda\colon \widetilde{\mathcal{C}\mathrm{Diff}_k} \bar{\Lambda}^n \to \mathcal{C}D^\bullet(\mathrm{Diff}_{k-1}^+ \bar{\Lambda}^n)$. This follows from the projectivity of the A-modules $\widetilde{\mathcal{C}\mathrm{Diff}_k} \bar{\Lambda}^n$, which, in turn, can be deduced from the representation $\mathcal{C}\mathrm{Diff}_k \bar{\Lambda}^i = A \otimes_{C^\infty(M)} \mathrm{Diff}_k \Lambda^i(M)$ valid in any special chart. This isomorphism is realized by the mapping $a \otimes \Delta \mapsto a\hat{\Delta}$, where $a \in A$, $\Delta \in \mathrm{Diff}_k \Lambda^i(M)$ (see Corollary 3.2.7). The $\mathcal{C}$-Green formula is of the form

$$\langle \Delta(p), q\rangle - \langle p, \Delta^* q\rangle = \bar{d}\mathcal{K}(\Delta(p,q)), \quad p \in P,\ q \in Q,\ \Delta \in \mathcal{C}\mathrm{Diff}(P, \hat{Q}).$$

The results of Subsection 1.3.5 are valid in the $\mathcal{C}$-theory without any changes. Using the same arguments as in Subsection 1.3.6 and taking $\bar{\omega} = dx_1 \wedge \cdots \wedge dx_n \in \bar{\Lambda}^n$, one can obtain coordinate representations of the splitting λ and the operator $\mathcal{K}_\lambda$:

$$\lambda(\Delta)_{\bar{\omega}} = \sum_\sigma a_\sigma D_{\bar{\sigma}} \otimes D_{i(\sigma)}, \qquad \Delta = \sum_{|\sigma|>0} a_\sigma D_\sigma, \quad a_\sigma \in A,$$

$$\mathcal{K}_\lambda(\Delta) = \sum_{|\sigma|>0} (-1)^{\bar{\sigma}} D_{\bar{\sigma}}(a_\sigma) \bar{\omega}_{i(\sigma)}, \qquad \bar{\omega}_i = (-1)^{i-1} D_i \lrcorner \bar{\omega}.$$

In the sequel, when referring to the $\mathcal{C}$-analogs of "usual" results and constructions, the letter $\mathcal{C}$ is added: e.g., Theorem 1.3.2 $\mathcal{C}$.

4.1.5. Transformations. Any morphism $F\colon A_1 \to A_2$ in the category DE (see Subsection 3.4.9) generates naturally a homomorphism $F_{\bar{\Lambda}} = F\colon \bar{\Lambda}^*(A_1) \to \bar{\Lambda}^*(A_2)$. If F is an isomorphism in the category DE, then, in addition, it generates, by the same formulas and under the same assumptions as in Subsection 1.2.5, the transformations $F_{\mathrm{Diff}}\colon \mathrm{Diff}(P_1, Q_1) \to \mathrm{Diff}(P_2, Q_2)$ (see Subsection 1.2.5). Moreover, it is easy to see that $F_{\mathrm{Diff}}(\mathcal{C}\mathrm{Diff}(P_1, Q_1)) \subset \mathcal{C}\mathrm{Diff}(P_2, Q_2)$, so it makes sense to define

$$F_{\mathcal{C}\mathrm{Diff}} = F_{\mathrm{Diff}}|_{\mathcal{C}\mathrm{Diff}(P_2,Q_2)}.$$

Then all definitions and formulas of Subsections 1.2.5 and 1.3.7 remain valid under the replacements $\Lambda^* \Rightarrow \bar{\Lambda}^*$, $\mathrm{Diff} \Rightarrow \mathcal{C}\mathrm{Diff}$, $F_{\Lambda} \Rightarrow F_{\bar{\Lambda}}$, $F_{\mathrm{Diff}} \Rightarrow F_{\mathcal{C}\mathrm{Diff}}$. It is interesting to note that in the situation under consideration one can take for F not only isomorphisms but also restriction mappings. To be more precise, let $\mathcal{Y}$ be a consequence of the equation $\mathcal{Y}'$, i.e., $i\colon \mathcal{Y}'_\infty \hookrightarrow \mathcal{Y}_\infty$. Let $A = \mathcal{F}(\mathcal{Y})$, $A' = \mathcal{F}(\mathcal{Y}')$, $F = i^*\colon A \to A'$, and let $I_{\mathcal{Y}}$, $I_{\mathcal{Y}'}$ be the ideals of the equations $\mathcal{Y}_\infty$ and $\mathcal{Y}'_\infty$ in $\mathcal{F}(E, n)$ (or in $\mathcal{F}(\pi)$), respectively. Since the ideal $I = I_{\mathcal{Y}'}/I_{\mathcal{Y}}$ is stable with respect to the action of $\mathcal{C}\mathrm{Diff}\, A$, from $F_{\bar{\Lambda}}(\bar{\omega}) = 0$, $\bar{\omega} \in \bar{\Lambda}^*(A)$, it follows that $F_{\bar{\Lambda}}(X(\bar{\omega})) = 0$ for any $X \in \mathcal{C}D(A)$. Besides, the restriction $F_{\mathcal{C}\mathrm{Diff}}\colon \mathcal{C}\mathrm{Diff}\, A \to \mathcal{C}\mathrm{Diff}\, A'$ is naturally defined and is an epimorphism (see Subsection 3.2.8). From these two facts it follows that the operation

$$\Delta[\bar{\omega}]_0 = F_{\bar{\Lambda}}(\square[\bar{\rho}])$$

with $\square \in \mathcal{C}\mathrm{Diff}\, A$, $\Delta = F_{\mathcal{C}\mathrm{Diff}}(\square)$, $\bar{\rho} \in \bar{\Lambda}^n(A)$, $\bar{\omega} = F_{\bar{\Lambda}}(\bar{\rho})$, is well defined and satisfies the axioms of action. Due to uniqueness of the action, one has $[\bullet]_0 = [\bullet]_{A'}$, i.e., the operation $[\bullet]$ is compatible with this kind of restrictions. This implies that

$$F_{\mathcal{C}\mathrm{Diff}}(\square^*) = F_{\mathcal{C}\mathrm{Diff}}(\square)^*, \quad \square \in \mathcal{C}\mathrm{Diff}\, A.$$

Obvious functorial arguments show that the homomorphism F generates homomorphisms of the Spencer $\mathcal{C}\mathrm{Diff}$-complexes over A to the corresponding Spencer $\mathcal{C}\mathrm{Diff}$-complexes over A'. One also has homomorphisms of complexes $\mathcal{S}_k P \to \mathcal{S}_k P'$, $\mathcal{S}_k PQ \to \mathcal{S}_k P'Q'$ in the $\mathcal{C}$-theory, where $P' = P|_{\mathcal{Y}'_\infty}$, $Q' = Q|_{\mathcal{Y}'_\infty}$. In particular, according to Theorem 1.3.2 $\mathcal{C}$, the Green formula over $\mathcal{Y}_\infty$ admits a restriction to $\mathcal{Y}'_\infty$ for any operator $\square \in \mathcal{C}\mathrm{Diff}(P, Q)$, and this restriction is the Green formula for $\Delta = F_{\mathcal{C}\mathrm{Diff}}(\square)$.

If $\mathcal{Y}' = L^{(k)} = \operatorname{im} j_k(L) \subset J^k(E, n)$ (resp., $\mathcal{Y}' = M_s^{(k)} \subset J^k(\pi)$, $s \in \Gamma_{\mathrm{loc}}(\pi)$), then, independently of k, $\mathcal{Y}_\infty = L^{(\infty)}$ (resp., $\mathcal{Y}_\infty = M_s^{(\infty)}$). For this reason, one can take $C^\infty(L)$ (resp., $C^\infty(\mathcal{U})$, where $\mathcal{U}$ is the domain of s) for A' and $j(L)^*$ (resp., $j(s)^*$) for F. Thus the theory of the adjoint operator and the formulas related to this theory are compatible with homomorphisms of the form $F = j(L)^*$ or $j(s)^*$. Let us emphasize that in this case $\mathcal{C}\Phi = 0$ for all $\Phi = \Phi(\mathcal{Y}'_\infty)$ and consequently $\bar{\Phi} = \Phi$. Hence F induces a homomorphism $F_{\bar{\Lambda}}\colon \bar{\Lambda}^*(A) \to \Lambda^*(L)$ (resp., $F_{\bar{\Lambda}}\colon \bar{\Lambda}^*(A) \to \Lambda^*(\mathcal{U})$). In other words, any $L \subset E$ (resp., $\mathcal{U} \subset M$) can be regarded as an object of the category DE if one sets $\mathcal{C} \equiv 0$. Then $j(L)$ (resp., $j(s)$) is a morphism in DE.

The theory of infinitesimal transformations in the situation under consideration is also somewhat richer than in the "usual" case. In fact, all definitions and formulas of Subsections 1.2.5 and 1.3.7 together with their proofs remain valid if for infinitesimal transformations one takes the fields $X \in \mathcal{C}D(A)$. Moreover, the class of such transformations can be extended to the fields of $D_{\mathcal{C}}(A)$. The proofs of corresponding formulas differ from those given in Subsections 1.2.5 and 1.3.7 and are presented below. For example, if $X \in D_{\mathcal{C}}(A) \setminus \mathcal{C}D(A)$, then the equality $[X, \Delta]^* = [X, \Delta^*]$ cannot be proved using the definition of the commutator on the left-hand side and properties of the operation $X \mapsto X^*$, since the operator X^* makes no sense in this case.

Thus below we assume that $X \in D_{\mathcal{C}}(A)$ and preserve the definitions of infinitesimal transformations of corresponding objects as given in Subsections 1.2.5 and 1.3.7. Let us emphasize that according to the definition of $D_{\mathcal{C}}(A)$, the equality $X(\bar{\omega}) = \overline{X(\omega)}$, $\omega \in \Lambda^k$, $\bar{\omega} = (\omega \bmod \mathcal{C}\Lambda^k) \in \bar{\Lambda}^k$, gives a well-posed definition of the Lie derivative of forms in $\bar{\Lambda}^*$ along fields $X \in D_{\mathcal{C}}(A)$.

1. The proof of the formula $\Delta[\bar{\omega}]_X = 0$, or $X(\Delta[\bar{\omega}]) = [X, \Delta][\bar{\omega}] + \Delta[X(\bar{\omega})]$.

The formula is obvious for $\Delta \in A = \mathcal{C}\mathrm{Diff}_0 A$, since $[X, \Delta] = X(\Delta)$ in this case. Now let $\Delta \in \mathcal{C}D(A)$. Then

$$[X, \Delta][\bar{\omega}] = -[X, \Delta](\bar{\omega}) = \Delta(X(\bar{\omega})) - X(\Delta(\bar{\omega})),$$
$$\Delta[X(\bar{\omega})] = -\Delta(X(\bar{\omega})), \quad X(\Delta[\bar{\omega}]) = -X(\Delta(\bar{\omega})).$$

Substituting these expressions into the desired formula one sees its validity. Finally, let $\Delta = \Box \circ \nabla$ and assume that the formula is valid for both $\Box$ and ∇. Then, first,

$$X(\Delta[\bar{\omega}]) = X((\Box \circ \nabla)[\bar{\omega}]) = X(\nabla[\Box[\bar{\omega}]]),$$
$$\Delta[X(\bar{\omega})] = (\Box \circ \nabla)[X(\bar{\omega})] = \nabla[\Box[X(\bar{\omega})]].$$

Second, using the relation $[X, \Box \circ \nabla] = \Box \circ [X, \nabla] + [X, \Box] \circ \nabla$, one obtains $[X, \Delta][\bar{\omega}] = [X, \nabla][\Box[\bar{\omega}]] + \nabla[[X, \Box][\bar{\omega}]]$. Since the desired formula is valid for the operators ∇ and $\Box$, one has

$$[X, \nabla][\Box[\bar{\omega}]] = X(\nabla[\Box[\bar{\omega}]]) - \nabla[X(\Box[\bar{\omega}])],$$
$$[X, \Box][\bar{\omega}] = X(\Box[\bar{\omega}]) - \Box[X(\bar{\omega})],$$

so that

$$[X, \Delta][\bar{\omega}] = X(\nabla[\Box[\bar{\omega}]]) - \nabla[\Box[X(\bar{\omega})]].$$

Putting together all these relations, we get the needed result.

Finally, taking into account that $\mathcal{C}\mathrm{Diff}\, A$ is generated by A and $\mathcal{C}D(A)$, the formula is proved by induction on $\mathrm{ord}\, \Delta$.

2. The formula $[X, \Delta]^* = [X, \Delta^*]$, where $\Delta \in \mathcal{C}\mathrm{Diff}\, \bar{\Lambda}^n$.

Let $\bar{\omega} \in \bar{\Lambda}^n$ be a local volume form. It is directly checked that

$$[X, \Delta]_{\bar{\omega}} = [X, \Delta_{\bar{\omega}}] + (\mathrm{div}_{\bar{\omega}}\, X) \cdot \Delta_{\bar{\omega}},$$

where $X(\bar{\omega}) = \mathrm{div}_{\bar{\omega}}\, X \cdot \bar{\omega}$. Then

$$[X, \Delta]^*(f) = [X, \Delta]_{\bar{\omega}}[f\bar{\omega}] = [X, \Delta_{\bar{\omega}}][f\bar{\omega}] + \Delta_{\bar{\omega}}[\mathrm{div}_{\bar{\omega}}\, X \cdot f\bar{\omega}].$$

But according to the formula already proved,

$$[X, \Delta_{\bar{\omega}}][f\bar{\omega}] = X(\Delta_{\bar{\omega}}[f\bar{\omega}]) - \Delta_{\bar{\omega}}[X(f\bar{\omega})]$$

and hence

$$[X, \Delta]^*(f) = X(\Delta_{\bar{\omega}}[f\bar{\omega}]) - \Delta_{\bar{\omega}}[X(f)\bar{\omega}].$$

The result is now obtained from the equalities

$$X(\Delta_{\bar{\omega}}[f\bar{\omega}]) = X(\Delta^*(f)), \quad \Delta_{\bar{\omega}}[X(f)\bar{\omega}] = \Delta^*(X(f)).$$

3. The formula $[X,\Delta]^* = [X,\Delta^*]$, where $\Delta \in \mathcal{C}\mathrm{Diff}(P,\hat{Q})$.
Recall that here two operators $X_P \in \operatorname{Der} P$ and $X_Q \in \operatorname{Der} Q$ covering X are assumed to be given and the commutators are understood as

$$[X,\Delta] = \hat{X}_Q \circ \Delta - \Delta \circ X_P \quad \text{and} \quad [X,\Delta^*] = \hat{X}_Q \circ \Delta^* - \Delta^* \circ X_P,$$

where for any $Y \in \operatorname{Der} R$ covering X the operator $\hat{Y} \in \operatorname{Der} \hat{R}$ is defined as

$$Y(\hat{r}) = X_{\bar{\Lambda}} \circ \hat{r} - \hat{r} \circ Y \colon R \to \bar{\Lambda}^n.$$

Below we omit the subscripts P and Q of X.

Recall that by definition

$$[X,\Delta^*](p,q) = ([X,\Delta](p,q))^* = (q^* \circ [X,\Delta] \circ p)^*$$

for any $p \in P$, $q \in Q$ (see Subsection 1.2.3). We use the following obvious relation:

$$\begin{aligned}[X, q^* \circ \Delta \circ p] &= [X,q^*] \circ \Delta \circ p + q^* \circ [X,\Delta] \circ p + q^* \circ \Delta \circ [X,p] \\ &= X(q)^* \circ \Delta \circ p + q^* \circ [X,\Delta] \circ p + q^* \circ \Delta \circ X(p).\end{aligned}$$

In other words,

$$q^* \circ [X,\Delta] \circ p = [X, q^* \circ \Delta \circ p] - X(q)^* \circ \Delta \circ p - q^* \circ \Delta \circ X(p).$$

But since $q^* \circ \Delta \circ p \in \mathcal{C}\mathrm{Diff}\,\bar{\Lambda}^n$, we find, by the formula proved above, that $[X, q^* \circ \Delta \circ p]^* = [X, p^* \circ \Delta^* \circ q]$. Therefore,

$$\begin{aligned}(q^* \circ [X,\Delta] \circ p)^* &= [X, p^* \circ \Delta^* \circ q] - p^* \circ \Delta^* \circ X(q) - X(p)^* \circ \Delta^* \circ q \\ &= p^* \circ [X,\Delta^*] \circ q = [X,\Delta^*](p,q).\end{aligned}$$

4. The formulas $X(\mu) = 0$ and $X(\mathcal{K}_\lambda) = \bar{d} \circ v$, $v \in \mathcal{C}\mathrm{Diff}(\mathcal{C}\mathrm{Diff}(\bar{\Lambda}^n, \bar{\Lambda}^{n-2}))$.
Using the results already obtained, these formulas can be proved in the same way as was done in Subsection 1.3.7.

4.2. Nonlinear Lagrangian formalism

In this section, we describe the basic notions of the nonlinear Lagrangian formalism in terms adequate for our purposes. This allows us to prove the most general version of the Noether theorem and, later, to put the Lagrangian formalism in the framework of the $\mathcal{C}$-spectral sequence.

4.2.1. Lagrangian densities. An element of the set $\Gamma_{\mathrm{loc}}(\pi)$ or an n-dimensional submanifold of $E = E^{n+m}$ is said to be *admissible* for a given variational problem if it satisfiess all additional conditions of this problem, e.g., boundary conditions or constraints. The set of such sections (submanifolds) will be denoted by $\Gamma_{\mathrm{adm}}(\pi)$ (resp., by $\mathcal{M}^n_{\mathrm{adm}}(E)$). We set

$$B' = \{j(\sigma)(x) \mid \sigma \in \Gamma_{\mathrm{adm}},\ x \in \operatorname{int} \mathcal{U}_\sigma\}$$

and, respectively,

$$B' = \{j(V)(x) \mid V \in \mathcal{M}^n_{\mathrm{adm}}(E),\ x \in \operatorname{int} V\}.$$

Boundary conditions determine in the space $J^\infty(\pi)$ (resp., $J^\infty(E,n)$), a "submanifold" ∂B of boundary conditions characterized by the property

$$j(\sigma)(\partial \mathcal{U}_\sigma) \subset \partial B, \quad \sigma \in \Gamma_{\mathrm{adm}}(\pi),$$

and, respectively,

$$j(V)(\partial V) \subset \partial B, \quad V \in \mathcal{M}^n_{\mathrm{adm}}(E).$$

Let $B = B' \cup \partial B$ and $\bar{\Lambda}^n(B, \partial B) = \{\omega \in \bar{\Lambda}^*(B) \mid \omega|_{\partial B} = 0\}$.

Usually an expression of the form $\mathcal{L} = \int L dx_1 \wedge \cdots \wedge dx_n$ is called a *Lagrangian*, and the function L depending on functions $u_1, \dots, u_m$ and their derivatives is called a *Lagrangian density*. But since the integrands are n-dimensional forms, a more invariant approach consists in treating the form $L dx_1 \wedge \cdots \wedge dx_n$ as a Lagrangian density. This form can be interpreted as a horizontal form on $J^\infty(\pi)$. Thus, taking into account the fact that Λ^n_Γ is naturally identified with $\bar{\Lambda}^n$ (see Subsection 3.2.2), the following definition is naturally motivated.

DEFINITION. A form $\omega \in \bar{\Lambda}^n = \bar{\Lambda}^n(B, \partial B)$ is called a *Lagrangian density*.

4.2.2. Lagrangians. A Lagrangian density ω over $\Gamma_{\mathrm{adm}}(\pi)$ (respectively, over $\mathcal{M}^n_{\mathrm{adm}}(N)$) determines the functional

$$\sigma \mapsto \int_{\mathcal{U}_\sigma, \partial \mathcal{U}_\sigma} j(\sigma)^*(\omega), \quad \sigma \in \Gamma_{\mathrm{adm}}(\pi),$$

or, respectively,

$$V \mapsto \int_{V, \partial V} j(V)^*(\omega), \quad V \in \mathcal{M}^n_{\mathrm{adm}}(E),$$

where $j(\sigma)^*$ (resp., $j(V)^*$) is understood as a homomorphism of $\bar{\Lambda}^n(B, \partial B)$ to $\Lambda^n(\mathcal{U}_\sigma, \partial \mathcal{U}_\sigma)$ (resp., to $\Lambda^n(V, \partial V)$) (see Subsection 4.1.5). It is assumed here that the domain $\mathcal{U}_\sigma$ (resp., the manifold V) is oriented. This assumption is made just to avoid an additional discussion of alternative de Rham-like complexes, which are to be used in the case of nonorientable manifolds or in more conceptually delicate situations.

A density $\omega \in \bar{\Lambda}^n(B)$ is said to be *trivial* if the functional defined by this density vanishes. In particular, for any form $\psi \in \bar{\Lambda}^{n-1}(B, \partial B)$ the density $\bar{d}\psi$ is trivial according to the Stokes formula. Hence, the functional corresponding to ω depends on the cohomology class $\langle \omega \rangle \in \bar{H}^n(B, \partial B)$ only. Here the symbol $\bar{H}^i$ denotes the i-th cohomology of the de Rham $\bar{d}$-complex. Keeping in mind that integration is an operation which assigns to any form of the highest dimension its cohomology class, it is natural to take the following

DEFINITION. A cohomology class $\mathcal{L} \in \bar{H}^n(B, \partial B)$ is called the *Lagrangian* of the variational problem.

Since, as was noted in Subsection 4.1.5, homomorphisms of the form $j(\sigma)^*$ (resp., $j(V)^*$) are morphisms in the category DE, they induce homomorphisms of $\bar{H}^n(B, \partial B)$ to the group $H^n(\mathcal{U}_\sigma, \partial \mathcal{U}_\sigma)$ (resp., to $H^n(V, \partial V)$). These homomorphisms will also be denoted by $j(\sigma)^*$ (resp., by $j(V)^*$). Thus the functional discussed above may be viewed as

$$\sigma \mapsto j(\sigma)^*(\mathcal{L}) \quad (\text{resp., } V \mapsto j(V)^*(\mathcal{L}))$$

and the corresponding variational problem consists in finding the extremum of this functional.

REMARK. The above discussion concerns only with the simplest variational problem. For instance, there are natural problems corresponding to Lagrangians of the form

$$\mathcal{L} = \mathcal{L}' + \mathcal{L}'' \quad \text{with} \quad \mathcal{L}' \in \bar{H}^n(B, \partial B), \ \mathcal{L}'' \in \bar{H}^{n-1}(\partial B),$$

to which the functionals

$$\sigma \mapsto j(\sigma)^*(\mathcal{L}') + j(\sigma|_{\partial\mathcal{U}_\sigma})^*(\mathcal{L}'')$$

or, respectively,

$$V \mapsto j(V)^*(\mathcal{L}') + (j(V)|_{\partial V})^*(\mathcal{L}'')$$

are associated.

This (see [**134**]) and other variational problems, including that of optimal control, can be treated along the same lines.

4.2.3. The first variation formula and the Euler operator. In this subsection, we assume $B' \subset J^\infty(\pi)$ to be open. This means, in particular, that the sections to be varied are not subject to any constraints in the form of differential equations.

Let $\sigma \in \Gamma_{\mathrm{adm}}(\pi)$, $\mathcal{L} = \langle\omega\rangle$, where $\omega \in \bar{\Lambda}^n(B)$. Consider a vertical $\mathcal{C}$-field $\mathcal{Э}_\chi$ corresponding to $\chi \in \varkappa(\pi)$. Then the velocity of the change of the Lagrangian $\mathcal{L}$ under the flow generated by this field equals $\mathcal{Э}_\chi(\mathcal{L}) = \langle\mathcal{Э}_\chi(\omega)\rangle = \langle\ell_\omega(\chi)\rangle$, whereas the velocity $\dot{\mathcal{L}}_\sigma$ with which the corresponding functional changes on σ is $j(\sigma)^*\langle\mathcal{Э}_\chi(\mathcal{L})\rangle = \int_{\mathcal{U}_\sigma,\partial\mathcal{U}_\sigma} j(\sigma)^*(\ell_\omega(\chi))$. According to the $\mathcal{C}$-Green formula one has

$$\ell_\omega(\chi) = \langle\chi, \ell_\omega^*(1)\rangle + \bar{d}\mathcal{K}_\lambda(\ell_\omega(\chi, 1)).$$

So, keeping in mind the Stokes formula, one has

$$\begin{aligned}\dot{\mathcal{L}}_\sigma &= \int_{\mathcal{U}_\sigma} j(\sigma)^*\langle\chi, \ell_\omega^*(1)\rangle + \int_{\partial\mathcal{U}_\sigma} j(\sigma)^*\big(\mathcal{K}_\lambda(\ell_\omega(\chi, 1))\big) \\ &= \int_{\mathcal{U}_\sigma} \langle j(\sigma)^*(\chi), j(\sigma)^*(\ell_\omega^*(1))\rangle + \int_{\partial\mathcal{U}_\sigma} \mathcal{K}_{\lambda(\sigma)}(\ell_{\omega,\sigma}(\chi, 1)),\end{aligned}$$

where $\lambda(\sigma)$ and $\ell_{\omega,\sigma}(\chi, 1)$ are images of the splitting λ and of the $\mathcal{C}$-differential operator $\ell_\omega(\chi, 1)$, respectively, under the morphism $j(\sigma)$ in the category DE (see Subsection 4.1.5). In the case of linear π, for $\mathcal{Э}_\chi = \mathcal{Э}_\varphi$, where $\varphi = \chi \lrcorner U_1(\pi)$, the equality above is called the *first variation formula* and can be rewritten in the form

$$\dot{\mathcal{L}}_\sigma = \int_{\mathcal{U}_\sigma} \langle\Delta_\varphi(\sigma), j(\sigma)^*(\ell_\omega^*(1))\rangle + \int_{\partial\mathcal{U}_\sigma} \mathcal{K}_{\lambda(\sigma)}(\ell_{\omega,\sigma}(\varphi, 1)).$$

Any section $\psi \in \Gamma(\pi|_{\mathcal{U}_\sigma})$ with a compact support lying in $\mathrm{int}\,\mathcal{U}_\sigma$ can be represented in the form $\psi = \Delta(\sigma)$, where the support of the operator Δ also lies in $\mathrm{int}\,\mathcal{U}_\sigma$. In this case, $\ell_{\omega,\sigma}(\varphi, 1) = \ell_{\omega,\sigma} \circ j(\sigma)^*(\varphi) = \ell_{\omega,\sigma} \circ \Delta(\sigma) = 0$ over $\partial\mathcal{U}_\sigma$, where $\varphi = \varphi_\Delta$. Therefore, $\dot{\mathcal{L}}_\sigma = \int_{\mathcal{U}_\sigma} \langle\psi, j(\sigma)^*(\ell_\omega^*(1))\rangle$. Hence, according to the Du Bois–Reymond lemma, the extremality condition $\dot{\mathcal{L}}_\sigma = 0$ implies $j(\sigma)^*(\ell_\omega^*(1)) = 0$. In other words, σ is a solution of the equation $\ell_\omega^*(1) = 0$.

For nonlinear bundles π, this result remains valid since any section $\sigma \in \Gamma_{\mathrm{loc}}(\pi)$ can be included in a subbundle with a linear structure. Thus we obtained the following

THEOREM. *A section $\sigma \in \Gamma_{\mathrm{adm}}(\pi)$ is extremal with respect to a Lagrangian $\mathcal{L} \in \bar{H}^n(B, \partial B)$, where B' is open in $J^\infty(\pi)$, if $j(\sigma)^*(\ell_\omega^*(1)) = 0$, i.e., extremals are solutions of the equation $\ell_\omega^*(1) = 0$.*

It is worth mentioning that the form $\ell^*_\omega(1) \in \bar{\Lambda}^n(B, \partial B)$ is uniquely determined by the Lagrangian $\mathcal{L}$. Indeed, $\ell_{\bar{d}\psi} = \bar{d} \circ \ell_\varphi$ so that

$$\ell^*_{\omega+\bar{d}\psi}(1) = \ell^*_\omega(1) + (\bar{d} \circ \ell_\psi)^*(1) = \ell^*_\omega(1) \pm \ell^*_\psi(\bar{d}(1)) = \ell^*_\omega(1),$$

since, according to Proposition 1.3.3 $\mathcal{C}$, $\bar{d}^* = \pm\bar{d}$.

The equation $\ell^*_\omega(1) = 0$ is called the *Euler–Lagrange equation* associated with the original variational problem. Rewritten in local coordinates, this equation coincides with the classical Euler–Lagrange equation. This fact can be checked directly using the coordinate formulas for ℓ_ω and for the adjoint operator (see Subsections 2.3.7 and 4.1.2). Thus the theorem above is the invariant global version of the classical Euler–Lagrange theorem. The operator

$$\mathcal{E} \colon \bar{H}^n(B, \partial B) \to \widehat{\varkappa(\pi)_B}, \quad \mathcal{L} \mapsto \ell^*_\omega(1)|_B, \; \mathcal{L} = \langle\omega\rangle,$$

where $\varkappa(\pi)_B = \varkappa(\pi)|_B$, is called the *Euler operator*.

The case $B \subset J^\infty(\pi)$ can be treated similarly. But in this situation the universal linearization operator ℓ_ω is no longer defined globally. Therefore, the Euler operator cannot be defined globally in the form $cL \mapsto \ell^*_\omega(1)$. By gluing together local representations of this type, it is possible to describe the Euler operator as $\mathcal{L} \mapsto \Delta^*_\mathcal{L}(1)$ for a suitable $\mathcal{C}$-differential operator $\Delta_\mathcal{L}$ on $J^\infty(E, n)$. But a more satisfactory (from the conceptual point of view) solution comes from the theory of the $\mathcal{C}$-spectral sequence, for which one of the differentials is naturally interpreted as the Euler operator (see Subsection 4.3.8). So, one may consider the above discussion as a motivation for this interpretation.

4.2.4. Conservation laws and the Noether theorem. Let $B \subset J^\infty(\pi)$ and $\mathcal{L} = \langle\omega\rangle$, where $\omega \in \Lambda^n_\Gamma(B) = \bar{\Lambda}^n(B)$. A field $X \in D_\mathcal{C}(B)$ is called an (*infinitesimal*) *symmetry* of the density ω if $X(\omega) = 0$ on $\mathcal{Y}_\infty$, where $\mathcal{Y} = \{\mathcal{E}(\mathcal{L}) = 0\}$. Below we assume that $\mathcal{Y}$ is quasi-integrable and consequently $X = \tilde{X}|_B$ for some $\tilde{X} \in D_\mathcal{C}(\pi)$.

Let $X = \mathcal{Э}_\chi + Y$ (see Subsection 2.3.12). Then

$$X(\omega) = \mathcal{Э}_\chi(\omega) + Y(\omega) = \ell_\omega(\chi) + \bar{d}(Y \lrcorner\, \omega).$$

Using the $\mathcal{C}$-Green formula for $\ell_\omega(\chi)$ and the fact that $\ell^*_\omega(1) = 0$, we obtain (on $\mathcal{Y}_\infty$)

$$X(\omega) = \bar{d}(\mathcal{K}(\ell_\omega \circ \chi) + X \lrcorner\, \omega) = \bar{d}c_\lambda(\omega, X),$$

where $c_\lambda(\omega, X) = \mathcal{K}(\ell_\omega \circ \chi) + X \lrcorner\, \omega$. Thus, if X is a symmetry of the density ω, then $\bar{d}c_\lambda(\omega, X) = 0$ on $\mathcal{Y}_\infty$ and we obtain a cohomology class $\langle c_{\lambda,\mathcal{Y}}(\omega, X)\rangle \in \bar{H}^{n-1}(\mathcal{Y}_\infty)$, where $c_{\lambda,\mathcal{Y}}(\omega, X) = c_\lambda(\omega, X)|_{\mathcal{Y}_\infty}$. Here $\bar{H}^i(\mathcal{Y}_\infty)$ denotes the i-th cohomology group of the de Rham $\bar{d}$-complex on $\mathcal{Y}_\infty$. The following terminology allows one to interpret appropriately the cohomology class obtained in such a way.

DEFINITION. A cohomology class $h \in \bar{H}^{n-1}(\mathcal{Y}_\infty)$ is called a (*local*) *conservation law* for the equation $\mathcal{Y}$. If $h = \langle\rho\rangle$, where $\rho \in \bar{\Lambda}^{n-1}(\mathcal{Y}_\infty)$, then ρ is called a *c-density*.

From the definition above one can see that for any solution $V \subset E$ (resp., for $\sigma \in \Gamma_{\mathrm{loc}}(\pi)$) of the equation $\mathcal{Y}$ the form $j(V)^*(\rho)$ (resp., $j(\sigma)^*(\rho)$) is closed on V

(resp., on $\mathcal{U}_\sigma$) and its cohomology class equals $j(V)^*(h)$ (resp., $j(\sigma)^*(h)$). Hence, the quantity

$$(j(V)^*(h), [W^{n-1}]) = \int_{W^{n-1}} j(V)^*(h)$$

or, respectively,

$$(j(\sigma)^*(h), [W^{n-1}]) = \int_{W^{n-1}} j(\sigma)^*(h)$$

is a conserved quantity for any oriented $(n-1)$-dimensional submanifold $W^{n-1} \subset V$ (resp., for $W^{n-1} \subset \mathcal{U}_\sigma$). This means that this quantity remains constant when W^{n-1} is changed by a homologous manifold. If one of independent variables in $\mathcal{Y}$ is the time t, then the surfaces $\{t = t_1\}$ and $\{t = t_2\}$ are homologous and the above integrals, provided they exist, taken along these surfaces coincide. This motivates the introduced terminology.

From the above-said one immediately obtains the following result which can be naturally called the *generalized Noether theorem.*

THEOREM (Generalized Noether theorem). *Let X be a symmetry of a Lagrangian density $\mathcal{L} = \langle\omega\rangle$. Then $c_{\lambda,\mathcal{Y}}(\omega, X)$ is a c-density for the corresponding Euler–Lagrange equation, and the class $\langle c_{\lambda,\mathcal{Y}}(\omega, X)\rangle$ is the corresponding conservation law which is independent of the choice of λ.*

If $X(\omega)|_{\mathcal{Y}_\infty} = \bar{d}\nu$, then the form $c_\lambda(\omega, X) - \nu$ is obviously a c-density for the Euler–Lagrange equation $\mathcal{Y}$. This fact turns out to be useful in practical search for conservation laws, though it can be reduced to the previous theorem. Indeed, if a field $Y \in \mathcal{C}D(B)$ is such that $(Y \lrcorner\, \omega)|_{\mathcal{Y}_\infty} = \nu$, then $Y(\omega)|_{\mathcal{Y}_\infty} = \bar{d}\nu$ and $c_\lambda(\omega, X - Y) = c_\lambda(\omega, X) - \nu$.

Symmetry transformations appearing in the classical Noether theorem are Lie fields. Thus the theorem above essentially extends the class of symmetries.

4.3. The $\mathcal{C}$-spectral sequence

In this section, the $\mathcal{C}$-spectral sequence $\{E_r^{p,q}, d_r^{p,q}\}$ is constructed. We establish some properties of this spectral sequence and give its first applications. In particular, it is shown that the differential $d_1^{0,n}$ coincides with the Euler operator which solves the problem of triviality of Lagrangians "in the large" as well as the inverse problem of the calculus of variations in the situation without nonholonomic constraints. In addition, we deduce "the infinitesimal Stokes formula" for the term E_1. This formula is a natural base for homotopical methods of computation of the term E_2, and moreover, allows one to obtain explicit formulas in the inverse problem of the calculus of variations, as well as in the problem of the Lagrangian triviality. The $\mathcal{C}$-spectral sequence is natural in the category DE. This fact, in particular, means that the Euler operator commutes with the morphisms in DE, generalizing the well-known property of Euler–Lagrange equations to be invariant under transformations of dependent and independent variables.

4.3.1. The $\mathcal{C}$-spectral sequence. Let $\mathcal{O}$ be an object of the category DE (see Subsection 3.4.9), $\Lambda^i = \Lambda^i(\mathcal{O})$, $\Lambda^* = \Lambda^*(\mathcal{O})$. Evidently, $\mathcal{C}\Lambda^* = \mathcal{C}\Lambda^*(\mathcal{O})$ is an ideal of the external algebra Λ^* stable with respect to the operator $\bar{d}$. Hence, all its powers $\mathcal{C}^k\Lambda^* = \mathcal{C}^k\Lambda^*(\mathcal{O})$ are also stable. In this way, the de Rham complex over $\mathcal{O}$ is supplied with a natural filtration

$$\Lambda^* = \mathcal{C}^0\Lambda^* \supset \mathcal{C}^1\Lambda^* = \mathcal{C}\Lambda^* \supset \cdots \supset \mathcal{C}^k\Lambda^* \supset \cdots,$$

which is called the *$\mathcal{C}$-filtration.* The spectral sequence

$$\{E_r^{p,q}, d_r^{p,q}\} = \{E_r^{p,q}(\mathcal{O}), d_r^{p,q}(\mathcal{O})\}$$

associated with this filtration is called the *$\mathcal{C}$-spectral sequence.* As usual, p denotes the filtration index and $p+q$ is the degree. Multiplication in Λ^* in a usual way induces multiplication in the terms E_r that satisfies $E_r^{p,q} \wedge E_r^{s,t} \subset E_r^{p+s,q+t}$ and

$$d_r(\omega_1 \wedge \omega_2) = d_r\omega_1 \wedge \omega_2 + (-1)^{p+q}\omega_1 \wedge d_r\omega_2, \quad \omega_1 \in E_r^{p,q}, \ \omega_2 \in E_r^{s,t}.$$

Below, $H^*(\mathcal{O})$ stands for the de Rham cohomology of $\mathcal{O}$. Recall that if $\mathcal{O} = \mathcal{Y}_\infty$, then $H^*(\mathcal{O})$ is the direct limit of the de Rham cohomology $H^*(\mathcal{Y}_k)$, $k \to \infty$. The cohomology of the horizontal de Rham complex $\{\bar{\Lambda}^i, \bar{d}\}$, $\bar{\Lambda}^i = \bar{\Lambda}^i(\mathcal{O})$, over $\mathcal{O}$ (see Subsection 3.1.4) is denoted by $\bar{H}^i = \bar{H}^i(\mathcal{O})$, $0 \le i < \infty$.

Recall also that a chain mapping of one filtered complex into another preserving the filtration generates a homomorphism of the associated spectral sequences, say $\{E'_r, d'_r\}$ and $\{E''_r, d''_r\}$, i.e., a series of chain mappings $h_r\colon E'_r \to E''_r$ for any $r \ge 0$.

PROPOSITION. *Let $\mathcal{O}_1$ and $\mathcal{O}_2$ be two objects of the category DE.*

1. *If $F\colon \mathcal{O}_1 \to \mathcal{O}_2$ is a morphism, then F^* induces a homomorphism of the $\mathcal{C}$-spectral sequence over $\mathcal{O}_2$ to the $\mathcal{C}$-spectral sequence over $\mathcal{O}_1$.*
2. *For any $q \ge 0$, one has $E_0^{0,q} = \bar{\Lambda}^q$ and $E_1^{0,q} = \bar{H}^q$.*
3. *For any $q < 0$ and $p \in \mathbb{Z}$, one has $E_0^{p,q} = 0$.*
4. *If $\bar{\Lambda}^{s+1} = 0$ for some $s \ge 0$, then $E_0^{p,q} = 0$ for all $q > s$.*
5. *The $\mathcal{C}$-spectral sequence converges and $\{E_\infty^{p,q}\}$ is adjoint to $H^*(\mathcal{O})$.*

PROOF. Statements (1) and (2) immediately follow from the definitions. To prove (3), note that $\mathcal{C}\Lambda^i = \mathcal{C}\Lambda^1 \wedge \Lambda^{i-1}$, which implies $\mathcal{C}^k\Lambda^i = 0$ for $k > i$.

Now let $\bar{\Lambda}^{s+1} = 0$. Then $\mathcal{C}\Lambda^{s+1} = \Lambda^{s+1}$ and, more generally, $\mathcal{C}\Lambda^i = \Lambda^i$ for $i > s$. In fact, $\mathcal{C}\Lambda^{s+1} \wedge \Lambda^{i-s-1} \subset \mathcal{C}\Lambda^i$ and $\mathcal{C}\Lambda^{s+1} \wedge \Lambda^{i-s-1} = \Lambda^{s+1} \wedge \Lambda^{i-s-1} = \Lambda^i$. In addition, from the equality $\mathcal{C}\Lambda^{m+1} = \mathcal{C}\Lambda^1 \wedge \Lambda^m$ it follows that $\mathcal{C}^k\Lambda^{m+k} = \underbrace{\mathcal{C}\Lambda^1 \wedge \cdots \wedge \mathcal{C}\Lambda^1}_{k \text{ times}} \wedge \Lambda^m$. Therefore, for $q > s$ one has

$$\begin{aligned} \mathcal{C}^p\Lambda^{p+q} &= \underbrace{\mathcal{C}\Lambda^1 \wedge \cdots \wedge \mathcal{C}\Lambda^1}_{p \text{ times}} \wedge \Lambda^q \\ &= \underbrace{\mathcal{C}\Lambda^1 \wedge \cdots \wedge \mathcal{C}\Lambda^1}_{p \text{ times}} \wedge \mathcal{C}\Lambda^q \subset \mathcal{C}^{p+1}\Lambda^{p+q} \subset \mathcal{C}^p\Lambda^{p+q}. \end{aligned}$$

In other words, $\mathcal{C}^p\Lambda^{p+q} = \mathcal{C}^{p+1}\Lambda^{p+q}$, which proves (4).

Finally, statement (5) follows from the fact that the $\mathcal{C}$-spectral sequence, according to (3), is situated in the first quadrant of the (p,q)-diagram. □

COROLLARY. *If $\mathcal{O} \subset J^\infty(E,n)$ or $\mathcal{O} \subset J^\infty(\pi)$, then $E_0^{p,q} = 0$ for all $q < 0$ and $q > n$.*

PROOF. Actually, in this case one can take $s = n$ (in the second case $n = \dim M$, where M is the base of the bundle π), where s is the number that appears in the forth statement of the previous proposition. □

4.3.2. Bicomplex structure in $\Lambda^*(J^\infty(\pi))$. To describe the term $E_0^{p,q}$ in a convenient way, it is useful to introduce in $\Lambda^* = \Lambda^*(J^\infty(\pi))$ the structure of a bicomplex. It can be done using the operation $\rceil$ (see Subsection 3.2.2) which gives the direct sum decomposition $\Lambda^1 = \mathcal{C}\Lambda^1 \oplus \Lambda^1_\Gamma$ and thus leads to the decomposition $\Lambda^* = \sum_{p,q\geq 0} \Lambda^{p,q}$, where

$$\Lambda^{p,q} = \underbrace{\mathcal{C}\Lambda^1 \wedge \dots \wedge \mathcal{C}\Lambda^1}_{p \text{ times}} \wedge \underbrace{\Lambda^1_\Gamma \wedge \dots \wedge \Lambda^1_\Gamma}_{q \text{ times}} = \underbrace{\mathcal{C}\Lambda^1 \wedge \dots \wedge \mathcal{C}\Lambda^1}_{p \text{ times}} \wedge \Lambda^q_\Gamma.$$

Since $\mathcal{C}^p\Lambda^* = \mathcal{C}\Lambda^1 \wedge \dots \wedge \mathcal{C}\Lambda^1 \wedge \Lambda^*$, the term $E_0^{p,q}$ is identified with $\Lambda^{p,q}$. To describe the operator d in these terms, we shall need the following

LEMMA. *$d(\hat{d}\varphi) \in \mathcal{C}\Lambda^1 \otimes \Lambda^1_\Gamma$ for any $\varphi \in A$.*

PROOF. Indeed, for any χ one has

$$\partial_\chi \lrcorner\, d(\hat{d}\varphi) = \partial_\chi(\hat{d}\varphi) - d(\partial_\chi \lrcorner\, \hat{d}\varphi) = \partial_\chi(\hat{d}\varphi) = \hat{d}(\partial_\chi(\varphi)) \in \Lambda^1_\Gamma$$

(we used here the identity $\partial_\chi \lrcorner\, \hat{d}\varphi = 0$ which is valid because ∂_χ is a vertical field and $\hat{d}\varphi$ is a horizontal form). Consequently, $d\hat{d}\varphi \in \mathcal{C}\Lambda^1 \wedge \Lambda^1_\Gamma \oplus \Lambda^1_\Gamma \wedge \Lambda^1_\Gamma$. But for any $u \in \Gamma_{\mathrm{loc}}(\pi)$ one has $j(u)^*(d(\hat{d}\varphi)) = d(j(u)^*(\hat{d}\varphi)) = d^2(j(u)^*(\varphi)) = 0$, so that $d\hat{d}\varphi \in \mathcal{C}\Lambda^2 = \mathcal{C}\Lambda^1 \wedge \Lambda^1 = \mathcal{C}\Lambda^1 \wedge \Lambda^1_\Gamma \oplus \mathcal{C}\Lambda^1 \wedge \mathcal{C}\Lambda^1$. □

COROLLARY. *For any p, q one has $d\Lambda^{p,q} \subset \Lambda^{p+1,q} \oplus \Lambda^{p,q+1}$.*

PROOF. First note that $dU_1(\varphi) = -d\hat{d}\varphi \in \Lambda^{1,1}$ and, obviously, $d\Lambda^1_\Gamma \subset \Lambda^{1,1} \oplus \Lambda^{0,2}$. According to the equality $\Lambda^q_\Gamma = \Lambda^1_\Gamma \wedge \dots \wedge \Lambda^1_\Gamma$ (q factors), the last embedding implies that $d\Lambda^q_\Gamma \subset \Lambda^{0,q+1} \oplus \Lambda^{1,q}$. On the other hand, elements of the form $\rho = U_1(\varphi_1) \wedge \dots \wedge U_1(\varphi_p) \wedge \omega$, $\omega \in \Lambda^q_\Gamma$, $\varphi_i \in A$, are additive generators of $\Lambda^{p,q}$, so that $d\rho \in \Lambda^{p+1,q} \wedge \Lambda^{p,q+1}$. □

Denote by $d' = d'_{p,q}$ the composition $\Lambda^{p,q} \xrightarrow{d} \Lambda^{p+1,q} \oplus \Lambda^{p,q+1} \to \Lambda^{p,q+1}$, where the second mapping is the natural projection. Set also $d''_{p,q} = d - d' \colon \Lambda^{p,q} \to \Lambda^{p+1,q}$. Then, under the above-described identification of $E_0^{p,q}$ with $\Lambda^{p,q}$, the operator $d_0^{p,q}$ is identified with $d'_{p,q}$. The formulas below immediately follow from comparison of bigradings:

$$\begin{aligned} &d'\big(U_1(\varphi_1) \wedge \dots \wedge U_1(\varphi_p) \wedge \omega_q\big) \\ &\qquad = \sum_{i=1}^{p} (-1)^{i-1} U_1(\varphi_1) \wedge \dots \wedge dU_1(\varphi_i) \wedge \dots \wedge U_1(\varphi_p) \wedge \omega_q \\ &\qquad\qquad + (-1)^p U_1(\varphi_1) \wedge \dots \wedge U_1(\varphi_p) \wedge \hat{d}\omega_q, \end{aligned} \tag{4.2}$$

$$d''\big(U_1(\varphi_1) \wedge \dots \wedge U_1(\varphi_p) \wedge \omega_q\big) = (-1)^p U_1(\varphi_1) \wedge \dots \wedge U_1(\varphi_p) \wedge U_1(\omega_q),$$

where $\varphi_i \in A$, $\omega_q \in \Lambda^q_\Gamma$.

REMARK (1). The results of this subsection remain valid for any $\mathcal{Y}_\infty \subset J^\infty(\pi)$, since the operators $d, \hat{d}, U_1$, and $\rceil$, as well as the construction of the $\mathcal{C}$-spectral sequence, commute with the operation of restriction to $\mathcal{Y}_\infty$.

REMARK (2). Consider the spectral sequence for the de Rham cohomology of the bundle $\pi_\infty \colon \mathcal{Y}_\infty \to M$. Then the complex $\{\Lambda^*, d''\}$ is naturally identified with the term E_0 of this spectral sequence.

4.3.3. The terms $E_0^{p,q}$ in the free case. Now we describe the terms $E_0^{p,q}$ in the case where $\mathcal{O} = J^\infty(\pi)$ or $J^\infty(E,n)$. We need the following

LEMMA. *Let $\mathcal{O}$ be an open domain in $J^\infty(\pi)$ or in $J^\infty(E,n)$. Consider a form $\omega \in \mathcal{C}\Lambda^1$ and the operator $\nabla_\omega \in \mathcal{C}\mathrm{Diff}(\varkappa, A)$ defined by $\nabla_\omega(\chi) = \chi \lrcorner \omega$. Then the correspondence $\omega \mapsto \nabla_\omega$ is an isomorphism of A-modules $\mathcal{C}\Lambda^1$ and $\mathcal{C}\mathrm{Diff}(\varkappa, A)$.*

PROOF. First note that $\nabla_{a\omega} = a\nabla_\omega$, i.e., the operation $\omega \mapsto \nabla_\omega$ is a homomorphism of A-modules. If $\omega = U_1(a)$, then $\nabla_\omega(\chi) = \chi \lrcorner U_1(a) = \ell_a(\chi)$, i.e., $\nabla_\omega = \ell_a$. Thus, if $\omega = \sum_i b_i U_1(a_i)$, then $\nabla_\omega = \sum_i b_i \ell_{a_i}$. Since $\mathcal{C}\mathrm{Diff}(\varkappa, P)$ as an A-module is generated by operators of the form ℓ_p, $p \in P$, while the forms $U_1(a)$, $a \in A$, generate $\mathcal{C}\Lambda^1$, we see that the mapping under consideration is an epimorphism.

If $\nabla_\omega = \sum_i b_i \nabla_{a_i} = 0$, then the form $\sum_i b_i U_1(a_i)$ also vanishes. In fact, in this case $\nabla_\omega(\chi) = \chi \lrcorner \left(\sum_i b_i U_1(a_i)\right) = 0$ for any $\chi \in \varkappa$. But any vertical tangent vector at a point $\theta \in J^\infty(\pi)$ can be realized as $\mathcal{Э}_\chi|_\theta$ for some χ. This can be seen, for example, from the coordinate representation (see Subsection 2.3.7). Hence, $\sum_i b_i U_1(a_i)$ is a horizontal form and consequently vanishes, because $\mathcal{C}\Lambda^1 \cap \Lambda^1_\Gamma = 0$.

Thus we have proved the lemma for domains in $J^\infty(\pi)$. Since $J^\infty(E,n)$ can be covered by the charts of the form $J^\infty(\pi)$, the result is valid in general. □

Let $\rho \in E_0^{p,q}$, $\chi \in \varkappa$. Note that the following inclusions hold:

$$X \lrcorner \mathcal{C}^p\Lambda^* \subset \mathcal{C}^{p-1}\Lambda^*,\ X \in D_\mathcal{C}, \qquad X \lrcorner \mathcal{C}^p\Lambda^* \subset \mathcal{C}^p\Lambda^*,\ X \in \mathcal{C}D.$$

This shows that the contraction $\chi \lrcorner \rho \in E_0^{p-1,q}$, $\rho \in E_0^{p,q}$, is well defined:

$$\chi \lrcorner \rho = X \lrcorner \omega \bmod \mathcal{C}^p\Lambda^{p+q-1}, \quad \chi = X \bmod \mathcal{C}D,\ \rho = \omega \bmod \mathcal{C}^{p+1}\Lambda^{p+q}.$$

For any $\rho \in E_0^{p,q}$, define a multidifferential operator $\nabla_\rho \in \mathcal{C}\mathrm{Diff}^{\mathrm{alt}}_{(p)}(\varkappa; \bar{\Lambda}^q)$ by setting

$$\nabla_\rho(\chi_1, \dots, \chi_p) = \chi_p \lrcorner (\dots(\chi_1 \lrcorner \rho)\dots) \in \bar{\Lambda}^q, \quad \chi_i \in \varkappa. \tag{4.3}$$

The required description of the terms $E_0^{p,q}$ is given by the following

PROPOSITION. *For any open domain $\mathcal{O}$ in $J^\infty(\pi)$ or in $J^\infty(E,n)$, the correspondence $\rho \mapsto \nabla_\rho$ establishes an isomorphism between the A-modules $E_0^{p,q}$ and $\mathcal{C}\mathrm{Diff}^{\mathrm{alt}}_{(p)}(\varkappa; \bar{\Lambda}^q)$.*

PROOF. We start with the case $\mathcal{O} \subset J^\infty(\pi)$, $p = 1$. Note that

$$E_0^{1,q} = \Lambda^{1,q} = \mathcal{C}\Lambda^1 \otimes \bar{\Lambda}^q,$$
$$\mathcal{C}\mathrm{Diff}^{\mathrm{alt}}_{(p)}(\varkappa, \bar{\Lambda}^q) = \mathcal{C}\,\mathrm{Diff}(\varkappa, \bar{\Lambda}^q) = \mathcal{C}\,\mathrm{Diff}(\varkappa, A) \otimes \bar{\Lambda}^q,$$
$$\nabla_{\lambda\otimes\omega} = \nabla_\lambda \otimes \omega, \quad \lambda \in \mathcal{C}\Lambda^1,\ \omega \in \bar{\Lambda}^q.$$

Hence, the result in this case follows from the lemma above. Since $J^\infty(E,n)$ can be covered by the charts of the form $J^\infty(\pi)$, the result also holds for $p = 1$ and $\mathcal{O} \subset J^\infty(E,n)$.

To prove the proposition in the general case, we introduce the bigraded associative algebras

$$\mathcal{C}\mathrm{Diff}^{\mathrm{alt}}\{P\} = \sum_{p,q} \mathcal{C}\mathrm{Diff}^{\mathrm{alt}}_{(p)}(P, \bar{\Lambda}^q),$$

where $\mathcal{C}\mathrm{Diff}^{\mathrm{alt}}_{(0)}(P, \bar{\Lambda}^q) = \bar{\Lambda}^q$, in which the multiplication operation is defined by

$$(\Delta_1 \wedge \Delta_2)(p_1, \dots, p_{l+k}) = \frac{(-1)^{lk}}{l!k!} \sum_{\sigma} (-)^{\sigma} \Delta_1(p_{\sigma(1)}, \dots, p_{\sigma(l)}) \wedge \Delta_2(p_{\sigma(l+1)}, \dots, p_{\sigma(l+k)}) \in \bar{\Lambda}^{q+r},$$

where $\Delta_1 \in \mathcal{C}\mathrm{Diff}^{\mathrm{alt}}_{(l)}(P, \bar{\Lambda}^q)$, $\Delta_2 \in \mathcal{C}\mathrm{Diff}^{\mathrm{alt}}_{(k)}(P, \bar{\Lambda}^r)$. Then the operation $\rho \mapsto \nabla_\rho$ is obviously a homomorphism of the algebra $E_0 = \sum E_0^{p,q}$ to the algebra $\mathcal{C}\mathrm{Diff}^{\mathrm{alt}}\{P\}$. Besides, one has

$$\mathcal{C}\mathrm{Diff}^{\mathrm{alt}}_{(p)}(P, \bar{\Lambda}^q) \wedge \mathcal{C}\mathrm{Diff}(P, \bar{\Lambda}^r) = \mathcal{C}\mathrm{Diff}^{\mathrm{alt}}_{(p+1)}(P, \bar{\Lambda}^{q+r})$$

and $E_0^{p,q} \wedge E_0^{1,r} = E_0^{p+1,q+r}$. Now the result can be proved by induction starting from the isomorphism just proved between $\mathcal{C}\mathrm{Diff}^{\mathrm{alt}}_{(i)}(\varkappa, \bar{\Lambda}^q)$ and $E_0^{i,q}$ for $i = 0, 1$. □

4.3.4. Bicomplex structure in $\mathcal{C}\mathrm{Diff}^{\mathrm{alt}}\{\varkappa\}$. Below, when defining the operators $l_{p,q}$, we work over $J^\infty(\pi)$ or over a chart in $J^\infty(E, n)$. In this setting, we identify $\bar{\Lambda}^i$ and Λ^i_Γ. This assumption is unnecessary in the definition of the operators $\bar{d}_{p,q}$. Thus, a bicomplex structure in $\mathcal{C}\mathrm{Diff}^{\mathrm{alt}}\{\varkappa\}$ introduced below is defined on $J^\infty(\pi)$ or in a special chart. The operators $l = l_{p,q} \colon \mathcal{C}\mathrm{Diff}^{\mathrm{alt}}_{(p)}(\varkappa; \bar{\Lambda}^q) \to \mathcal{C}\mathrm{Diff}^{\mathrm{alt}}_{(p+1)}(\varkappa; \bar{\Lambda}^q)$ are defined by the "standard" formula

$$\begin{aligned} l(\Delta)(\chi_1, \dots, \chi_{p+1}) &= \sum_i (-1)^{i-1} \mathcal{E}_{\chi_i}(\Delta(\chi_1, \dots, \hat{\chi}_i, \dots, \chi_{p+1})) \\ &+ \sum_{i<j} (-1)^{i+j} \Delta([\chi_i, \chi_j], \chi_1, \dots, \hat{\chi}_i, \dots, \hat{\chi}_j, \dots, \chi_{p+1}). \end{aligned} \tag{4.4}$$

Directly from the definition we get

$$l_{p+1,q} \circ l_{p,q} = 0$$

and

$$l_{p+k,q+s}(\Delta_1 \wedge \Delta_2) = l_{p,q}(\Delta_1) \wedge \Delta_2 + (-1)^{p+q} \Delta_1 \wedge l_{k,s}(\Delta_2),$$

where $\Delta_1 \in \mathcal{C}\mathrm{Diff}^{\mathrm{alt}}_{(p)}(\varkappa; \bar{\Lambda}^q)$, $\Delta_1 \in \mathcal{C}\mathrm{Diff}^{\mathrm{alt}}_{(k)}(\varkappa; \bar{\Lambda}^s)$. Consider also the operators

$$\bar{d} = \bar{d}_{p,k} \colon \mathcal{C}\mathrm{Diff}^{\mathrm{alt}}_{(p)}(Q; \bar{\Lambda}^k) \to \mathcal{C}\mathrm{Diff}^{\mathrm{alt}}_{(p)}(Q; \bar{\Lambda}^{k+1})$$

defined by the relation

$$\bar{d}_{p,k}(\Delta)(q_1, \dots, q_p) = (-1)^p \bar{d}(\Delta(q_1, \dots, q_p)), \quad q_i \in Q. \tag{4.5}$$

Since the Lie derivative along $\chi \in \varkappa$ commutes with the operators $\bar{d} \colon \bar{\Lambda}^q \to \bar{\Lambda}^{q+1}$, it follows immediately from definitions that for $Q = \varkappa$ one has

$$\bar{d}_{p+1,q} \circ l_{p,q} + l_{p,q} \circ \bar{d}_{p,q} = 0.$$

In other words, the systems of operators $\{l_{p,q}\}$ and $\{\bar{d}_{p,q}\}$ supply $\mathcal{C}\mathrm{Diff}^{\mathrm{alt}}\{\varkappa\}$ with a bicomplex structure.

Note the following property of the operators $\bar{d}_{p,q}$:

$$\bar{d}_{p+k,q+l}(\Delta_1 \wedge \Delta_2) = \bar{d}_{p,q}(\Delta_1) \wedge \Delta_2 + (-1)^{p+q} \Delta_1 \wedge \bar{d}_{k,l}(\Delta_2), \tag{4.6}$$

where $\Delta_1 \in \mathcal{C}\mathrm{Diff}^{\mathrm{alt}}_{(p)}(\varkappa; \bar{\Lambda}^q)$, $\Delta_2 \in \mathcal{C}\mathrm{Diff}^{\mathrm{alt}}_{(k)}(\varkappa; \bar{\Lambda}^l)$.

PROPOSITION. *Under the isomorphism of Proposition* 4.3.3, *the operator* $d_0^{p,q}$ *is identified with* $\bar{d}_{p,q}$. *Moreover, the operators* $d''_{p,q}$ *and* $d'_{p,q}$ *are identified with* $l_{p,q}$ *and* $\bar{d}_{p,q}$, *respectively, under the identifications of the modules* $\Lambda^{p,q}$ *and* $\mathcal{C}\mathrm{Diff}_{(p)}(\varkappa; \bar{\Lambda}^q)$ *established in Proposition* 4.3.3.

PROOF. Since $d_0 = \{d_0^{p,q}\}$ and $\bar{d} = \{\bar{d}_{p,q}\}$ are derivations of the algebras E_0 and $\mathcal{C}\mathrm{Diff}^{\mathrm{alt}}\{\varkappa\}$, respectively, it is sufficient to check the first statement for $p = 0$ and $p = 1$. The case $p = 0$ is trivial.

Consider the case $p = 1$. Since d_0 and $\bar{d}$ are derivations of the A-algebras $\sum_q E_0^{1,q}$ and $\sum_q \mathcal{C}\mathrm{Diff}^{\mathrm{alt}}_{(1)}(\varkappa; \bar{\Lambda}^q)$, respectively, it suffices to check the assertion on the elements $U_1(\omega) \in E_0^{1,q}$, $\omega \in \Lambda^q_\Gamma$, over a special chart. But in $\bar{\Lambda}^*_\Gamma$ one has $U_1 = d - \hat{d}$, while $d''U_1(\omega) = 0$ according to (4.2). Therefore,

$$d'U_1(\omega) = dU_1(\omega) = d(d - \hat{d})(\omega) = -(d - \hat{d})(\hat{d}\omega) = -U_1(\hat{d}\omega).$$

But from the relations $\nabla_{U_1(\omega)} = l_\omega$ and $\nabla_{U_1(\hat{d}\omega)} = \ell_{\hat{d}\omega} = \hat{d} \circ \ell_\omega$ one gets $\nabla_{d'U_1(\omega)} = -\hat{d} \circ \nabla_{U_1(\omega)}$.

In a similar way, it suffices to verify the coincidence of the operators $d''_{p,q}$ and $l_{p,q}$ for the elements of the form $\rho = U_1(\varphi) \wedge \omega$, $\varphi \in A$, $\omega \in \Lambda^q_\Gamma$. But $\nabla_\rho(\ni_f) = \ni_f(\varphi)\omega$, $f \in \varkappa$. Therefore, due to the equality $[\ni_f, \ni_g] = \ni_{\{f,g\}}$, one has

$$\begin{aligned} l(\nabla_\rho)(f,g) &= \ni_f(\ni_g(\varphi) \cdot \omega) - \ni_g(\ni_f(\varphi) \cdot \omega) - \ni_{\{f,g\}}(\varphi) \cdot \omega \\ &= [\ni_f, \ni_g](\varphi) \cdot \omega + \ni_g(\varphi)\ni_f(\omega) - \ni_f(\varphi)\ni_g(\omega) - \ni_{\{f,g\}}(\varphi) \cdot \omega \\ &= \ni_g(\varphi)\ni_f(\omega) - \ni_f(\varphi)\ni_g(\omega). \end{aligned}$$

On the other hand, according to (4.2), $d''\rho = -U_1(\varphi) \wedge U_1(\omega)$ and $\nabla_{d''\rho}(\ni_f, \ni_g) = \ni_g(\varphi)\ni_f(\omega) - \ni_f(\varphi)\ni_g(\omega)$ in an obvious way. □

REMARK. The complex $\{\mathcal{C}\mathrm{Diff}^{\mathrm{alt}}\{\varkappa\}, l\}$ can be identified with the term E_0 of the spectral sequence for the de Rham cohomology of the bundle $\pi_\infty \colon J^\infty(\pi) \to M$. This fact follows from Remark 2 of Subsection 4.3.2 and Proposition 4.3.3.

4.3.5. The terms $E_1^{p,q}$ in the free case. Using the isomorphism of Propositions 4.3.3 and 4.3.4, we compute the term E_1 of the $\mathcal{C}$-spectral sequence as the cohomology of the complex $\{\mathcal{C}\mathrm{Diff}^{\mathrm{alt}}\{\varkappa\}, \bar{d}\}$. Since the complex $\{\mathcal{C}\mathrm{Diff}^{\mathrm{alt}}\{\varkappa\}, \bar{d}\}$ is the direct sum of the complexes $\{\mathcal{C}\mathrm{Diff}^{\mathrm{alt}}_{(p)}(\varkappa), \bar{d}\}$, where, by definition, $\mathcal{C}\mathrm{Diff}^{\mathrm{alt}}_{(p)}(\varkappa) = \sum_q \mathcal{C}\mathrm{Diff}^{\mathrm{alt}}_{(p)}(\varkappa; \bar{\Lambda}^q)$, it suffices to consider only the complex $\{\mathcal{C}\mathrm{Diff}^{\mathrm{alt}}_{(p)}(\varkappa), \bar{d}\}$. The cohomology of this complex is computed using the following considerations.

Let $Q_1, \dots, Q_k$ be A-modules. Consider the module

$$\mathcal{C}\mathrm{Diff}\{Q_1, \dots, Q_k\} = \sum_q \mathcal{C}\mathrm{Diff}(Q_1, \dots, Q_k; \bar{\Lambda}^q),$$

where

$$\mathcal{C}\mathrm{Diff}(Q_1, \dots, Q_k; P) = \mathcal{C}\mathrm{Diff}(Q_1, \mathcal{C}\mathrm{Diff}(Q_2, \dots, Q_k; P))$$

and $\mathcal{C}\mathrm{Diff}(Q; P) = \mathcal{C}\mathrm{Diff}(Q, P)$. One can see that $\mathcal{C}\mathrm{Diff}\{Q_1, \dots, Q_k\}$ becomes a complex with the differential $\bar{d}$ defined by

$$(\bar{d}\nabla)(q_1, \dots, q_k) = (-1)^k d(\nabla(q_1, \dots, q_k)), \quad q_i \in Q_i,$$

where $\nabla(q_1, \dots, q_k) = (\dots(\nabla(q_1))(q_2)\dots)(q_k)$. Starting with the operator $\mu = \mu_1 \colon \mathcal{C}\mathrm{Diff}(Q_1, \bar{\Lambda}^n) \to \hat{Q}_1$ (see Subsection 1.3.2), let us define by induction the operators

$$\mu = \mu_k \colon \mathcal{C}\mathrm{Diff}(Q_1, \dots, Q_k; \bar{\Lambda}^n) \to \mathcal{C}\mathrm{Diff}(Q_1, \dots, Q_k; \hat{Q}_k)$$

by setting $\mu_k(\nabla) = \mu_{k-1} \circ \nabla$ for $\nabla \colon Q_1 \to \mathcal{C}\mathrm{Diff}(Q_2, \dots, Q_k; \bar{\Lambda}^n)$.

PROPOSITION. *Let $A = \mathcal{F}(\mathcal{Y})$, and let A-modules Q_i be projective, $2 \le i \le k$. Then the cohomology of the complex $\{\mathcal{C}\mathrm{Diff}\{Q_1, \dots, Q_k\}, \bar{d}\}$ is trivial in all dimensions except dimension n, the latter being isomorphic to $\mathcal{C}\mathrm{Diff}(Q_1, \dots, Q_{k-1}; \hat{Q}_k)$.*

PROOF. For $k = 1$ this statement is equivalent to the exactness of the complex (4.1) (see Subsection 4.3.5). Let us extend the complex $\mathcal{C}\mathrm{Diff}\{Q_1, \dots, Q_k\}$ with the operator μ_k. Then the proposition, being equivalent to acyclicity of the extended complex, is proved by induction on k in just the same way as the acyclicity of the complex $\mathcal{S}PQ$ was proved in Subsection 1.4.1. □

The permutation group S_k acts on the complex $\{\mathcal{C}\mathrm{Diff}_{(k)}\{Q\}, \bar{d}\}$, where

$$\mathcal{C}\mathrm{Diff}_{(k)}\{Q\} = \mathcal{C}\mathrm{Diff}\{\underbrace{Q, \dots, Q}_{k \text{ times}}\}.$$

Namely, any $\Delta \in \mathcal{C}\mathrm{Diff}_{(k)}\{Q\}$ can be understood as a multidifferential operator (see Subsection 1.4.2):

$$(q_1, \dots, q_k) \mapsto \Delta(q_1, \dots, q_k) = (\dots(\Delta(q_1))(q_2)\dots)(q_k) \in \bar{\Lambda}^*, \quad q_i \in Q.$$

Then for $\tau \in S_k$ we set $\tau(\Delta)(q_1, \dots, q_k) = \Delta(q_{\tau(1)}, \dots, q_{\tau(k)})$. Note that the skew-symmetrization operation realizes the complex $\{\mathcal{C}\mathrm{Diff}^{\mathrm{alt}}_{(k)}\{\varkappa\}, \bar{d}\}$ as a direct summand in $\{\mathcal{C}\mathrm{Diff}_{(k)}\{\varkappa\}, \bar{d}\}$. This shows that the cohomology of this complex is trivial in all dimensions except dimension n, where it is isomorphic to the skew-symmetric part $L_k(\varkappa)$ of the A-module $\mathcal{C}\mathrm{Diff}_{(k-1)}(\varkappa; \hat{\varkappa})$ with respect to the induced action of the permutation group described below.

4.3.6. Action of S_k on $\mathcal{C}\mathrm{Diff}_{(k-1)}(P; \hat{P})$. Now we are going to describe the action of the permutation group S_k on $\mathcal{C}\mathrm{Diff}_{(k-1)}(P; \hat{P})$, i.e., on the n-th cohomology of the complex $\{\mathcal{C}\mathrm{Diff}_{(k)}\{P\}, \bar{d}\}$. The inclusion

$$i_1 \colon \mathcal{C}\mathrm{Diff}_{(0)}(P; \hat{P}) = \hat{P} = \mathrm{Hom}_A(P, \bar{\Lambda}^n) \hookrightarrow \mathcal{C}\mathrm{Diff}(P, \bar{\Lambda}^n)$$

induces the following series of inclusions:

$$\begin{aligned} i_2 \colon \mathcal{C}\mathrm{Diff}_{(1)}(P; \hat{P}) &= \mathcal{C}\mathrm{Diff}(P, \mathrm{Hom}(P, \bar{\Lambda}^n)) \\ &\to \mathcal{C}\mathrm{Diff}(P, \mathcal{C}\mathrm{Diff}(P, \bar{\Lambda}^n)) = \mathcal{C}\mathrm{Diff}_{(2)}(P; \bar{\Lambda}^n), \end{aligned}$$

....

$$i_k \colon \mathcal{C}\mathrm{Diff}_{(k-1)}(P; \hat{P}) \to \mathcal{C}\mathrm{Diff}_{(k)}(P; \bar{\Lambda}^n),$$

which satisfy the identity $\mu_k \circ i_k = \mathrm{id}$. Therefore, if $\Delta \in \mathrm{Diff}_{(k-1)}(P; \hat{P})$, then $i_k(\Delta)$ is an n-cocycle of the complex $\mathcal{C}\mathrm{Diff}_{(k)}\{P\}$ whose cohomology class is Δ. For this reason, the induced action of a permutation $\tau \in S_k$ on Δ is the cohomology class of the multidifferential operator $\tau(i_k(\Delta)) \in \mathcal{C}\mathrm{Diff}_{(k)}(P; \bar{\Lambda}^n)$. This shows that if $\nabla \in \mathcal{C}\mathrm{Diff}_{(k-1)}(P; \hat{P})$ is such that $i_k(\nabla) - \tau(i_k(\nabla)) = \bar{d} \circ \square$ with $\square \in \mathcal{C}\mathrm{Diff}_{(k)}(P, \hat{P})$, then $\bar{\tau}(\Delta) = \nabla$, $\bar{\tau}$ being the induced action of τ. Since, obviously, $\mathrm{im}\, i_k$ is invariant with respect to the subgroup $S_{k-1} \subset S_k$ leaving the k-th argument invariant, the induced action of a permutation $\tau \in S_{k-1} \subset S_k$ is just a permutation of the arguments in Δ, i.e.,

$$\bar{\tau}(\Delta)(p_1, \dots, p_{k-1}) = \Delta(p_{\tau(1)}, \dots, p_{\tau(k-1)}).$$

Now we describe the induced action of the element $\tau \in S_k$ transposing the i-th and the k-th arguments, $i < k$.

Fix elements $p_1, \dots, p_{i-1}, p_{i+1}, \dots, p_{k-1} \in P$ and consider the operator $\bar{\Box} = i_2(\Box)$ defined by $\bar{\Box}(p_i)(p_k) = \bar{\Delta}(p_1, \dots, p_k)$, where $\bar{\Delta} = i_k(\Delta)$. Let $\tilde{\Box}(p_i)(p_k) = \bar{\Box}(p_k)(p_i)$. Then from the $\mathcal{C}$-Green formula it follows that

$$\tilde{\Box}(p_i)(p_k) = (\Box(p_k), p_i) = (p_k, \Box^*(p_i)) + \bar{d}\nu(p_1, \dots, p_k),$$

where ν is a multidifferential operator. The multidifferential operator

$$\Delta' \colon (p_1, \dots, p_k) \mapsto (p_k, \Box^*(p_i)),$$

being A-linear with respect to p_k, lies in the image of i_k and $\Delta' = i_k(\Box^*)$. Since $\tau(\bar{\Delta})(p_1, \dots, p_k) = \tilde{\Box}(p_i)(p_k)$, we see that $\tau(\bar{\Delta}) - \Delta' = \bar{d} \circ \nu$ and therefore $\Delta' = i_k(\bar{\tau}(\Delta))$. On the other hand, $\Delta' = i_k(\Box^*)$, which implies $\bar{\tau}(\Delta) = \Box^*$. Thus we have proved the following

PROPOSITION. *Let $\tau \in S_k$ be the transposition of the i-th and the k-th elements, $i < k$, and $\nabla \in \mathcal{C}\mathrm{Diff}_{(k-1)}(P; \hat{P})$. Then*

$$\tau(\nabla)(p_1, \dots, p_{k-1}) = \Box_i(p_1, \dots, p_{k-1})^*(p_i),$$

where $\Box_i(p_1, \dots, p_{k-1}) \in \mathcal{C}\mathrm{Diff}(P, \hat{P})$ is defined by

$$\Box_i(p_1, \dots, p_{k-1})(p) = \nabla(p_1, \dots, p_{i-1}, p, p_{i+1}, \dots, p_{k-1}).$$

COROLLARY. *The set $L_k(P)$ consists of operators $\nabla \in \mathcal{C}\mathrm{Diff}^{\mathrm{alt}}(P; \hat{P})$ such that*

$$\Box_i(p_1, \dots, p_{k-1})^* = -\Box_i(p_1, \dots, p_{k-1})$$

for all $1 \le i \le k-1$, $p_1, \dots, p_{k-1} \in P$.

4.3.7. Description of $\{E_r^{p,q}, d_r\}$ in the free case.

THEOREM. *Let $\mathcal{O}$ be open in $J^\infty(\pi)$ or in $J^\infty(E, n)$. Then*

1. *$E_1^{0,q} = \bar{H}^q(\mathcal{O})$ and $E_1^{p,q} = 0$ if $p > 0$ and $q \neq n$. If $q = n$, then $E_1^{p,n} = L_p(\varkappa)$ for all $p > 0$ assuming that $L_1(\varkappa) = \hat{\varkappa}$.*
2. *$E_1^{0,q} = E_\infty^{0,q}$ if $q < n$; $E_2^{p,n} = E_\infty^{p,n}$ if $p \ge 0$. If $2 \le r \le \infty$, then $E_r^{p,q} = 0$ for $p > 0$, $q \neq n$ and for $p = 0$, $q > n$.*

PROOF. The proof immediately follows from the above description of the terms E_0 and E_1. □

Set $L_0(\varkappa) = \bar{H}^n(\mathcal{O})$ and introduce the operators $\mathcal{E}_i \colon L_i(\varkappa) \to L_{i+1}(\varkappa)$ coinciding with $d_1^{i,n}$ under the above-described identification of $E_1^{i,n}$ with $L_i(\varkappa)$.

COROLLARY. *The following facts hold:*

1. *If $q < n$, then $H^q(\mathcal{O}) = \bar{H}^q(\mathcal{O})$. If $q \ge n$, then $H^q(\mathcal{O}) = E_2^{q-n,n}$, or, which is the same, $H^q(\mathcal{O})$ is isomorphic to the $(q-n)$-th cohomology of the complex $\{L_i, \mathcal{E}_i\}$.*
2. *If $\mathcal{O} = J^\infty(\pi)$, then $H^q(J^\infty(\pi)) = H^q(J^0(\pi))$. If $\mathcal{O} = J^\infty(E, n)$, then $H^q(J^\infty(E, n)) = H^q(J^1(E, n))$.*

PROOF. Statement (1) immediately follows from the theorem above. To prove the second statement, it suffices to note the following. First, the fibers of the projection $\pi_{k,k-1}\colon J^k(\pi) \to J^{k-1}(\pi)$ are contractible for $k \geq 1$, whereas the fibers of $\pi_{k,k-1}\colon J^k(E,n) \to J^{k-1}(E,n)$ are contractible for $k \geq 2$ (see, for example, [**71**]). Second, Λ^* is the direct limit of the sequence of complexes $\pi_{\infty,k}(\Lambda^*(\mathcal{O}_k))$, where $\mathcal{O}_k$ is either $J^k(\pi)$ or $J^k(E,n)$, while $\pi^*_{\infty,k}$ are monomorphisms. □

REMARK. The manifold $J^1(E,n)$ is the manifold of all n-dimensional tangent subspaces of the manifold M. Therefore, the fibers of the projection $\pi_{1,0}\colon J^1(E,n) \to J^0(E,n) = E$ are Grassmann manifolds $G_{m+n,n}$, and the cohomology of $J^1(E,n)$ can be computed using the standard techniques of algebraic topology (the Leray–Hirsch theorem, etc.).

4.3.8. Relations with the calculus of variations and computation of the operators $\mathcal{E}_k$. In this subsection, the differentials $d_1^{i,n}$ (or, which is the same, $\mathcal{E}_i$) are described explicitly. The $\mathcal{C}$-spectral sequence and the calculus of variations are related by the fact that the differential $d_1^{0,n}\colon E_1^{0,n} = \bar{H}^n(\mathcal{Y}) \to E_1^{1,n} = \hat{\varkappa}$ in any special chart coincides with the Euler operator $\mathcal{E}$ (see Subsection 4.2.3). Thus $d_1^{0,n}$ can be viewed as the "global" Euler operator. The following proposition is a direct consequence of the above remark and the naturality of the $\mathcal{C}$-spectral sequence in the category DE.

PROPOSITION. *The Euler operator is natural in the category DE, i.e., commutes with morphisms in DE.*

REMARK. Let Δ be a nonlinear differential operator acting from $\Gamma(\pi_2)$ to $\Gamma(\pi_1)$, where $\Gamma(\pi_i)\colon E_{\pi_i} \to M$, $i = 1,2$, are smooth bundles. This operator generates in a natural way a mapping $F\colon J^\infty(\pi_1) \to J^\infty(\pi_2)$, and $F^*\colon \mathcal{F}(\pi_2) \to \mathcal{F}(\pi_1)$ is a morphism in the category DE. In a similar manner, a nonlinear differential operator Δ acting on n-dimensional submanifolds of a manifold E_2 and taking values in n-dimensional submanifolds of a manifold E_1 generates a morphism $F^*\colon \mathcal{F}(E_2,n) \to \mathcal{F}(E_1,n)$ in the category DE, where $\dim E_i = n + m_i$. Then Proposition 4.3.8 means that

$$\mathcal{E} \circ F^*_{\bar{H}} = F^*_{\hat{\varkappa}} \circ \mathcal{E},$$

where $F^*_{\bar{H}}\colon \bar{H}^*(\pi_2) \to \bar{H}^*(\pi_1)$ (resp., $F^*_{\bar{H}}\colon \bar{H}^*(J^\infty(E_2,n)) \to \bar{H}^*(J^\infty(E_1,n))$) and $F^*_{\hat{\varkappa}}\colon \hat{\varkappa}(\pi_1) \to \hat{\varkappa}(\pi_2)$ (resp., $F^*_{\hat{\varkappa}}\colon \hat{\varkappa}(J^\infty(E_2,n)) \to \hat{\varkappa}(J^\infty(E_1,n))$) are the mappings induced by F. This shows that Proposition 4.3.8 is a generalization of the well-known fact about the invariance of the Euler operator under transformations of dependent and independent variables. These transformations correspond to the zero-order operators in the situation above.

From Propositions 4.3.3 and 4.3.4 and from the general theory of bicomplexes (see [**91**]) it follows that under the above identifications the composition

$$E_0^{0,n} = \bar{\Lambda}^n \to E_1^{0,n} = \bar{H}^n(\mathcal{O}) \xrightarrow{d_1^{0,n}} E_1^{1,n} = \hat{\varkappa}$$

coincides with $\mu \circ l_{0,n}$, while $d_1^{p,n} - \mathcal{E}_p$ coincides with $\mu_{p+1} \circ l_{p,n} \circ i_p^{\text{alt}}$, where i_p^{alt} is the composition of i_p with the alteration operation (see Subsection 4.3.5).

Below we work over $J^\infty(\pi)$ or, equivalently, in a special chart in $J^\infty(E,n)$. In agreement with Subsection 3.3.5, we identify $\varkappa = \varkappa(\pi)$ with $\mathcal{F}(\pi,\pi)$. If ∇

is a multidifferential operator on $\varkappa = \varkappa(\pi)$, we write $\nabla(f_1, \dots, f_p)$ instead of $\nabla(\ni_{f_1}, \dots, \ni_{f_m})$.

As was noted before, $l_{0,k}(\omega) = \ell_\omega$, $\omega \in \bar{\Lambda}^k$. Hence, $(\mu \circ l_{0,n})(\omega) = \ell^*_\omega(1) = \mathcal{E}(\mathcal{L})$, where $\mathcal{L} = \langle\omega\rangle$. Thus we have proved the coincidence of $d_1^{0,n}$ with $\mathcal{E}$. To compute the operators $d_1^{p,n}$, $p > 0$, we shall use the following

LEMMA. *Let* $\square \in \mathcal{C}\mathrm{Diff}^{\mathrm{alt}}_{(p)}(\varkappa; \bar{\Lambda}^n)$. *Then*

(4.7)
$$(\mu_{p+1} \circ l_{p,n})(\square)(f_1, \dots, f_p) = (-1)^p \ell^*_{\square(f_1,\dots,f_p)}(1) + \sum_{s\le p}(-1)^{s-1} \ni_{f_s}(\square^*_s(1))$$
$$+ \sum_{s\le p}(-1)^{s-1}\ell^*_{f_s}(\square^*_s(1)) + \sum_{i<j\le p}(-1)^{i+j}\square^*_{ij}(1),$$

where the operators $\square_s$ *and* $\square_{ij} \in \mathcal{C}\mathrm{Diff}(\varkappa, \bar{\Lambda}^n)$ *are defined by the following relations:*

$$\square_s(f) = \square(f_1, \dots, \hat{f}_s, \dots, f_p, f)$$

and

$$\square_{ij}(f) = \square(\{f_i, f_j\}, f_1, \dots, \hat{f}_i, \dots, \hat{f}_j, \dots, f_p, f).$$

PROOF. According to (4.4), one has

$$\begin{aligned} l_{p,n}(\square)(f_1, \dots, f_{p+1}) &= \sum_{s\le p}(-1)^{s-1}\ni_{f_s}\square(f_1, \dots, \hat{f}_s, \dots, f_{p+1}) \\ &+ (-1)^p \ni_{f_{p+1}}\square(f_1, \dots, f_p) \\ &+ \sum_{i<j\le p}(-1)^{i+j}\square(\{f_i, f_j\}, f_1, \dots, \hat{f}_i, \dots, \hat{f}_j, \dots, f_{p+1}) \\ &+ \sum_{s\le p}(-1)^{s+p+1}\square(\{f_s, f_{p+1}\}, f_1, \dots, \hat{f}_s, \dots, f_p) \\ &= \sum_{s\le p}(-1)^{s-1}\left([\ni_{f_s}, \square_s](f_{p+1}) + \square_s(\ni_{f_s}(f_{p+1}))\right) \\ &+ (-1)^p \ell_{\square(f_1,\dots,f_p)}(f_{p+1}) + \sum_{i<j\le l}(-1)^{i+j}\square_{ij}(f_{p+1}) \\ &+ \sum_{s\le p}(-1)^s\left(\square_s(\ni_{f_s}(f_{p+1})) - \square_s(\ni_{f_{p+1}}(f_s))\right) \\ &= \sum_{s\le p}(-1)^{s-1}[\ni_{f_s}, \square_s](f_{p+1}) + (-1)^p \ell_{\square(f_1,\dots,f_p)}(f_{p+1}) \\ &+ \sum_{i<j\le p}(-1)^{ij}\square_{ij}(f_{p+1}) + \sum_{s\le p}(-1)^{s-1}\square_s(\ell_{f_s}(f_{p+1})). \end{aligned}$$

Consider the operator

$$\nabla = \nabla_{f_1,\dots,f_p} = \sum_{s\le p}(-1)^{s-1}[\ni_{f_s}, \square_s]$$
$$+ (-1)^p \ell_{\square(f_1,\dots,f_p)} + \sum_{i<j\le p}(-1)^{i+j}\square_{ij} + \sum_{s\le p}(-1)^{s-1}\square_s \circ \ell_{f_s}.$$

Then the result of the previous computations can be represented in the form $l_{p,n}(\square)(f_1,\dots,f_{p+1}) = \nabla(f_{p+1})$. But

$$\nabla(f_{p+1}) = (\nabla^*(1), f_{p+1}) + \bar{d}\mathcal{K}_\lambda(\nabla \circ f_{p+1}),$$

and since the multidifferential operator $(f_1,\dots,f_{p+1}) \mapsto (\nabla^*(1), f_{p+1})$ is an A-homomorphism with respect to the last argument, it belongs to the image of i_{p+1}. Hence,

$$(\mu_{p+1} \circ l_{p,n})(\square)(f_1,\dots,f_p) = \nabla^*_{f_1,\dots,f_p}(1).$$

□

Now let $\Delta \in L_p(\varkappa)$, $p \geq 1$. If $\square = i_p^{\mathrm{alt}}(\square)$, then

$$\square(f_1,\dots,f_p) = \frac{1}{p}\sum_{i\leq p}(-1)^{p-i}\Delta(f_1,\dots,\hat{f}_i,\dots,f_p)(f_i). \tag{4.8}$$

To compute $\mathcal{E}_p(\Delta)$, we have to express $\square_s^*(1)$, $\square_{ij}^*(1)$, and $\ell^*_{\square(f_1,\dots,f_p)}(1)$ in terms of Δ and then use the lemma above. Let us introduce operators $\Delta_{is} \in \mathcal{C}\mathrm{Diff}(\varkappa,\hat{\varkappa})$ by setting $\Delta_{is}(f_s) = \Delta(f_1,\dots,\hat{f}_i,\dots,f_p)$. Then directly from definitions one gets

$$\square_s = \frac{1}{p}\sum_{i\leq p}(-1)^{i+s} f_i^* \circ \Delta_{is}.$$

Since $\Delta \in L_p(\varkappa)$, $\Delta^*_{is} = -\Delta_{is}$ by Corollary 4.3.6, and we obtain

$$\begin{aligned}\square_s^*(1) &= \frac{1}{p}\sum_{i\leq p}(-1)^{i+s}(\Delta^*_{is}\circ f_i)(1) = \frac{1}{p}\sum_{i\leq p}(-1)^{i+s}\Delta^*_{is}(f_i)\\ &= \frac{1}{p}\sum_{i\leq p}(-1)^{i+s-1}\Delta_{is}(f_i) = \Delta(f_1,\dots,\hat{f}_s,\dots,f_p).\end{aligned}$$

In a similar way,

$$\square_{ij}^*(1) = \Delta(\{f_i,f_j\}, f_1,\dots,\hat{f}_i,\dots,\hat{f}_j,\dots,f_p).$$

Further, since the operator $\varphi \mapsto \ell_\varphi$ is a derivation (see Subsection 2.3.4), for any $\varphi \in P$ and $\psi \in \hat{P}$ one has $\ell_{(\varphi,\psi)} = \varphi^* \circ \ell_\psi + \psi^* \circ \ell_\varphi$ and $\ell^*_{(\varphi,\psi)}(1) = \ell^*_\psi(\varphi) + \ell^*_\varphi(\psi)$. Applying this equality to (4.8), we find that

$$\ell^*_{\square(f_1,\dots,f_p)}(1) = \frac{1}{p}\sum_{i\leq p}(-1)^{p-i}\Big(\ell^*_{\Delta(f_1,\dots,\hat{f}_i,\dots,f_p)}(f_i) + \ell^*_{f_i}(\Delta(f_1,\dots,\hat{f}_i,\dots,f_p))\Big).$$

Substituting these expressions into (4.7), one gets for $p > 0$

(4.9)

$$\begin{aligned}\mathcal{E}_p(\Delta)(f_1,\dots,f_p) &= \sum_{s\leq p}(-1)^{s-1}\mathfrak{I}_{f_s}(\Delta(f_1,\dots,\hat{f}_s,\dots,f_p))\\ &+ \sum_{s<k\leq p}(-1)^{s+k}\Delta(\{f_s,f_k\}, f_1,\dots,\hat{f}_s,\dots,\hat{f}_k,\dots,f_p)\\ &+ \frac{1}{p}\sum_{s\leq p}(-1)^{s-1}\Big((p-1)\ell^*_{f_s}(\Delta(f_1,\dots,\hat{f}_s,\dots,f_p)) - \ell^*_{\Delta(f_1,\dots,\hat{f}_s,\dots,f_p)}(f_s)\Big).\end{aligned}$$

For $p = 1$ this formula reduces to $\mathcal{E}_1(\varphi)(f) = \partial_f(\varphi) - \ell^*_\varphi(f) = (\ell_\varphi - \ell^*_\varphi)(f)$, where $\varphi \in \hat{\varkappa}$, i.e.,

$$\mathcal{E}_1(\varphi) = \ell_\varphi - \ell^*_\varphi. \tag{4.10}$$

The last summation in formula (4.9) can be rewritten as

$$\left(\sum_{s<p}(-1)^{s-1}\ell^*_{f_s} \circ \Delta_{f_1,\dots,\hat{f}_s,\dots,f_{p-1}} + (-1)^p \ell^*_{\Delta(f_1,\dots,f_{p-1})}\right)(f_p),$$

where $\Delta_{g_1,\dots,g_{p-2}} = \Delta(g_1, \dots, g_{p-2}, \cdot) \in \mathcal{C}\,\mathrm{Diff}(\varkappa, \hat{\varkappa})$. This results from the identity

$$\begin{aligned} &\ell^*_{\Delta(\chi,\chi_1,\dots,\hat{\chi}_i,\dots,\chi_{p-1})}(\chi_i) + \ell^*_{\chi_i}(\Delta(\chi, \chi_1, \dots, \hat{\chi}_i, \dots, \chi_{p-1})) \\ &\qquad = (-1)^i \left(\ell^*_{\Delta(\chi_1,\dots,\chi_{p-1})}(\chi) + \ell^*_\chi(\Delta(\chi_1, \dots, \chi_{p-1}))\right) \end{aligned} \tag{4.11}$$

valid for any $\Delta \in L_p(\varkappa)$. Obviously, this identity is a particular case of the identity

$$\ell^*_g(\square(f)) + \ell_{\square(g)}(f) = -\ell^*_f(\square(g)) - \ell^*_{\square(g)}(f), \tag{4.12}$$

which holds for any skew-adjoint operator $\square\colon \varkappa \to \hat{\varkappa}$ and is specified for $\square = \Delta(\chi_1, \dots, \hat{\chi}_i, \dots, \chi_{p-1})$, $f = \chi_i$, $g = \chi$. In turn, (4.12) is a consequence of the $\mathcal{C}$-Green formula

$$(\square(f), g) + (f, \square(g)) = \bar{d}\mathcal{K}(\square(f, g)),$$

which implies

$$\ell^*_{(\square(f),g)}(1) = -\ell^*_{(f,\square(g))}(1).$$

But as we have already noticed earlier,

$$\ell_{(\square(f),g)} = \square(f)^* \circ \ell_g + g^* \circ \ell_{\square(f)},$$

so that

$$\ell^*_{(\square(f),g)}(1) = \ell^*_g(\square(f)) + \ell^*_{\square(f)}(g).$$

Similarly, $\ell^*_{(f,\square(g))}(1) = \ell^*_f(\square(g)) + \ell^*_{\square(g)}(f)$, which proves (4.12).

By putting

$$\begin{aligned} \partial_f(\Delta)(\chi_1, \dots, \chi_{p-1}) = &- \partial_f(\Delta(\chi_1, \dots, \chi_{p-1})) \\ &+ \sum_{s \le p-1} \Delta(\chi_1, \dots, \partial_f(\chi_s), \dots, \chi_{p-1}) \end{aligned}$$

and taking (4.11) into account, we obtain a more compact description of $\mathcal{E}_p$:

$$\begin{aligned} \mathcal{E}_p(\Delta)(f_1, \dots, f_p) = &\sum_{i \le p}(-1)^i \partial_{f_i}(\Delta)(f_1, \dots, \hat{f}_i, \dots, f_p) \\ &+ \left(\sum_{s<p} \ell^*_{f_s} \circ \Delta_{f_1,\dots,\hat{f}_i,\dots,f_{p-1}} + (-1)^p \ell^*_{\Delta(f_1,\dots,\dots,f_{p-1})}\right)(f_p). \end{aligned} \tag{4.13}$$

Note that the second summation in (4.13) is not manifestly skew-symmetric with respect to $f_1, \dots, f_p$. Its hidden skew-symmetry leads to a typical identity occurring in the context we deal with. Other identities of this kind can be obtained from the equality $\mathcal{E}_{p+1} \circ \mathcal{E}_p = 0$ by rewriting it in accordance with (4.9) or (4.13).

4.3.9. A resolution for the Euler operator. If $\mathcal{O}$ is open in $J^\infty(E,n)$ (or in $J^\infty(\pi)$), then the description of the $\mathcal{C}$-spectral sequence given in Subsection 4.3.7 can be reformulated in the following way. Consider the complex

$$0 \to A = \bar{\Lambda}^0 \xrightarrow{\bar{d}} \cdots \xrightarrow{\bar{d}} \bar{\Lambda}^n \xrightarrow{\mathcal{E}'} L_1 \xrightarrow{\mathcal{E}_1} L_2 \xrightarrow{\mathcal{E}_2} \cdots \tag{4.14}$$

called the *variational complex*. Here $\mathcal{E}'$ is the composition of the natural projection $\bar{\Lambda}^n \to \bar{H}^n(\mathcal{O})$ and of $\mathcal{E}_0 = d_1^{0,n}$. According to Theorem 4.3.7, the k-th cohomology of this complex (if one counts from the left) coincides with $H^k(\mathcal{O})$. Therefore, the fragment of this complex lying to the left of $\bar{\Lambda}^n$, as well as the complex $\{L_i, \mathcal{E}_i\}$, is a resolution for the Euler operator in the case $H^k(\mathcal{O}) = 0$, $k > n$. This is always true locally. Note the following particular case.

COROLLARY. *If π is a vector bundle, then $\{L_i, \mathcal{E}_i\}$ is a resolution for the Euler operator on $J^\infty(\pi)$.*

A Lagrangian $\mathcal{L} \in \bar{H}^n(\mathcal{O})$ is said to be *trivial* if $\mathcal{E}(\mathcal{L}) = 0$. The vector space of trivial Lagrangians obviously coincides with $E_2^{0,n}$ and therefore, according to Theorem 4.3.7, one obtains

PROPOSITION. *The vector space of trivial Lagrangians is isomorphic to $H^n(\mathcal{O})$. In particular, if $H^n(\mathcal{O}) = 0$, then a Lagrangian is trivial if and only if its density is trivial, i.e., is of the form $\bar{d}\rho$, $\rho \in \bar{\Lambda}^{n-1}(\mathcal{O})$.*

One of the simplest applications of the variational complex is to the *inverse problem* of the calculus of variations, i.e., to the problem whether a given system of differential equations can be derived from the variational principle. There are various settings for this problem.

The *weak inverse problem* asks whether a given representation $\varphi = 0$ of a given system of partial differential equations $\mathcal{Y} \subset J^k(E,n)$ or $\subset J^k(\pi)$ is in the Euler–Lagrange form, i.e., whether $\varphi = \mathcal{E}(\mathcal{L})$ for a Lagrangian $\mathcal{L} \in \bar{H}^n(\mathcal{O})$. The results obtained previously prove

THEOREM. *The weak inverse problem for a system $\varphi = 0$, $\varphi \in \hat{\varkappa}$, admits a solution if and only if $\mathcal{E}_1(\varphi) = 0$ (i.e., φ is a cocycle of the variational complex) and the corresponding cohomology class $[\varphi] \in E_1^{1,n}(\mathcal{O}) = H^{n+1}(\mathcal{O})$ is trivial.*

It is worth noting the following important particular case.

COROLLARY. *Let $\pi\colon E_\pi \to M$ be a fiber bundle such that $H^{n+1}(E_\pi) = 0$. Then a system $\varphi = 0$, $\varphi \in \hat{\pi}$, is an Euler–Lagrange system if and only if $\ell_\varphi = \ell_\varphi^*$.*

PROOF. The result follows from the theorem and equation (4.10). □

It is natural to say that a nonlinear system of equation is *selfadjoint* if it is of the form $\varphi = 0$ with $\ell_\varphi = \ell_\varphi^*$. Using this terminology, one can observe a direct connection between the theorem above and the results of Subsection 1.4.3.

The *strong inverse problem* is to determine whether a given system of partial differential equations $\mathcal{Y}$ represented by $\varphi = 0$ admits another representation, say $\psi = 0$, such that $\psi = \mathcal{E}(\mathcal{L})$.

In this situation $\psi = F(\varphi)$, F being an automorphism of the $\mathcal{F}_k$-module $\hat{\varkappa}_k$ (k-th filtration of $\hat{\varkappa}$) and the problem may be viewed as that of solvability of the equation $\mathcal{E}_1(F(\varphi))$ imposed on F. If $\mathcal{Y} \subset J^k(\pi)$, or if we are in a local situation, the last equation takes the form $\ell_{F(\varphi)} = \ell_{F(\varphi)}^*$. Since the operation $\theta \mapsto \ell_\theta$ is a derivation, the restriction $\bar{\ell}_{F(\varphi)}$ of the $\mathcal{C}$-differential operator $\ell_{F(\varphi)}$ to $\mathcal{Y}_\infty$ is equal to $\bar{F} \circ \bar{\ell}_{F(\varphi)}$, where the bars refer to the restrictions to $\mathcal{Y}_\infty$. Therefore, a necessary condition for solvability of the strong problem is the solvability with respect to $\bar{F}$ of the equation

$$\bar{F} \circ \bar{\ell}_\varphi = \bar{\ell}^*_\varphi \circ \bar{F}^*. \tag{4.15}$$

This equality shows that if ξ is a higher symmetry of the equation $\mathcal{Y}$, then $\bar{F}^*(\xi)$ is the generating function of a conservation law of $\mathcal{Y}$. Note also that the condition (4.15) is very close to being sufficient.

In a more adequate version of the strong problem, F is supposed to be an invertible $\mathcal{C}$-differential operator. Condition (4.15) also remains valid in this case provided $\bar{F}$ is assumed to be an invertible $\mathcal{C}$-differential operator.

4.3.10. Infinitesimal Stokes formula in the term E_1. Let $\Delta \in L_i$ and assume that $\mathcal{E}_i(\Delta) = 0$. Then, if there are no topological obstructions, i.e., if $H^i(\mathcal{O}) = 0$, one has $\Delta = \mathcal{E}_{i-1}(\Delta')$. Our goal is to deduce explicit ("inversion") formulas for Δ'. These formulas are consequences of "the infinitesimal Stokes formula", which, as will be shown below, can be adapted to the term E_1 of the $\mathcal{C}$-spectral sequence.

Let an algebra A be an object of the category DE. Then any operator of the Lie derivative along a field $X \in D_{\mathcal{C}}(A)$ preserves the ideals $\mathcal{C}^k\Lambda^*$ and consequently generates an endomorphism of the $\mathcal{C}$-spectral sequence. This endomorphism is also called the Lie derivative (on $E_r^{p,q}$) and is denoted by $e \mapsto X\{e\}$, $e \in E_r^{p,q}$.

Let $X \in \mathcal{C}D(A)$. Then

$$X \lrcorner \mathcal{C}\Lambda^* = X \lrcorner (\mathcal{C}\Lambda^1 \wedge \Lambda^*) \subset \mathcal{C}\Lambda^1 \wedge \Lambda^* = \mathcal{C}\Lambda^*$$

and therefore $X \lrcorner \mathcal{C}^k\Lambda^* \subset \mathcal{C}^k\Lambda^*$. In a similar way, for $X \in D(A)$ one has $X \lrcorner \mathcal{C}^k\Lambda^* \subset \mathcal{C}^{k-1}\Lambda^*$. If $\omega \in \mathcal{C}^p\Lambda^*$ and $d\omega \in \mathcal{C}^{p+k}\Lambda^*$, then from the formula $X(\omega) = X \lrcorner d\omega + d(X \lrcorner \omega)$ it follows that $X(\omega) \equiv d\rho \bmod \mathcal{C}^{p+k}\Lambda^*$, where $\rho \in \mathcal{C}^p\Lambda^*$ and $X \in \mathcal{C}D(A)$. In other words, $X\{e\} = 0$ for $k > 0$ under the assumption that the element $e \in E_k$ is represented by the form ω. Thus the Lie derivative along $X \in \mathcal{C}D(A)$ is always trivial in the $\mathcal{C}$-spectral sequence for $k > 0$. It means that the action of elements of the quotient space $D_{\mathcal{C}}(A)/\mathcal{C}D(A)$ on $E_r^{p,q}$, $r > 0$, is well defined. This action is again called the Lie derivative. Thus, if A is $\mathcal{F}(\mathcal{Y}_\infty)$, $\mathcal{F}(E, n)$, or $\mathcal{F}(\pi)$, then the Lie derivative $\chi\{e\}$ is defined, where $e \in E_r^{p,q}$, while χ lies in $\operatorname{sym}\mathcal{Y}$, $\varkappa(J^\infty(E, n))$, or $\varkappa(\pi)$, respectively.

For $X \in D_{\mathcal{C}}(A)$ the insertion operation $X \lrcorner e \in E_1^{p-1,q}$, where $e \in E_1^{p,q}$, is also well defined. Indeed, let $\omega \in \mathcal{C}^p\Lambda^{p+q}$ be a d_0-cocycle representing e, i.e., $e = \omega + d(\mathcal{C}^p\Lambda^{p+q-1}) + \mathcal{C}^{p+1}\Lambda^{p+q}$. Then $X \lrcorner \omega \in \mathcal{C}^{p-1}\Lambda^{p+q-1}$ and $d(X \lrcorner \omega) = X(\omega) - X \lrcorner d\omega \in \mathcal{C}^p\Lambda^{p+q}$, i.e., $X \lrcorner \omega$ is a d_0-cocycle. If $\rho \in \mathcal{C}^p\Lambda^{p+q-1}$ and $\lambda \in \mathcal{C}^{p+1}\Lambda^{p+q}$, then

$$X \lrcorner (d\rho + \lambda) = X(\rho) - d(X \lrcorner \rho) + X \lrcorner \lambda \in \mathcal{C}^p\Lambda^{p+q-1} + d\mathcal{C}^{p-1}\Lambda^{p+q-2}.$$

Hence, the d_0-cohomology class of the d_0-cocycle $X \lrcorner \omega$ is determined by the class e. Denote it by $X \lrcorner e$.

If in the previous considerations we assume that $X \in \mathcal{C}D(A) \subset D_{\mathcal{C}}(A)$, then according to the inclusion $X \lrcorner \mathcal{C}^p\Lambda^* \subset \mathcal{C}^p\Lambda^*$, we find that $X \lrcorner e = 0$. This allows one to define the insertion operation $\chi \lrcorner e \in E_1^{p-1,q}$, where $e \in E_1^{p,q}$ and $\chi \in D_{\mathcal{C}}(A)/\mathcal{C}D(A)$. Taking now the d_0-cohomology classes of each summand in the infinitesimal Stokes formula $X(\omega) = X \lrcorner\, d\omega + d(X \lrcorner\, \omega)$, we arrive at the following *infinitesimal Stokes formula* in E_1:

$$\chi\{e\} = d_1^{p-1,q}(\chi \lrcorner\, e) + \chi \lrcorner\, d_1^{p,q}, \quad e \in E_1^{p,q}, \ \chi \in D_{\mathcal{C}}(A)/\mathcal{C}D(A). \tag{4.16}$$

If $\mathcal{O}$ is an open set in $J^\infty(\pi)$ or in $J^\infty(E,n)$, the operations of insertion and of Lie derivative can be carried over from $E_1^{p,q}(\mathcal{O})$ to $L_p(\varkappa)$ using the identifications of Subsection 4.3.6. Preserving the same notations for these operations, we obtain from (4.16) that

$$\chi\{\Delta\} = \mathcal{E}_{p-1}(\chi \lrcorner\, \Delta) + \chi \lrcorner\, \mathcal{E}_p(\Delta), \quad \chi \in \varkappa(\mathcal{O}), \ \Delta \in L_p(\varkappa). \tag{4.17}$$

Now we compute the operations $\chi \lrcorner$ and $\chi\{\bullet\}$ explicitly. Let $\Delta \in L_p(\varkappa)$ and assume that the operator $\square \in \mathcal{C}\mathrm{Diff}^{\mathrm{alt}}_{(p)}(\varkappa; \bar\Lambda^n)$ is defined by (4.8). Then from (4.3), according to identification of L_p with $E_1^{p,n}$, one obtains

$$\chi \lrcorner\, \Delta = \mu_{p-1}(\chi \lrcorner\, \square),$$

where $\chi \lrcorner\, \square \in \mathcal{C}\mathrm{Diff}^{\mathrm{alt}}_{(p-1)}(\varkappa; \bar\Lambda^n)$ and

$$(\chi \lrcorner\, \square)(\chi_1, \dots, \chi_{p-1}) = \square(\chi, \chi_1, \dots, \chi_{p-1}).$$

But in view of (4.8),

$$\begin{aligned}
p \cdot (\chi \lrcorner\, \square)&(\chi_1, \dots, \chi_{p-1}) \\
&= (-1)^{p-1}\Delta(\chi_1, \dots, \chi_{p-1})(\chi) + \sum_{i=2}^{p-2}(-1)^{p-i-1}\Delta(\chi, \chi_1, \dots, \hat\chi_i, \dots, \chi_{p-1})(\chi_i) \\
&\quad + \Delta(\chi, \chi_1, \dots, \chi_{p-2})(\chi_{p-1}) \\
&= (-1)^{p-1}(\chi^* \circ \Delta_{\chi_1,\dots,\chi_{p-2}})(\chi_{p-1}) \\
&\quad + \sum_{i=2}^{p-2}(-1)^{p-i-1}(\chi_i^* \circ \Delta_{\chi,\chi_1,\dots,\hat\chi_i,\dots,\chi_{p-2}})(\chi_{p-1}) + \Delta_{\chi,\chi_1,\dots,\chi_{p-2}}(\chi_{p-1}),
\end{aligned}$$

where

$$\begin{aligned}
\Delta_{\chi_1,\dots,\chi_{p-2}}(\chi_{p-1}) &= \Delta(\chi_1, \dots, \chi_{p-1}), \\
\Delta_{\chi,\chi_1,\dots,\hat\chi_i,\dots,\chi_{p-2}}(\chi_{p-1}) &= \Delta(\chi, \chi_1, \dots, \hat\chi_i, \dots, \chi_{p-1}).
\end{aligned}$$

Using now the $\mathcal{C}$-Green formula and Corollary 4.3.6, we get

$$\begin{aligned}
(-1)^{p-1}&(\chi^* \circ \Delta_{\chi_1,\dots,\chi_{p-2}})(\chi_{p-1}) \\
&= (-1)^{p-1}(\Delta^*_{\chi_1,\dots,\chi_{p-2}}(\chi), \chi_{p-1}) + \bar d\nu(\chi, \chi_1, \dots, \chi_{p-1}) \\
&= (-1)^p(\Delta_{\chi_1,\dots,\chi_{p-2}}(\chi))(\chi_{p-1}) + \bar d\nu = \Delta(\chi, \chi_1, \dots, \chi_{p-2})(\chi_{p-1}) + \bar d\nu,
\end{aligned}$$

where $\nu \in \mathcal{C}\mathrm{Diff}^{\mathrm{alt}}_{(p-1)}(\varkappa; \bar\Lambda^{n-1})$. Similarly,

$$(-1)^{p-i-1}(\chi_i^* \circ \Delta_{\chi,\chi_1,\dots,\hat\chi_i,\dots,\chi_{p-2}})(\chi_{p-1}) = \Delta(\chi, \chi_1, \dots, \chi_{p-2})(\chi_{p-1}) + \bar d\nu_i.$$

Hence,

$$(\chi \lrcorner\, \Delta)(\chi_1, \dots, \chi_{p-2}) = \Delta(\chi, \chi_1, \dots, \chi_{p-2}). \tag{4.18}$$

Now an explicit formula for $\chi\{\Delta\}$ directly follows from (4.17), (4.18), and (4.9):

(4.19)
$$\chi\{\Delta\}(\chi_1,\dots,\chi_{p-1}) = \mathrm{Э}_\chi(\Delta(\chi_1,\dots,\chi_{p-1})) + \sum_i \Delta(\chi_1,\dots,\{\chi_i,\chi\},\dots,\chi_{p-1})$$
$$+ \frac{1}{p(p-1)}\sum_i (-1)^i \Big(\ell^*_{\Delta(\chi,\chi_1,\dots,\hat{\chi}_i,\dots,\chi_{p-1})}(\chi_i) + \ell^*_{\chi_i}\Delta(\chi,\chi_1,\dots,\hat{\chi}_i,\dots,\chi_{p-1})\Big)$$
$$+ \frac{p-1}{p}\ell^*_\chi(\Delta(\chi_p,\dots,\chi_{p-1})) - \frac{1}{p}\ell^*_{\Delta(\chi_1,\dots,\chi_{p-1})}(\chi).$$

Modifying the second summation in (4.19) using identity (4.11), we find a more satisfactory description of $\chi\{\Delta\}$:

$$\begin{aligned}\chi\{\Delta\}(\chi_1,\dots,\chi_{p-1}) &= \mathrm{Э}_\chi(\Delta(\chi_1,\dots,\chi_{p-1}))\\ &+ \sum_i \Delta(\chi_1,\dots,\{\chi_i,\chi\},\dots,\chi_{p-1}) + \ell^*_\chi(\Delta(\chi_1,\dots,\chi_{p-1})).\end{aligned} \tag{4.20}$$

By introducing the action of evolutionary derivation on $L_p(\chi)$,

$$\mathrm{Э}_\chi[\Delta] = \mathrm{Э}_\chi(\Delta(\chi_1,\dots,\chi_{p-1})) + \sum_i \Delta(\chi_1,\dots,\{\chi_i,\chi\},\dots,\chi_{p-1}),$$

the last formula may be rewritten as

$$\mathrm{Э}\{\Delta\} = \mathrm{Э}_\chi[\Delta] + \ell^*_\chi \circ \Delta. \tag{4.21}$$

Since in the last formula, p is assumed to be > 1, it is necessary to consider the case $p = 1$ separately. Let $\varphi \in \hat{\varkappa} = L_1(\varkappa)$. Then $\chi \lrcorner \varphi = (\varphi, \chi)$ and, as it was shown earlier (see Subsection 4.3.8), $\ell_{(\varphi,\chi)} = \varphi^* \circ \ell_\chi + \chi^* \circ \ell_\varphi$. Therefore, due to the equality $\mathcal{E}_0(\langle\omega\rangle) = \ell^*_\omega(1)$, formulas (4.17), (4.10) imply

$$\chi\{\varphi\} = \ell_\varphi(\chi) + \ell^*_\chi(\varphi) = \mathrm{Э}_\chi(\varphi) + \ell^*_\chi(\varphi). \tag{4.22}$$

Finally, note that due to (4.22) formula (4.20) may be represented as

(4.23)
$$\chi\{\Delta\}(\chi_1,\dots,\chi_{p-1}) = \chi\{\Delta(\chi_1,\dots,\chi_{p-1})\} + \sum_i \Delta(\chi_1,\dots,\{\chi_i,\chi\},\dots,\chi_{p-1}).$$

It should be noted that there exists a natural action of $\chi \in \varkappa$ on the elements of $L_i(\varkappa)$, with the latter regarded as multidifferential operators. This action is easily defined using the results of Subsection 1.2.5 $\mathcal{C}$ and it differs from the action considered above.

4.3.11. Inversion formulas. Let $G_t\colon \mathcal{O} \to \mathcal{O}$ be a smooth family of automorphisms of an object $\mathcal{O}$, $t \in [0,1]$, and let $X_t \in D_\mathcal{C}(\mathcal{O})$ be the corresponding family of $\mathcal{C}$-fields, i.e.,

$$\frac{d}{dt}(G_t^*) = G_t^* \circ X_t.$$

If $d^{p,q}(e) = 0$ and $\chi_t = X_t \bmod \mathcal{C}D(\mathcal{O})$, then, according to (4.16), we get

$$\frac{d}{dt}(G_t^*(e)) = G_t^*(\chi_t\{e\}) = G_t^*(d_1(\chi_t \lrcorner e)) = d_1 G_t^*(\chi_t \lrcorner e).$$

Suppose that $G_0 = \mathrm{id}$. Integrating this equality with respect to t we obtain

$$(4.24)\qquad e = G_1^*(e) - d_1\left(\int_0^1 G_t^*(\chi_t \rfloor e)dt\right), \quad e \in E_1^{p,q},$$

or, identifying $E_1^{p,n}$ with $L_p(\varkappa)$,

$$(4.25)\qquad \Delta = G_1^*\{\Delta\} - \mathcal{E}_{p-1}\left(\int_0^1 G_t^*\{\chi_t \rfloor \Delta\}dt\right), \quad \Delta \in L_p(\varkappa),$$

where $G_t^*\{\square\} = i^{-1}\left(G_t^*(i(\square))\right)$, $\square \in L_p(\varkappa)$, and $i\colon L_p \to E_1^{p,n}$ is the identification mapping. Formulas (4.24) and (4.25) obviously remain valid if G_t are automorphisms of the object $\mathcal{O}$ for $t \in [0,1)$ only. In particular, if $\operatorname{im} G_1 \subset \mathcal{O}'$, where $\mathcal{O}'$ is a subobject in $\mathcal{O}$ such that $E_1^{p,q}(\mathcal{O}') = 0$, these formulas give us the desired inversion, since in such a case the term $G_1^*(e)$ (resp., $G_1^*(\Delta)$) vanishes.

Let $\pi\colon E_\pi \to M$ be a vector bundle, and $G_t^{(0)}\colon E_\pi \to E_\pi$ its fiberwise homothety with the coefficient $1-t$. Then $G_t = G_t^{(\infty)}$ is a Lie transformation (see Subsection 3.4.5) for any $t \in [0,1)$ and $\operatorname{im} G_1 = \operatorname{im} j(0)$, $0 \in \Gamma(\pi)$. Therefore, $G_1^*(E_1^{p,q}) = 0$, $p > 0$.

In special coordinates one has

$$G_t^*(x_k) = x_k,\ G_t^*(p_\sigma^i) = (1-t)p_\sigma^i, \quad t \in [0,1],$$
$$X_t = \frac{1}{t-1}\sum_{i,\sigma} p_\sigma^i \frac{\partial}{\partial p_\sigma^i} = \frac{1}{t-1}\mathfrak{Э}_u, \quad u = \rho_0(\pi).$$

Identifying $\varkappa(\pi)$ and $\mathcal{F}(\pi,\pi)$ according to Subsection 3.3.5, one sees that $\chi_t = (1/(t-1))\mathfrak{Э}_u$. Therefore, (4.18) shows that

$$(4.26)\qquad (\chi_t \rfloor \Delta)(f_1,\dots,f_{p-1}) = \frac{1}{t-1}\Delta(u, f_1,\dots,f_{p-1}),$$

where $f_i \in \mathcal{F}(\pi,\pi)$, $\Delta \in L_{p+1}(\varkappa)$.

LEMMA. *If $\Delta \in L_p(\varkappa)$, then*

$$G_t^*\{\Delta\} = (1-t)^p (G_t^*)^{-1} \circ \Delta \circ G_t^* = (1-t)^p G_t(\Delta).$$

PROOF. Since G_t^0 is an automorphism of the bundle π, one has $G_t(\mathfrak{Э}_f) = (G_t^*)^{-1} \circ \mathfrak{Э}_f \circ G_t^* = \mathfrak{Э}_{\tilde f}$ for some $\tilde f$, or, equivalently, $G_t^* \circ \mathfrak{Э}_{\tilde f} = \mathfrak{Э}_f \circ G_t^*$. Applying this equality to the element $u = \rho_0(\pi)$ and using the identities $\mathfrak{Э}_g(u) = g$ and $G_t^*(u) = (1-t)u$, we obtain $\tilde f = (1-t)(G_t^*)^{-1}(f)$. Thus

$$(4.27)\qquad \mathfrak{Э}_f \circ G_t^* = (1-t)G_t^* \circ \mathfrak{Э}_{(G_t^*)^{-1}(f)}.$$

By the definition of evolutionary derivations (Subsection 2.3.4), this is equivalent to the equality

$$(4.28)\qquad \ell_{G_t^*(\varphi)} = (1-t)G_t^* \circ \ell_\varphi \circ (G_t^*)^{-1}.$$

According to Subsection 1.2.5, (4.28) implies

$$(4.29)\qquad \ell^*_{G_t^*(\varphi)} = (1-t)G_t^* \circ \ell^*_\varphi \circ (G_t^*)^{-1}.$$

Since $L_p(\varkappa)$ as an A-module is generated by $\operatorname{im}\mathcal{E}_{p-1}$, the required statement results from

$$\mathcal{E}_{p-1}((G_t^*)^{-1} \circ \Delta \circ G_t^*) = (1-t)(G_t^*)^{-1} \circ \mathcal{E}_{p-1}(\Delta) \circ G_t^*, \quad p > 1,$$

and

$$\mathcal{E}(G_t^*(h)) = (1-t)G_t^*(\mathcal{E}(h)).$$

The last equality can be rewritten as

$$\ell^*_{G_t^*(\omega)}(1) = (1-t)G_t^*(\ell^*_\omega(1)), \quad h = \langle\omega\rangle,\ \omega \in \bar{\Lambda}^n,$$

and immediately follows from (4.29). For the case $p > 1$ one must use equalities (4.13) and (4.27)–(4.29) together with

$$-\{G_t^*(f), G_t^*(g)\} = (1-t)G_t^*\{f, g\}.$$

But due to (4.27), $\mathbf{\partial}_{G_t^*(f)}(G_t^*(g))$ equals $(1-t)G_t^*(\mathbf{\partial}_f(g))$ and one only needs to use (2.15). □

Finally, substituting into (4.25) the above data about the family of Lie transformations generated by homotheties, we obtain the desired formula:

$$\Delta = \mathcal{E}_{p-1}\left(\int_0^1 \left(G_t^* \circ (\mathbf{\partial}_u \lrcorner \Delta) \circ (G_t^*)^{-1}\right)(1-t)^{p-1}dt\right) \tag{4.30}$$

assuming that $\Delta \in L_p(\varkappa)$, $p > 1$, $\mathcal{E}_p(\Delta) = 0$.

The case $p = 1$ is somewhat different from the general case since $G_t^*\{\alpha\} = G_t^*(\alpha)$, $\alpha \in \bar{H}^n$, and $\chi \lrcorner h = \langle h(\chi)\rangle \in \bar{H}^n$, where $h \in \hat{\varkappa} = L_1(\varkappa)$. Consequently, (4.25) implies in this case that

$$h = \mathcal{E}\left(\int_0^1 G_t^*(h(u))\frac{dt}{1-t}\right), \quad h(u) = h(\mathbf{\partial}_u). \tag{4.31}$$

4.3.12. "Potentials" of trivial Lagrangians. If a Lagrangian $\mathcal{L} \in \bar{H}^n(B)$ is trivial and $H^n(B) = 0$, then according to Proposition 4.3.9, $\omega = \bar{d}\rho$, where $\omega \in \bar{\Lambda}^n(B)$ is such that $\mathcal{L} = \langle\omega\rangle$. In this subsection, we deduce an explicit formula for the "potential" ρ of the form ω.

For any Lie field X_f, consider the decomposition

$$X_f = \mathbf{\partial}_f + Y_f, \quad Y_f \in \mathcal{C}D.$$

If $\omega \in \bar{\Lambda}^n$, then, according to Subsection 4.2.4, we have

$$X_f(\omega) = \mathbf{\partial}_f(\omega) + \bar{d}(Y_f \lrcorner \omega) = \mathbf{\partial}_f(\omega) + \bar{d}(X_f \lrcorner \omega).$$

If $\mathcal{E}\langle\omega\rangle = \ell^*_\omega(1) = 0$, then by the $\mathcal{C}$-Green formula we have $\mathbf{\partial}_f(\omega) = \ell_\omega(f) = \bar{d}\mathcal{K}_\lambda(\ell_\omega \circ f)$ and thus

$$X_f(\omega) = \bar{d}(\mathcal{K}_\lambda(\ell_\omega \circ f) + X_f \lrcorner \omega).$$

Let G_t be a family of automorphisms of the domain B considered as an object of the category DE and $G_0 = \mathrm{id}$. Then for the corresponding family $X_t = X_{f_t}$ of $\mathcal{C}$-fields one gets

$$\begin{aligned} G_1^*(\omega) - \omega &= \int_0^1 \frac{d}{dt}G_t^*(\omega)dt = \int_0^1 G_t^*(X_t(\omega))dt \\ &= \bar{d}\left(\int_0^1 G_t^*(\mathcal{K}_\lambda(\ell_\omega \circ f_t) + X_{f_t} \lrcorner \omega)dt\right). \end{aligned}$$

Thus if $\mathcal{E}(\omega) = 0$, then

$$\omega = G_1^*(\omega) + \bar{d}\rho', \qquad \rho' = -\int_0^1 (\mathcal{K}_\lambda(\ell_\omega \circ f_t) + X_{f_t} \lrcorner \omega)dt. \tag{4.32}$$

If $G_1^*(\omega) = 0$, this formula gives the desired expression for ρ. In particular, for the family of homotheties described in the previous subsection one has

$$\omega = \pi_\infty^*(\omega|_M) + \bar{d}\rho', \qquad \rho' = \int_0^1 \mathcal{K}_\lambda(\ell_\omega \circ u)\frac{dt}{1-t}, \tag{4.33}$$

since $G_1^*(\omega) = \pi_\infty^*(j(0)^*(\omega)) = \pi_\infty^*(\omega|_M)$, $f_t = u/(t-1)$, and $Y_{f_t} = 0$. Therefore, under the condition

$$\int_M \omega|_M = 0 \Leftrightarrow \omega|_M = d\rho_0, \quad \rho_0 \in \Lambda^{n-1}(M),$$

formula (4.33) gives the required expression for the potential ρ:

$$\rho = \rho' + \pi_\infty^*(\rho_0). \tag{4.34}$$

4.4. The $\mathcal{C}$-spectral sequence of infinitely prolonged equations

In this section, we study the $\mathcal{C}$-spectral sequence of infinitely prolonged equations. Our attention is concentrated mostly on how to describe the term E_1. The role of the latter in the category DE is completely similar to the role of the de Rham complex in the category of smooth manifolds. To achieve this goal, the complexes $E_0^{p,*}$ are represented as cokernels of the mappings acting on some auxiliary complexes which are analogs of the Spencer complexes for multidifferential operators. To compute these cokernels, we construct another spectral sequence for which the term E_0 is similar to the Spencer δ-complex. As a result, it turns out to be possible to obtain a satisfactory description of the $\mathcal{C}$-spectral sequence for nonoverdetermined systems. The methods of this section allow one to obtain useful estimates in the overdetermined case as well.

At the end of the section, we prove "the homotopy theorem" which is an analog of the similar theorem for the de Rham cohomology. In many cases, this theorem makes it possible to compute efficiently the term E_2 of the $\mathcal{C}$-spectral sequence. Here we give one of the simplest consequences of this theorem — an analog of the Poincaré lemma in the category DE.

4.4.1. Let $\mathcal{Y} \subset J^k(E,n)$ be an equation, with $i = i_\mathcal{Y} \colon \mathcal{Y}_\infty \to J^\infty(E,n)$ being the corresponding embedding. Denote by $\{E_r^{p,q}, d_r^{p,q}\}$ (resp., by $\{E_r^{p,q}(\mathcal{Y}), d_r^{p,q}\}$) the $\mathcal{C}$-spectral sequence on $J^\infty(E,n)$ (resp., on $\mathcal{Y}_\infty$). Since i is a morphism in the category DE, it induces homomorphisms of $\mathcal{C}$-spectral sequences, i.e., we obtain the system of mappings $i^* = i_r^{p,q} \colon E_r^{p,q} \to E_r^{p,q}(\mathcal{Y})$ such that $i^* \circ d_r^{p,q} = d_r^{p,q} \circ i^*$. Evidently, $\operatorname{coker} i_0^{p,q} = 0$. To describe $\{E_r^{p,q}(\mathcal{Y}), d_r^{p,q}\}$, we need to describe $\ker i_0^{p,q}$ first of all.

We start with a special chart or, equivalently, assume that $\mathcal{Y} \subset J^k(\pi)$ and $\mathcal{Y}_\infty \subset J^\infty(\pi)$. According to Subsection 4.3.2, $E_0^{p,q} = \Lambda^{p,q}$, $E_0^{p,q}(\mathcal{Y}) = \Lambda^{p,q}(\mathcal{Y})$ and the mapping $i_0^{p,q} \colon \Lambda^{p,q} \to \Lambda^{p,q}(\mathcal{Y})$ is the natural restriction of differential forms. But as it follows from the general theory of differential forms, the kernel of the mapping $\Lambda^k \to \Lambda^k(\mathcal{Y}_\infty)$ coincides with $\mathcal{I}_\mathcal{Y} \cdot \Lambda^k + \Lambda^{k-1} \wedge d\mathcal{I}_\mathcal{Y}$, where $\mathcal{I}_\mathcal{Y}$ is the ideal of the submanifold $\mathcal{Y}_\infty \subset J^\infty(\pi)$. The (p,q)-component of this kernel is

$$\ker i_0^{p,q} = \mathcal{I}_\mathcal{Y} \cdot \Lambda^{p,q} + \mathcal{F} \cdot U_1(\mathcal{I}_\mathcal{Y}) \wedge \underbrace{\mathcal{C}\Lambda^1 \wedge \dots \wedge \mathcal{C}\Lambda^1}_{(p-1)\text{ times}} \wedge \Lambda_\Gamma^q$$
$$+ \underbrace{\mathcal{C}\Lambda^1 \wedge \dots \wedge \mathcal{C}\Lambda^1}_{p\text{ times}} \wedge \hat{d}\mathcal{I}_\mathcal{Y} \wedge \Lambda_\Gamma^{q-1},$$

where $\mathcal{F} = \mathcal{F}(\pi)$. But obviously $\hat{d}\mathcal{I}_{\mathcal{Y}} \subset \mathcal{I}_{\mathcal{Y}} \cdot \Lambda^1$. Consequently, $\hat{d}\mathcal{I}_{\mathcal{Y}} \subset \mathcal{I}_{\mathcal{Y}} \cdot \Lambda^1_{\Gamma}$ and

$$\ker i_0^{p,q} = \mathcal{I}_{\mathcal{Y}} \cdot \Lambda^{p,q} + \mathcal{F} \cdot U_1(\mathcal{I}_{\mathcal{Y}}) \wedge \Lambda^{p-1,q}.$$

In particular, $\ker i_0^{1,0} = \mathcal{I}_{\mathcal{Y}} \cdot \mathcal{C}\Lambda^1 + \mathcal{F} \cdot U_1(\mathcal{I}_{\mathcal{Y}})$. From the last equality it follows that in spite of the dependence of U_1 on the choice of a special chart, the image of the submodule $\mathcal{F} \cdot U_1(\mathcal{I}_{\mathcal{Y}})$ in $\mathcal{C}\Lambda^1/\mathcal{I}_{\mathcal{Y}} \cdot \mathcal{C}\Lambda^1$ is uniquely determined. Thus, denoting by $[P] = P/\mathcal{I}_{\mathcal{Y}} \cdot P$ the restriction of an $\mathcal{F}$-module P to $\mathcal{Y}_\infty$, we obtain

$$[\ker i_0^{p,q}] = [\mathcal{F} \cdot U_1(\mathcal{I}_{\mathcal{Y}})] \wedge \left[E_0^{p-1,q}\right], \tag{4.35}$$

where $[\mathcal{F} \cdot U_1(\mathcal{I}_{\mathcal{Y}})]$ denotes the submodule in $\left[\mathcal{C}\Lambda^1\right]$ generated in a special chart by the set $U_1(\mathcal{I}_{\mathcal{Y}})$. The operator U_1 is considered in this chart too. Note that

$$[\mathcal{F} \cdot U_1(\mathcal{I}_{\mathcal{Y}})] = \left[\ker i_0^{1,0}\right].$$

REMARK. It is worth noting the difference between the two kinds of restrictions we deal with: the restriction of representing objects, say $\Lambda^k \to \Lambda^k(\mathcal{Y})$, and that of modules, say $\Lambda^k \to \left[\Lambda^k\right]$. In the first case, Λ^k is quotiened by $\mathcal{I}_{\mathcal{Y}} \cdot + \Lambda^{k-1} \wedge d\mathcal{I}_{\mathcal{Y}}$ while in the second case by $\mathcal{I}_{\mathcal{Y}} \cdot \Lambda^k$ only.

4.4.2. Now we are going to describe the term $E_0^{p,q}(\mathcal{Y})$, just like it was done for the term $E_0^{p,q}$ in Subsection 4.3.3. For this purpose, let us associate to any element $\varphi \in [E_0^{p,q}]$ an operator $\nabla_\varphi \in \mathcal{C}\mathrm{Diff}^{\mathrm{alt}}_{(p)}([\varkappa]\,; \bar{\Lambda}^q(\mathcal{Y}_\infty))$ by setting

$$\nabla_\varphi([\chi_1], \dots, [\chi_p]) = [\chi_p \lrcorner (\dots(\chi_1 \lrcorner \rho)\dots)], \tag{4.36}$$

where $\chi_i \in \varkappa$, $[\chi_i]$ denotes the image of χ_i in $[\varkappa]$, and $\varphi = [\rho]$. It should be noted that the right-hand side of (4.36) lies in $\left[E_0^{0,q}\right] = \left[\bar{\Lambda}^q\right]$ and obviously $\left[\bar{\Lambda}^q\right] = \bar{\Lambda}^q(\mathcal{Y}_\infty)$. The proof of the following statement is the same as that of Proposition 4.3.3.

PROPOSITION. *The correspondence $\varphi \mapsto \nabla_\varphi$ establishes an isomorphism of $\mathcal{F}(\mathcal{Y})$-modules $[E_0^{p,q}]$ and $\mathcal{C}\mathrm{Diff}^{\mathrm{alt}}_{(p)}([\varkappa]\,; \bar{\Lambda}^q(\mathcal{Y}_\infty))$.*

We note that if $\rho \in E_0^{p,q}$, then $\nabla_{[\rho]} = [\nabla_\rho]$, where $[\Delta]$ denotes the restriction of a $\mathcal{C}$-differential operator Δ from $J^\infty(E, n)$ (or from $J^\infty(\pi)$) to $\mathcal{Y}_\infty$ (see Subsection 3.2.8).

Unlike the algebra $\mathcal{C}\mathrm{Diff}^{\mathrm{alt}}\{\varkappa\}$, the bigraded algebra

$$\mathcal{C}\mathrm{Diff}^{\mathrm{alt}}\{[\varkappa]\} = \sum_{p,q} \mathcal{C}\mathrm{Diff}^{\mathrm{alt}}_{(p)}([\varkappa]\,; \bar{\Lambda}^q(\mathcal{Y}_\infty))$$

over $\mathcal{F}(\mathcal{Y})$, is not a bicomplex. One can define the differentials

$$\bar{d} = d_{p,q} \colon \mathcal{C}\mathrm{Diff}^{\mathrm{alt}}_{(p)}([\varkappa]\,; \bar{\Lambda}^q(\mathcal{Y}_\infty)) \to \mathcal{C}\mathrm{Diff}^{\mathrm{alt}}_{(p)}([\varkappa]\,; \bar{\Lambda}^{q+1}(\mathcal{Y}_\infty))$$

using (4.5), but formula (4.4) which defines the differentials $l_{p,q}$ makes no sense in this situation.

4.4.3. Consider $\mathcal{F}(\mathcal{Y})$-modules

$$K_0^{p,q} = \{\nabla_\varphi \mid \varphi \in [\ker i_0^{p,q}]\} \subset \mathcal{C}\mathrm{Diff}^{\mathrm{alt}}_{(p)}([\varkappa]\,; \bar{\Lambda}^q(\mathcal{Y}_\infty)),$$

$$K_0 = \sum_{p,q} K_0^{p,q} \subset \mathcal{C}\mathrm{Diff}^{\mathrm{alt}}\{[\varkappa]\}.$$

From (4.35) and Proposition 4.4.2 it follows that K_0 is a bigraded ideal of the algebra $\mathcal{C}\mathrm{Diff}^{\mathrm{alt}}\{[\varkappa]\}$ generated by $K_0^{1,0}$, i.e.,

$$K_0^{p,q} = K_0^{1,0} \wedge \mathcal{C}\mathrm{Diff}^{\mathrm{alt}}_{(p-1)}([\varkappa]\,;\bar{\Lambda}^q(\mathcal{Y}_\infty)). \tag{4.37}$$

Obviously, $\bar{d}_{1,0}(K_0^{1,0}) \subset K_0^{1,1}$. This inclusion, together with (4.37) and (4.6), shows that $\bar{d}(K_0) \subset K_0$.

Consider the bigraded quotient algebra

$$K = \mathcal{C}\mathrm{Diff}^{\mathrm{alt}}\{[\varkappa]\}/K_0 = \sum K^{p,q}, \qquad K^{p,q} = \mathcal{C}\mathrm{Diff}^{\mathrm{alt}}_{(p)}([\varkappa]\,;\bar{\Lambda}^q)/K_0^{p,q}.$$

By the above, this quotient algebra inherits the differentials $\bar{d}_{p,q}$. For these quotient differentials the same notation will be used:

$$\bar{d} = \bar{d}_{p,q} \colon K^{p,q} \to K^{p,q+1}.$$

Since $E_0^{p,q}(\mathcal{Y}) = [E_0^{p,q}]\,/\,[\ker i_0^{p,q}]$, the isomorphism from Proposition 4.4.2 induces isomorphisms $\mathit{И} = \mathit{И}^{p,q} \colon K^{p,q} \to E_0^{p,q}(\mathcal{Y})$, $\mathit{И} \colon K \to E_0(\mathcal{Y})$. Then Proposition 4.3.4 shows that the differentials $\bar{d}_{p,q}$ are identified with $d_0^{p,q} = d_0^{p,q}(\mathcal{Y})$.

If $\mathcal{Y}_\infty \subset J^\infty(\pi)$, then, according to the construction of Subsection 4.3.2, $E_0^{p,q}(\mathcal{Y})$ is identified with $\Lambda^{p,q}(\mathcal{Y}_\infty)$. Therefore, the isomorphisms $\mathit{И}^{p,q}$ allow one to identify $K^{p,q}$ with $\Lambda^{p,q}(\mathcal{Y}_\infty)$ and K with $\Lambda^*(\mathcal{Y}_\infty)$. Then the differentials $\bar{d}_{p,q}$ are identified with $d'_{p,q}$. Let us describe now the differentials $K^{p,q} \to K^{p,q+1}$ which are identified with $d''_{p,q}$.

Note first that if $X \in D_{\mathcal{C}}$ and $X = 0$ on $\mathcal{Y}_\infty$, then $X(\omega) = 0$ on $\mathcal{Y}_\infty$ for any $\omega \in \bar{\Lambda}^q$. In fact, if $X = \mathit{Э}_\varphi$, then $\varphi = 0$ on $\mathcal{Y}_\infty$ and $\mathit{Э}_\varphi(\omega) = \ell_\omega(\varphi) = 0$ on $\mathcal{Y}_\infty$, since the ideal $\mathcal{I}_{\mathcal{Y}}$ is differentially closed. If $X \in \mathcal{C}D$, then $X(\omega) = \bar{d}(X \lrcorner\, \omega) + X \lrcorner\, \bar{d}\omega$. But $(X \lrcorner\, \omega)|_{\mathcal{Y}_\infty} = 0$ and $(X \lrcorner\, \bar{d}\omega)|_{\mathcal{Y}_\infty} = 0$ if $X = 0$ on $\mathcal{Y}_\infty$, while $\bar{d}(X \lrcorner\, \omega)|_{\mathcal{Y}_\infty} = 0$ because $\mathcal{I}_{\mathcal{Y}}$ is differentially closed.

Therfore, the action of the $\mathcal{F}(\mathcal{Y})$-module $[\varkappa]$ on $\mathcal{F}$-modules $\bar{\Lambda}^q$ with values in $[\bar{\Lambda}^q]$,

$$([\chi]\,,\omega) \mapsto [\chi(\omega)]\,, \quad \chi \in \varkappa,\ \omega \in \bar{\Lambda}^q,$$

is well defined. Moreover, since $\chi(\omega) = \chi \lrcorner\, U_1(\omega)$, the differential operator $\square \in \mathcal{C}\mathrm{Diff}([\varkappa]\,,\bar{\Lambda}^q)$ defined by $\square([\chi]) = [\chi(\omega)]$, belongs to K_0 if $\omega|_{\mathcal{Y}_\infty} = 0$. It should also be noted that since $\mathcal{I}_{\mathcal{Y}}$ is differentially closed, $\{\varphi, \psi\} = \ell_\psi(\varphi) - \ell_\varphi(\psi) = 0$ on $\mathcal{Y}_\infty$ if $\varphi|_{\mathcal{Y}_\infty} = \psi|_{\mathcal{Y}_\infty} = 0$ and $\Delta(\chi_1, \dots, \chi_p) = 0$ on $\mathcal{Y}_\infty$ for any $\mathcal{C}$-multidifferential operator Δ and χ_i vanishing on $\mathcal{Y}_\infty$.

This shows that (4.4) induces the operators $l = l_{p,q}$ in K. Thus we have proved

PROPOSITION. *The algebra K equipped with the differential $\bar{d}$ is isomorphic to $\{E_0(\mathcal{Y}), d_0\}$. If $\mathcal{Y}_\infty \subset J^\infty(\pi)$, then K, together with the differentials $\bar{d}$ and l, form a bicomplex isomorphic to $\{\Lambda^*(\mathcal{Y}_\infty), d', d''\}$.*

REMARK. Let $\mathcal{Y}_\infty \subset J^\infty(\pi)$. Then (K, l) is isomorphic to the 0-th term of the spectral sequence for the de Rham cohomology of the mapping $\mathcal{Y}_\infty \to M$ (see Remark 2 in Subsection 4.3.2).

4.4.4. In the "regular" situation described below, the complexes

$$K^{p,*} = \sum_q K^{p,q} \subset K$$

with the differential $\bar{d}$ can be represented as the cokernel of a mapping of Spencer type complexes. This gives an approach to compute the terms $E_1^{p,q}(\mathcal{Y})$, since $H^q(K^{p,*}) = E_1^{p,q}(\mathcal{Y})$.

An equation $\mathcal{Y} \subset J^k(\pi)$ is said to be *regular* if $\mathcal{Y} = \{\varphi = 0\}$, where $\varphi \in \mathcal{F}_k(\pi, \xi)$ for some bundle $\xi \colon E_\xi \to M$ and

$$\mathcal{I}_\mathcal{Y} = \mathcal{C}\mathrm{Diff}(P, \mathcal{F})(\varphi), \quad P = \mathcal{F}(\pi, \xi).$$

For regular equations the following simple but very important lemma implies the desired interpretation of the complexes $K^{p,*}$:

LEMMA. *If $\mathcal{Y}$ is a regular equation, then*

$$K_0^{1,0} = \{\square \in \mathcal{C}\mathrm{Diff}([\varkappa], \mathcal{F}(\mathcal{Y})) \mid \square = \Delta \circ [\ell_\varphi], \ \Delta \in \mathcal{C}\mathrm{Diff}([P], \mathcal{F}(\mathcal{Y}))\},$$

where $[\bullet]$ denotes the restriction of a $\mathcal{C}$-differential operator to $\mathcal{Y}_\infty$.

PROOF. According to Subsection 4.4.1 and Proposition 4.4.2, $K^{1,0}$ is generated as an $\mathcal{F}(\mathcal{Y})$-module by the operators $[\ell_f]$, $f \in \mathcal{I}_\mathcal{Y}$. Regularity of the equation $\mathcal{Y}$ implies $f = \nabla(\varphi)$, $\nabla \in \mathcal{C}\mathrm{Diff}(P, \mathcal{F})$, and according to Subsection 3.2.7 it follows that $\nabla = \sum_i a_i \hat{\nabla}_i$, $a_i \in \mathcal{F}$. But $\ell_{\hat{\square}(\psi)} = \hat{\square} \circ \ell_\psi$ and since the operation $g \mapsto \ell_g$ is a derivation, we find

$$\ell_f = \sum_i a_i \hat{\nabla}_i \circ \ell_\varphi + \sum_i \hat{\nabla}_i(\varphi) \ell_{a_i}.$$

From $\hat{\nabla}_i(\varphi) \in \mathcal{I}_\mathcal{Y}$ it follows that $[\ell_\varphi] = \sum \left[a_i \hat{\nabla}_i\right] \circ [\ell_\varphi]$. □

Projectivity of $\mathcal{F}(\mathcal{Y})$-modules $[P]$ and $[\varkappa]$ implies

$$\mathcal{C}\mathrm{Diff}([P], \bar{\Lambda}^q(\mathcal{Y}_\infty)) = \mathcal{C}\mathrm{Diff}([P], \mathcal{F}(\mathcal{Y})) \otimes_{\mathcal{F}(\mathcal{Y})} \bar{\Lambda}^q(\mathcal{Y}_\infty)$$

and a similar relation for $[\varkappa]$. This fact and the lemma above imply that

$$K_0^{1,q} = \{\square \in \mathcal{C}\mathrm{Diff}([\varkappa], \bar{\Lambda}^q) \mid \square = \Delta \circ [\ell_\varphi], \ \Delta \in \mathcal{C}\mathrm{Diff}([P], \bar{\Lambda}^q)\},$$

where $\bar{\Lambda}^q = \bar{\Lambda}^q(\mathcal{Y}_\infty)$. This shows that the complex $K_0^{1,*} = \sum_q K_0^{1,q}$ is the image of the mapping

$$\mathcal{S}_{[\ell_\varphi]} \colon \widetilde{\mathcal{S}[P]} \to \widetilde{\mathcal{S}[\varkappa]}$$

(see Subsection 1.5.2) and hence

$$K^{1,*} = \operatorname{coker} \mathcal{S}_{[\ell_\varphi]}.$$

4.4.5. Now let $\mathcal{Y} \subset J^k(E, n)$ and $\mathcal{Y} = \{\varphi = 0\}$, $\varphi \in P$, $P = \mathcal{F}(J^\infty(E, n), \xi)$. The definition of regularity given in the previous subsection can be literally carried over to this case. But now Lemma 4.4.4 makes no sense since the operator ℓ_φ is not defined globally. To compensate for this, we introduce the following terminology:

DEFINITION. Let $\mathcal{Y} \subset J^k(E, n)$ be a differential equation.

1. The equation $\mathcal{Y}$ is said to be *locally regular* if there exists a covering $\{\mathcal{U}_\alpha\}$ of the space $J^\infty(E, n)$ by special charts such that the equation $\mathcal{Y}$ is regular in each of these charts.
2. The equation $\mathcal{Y}$ is said to be *regular with respect to an operator* $\nabla \in \mathcal{C}\mathrm{Diff}([\varkappa], [P])$ if

$$K_0^{0,1} = \{\square \in \mathcal{C}\mathrm{Diff}([\varkappa], \mathcal{F}(\mathcal{Y})) \mid \square = \Delta \circ \nabla, \ \Delta \in \mathcal{C}\mathrm{Diff}([P], \mathcal{F}(\mathcal{Y}))\}.$$

REMARK. From Lemma 4.4.4 it follows that a regular equation $\mathcal{Y} = \{\varphi = 0\} \subset J^k(\pi)$ is also regular with respect to the operator $[\ell_\varphi]$.

PROPOSITION. *Let $\mathcal{Y} \subset J^k(E, n)$ be a differential equation.*

1. *If the equation $\mathcal{Y}$ is regular with respect to an operator ∇, then $K^{1,*} =$ coker $\mathcal{S}_\nabla$.*
2. *If an equation $\mathcal{Y}$ is locally regular, then there exists an operator ∇ such that $\mathcal{Y}$ is regular with respect to ∇.*

PROOF. The first statement is obvious. To prove the second, we use the formula showing how the operator ℓ_φ changes when passing from one special chart to another. This formula is a direct consequence of transformation properties of the operator U_1 (see [**71, 124**]) and its relationship with the operation ℓ (Subsection 4.3.3). Thus we have

$$\tilde{l}_\varphi = \ell_\varphi + \sum_{i,k} D_k(\varphi_i)\tilde{\ell}_{x_i\psi_k} + \sum_i \varphi_i \tilde{\ell}_{\psi_i},$$

where (x_i, p^j_σ) are coordinates in the first chart, ψ_i is a basis of the module P in this chart, $D_k = \widehat{\partial/\partial x_k}$, and $\tilde{\ell}$ is the operation of universal linearization in the second chart. This formula immediately implies that $[\tilde{\ell}_\varphi] = [\ell_\varphi]$ in the intersection of the two charts. Therefore, one has a well-defined operator ∇ such that its restriction to any local chart is of the form $[\ell_\varphi]$, where the operation ℓ is considered in the sense of this chart. To prove that $\mathcal{Y}$ is regular with respect to ∇ we only need to use Lemma 4.4.4 and the standard techniques of the partition of unity. □

4.4.6. Representation of the complex $K^{1,*}$ as the cokernel of the operator $\mathcal{S}_\nabla$ is crucial for our consideration because it reduces the computation of the terms $E_1^{1,q}$ to the computation of the Spencer cohomology for the operator ∇. The latter turns out to be very efficient in many cases. That is why it is useful to have a similar representation for the complexes $K^{p,*}$, $p > 1$, as well. We describe such a representation in this subsection.

Let

$$\mathcal{C}\mathrm{Diff}_{[k]}(Q, Q'; \bar{\Lambda}^q(\mathcal{Y}_\infty)) = \mathcal{C}\mathrm{Diff}(\underbrace{Q, \dots, Q}_{k \text{ times}}, Q'; \bar{\Lambda}^q(\mathcal{Y}_\infty)),$$

$$\mathcal{C}\mathrm{Diff}_{[k]}\{Q, Q'\} = \sum_q \mathcal{C}\mathrm{Diff}_{[k]}(Q, Q'; \bar{\Lambda}^q(\mathcal{Y}_\infty)),$$

where Q and Q' are $\mathcal{F}(\mathcal{Y})$-modules. Recall that in Subsection 4.3.5 it was shown that $\mathcal{C}\mathrm{Diff}_{[k]}\{Q, Q'\}$ is a complex with respect to the differential $\bar{d}$. The permutation group S_k acts naturally on this complex by interchanging the first k arguments. The complex

$$\mathcal{C}\mathrm{Diff}^{\mathrm{alt}}_{[k]}\{Q, Q'\} = \sum_q \mathcal{C}\mathrm{Diff}^{\mathrm{alt}}_{[k]}(Q, Q'; \bar{\Lambda}^q(\mathcal{Y}_\infty))$$

is invariant with respect to this action. Note the following isomorphisms used in the sequel without special mention:

$$\begin{aligned} \mathcal{C}\mathrm{Diff}^{\mathrm{alt}}_{[k]}(Q, Q'; \bar{\Lambda}^q) &= \mathcal{C}\mathrm{Diff}(Q', \bar{\Lambda}^q) \otimes_{\mathcal{F}(\mathcal{Y})} \mathcal{C}\mathrm{Diff}^{\mathrm{alt}}_{(k)}(Q; \mathcal{F}(\mathcal{Y})), \\ \mathcal{C}\mathrm{Diff}^{\mathrm{alt}}_{[k]}\{Q, Q'\} &= \widetilde{\mathcal{S}Q}' \otimes_{\mathcal{F}(\mathcal{Y})} \mathcal{C}\mathrm{Diff}^{\mathrm{alt}}_{(k)}(Q; \mathcal{F}(\mathcal{Y})), \end{aligned} \tag{4.38}$$

where $\bar{\Lambda}^q = \bar{\Lambda}^q(\mathcal{Y}_\infty)$. For $\widetilde{\mathcal{S}Q}'$ see Subsection 1.5.2. Of course, the superscript "alt" in (4.38) can be omitted.

Consider now the composition $v_\nabla^{(p)}$ of the following mappings:

$$\widetilde{\mathcal{S}[P]} \otimes_{\mathcal{F}} \mathcal{C}\mathrm{Diff}^{\mathrm{alt}}_{(p-1)}([\varkappa]\,;\mathcal{F}) \xrightarrow{\mathcal{S}_\nabla \otimes \mathrm{id}} K_0^{1,*} \otimes \mathcal{C}\mathrm{Diff}^{\mathrm{alt}}_{(p-1)}([\varkappa]\,;\mathcal{F})$$
$$\to K_0^{1,*} \wedge \mathcal{C}\mathrm{Diff}^{\mathrm{alt}}_{(p-1)}([\varkappa]\,;\mathcal{F}) = K_0^{p,*},$$

where $\mathcal{F} = \mathcal{F}(\mathcal{Y})$. For the mappings $\mathcal{S}_\nabla$ see Subsection 1.5.2. Recall also that according to Subsections 4.4.4 and 4.4.5 one has $K_0^{1,*} = \operatorname{im} \mathcal{S}_\nabla \subset \tilde{\mathcal{S}}\,[\varkappa]$. Obviously, $v_\nabla^{(p)}$ is an epimorphism. Thus we have the following

PROPOSITION. *If an equation $\mathcal{Y}$ is regular with respect to an operator ∇, then $K^{p,*} = \operatorname{coker} v_\nabla^{(p)}$ for any $p \geq 1$.*

It is useful to note that the composition $\bar{v}_\nabla^{(p)}$ of $v_\nabla^{(p)}$ with the embedding $K_0^{p,*} \subset \mathcal{C}\mathrm{Diff}^{\mathrm{alt}}_{(p)}\{[\varkappa]\}$ coincides with the composition

$$\mathcal{C}\mathrm{Diff}^{\mathrm{alt}}_{[p-1]}\{[\varkappa]\,;[P]\} \xrightarrow{\mathcal{C}\mathrm{Diff}\{\mathrm{id};\nabla\}} \mathcal{C}\mathrm{Diff}^{\mathrm{alt}}_{[p-1]}\{[\varkappa]\,;[\varkappa]\} \xrightarrow{\mathrm{alt}} \mathcal{C}\mathrm{Diff}^{\mathrm{alt}}_{(p)}\{[\varkappa]\}. \tag{4.39}$$

4.4.7. Now we begin to study the cohomology of the complexes $K^{p,*}$, $p \geq 1$. First recall (see Subsection 4.3.5) that the complex $\mathcal{C}\mathrm{Diff}_{[k]}\{Q;Q'\}$ is acyclic in all dimensions except for $i = n$ and

$$H^n(\mathcal{C}\mathrm{Diff}_{[k]}\{Q;Q'\}) = \mathcal{C}\mathrm{Diff}_{(k)}\{Q;\hat{Q}'\}$$

for $i = n$. Therefore, since the complex $\mathcal{C}\mathrm{Diff}^{\mathrm{alt}}_{(k)}\{Q;Q'\}$ is a direct summand in $\mathcal{C}\mathrm{Diff}_{(k)}\{Q;Q'\}$ invariant with respect to the action of the permutation group S_k, it is also acyclic in dimensions $i \neq n$ and

$$H^n(\mathcal{C}\mathrm{Diff}^{\mathrm{alt}}_{[k]}\{Q;Q'\}) = \mathcal{C}\mathrm{Diff}^{\mathrm{alt}}_{(k)}\{Q;\hat{Q}'\}. \tag{4.40}$$

Consider the following short sequences of complexes:

$$0 \to \ker v_\nabla^{(p)} \to \widetilde{\mathcal{S}[P]} \otimes_{\mathcal{F}(\mathcal{Y})} \mathcal{C}\mathrm{Diff}^{\mathrm{alt}}_{(p-1)}([\varkappa]\,;\mathcal{F}(\mathcal{Y}))$$
$$= \mathcal{C}\mathrm{Diff}^{\mathrm{alt}}_{[p-1]}\{[\varkappa]\,;[P]\} \xrightarrow{v_\nabla^{(p)}} K_0^{p,*} \to 0,$$

and

$$0 \to K_0^{p,*} \to \mathcal{C}\mathrm{Diff}^{\mathrm{alt}}_{(p)}\{[\varkappa]\} \to K^{p,*} \to 0.$$

Using the results of subsection 4.3.5 and (4.40), from the corresponding exact cohomology sequences one finds that

$$H^i(K^{p,*}) = H^{i+1}(K_0^{p,*}) = H^{i+2}(\ker v_\nabla^{(p)}), \quad i < n-2,$$
$$H^{n-2}(K^{p,*}) = H^{n-1}(K_0^{p,*}),$$

and

$$0 \to H^{n-1}(K_0^{p,*}) \to H^n(\ker v_\nabla^{(p)}) \xrightarrow{I_p} \mathcal{C}\mathrm{Diff}^{\mathrm{alt}}_{(p-1)}([\varkappa]\,;[\hat{P}]) \to H^n(K_0^{p,*}) \to 0,$$

(4.41)

$$0 \to H^{n-1}(K^{p,*}) \to H^n(K_0^{p,*}) \to L_p([\varkappa]) \to H^n(K^{p,*}) \to 0.$$

Recall that the cohomology of the complexes under consideration vanishes for $i < 0$ and $i > n$.

From (4.41) it follows that to compute the cohomology of the complex $K^{p,*}$ it suffices to compute that of the complex $\ker v_\nabla^{(p)}$. In particular, if this complex is acyclic for $i \neq n$ and the mapping $I_p\colon H^n(\ker v_\nabla^{(p)}) \to \mathcal{C}\mathrm{Diff}^{\mathrm{alt}}_{(p-1)}([\varkappa]\,;[\hat{P}])$ is a monomorphism, then $H^i(K^{p,*}) = 0$ for $i \neq n-1, n$. Besides, one always has

$$H^n(K^{p,*}) = \operatorname{coker} \nabla_{(p)},$$

where $\nabla_{(p)}\colon \mathcal{C}\mathrm{Diff}^{\mathrm{alt}}_{(p-1)}([\varkappa]\,;[\hat{P}]) \to L_p([\varkappa])$ is the mapping of the n-dimensional cohomology of the complex $\mathcal{C}\mathrm{Diff}^{\mathrm{alt}}_{[p-1]}\{[\varkappa]\,;[P]\}$ to the cohomology

$$H^n(\mathcal{C}\mathrm{Diff}^{\mathrm{alt}}_{(p)}([\varkappa])) = L_p([\varkappa])$$

corresponding to $v_\nabla^{(p)}$. This mapping is defined by the following commutative diagram:

$$\begin{array}{ccc}
\mathcal{C}\mathrm{Diff}^{\mathrm{alt}}_{[p-1]}([\varkappa],[P];\bar{\Lambda}^n(\mathcal{Y}_\infty)) & \xrightarrow{\bar{v}_\nabla^{(p)}} & \mathcal{C}\mathrm{Diff}^{\mathrm{alt}}_{(p)}([\varkappa];\bar{\Lambda}^n(\mathcal{Y}_\infty)) \\
\downarrow & & \downarrow \\
\mathcal{C}\mathrm{Diff}^{\mathrm{alt}}_{[p-1]}([\varkappa],[P];\bar{\Lambda}^n(\mathcal{Y}_\infty))/\operatorname{im}\bar{d} & \xrightarrow{\nabla_{(p)}} & \mathcal{C}\mathrm{Diff}^{\mathrm{alt}}_{(p)}([\varkappa];\bar{\Lambda}^n(\mathcal{Y}_\infty))/\operatorname{im}\bar{d} \\
\| & & \| \\
\mathcal{C}\mathrm{Diff}^{\mathrm{alt}}_{(p-1)}([\varkappa];[\hat{P}]) & \xrightarrow{\nabla_{(p)}} & L_p([\varkappa])
\end{array}$$

Directly from the definitions and the results of Subsection 4.3.5 it follows that $\nabla_{(p)}$ is the composition of the mappings

$$\mathcal{C}\mathrm{Diff}^{\mathrm{alt}}_{(p-1)}([\varkappa];[\hat{P}]) \xrightarrow{\alpha_p} \mathcal{C}\mathrm{Diff}^{\mathrm{alt}}_{(p-1)}([\varkappa];[\hat{\varkappa}]) \xrightarrow{\beta} L_p([\varkappa]),$$

where α_p is defined by

$$\alpha_p(\Delta)(\chi_1,\dots,\chi_{p-1}) = \nabla^*(\Delta(\chi_1,\dots,\chi_{p-1})), \quad \chi_i \in [\varkappa],$$

and β_p is the skew-symmetrization with respect to the action of the permutation group S_p in $\mathcal{C}\mathrm{Diff}_{(p-1)}([\varkappa];[\hat{\varkappa}])$ (see Subsection 4.3.6). Note that $\nabla_{(1)} = \nabla^*$.

By definition, $\nabla_{(p)}$ is the mapping in the n-th cohomology corresponding to the composition

$$\mathcal{C}\mathrm{Diff}_{[p-1]}\{[\varkappa],[P]\} \xrightarrow{v_\nabla^{(p)}} K_0^{p,*} \hookrightarrow \mathcal{C}\mathrm{Diff}_{(p)}\{[\varkappa]\}.$$

Therefore, it is the composition of the induced mappings of n-th cohomology

$$\mathcal{C}\mathrm{Diff}^{\mathrm{alt}}_{(p-1)}([\varkappa];[\hat{P}]) \to H^n(K_0^{p,*}) \to L_p([\varkappa]).$$

Due to (4.41), this fact leads to the following exact sequence:

$$0 \to H^{n-1}(K_0^{p,*}) \to H^n(\ker v_\nabla^{(p)}) \xrightarrow{I_p} \ker \nabla_{(p)} \to H^{n-1}(K^{p,*}) \to 0. \tag{4.42}$$

Putting together the above facts and taking into account Proposition 4.4.3, we obtain the following result:

THEOREM (The Two Line Theorem). *Let an equation $\mathcal{Y}$ be regular with respect to an operator ∇, and let*

(a) *the complexes $K_0^{p,*}$, $p \geq 0$, be acyclic in dimensions $i \neq n$, or equivalently,*
(b) *the complexes $\ker v_\nabla^{(p)}$ be acyclic in dimensions $i \neq n$ and I_p be monomorphisms.*

Then

1. $E_1^{p,q}(\mathcal{Y}) = 0$ *for* $p \geq 1$ *and* $q \neq n-1, n$.
2. $E_1^{p,n}(\mathcal{Y}) = \operatorname{coker} \nabla_{(p)}$.
3. *The following sequence is exact:*

$$0 \to H^n(\ker v_\nabla^{(p)}) \to \ker \nabla_{(p)} \to E_1^{p,n-1}(\mathcal{Y}) \to 0.$$

Recall that $E_1^{0,q} = \bar{H}^Q(\mathcal{Y})$ (Proposition 4.3.1). Therefore, under the above assumptions, nonvanishing terms $E_1^{p,q}(\mathcal{Y})$, except for those in the interval $0 \leq q \leq n$ of the column $p = 0$, can be situated in two lines only: $q = n-1$ and $q = n$. They form two complexes,

$$0 \to \bar{H}^{n-1}(\mathcal{Y}_\infty) \xrightarrow{d_1^{0,n-1}} E_1^{1,n-1} \xrightarrow{d_1^{1,n-1}} \cdots \xrightarrow{d_1^{p-1,n-1}} E_1^{p,n-1} \xrightarrow{d_1^{p,n-1}} \cdots \tag{4.43}$$

and

(4.44)

$$0 \to \bar{H}^n(\mathcal{Y}_\infty) \xrightarrow{d_1^{0,n}} E_1^{1,n} = \operatorname{coker} \nabla_{(1)} \xrightarrow{d_1^{1,n}} \cdots \xrightarrow{d_1^{p-1,n}} \operatorname{coker} \nabla_{(p)} \xrightarrow{d_1^{p,n}} \cdots .$$

The following result is a direct consequence of the Two Line Theorem:

COROLLARY. *Under the assumptions of the Two Line Theorem, one has:*

1. $E_r^{p,q}(\mathcal{Y}) = 0$ *for* $p \geq 1$ *and* $1 \leq r < \infty$ *if* $q \neq n-1, n$.
2. $E_3^{p,q}(\mathcal{Y}) = E_\infty^{p,q}(\mathcal{Y})$.
3. $\bar{H}^q(\mathcal{Y}_\infty) = E_1^{0,q}(\mathcal{Y}) = H^q(\mathcal{Y}_\infty)$ *if* $q \leq n-2$.
4. $E_2^{0,n-1}(\mathcal{Y}) = E_\infty^{0,n-1}(\mathcal{Y}) = H^{n-1}(\mathcal{Y}_\infty)$, $E_2^{1,n-1}(\mathcal{Y}) = E_\infty^{1,n-1}(\mathcal{Y})$.

This corollary shows that to compute the $\mathcal{C}$-spectral sequence it is necessary to know the groups $H^q(\mathcal{Y}_\infty)$. Usually their description presents no difficulty (see Subsection 4.5.1).

The assumptions of the Two Line Theorem are usually satisfied for nonoverdetermined systems (see below). Thus the theorem describes the structure of the $\mathcal{C}$-spectral sequence for such equations and this fact explains its significance.

REMARK. It is convenient to use the assumptions of the theorem in the form (b), when the kernel $\ker v_\nabla^{(p)}$ is "small", i.e., when the equation $\mathcal{Y}$ is not "too overdetermined". In the sequel, in our general arguments, we will use them in the form (a).

4.4.8. Let us show first that for "almost all" nonoverdetermined systems $\ker v_\nabla^{(1,q)} = 0$. Since $v_\nabla^{(1,q)} = \mathcal{S}_\nabla|_{\mathcal{C}\mathrm{Diff}([P],\bar{\Lambda}^q(\mathcal{Y}))}$, we must find reasonable conditions implying $\ker \mathcal{S}_\nabla = 0$. To do this, we use the arguments of Subsection 1.5.3 in the framework of the $\mathcal{C}$-theory.

Consider an equation $\mathcal{Y}$ and its infinite prolongation $\mathcal{Y}_\infty \subset J^\infty(\pi)$. Then the maximal spectrum of the algebra $\mathcal{C}\operatorname{smbl}\mathcal{F}(\mathcal{Y})$ (see Subsection 3.2.9) contains the total space of the bundle $T^*(M, \mathcal{Y}_\infty) \to \mathcal{Y}_\infty$, which is the pullback of the cotangent bundle $T^*(M) \to M$ along the projection $\mathcal{Y}_\infty \to M$. Thus any element of $\mathcal{C}\operatorname{smbl}\mathcal{F}(\mathcal{Y})$ can be viewed as a function on $T^*(M, \mathcal{Y}_\infty)$.

Let $\xi_i \colon E_{\xi_i} \to M$, $i = 1, 2$, be vector bundles, $Q_i = \mathcal{F}(\mathcal{Y}, \xi_i)$, and $\tilde{\xi}_i \colon \tilde{E}_i \to T^*(M, \mathcal{Y}_\infty)$ be their pullbacks along the natural projection $T^*(M, \mathcal{Y}_\infty) \to M$. Then, since

$$\mathcal{C}\operatorname{smbl}(Q_1, Q_2) = \operatorname{Hom}_{\mathcal{F}(\mathcal{Y})}(Q_1, \mathcal{C}\operatorname{smbl}(Q_2))$$

(see Subsection 3.2.9), any element $s \in \mathcal{C}\,\mathrm{smbl}(Q_1, Q_2)$ can be understood as a homomorphism $h(s)\colon \tilde{\xi}_1 \to \tilde{\xi}_2$.

Recall that $\mathcal{C}\,\mathrm{smbl}(Q_1, Q_2)$ is a graded module associated with the filtered module $\mathcal{C}\mathrm{Diff}(Q_1, Q_2) = \{\mathcal{C}\mathrm{Diff}_i(Q_1, Q_2)\}_{i\geq 0}$. Consequently, the homomorphism $\mathcal{S}_\nabla\colon \mathcal{C}\mathrm{Diff}([P], \bar{\Lambda}^q(\mathcal{Y}_\infty)) \to \mathcal{C}\mathrm{Diff}([\varkappa], \bar{\Lambda}^q(\mathcal{Y}_\infty))$ induces a homomorphism

$$\mathcal{S}_\nabla^{\mathrm{gr}}\colon \mathcal{C}\,\mathrm{smbl}([P], \bar{\Lambda}^q(\mathcal{Y}_\infty)) \to \mathcal{C}\,\mathrm{smbl}([\varkappa], \bar{\Lambda}^q(\mathcal{Y}_\infty))$$

defined by

$$\mathcal{S}_\nabla^{\mathrm{gr}}(\mathcal{C}\,\mathrm{smbl}_i \square) = \mathcal{C}\,\mathrm{smbl}_{k+i}(\square \circ \nabla), \qquad k = \mathrm{ord}\,\nabla.$$

Obviously, $\ker \mathcal{S}_\nabla = 0$ when $\ker \mathcal{S}_\nabla^{\mathrm{gr}} = 0$. Hence, it suffices to describe conditions under which $\ker \mathcal{S}_\nabla^{\mathrm{gr}} = 0$.

Let $\xi_i\colon E_{\xi_i} \to M$, $i = 1, 2, 3$, be vector bundles such that $\mathcal{F}(\mathcal{Y}, \xi_1) = [\varkappa]$, $\mathcal{F}(\mathcal{Y}, \xi_2) = [P]$, and $\mathcal{F}(\mathcal{Y}, \xi_3) = \bar{\Lambda}^q(\mathcal{Y})$. Then

$$h\left(\mathcal{S}_\nabla^{\mathrm{gr}}(\mathcal{C}\,\mathrm{smbl}_i \square)\right) = h(\mathcal{C}\,\mathrm{smbl}_i \square) \circ h_\nabla, \qquad h_\nabla = h(\mathcal{C}\,\mathrm{smbl}\,\nabla).$$

Hence, $\ker \mathcal{S}_\nabla^{\mathrm{gr}} = 0$ provided that the set

$$\mathrm{char}\,\nabla = \{\theta \in T^*(M, \mathcal{Y}_\infty) \mid \mathrm{coker}\, h_{\nabla,\theta} \neq 0\}$$

is nowhere dense in $T^*(M, \mathcal{Y}_\infty)$. Thus we have

PROPOSITION. *Let $\mathcal{Y}$ be a regular equation with respect to an operator ∇, and let the set* $\mathrm{char}\,\nabla$ *be nowhere dense. Then* $\ker v_\nabla^{(1,q)} = 0$.

PROOF. Previous considerations prove the statement in any special chart, from which it follows in full generality. □

Recall that $[P]$ and $[\varkappa]$ are projective modules and suppose that the dimension of $[P]$ does not exceed that of $[\varkappa]$. Then in a generic situation the assumptions of Theorem 1.5.3 hold. In fact, if $m = \dim[\varkappa]$ and $l = \dim[P]$, then in local coordinates ∇ is given by an $l \times m$-matrix $\|\nabla_{ij}\|$, $\nabla_{ij} \in \mathcal{C}\mathrm{Diff}\,\mathcal{F}(\mathcal{Y})$. Therefore, the homomorphism h_∇ is given by the matrix $\|\mathcal{C}\,\mathrm{smbl}_k \nabla_{ij}\| = s(\nabla)$. Thus, when $l \leq m$, the condition $\theta \in \mathrm{char}\,\nabla$ means that all l-th order minors of the matrix $s(\nabla)$ vanish at the point θ. Hence, in this situation $\mathrm{codim}(\mathrm{char}\,\nabla) \geq 1$ if $s(\nabla)$ is generic.

This gives a simple way to describe the set $\mathrm{char}\,\nabla$. To do this, one needs to write down the elements of the matrix $s(\nabla)$ in the coordinates introduced in Subsection 3.2.10 and to impose the condition that all its $l \times l$-minors equal zero.

4.4.9. Now we consider some methods allowing us to estimate and compute the cohomology of the complexes $K^{p,*}$, $p > 1$. In particular, we present easily verifyable conditions under which the Two Line Theorem holds. These methods are also rather efficient in the study of concrete equations. In this subsection, the cohomology of some auxiliary complexes is studied. All considerations below are carried out over an object of the category DE. The function algebra of this object is denoted by $\mathcal{F}$.

Let Q_i, $i = 1, \ldots, k$, be projective $\mathcal{F}$-modules. We identify the $\mathcal{F}$-module $\mathcal{C}\mathrm{Diff}(Q_1, \ldots, Q_k; \bar{\Lambda}^q)$ with the tensor product (over $\mathcal{F}$)

$$\mathcal{C}\mathrm{Diff}(Q_1, \mathcal{F}) \otimes \cdots \otimes \mathcal{C}\mathrm{Diff}(Q_k, \mathcal{F}) \otimes \bar{\Lambda}^q.$$

Let us fix an integer i, $1 \le i \le k$, and introduce the *i-th* (*increasing*) *filtration* in the complex $\mathcal{C}\mathrm{Diff}\{Q_1, \dots, Q_k\}$ with respect to which the subcomplex of filtration p is defined to be

$$A^{p,q} = \mathcal{C}\mathrm{Diff}(Q_1, \mathcal{F}) \otimes \cdots \otimes \mathcal{C}\mathrm{Diff}(Q_{i-1}, \mathcal{F}) \\ \otimes \mathcal{C}\mathrm{Diff}_{p+q}(Q_i, \mathcal{F}) \otimes \mathcal{C}\mathrm{Diff}(Q_{i+1}, \mathcal{F}) \otimes \cdots \otimes \bar{\Lambda}^q.$$

In the case $i = k = 1$, this filtration coincides with the filtration of the complex $\widetilde{\mathcal{S}Q}$, $Q = Q_1$, by subcomplexes $\widetilde{\mathcal{S}Q}_p$ (see Subsection 1.5.3). Consider the spectral sequence associated with this filtration. The components $B_0^{p,q}$ of filtration p and degree q in the zero term of this spectral sequence are

$$B_0^{p,q} = A^{p,q}/A^{p+1,q} \\ = \mathcal{C}\mathrm{Diff}(Q_1, \dots, Q_{i-1}, Q_{i+1}, \dots, Q_k, \mathcal{F}) \otimes \mathcal{C}\,\mathrm{smbl}_{p+q}(Q_i, \bar{\Lambda}^q),$$

while the differential taking $B_0^{p,q}$ to $B_0^{p,q+1}$ is of the form $\mathrm{id} \otimes \delta_{p,q}$, where

$$\delta_{p,q} \colon \mathcal{C}\,\mathrm{smbl}_{p+q}(Q_i, \bar{\Lambda}^q) \to \mathcal{C}\,\mathrm{smbl}_{p+q+1}(Q_i, \bar{\Lambda}^{q+1})$$

is the similar differential in the spectral sequence for the complex $\widetilde{\mathcal{S}Q} = \mathcal{C}\mathrm{Diff}_{(1)}\{Q\}$, $Q = Q_i$, with respect to the filtration described above. The operators $\delta = \delta_{p,q}$ are versions of the Spencer δ-operators (see [**71, 98, 103**]) and are obviously homomorphisms of $\mathcal{F}$-modules. The complex

$$0 \to \mathcal{C}\,\mathrm{smbl}_p(Q, \bar{\Lambda}^0) \xrightarrow{\delta_{p,0}} \mathcal{C}\,\mathrm{smbl}_{p+1}(Q, \bar{\Lambda}^1) \\ \xrightarrow{\delta_{p,1}} \cdots \xrightarrow{\delta_{p,n-1}} \mathcal{C}\,\mathrm{smbl}_{p+n}(Q, \bar{\Lambda}^n) \to 0$$

is acyclic for $p > -n$ (we assume that $\mathcal{C}\,\mathrm{smbl}_p = 0$ for $p < 0$) and consequently all homogeneous components of the first term of the above spectral sequence are trivial except for that of filtration 0 and degree n, which implies

$$\mathcal{C}\mathrm{Diff}(Q_1, \dots, Q_{i-1}, Q_{i+1}, \dots, Q_k) \otimes \mathcal{C}\,\mathrm{smbl}_0(Q_i, \bar{\Lambda}^n) \\ = \mathcal{C}\mathrm{Diff}(Q_1, \dots, Q_{i-1}, Q_{i+1}, \dots, Q_k; \mathcal{F}) \otimes \hat{Q}_i \\ = \mathcal{C}\mathrm{Diff}(Q_1, \dots, Q_{i-1}, Q_{i+1}, \dots, Q_k; \hat{Q}_i) = H^n(\mathcal{C}\mathrm{Diff}\{Q_1, \dots, Q_k\}).$$

Below we are interested in subcomplexes of the complex $\mathcal{C}\mathrm{Diff}\{Q_1, \dots, Q_k\}$ of the following form. Let

$$\mathcal{O}' \subset \mathcal{C}\mathrm{Diff}(Q_1, \mathcal{F}) \otimes \cdots \otimes \mathcal{C}\mathrm{Diff}(Q_{i-1}, \mathcal{F}), \\ \mathcal{O} \subset \mathcal{C}\mathrm{Diff}(Q_i, \mathcal{F}), \\ \mathcal{O}'' \subset \mathcal{C}\mathrm{Diff}(Q_{i+1}, \mathcal{F}) \otimes \cdots \otimes \mathcal{C}\mathrm{Diff}(Q_k, \mathcal{F})$$

be submodules such that $\mathcal{O}' \otimes \bar{\Lambda}^*$, $\mathcal{O} \otimes \bar{\Lambda}^*$, and $\mathcal{O}'' \otimes \bar{\Lambda}^*$ are subcomplexes in $\mathcal{C}\mathrm{Diff}\{Q_1, \dots, Q_{i-1}\}$, $\mathcal{C}\mathrm{Diff}_{(1)}\{Q_i\}$, and $\mathcal{C}\mathrm{Diff}\{Q_{i+1}, \dots, Q_k\}$, respectively. Then $\mathcal{O}' \otimes \mathcal{O} \otimes \mathcal{O}'' \otimes \bar{\Lambda}^*$ is a subcomplex in $\mathcal{C}\mathrm{Diff}\{Q_1, \dots, Q_k\}$ which is denoted by $\{\mathcal{O}', \mathcal{O}, \mathcal{O}''\}$. In the case where submodules $\mathcal{O}_j \subset \mathcal{C}\mathrm{Diff}(Q_j, \mathcal{F})$ are such that $\mathcal{O}' = \mathcal{O}_1 \otimes \cdots \otimes \mathcal{O}_{i-1}$, $\mathcal{O} = \mathcal{O}_i$, and $\mathcal{O}'' = \mathcal{O}_{i+1} \otimes \cdots \otimes \mathcal{O}_k$, the complex $\{\mathcal{O}', \mathcal{O}, \mathcal{O}''\}$ is denoted by $\{\mathcal{O}_1, \dots, \mathcal{O}_k\}$.

The i-th filtration in the complex $\mathcal{C}\mathrm{Diff}\{Q_1, \dots, Q_k\}$ induces a filtration, also called i-th, in the subcomplex $\{\mathcal{O}', \mathcal{O}, \mathcal{O}''\}$ (resp., $\{\mathcal{O}_1, \dots, \mathcal{O}_k\}$) and one can consider the spectral sequence associated with this filtration. If the modules $\mathcal{O}'$ and $\mathcal{O}''$ (resp., $\mathcal{O}_s, s \neq i$) are projective, then, as above, the components $B_0^{p,q} =$

$B_0^{p,q}(\mathcal{O}',\mathcal{O},\mathcal{O}'')$ (resp., $B_0^{p,q} = B_0^{p,q}(\mathcal{O}_1,\dots,\mathcal{O}_k)$) of filtration p and degree q of this spectral sequence are of the form

$$\begin{aligned} B_0^{p,q}(\mathcal{O}',\mathcal{O},\mathcal{O}'') &= \mathcal{O}'\otimes\mathcal{O}''\otimes \mathrm{sm}_{p+q}(\mathcal{O},\bar\Lambda^q),\\ B_0^{p,q}(\mathcal{O}_1,\dots,\mathcal{O}_k) &= \mathcal{O}_1\otimes\dots\otimes\mathcal{O}_{i-1}\otimes\mathcal{O}_{i+1}\otimes\dots\otimes\mathcal{O}_k\otimes \mathrm{sm}_{p+q}(\mathcal{O}_i,\bar\Lambda^q)\\ &= \mathcal{O}^{(i)}\otimes \mathrm{sm}_{p+q}(\mathcal{O}_i,\bar\Lambda^q), \end{aligned}$$

where

$$\mathrm{sm}_{p+q}(\mathcal{O},\bar\Lambda^q) = \frac{(\mathcal{O}\otimes\bar\Lambda^q)\cap \mathcal{C}\mathrm{Diff}_{p+q}(Q,\bar\Lambda^q)}{(\mathcal{O}\otimes\bar\Lambda^q)\cap \mathcal{C}\mathrm{Diff}_{p+q-1}(Q,\bar\Lambda^q)}.$$

Note that $\mathrm{sm}_j(\mathcal{O},\bar\Lambda^q)\subset \mathcal{C}\,\mathrm{smbl}_j(Q_j,\bar\Lambda^q)$ and

$$\delta_{p,q}(\mathrm{sm}_j(\mathcal{O},\bar\Lambda^q))\subset \mathrm{sm}_{p+q+1}(\mathcal{O},\bar\Lambda^{q+1}).$$

The differential in the zero term of the spectral sequence under consideration takes $B_0^{p,q}$ to $B_0^{p,q+1}$ and is of the form $\mathrm{id}_{\mathcal{O}'\otimes\mathcal{O}''}\otimes\delta_{p,q}$ (resp., $\mathrm{id}_{\mathcal{O}^{(i)}}\otimes\delta_{p,q}$). Therefore, the first term $B_1^{p,q}(\mathcal{O}',\mathcal{O},\mathcal{O}'')$ (resp., $B_1^{p,q}(\mathcal{O}_1,\dots,\mathcal{O}_k)$) of this spectral sequence is of the form

$$\begin{aligned} B_1^{p,q}(\mathcal{O}',\mathcal{O},\mathcal{O}'') &= \mathcal{O}'\otimes\mathcal{O}''\otimes h^{p,q}(\mathcal{O}),\\ B_1^{p,q}(\mathcal{O}_1,\dots,\mathcal{O}_k) &= \mathcal{O}_1\otimes\dots\otimes\mathcal{O}_{i-1}\otimes\mathcal{O}_{i+1}\otimes\dots\otimes\mathcal{O}_k\otimes h^{p,q}(\mathcal{O}_i)\\ &= \mathcal{O}^{(i)}\otimes h^{p,q}(\mathcal{O}_i), \end{aligned}\tag{4.45}$$

where $h^{p,q}(\mathcal{O})$ is the q-th cohomology module of the complex $\mathrm{sm}_p(\mathcal{O})$:

$$\begin{aligned} 0\to \mathrm{sm}_p(\mathcal{O},\bar\Lambda^0)\xrightarrow{\delta_{p,0}}\cdots\xrightarrow{\delta_{p,q-1}} \mathrm{sm}_{p+q}(\mathcal{O},\bar\Lambda^q)\\ \xrightarrow{\delta_{p,q}}\cdots\xrightarrow{\delta_{p,n-1}} \mathrm{sm}_{p+n}(\mathcal{O},\bar\Lambda^n)\to 0. \end{aligned}$$

Note the following immediate consequence of the spectral sequence introduced above:

PROPOSITION. *Consider projective modules $\mathcal{O}'$, $\mathcal{O}''$ and assume that the complexes $\mathrm{sm}_p(\mathcal{O})$ are acyclic in all dimensions $< s$. Then the complex $\{\mathcal{O}',\mathcal{O},\mathcal{O}''\}$ is also acyclic in the same dimensions.*

PROOF. The assumption of the proposition means that $h^{p,q}(\mathcal{O}) = 0$ for $q < s$. Hence, from (4.45) it follows that $B_1^{p,q}(\mathcal{O}',\mathcal{O},\mathcal{O}'') = 0$ for $q < s$. Consequently, the components of filtration p and of degree q in the limit term are also trivial if $q < s$. □

COROLLARY. *If modules $\mathcal{O}'$ and $\mathcal{O}''$ are projective and there exists a number l_0 such that $h^{l,q}(\mathcal{O}) = 0$ for $(l,q)\neq(l_0,n)$, then the complex $\{\mathcal{O}',\mathcal{O},\mathcal{O}''\}$ is acyclic in all dimensions different from n and*

$$H^n(\{\mathcal{O}',\mathcal{O},\mathcal{O}''\}) = \mathcal{O}'\otimes\mathcal{O}''\otimes h^{l_0,n}(\mathcal{O}).$$

REMARK. The assumptions of the corollary are obviously satisfied if $\mathcal{O} = \mathcal{C}\mathrm{Diff}(Q_i,\mathcal{F})$.

4.4.10. Below, for the submodules $\mathcal{O} = \mathcal{O}_i \subset \mathcal{C}\mathrm{Diff}(Q_i, \mathcal{F})$ we take the images of the mappings

$$\mathcal{C}\mathrm{Diff}(\nabla, \mathrm{id}_{\mathcal{F}})\colon \mathcal{C}\mathrm{Diff}(Q_i', \mathcal{F}) \to \mathcal{C}\mathrm{Diff}(Q_i, \mathcal{F}), \quad \square \mapsto \square \circ \nabla,$$

where $\nabla \in \mathcal{C}\mathrm{Diff}(Q_i, Q_i')$, $\square \in \mathcal{C}\mathrm{Diff}(Q_i', \mathcal{F})$. In this case, the subcomplex $\mathcal{O} \otimes \bar{\Lambda}^* \subset \widetilde{\mathcal{S}Q}_i$ can be equipped with a filtration which is called the *left filtration* (as opposed to the one introduced in the previous subsection and called the *right filtration*). To do this, note that $\mathcal{O} \otimes \bar{\Lambda}^*$ is the image of the complex $\widetilde{\mathcal{S}Q}_i'$ under the mapping $\mathcal{S}_\nabla$ (see Subsection 1.5.2). Hence, the filtration of the complex $\widetilde{\mathcal{S}Q}_i'$ by the subcomplexes $\widetilde{\mathcal{S}_k Q}_i'$ is carried over to $\mathcal{O} \otimes \bar{\Lambda}^*$ by the mapping $\mathcal{S}_\nabla$. Obviously, the elements of degree q and filtration $\leq p$ form the submodule

$$\mathcal{O}_{(p+q)} \otimes \bar{\Lambda}^* = \{\square \circ \nabla \mid \square \in \mathcal{C}\mathrm{Diff}_{p+q}(Q_i', \mathcal{F})\}$$

in $\mathcal{O} \otimes \bar{\Lambda}^*$. If $\operatorname{ord} \nabla = k$, then

$$\mathcal{O}_{(p+q)} \otimes \bar{\Lambda}^q \subset \{\mathcal{C}\mathrm{Diff}_{p+q+k}(Q_i, \mathcal{F}) \cap \mathcal{O}\} \otimes \bar{\Lambda}^q,$$

and the last submodule consists of the elements of degree q and right filtration $\leq p + k$, i.e., the left filtration is embedded in the right one with increase of the filtration index by k.

Consider the spectral sequence generated by the left filtration in $\mathcal{O} \otimes \bar{\Lambda}^q$. Then the component of its zero term, consisting of the elements of degree q and of filtration p, is of the form

$$D_0^{p,q} = D_0^{p,q}(\mathcal{O}) = \frac{\mathcal{O}_{(p+q)}}{\mathcal{O}_{(p+q-1)}} \otimes \bar{\Lambda}^q.$$

Zero differential of this spectral sequence, denoted by $\partial = \partial_{p,q}$, takes $D_0^{p,q}$ to $D_0^{p,q+1}$, and thus the zero term of this spectral sequence splits into the complexes

$$0 \to D_0^{p,0} \xrightarrow{\partial_{p,0}} \cdots \xrightarrow{\partial_{p,q-1}} D_0^{p,q} \xrightarrow{\partial_{p,q}} \cdots \xrightarrow{\partial_{p,n-1}} D_0^{p,n} \to 0,$$

which are denoted by $\mathrm{Sml}_p(\mathcal{O})$. Obviously, $\partial_{p,q}$ is an $\mathcal{F}$-module homomorphism. The cohomology of the complex $\mathrm{Sml}_p(\mathcal{O})$ at the term $D_0^{p,q}$ will be denoted by $H^{p,q}(\mathcal{O})$. Thus, $H^{p,q}(\mathcal{O}) = D_1^{p,q}$, where $D_1^{p,q}$ is the component of degree q and filtration p in the first term of the spectral sequence under consideration.

The left filtration in the complex $\mathcal{O} \otimes \bar{\Lambda}^*$ induces a filtration in the complex $\{\mathcal{O}', \mathcal{O}, \mathcal{O}''\}$ (resp., $\{\mathcal{O}_1, \dots, \mathcal{O}_k\}$) which is called the *i-th lcft filtration.* The elements of degree q and filtration $\leq p$ form the submodule

$$\mathcal{O}' \otimes \mathcal{O}_{(p+q)} \otimes \mathcal{O}'' \otimes \bar{\Lambda}^q,$$

or, respectively,

$$\mathcal{O}_1 \otimes \cdots \otimes \mathcal{O}_{i-1} \otimes \mathcal{O}_{(p+q)} \otimes \mathcal{O}_{i+1} \cdots \otimes \mathcal{O}_k \otimes \bar{\Lambda}^q.$$

Consider the spectral sequence generated by this filtration in the situation where the modules $\mathcal{O}'$, $\mathcal{O}''$ (resp., $\mathcal{O}_s$, $s \neq i$) are projective. Then the elements of degree q and filtration p are of the form

$$\begin{aligned} D_0^{p,q}(\mathcal{O}', \mathcal{O}, \mathcal{O}'') &= \mathcal{O}' \otimes \mathcal{O}'' \otimes D_0^{p,q}(\mathcal{O}), \\ D_0^{p,q}(\mathcal{O}_1, \dots, \mathcal{O}_k) &= \mathcal{O}_1 \otimes \cdots \otimes \mathcal{O}_{i-1} \otimes \mathcal{O}_{i+1} \otimes \cdots \otimes \mathcal{O}_k \otimes D_0^{p,q}(\mathcal{O}) \\ &= \mathcal{O}^{(i)} \otimes D_0^{p,q}(\mathcal{O}). \end{aligned}$$

As before, the differentials at the zero term of this spectral sequence are of the form $\mathrm{id}_{\mathcal{O}'\otimes\mathcal{O}''}\otimes\partial_{p,q}$ (resp., $\mathrm{id}_{\mathcal{O}^{(i)}}\otimes\partial_{p,q}$). Hence, the component of degree q and filtration p is of the form $D_1^{p,q}(\mathcal{O}',\mathcal{O},\mathcal{O}'') = \mathcal{O}'\otimes\mathcal{O}''\otimes H^{p,q}(\mathcal{O})$ (resp., $D_1^{p,q}(\mathcal{O}_1,\dots,\mathcal{O}_k) = \mathcal{O}^{(i)}\otimes H^{p,q}(\mathcal{O})$).

Similarly to the previous subsection, we have

PROPOSITION. *Let the modules $\mathcal{O}'$, $\mathcal{O}''$ be projective and the complexes $\mathrm{Sml}_p(\mathcal{O})$ be acyclic in the dimensions $\le s$. Then the complex $\{\mathcal{O}',\mathcal{O},\mathcal{O}''\}$ is also acyclic in the same dimensions.*

COROLLARY. *If the modules $\mathcal{O}'$ and $\mathcal{O}''$ are projective and there exists a number l_0 such that $H^{l,q}(\mathcal{O}) = 0$ for $(l,q)\neq(l_0,n)$, then the complex $\{\mathcal{O}',\mathcal{O},\mathcal{O}''\}$ is acyclic in all dimensions different from n, while*

$$H^n(\{\mathcal{O}',\mathcal{O},\mathcal{O}''\}) = \mathcal{O}'\otimes\mathcal{O}''\otimes H^{l_0,n}(\mathcal{O}).$$

REMARK. The assumptions of the previous corollary are obviously satisfied provided $\ker\mathcal{S}_\nabla = 0$. In this case $l_0 = -n$.

As we shall see later, it is convenient to use the left (resp., right) filtration in the case where the kernel (resp., the image) of the mapping $\mathcal{S}_\nabla$ is sufficiently simple.

4.4.11. Let us return to the situation we are dealing with and set $R = K_0^{-1,0}$. Recall that the module $K_0^{-1,0}$ is the image of the mapping $\mathcal{C}\mathrm{Diff}(\nabla,\mathrm{id}_{\mathcal{F}})$ (see Subsection 4.4.5). Introduce the following notation.

Let Π be the set of sequences consisting of zeros and units. We include in Π the empty sequence of length zero. We say that a sequence $\sigma\in\Pi$ is *trivial* if it is either empty or consists of zeros only. If $\sigma,\tau\in\Pi$, $\sigma=(\varepsilon_1,\dots,\varepsilon_k)$, $\tau=(\varepsilon'_1,\dots,\varepsilon'_l)$, we set $|\sigma| = k$, $|\tau| = l$, $\sigma+\tau = (\varepsilon_1,\dots,\varepsilon_k,\varepsilon'_1,\dots,\varepsilon'_l)$, and if $|\sigma| = |\tau|$, then we set $\sigma\tau = (\varepsilon_1\cdot\varepsilon'_1,\dots,\varepsilon_k\cdot\varepsilon'_k)$. With any sequence $\sigma = (\varepsilon_1,\dots,\varepsilon_k)\in\Pi$ we associate a subcomplex $C(\sigma) = C(\sigma;R)$ of the complex $\mathcal{C}\mathrm{Diff}_{(k)}\{[\varkappa]\}$ by setting $C(\sigma) = \{\mathcal{O}_1,\dots,\mathcal{O}_k\}$, where

$$\mathcal{O}_i = \begin{cases} \mathcal{C}\mathrm{Diff}([\varkappa],\mathcal{F}) & \text{if } \varepsilon_i = 1,\\ R & \text{if } \varepsilon_i = 0.\end{cases}$$

If R can be represented as a direct summand in $\mathcal{C}\mathrm{Diff}([\varkappa],\mathcal{F})$ and $|\sigma| = |\tau|$, then obviously $C(\sigma)\cap C(\tau) = C(\sigma\tau)$.

Let $\sigma_{i,p} = (\varepsilon_1,\dots,\varepsilon_p)\in\Pi$ be the sequence such that $\varepsilon_s = 1$ for $s\neq i$ and $\varepsilon_i = 0$. Consider the subcomplex

$$C_p(R) = \sum_{i=1}^{p} C(\sigma_{i,p})$$

of the complex $\mathcal{C}\mathrm{Diff}_{(p)}([\varkappa])$. This subcomplex is evidently invariant with respect to the action of the permutation group S_p in $\mathcal{C}\mathrm{Diff}_{(p)}([\varkappa])$ and its image under skew-symmetrization coincides with $K_0^{p,*}$. Hence, $K_0^{p,*}$ is represented as a direct summand in $C_p(R)$ and

$$H^s(K_0^{p,*}) = H^s_{\mathrm{alt}}(C_p(R)), \tag{4.46}$$

where $H^s_{\mathrm{alt}}(C_p(R))$ is the skew-symmetric part of $H^s(C_p(R))$ with respect to the induced action of S_p in $H^*(C_p(R))$. In particular, the acyclicity of the complex

$K_0^{p,*}$ in a certain dimension follows from that of $C_p(R)$. Therefore, we study the cohomology of the latter. We say that a complex is of the type s, $s \le n$, if it is acyclic in all dimensions $< s$ and $> n$.

PROPOSITION. *The following assertions about* $K_0^{1,0}$ *hold:*

1. *Let* $h^{l,q}(K_0^{1,0}) = 0$ *(resp.,* $H^{l,q}(K_0^{1,0}) = 0$*) for* $q \le s$*, and let* $K_0^{1,0}$ *be a direct summand in* $\mathcal{C}\mathrm{Diff}([\varkappa], \mathcal{F}(\mathcal{Y}))$. *Then the complexes* $C_p(K_0^{1,0})$, $p > 1$, *are of the type* s.

2. *Let* $K_0^{1,0}$ *be a direct summand in* $\mathcal{C}\mathrm{Diff}([\varkappa], \mathcal{F}(\mathcal{Y}))$, *and let there exist a number* l_0 *such that* $h^{l,q}(K_0^{1,0}) = 0$ *(resp.,* $H^{l,q}(K_0^{1,0}) = 0$*) for* $(l, q) \ne (l_0, n)$. *Suppose that the natural embedding* $C((0,0)) \to C((1,0))$ *induces a monomorphism* $H^n(C((0,0))) \to H^n(C((1,0)))$. *Then the complexes* $C_p(K_0^{1,0})$, $p > 1$, *are of the type* n.

PROOF. We prove the assertion using the assumptions about $h^{p,q}$. However, the proof does not change when we pass to the assumptions about $H^{p,q}$. The only modification to be made is to replace the references to Proposition 4.4.9 and its corollary with references to Proposition 4.4.10.

Consider the complexes

$$C_{p,\tau} = C_{p,\tau}(R) = \sum_i C(\sigma_{i,p} + \tau; R) \subset \mathcal{C}\mathrm{Diff}_{(p+|\tau|)}\{[\varkappa]\}.$$

Then $C_{p,\tau} = C_p(R)$ for $\tau = \emptyset$. Denote by $[l]$ the sequence of length l consisting of ones. We shall also write $\varepsilon + \tau$ instead of $(\varepsilon) + \tau$, if $\varepsilon = 0$ or 1. In this notation,

$$C_{p+1,\tau} = C_{p,1+\tau} + C([p] + 0 + \tau), \quad C_{p,1+\tau} \cap C([p] + 0 + \tau) = C_{p,0+\tau}. \tag{4.47}$$

All the complexes in (4.47) are regarded as subcomplexes of $\mathcal{C}\mathrm{Diff}_{(p+1+|\tau|)}\{[\varkappa]\}$. The second equality is valid because $R = K_0^{1,0}$ is a direct summand in $\mathcal{C}\mathrm{Diff}([\varkappa], \mathcal{F})$ and consequently $C(\sigma) \cap C(\sigma') = C(\sigma\sigma')$.

The complex $C_{p,\varepsilon+\tau}$, where $\varepsilon = 0$ or 1, is of the form $\{\mathcal{O}', \mathcal{O}, \mathcal{O}''\}$, where

$$\begin{aligned} \mathcal{O}' &= \sum_{i=1}^{p} A_1 \otimes \cdots \otimes A_{i-1} \otimes R \otimes A_{i+1} \otimes \cdots \otimes A_p, \\ A_j &= \mathcal{C}\mathrm{Diff}([\varkappa], \mathcal{F}(\mathcal{Y})), \\ \mathcal{O}'' &= B_1 \otimes \cdots \otimes B_{|\tau|}, \quad \tau = (\varepsilon_1, \dots, \varepsilon_{|\tau|}), \end{aligned}$$

and

$$B_j = \begin{cases} \mathcal{C}\mathrm{Diff}([\varkappa], \mathcal{F}(\mathcal{Y})) & \text{if } \varepsilon_j = 1, \\ R & \text{if } \varepsilon_j = 0. \end{cases}$$

In this case

$$\mathcal{O} = \begin{cases} \mathcal{C}\mathrm{Diff}([\varkappa], \mathcal{F}(\mathcal{Y})) & \text{if } \varepsilon = 1, \\ R & \text{if } \varepsilon = 0. \end{cases}$$

Hence, Proposition and Corollary 4.4.9 imply that the complex $C_{p,0+\tau}$ is of the type $s+1$ if assumption (1) holds, while $C_{p,1+\tau}$ is of the type n. Note now that

$$\frac{C_{p+1,\tau}}{C_{p,1+\tau}} = \frac{C([p] + 0 + \tau)}{C_{p,1+\tau} \cap C([p] + 0 + \tau)} = \frac{C([p] + 0 + \tau)}{C_{p,0+\tau}} = C'_{p,\tau}. \tag{4.48}$$

If assumption (1) holds, the exact cohomology sequence of the pair $C_{p,0+\tau} \subset C([p]+0+\tau)$ shows that the quotient complex $C'_{p,\tau}$ is of the type s, while according to (4.48) the exact sequence of the pair $C_{p,1+\tau} \subset C_{p+1,\tau}$ implies that $C_{p+1,\tau}$ is of the type s. This proves (1). Statement (2) will be proved by induction on $p \geq 1$.

Taking into account (4.48), consider the following commutative diagram:

$$\begin{array}{ccccccccc} 0 \longrightarrow & C_{p,1+\tau} & \longrightarrow & C_{p+1,\tau} & \longrightarrow & \dfrac{C_{p+1,\tau}}{C_{p,1+\tau}} & = & \dfrac{C([p]+0+\tau)}{C_{p,0+\tau}} & \longrightarrow 0 \\ & \Big\| & & \Big\downarrow{\scriptstyle\alpha} & & \Big\downarrow{\scriptstyle\beta} & & & \\ 0 \longrightarrow & C_{p,1+\tau} & \longrightarrow & C([p+1]+\tau) & \longrightarrow & \dfrac{C([p+1]+\tau)}{C_{p,1+\tau}} & \longrightarrow & 0 & \end{array} \tag{4.49}$$

where $\alpha\colon C_{p+1,\tau} \hookrightarrow C_{([p+1]+\tau)}$ is the natural embedding. The following sequence is also exact:

$$0 \to \frac{C([p]+0+\tau)}{C_{p,0+\tau}} \to \frac{C([p+1]+\tau)}{C_{p,1+\tau}} \to \frac{C([p+1]+\tau)}{C([p]+0+\tau)} \to 0. \tag{4.50}$$

Let us show by induction that the complexes $C_{p,\sigma}$ are of the type n, while the homomorphisms $H^n(C_{p,\sigma}) \to H^n(C([p]+\sigma))$ induced by the inclusions $C_{p,\sigma} \hookrightarrow C([p]+\sigma)$ are injective ($|\sigma| > 0$ if $p = 1$).

First of all note that $C_{1,\sigma} = C(0+\sigma)$ and the complexes $C(\tau)$, $|\tau| > 0$, are of the type n (Remark 4.4.9). On the other hand, in view of Corollary 4.4.9, injectivity of the homomorphisms $H^n(C(0,\varepsilon)) \to H^n(C(1,\varepsilon))$, $\varepsilon = 0, 1$, implies injectivity of

$$H^n(C(0+\varepsilon+\tau)) \to H^n(C(1+\varepsilon+\tau)), \quad \tau \in \Pi.$$

Thus we can start the induction.

Suppose the induction hypothesis holds for some p and consider the exact cohomology sequence for the pair $C_{p,0+\tau} \subset C([p]+0+\tau)$. Then the quotient complex $C([p]+0+\tau)/C_{p,0+\tau}$ is of the type n. This fact, combined with the exact cohomology sequence corresponding to the upper line of diagram (4.49), implies that the complex $C_{p+1,\tau}$ is of the type n.

According to the above-said, the exact cohomology sequences corresponding to the lines of diagram (4.49) lead to the commutative diagram

$$\begin{array}{ccccccc} 0 \longrightarrow & H^n(C_{p,1+\tau}) & \longrightarrow & H^n(C_{p+1,\tau}) & \longrightarrow & H^n\left(\dfrac{C([p]+0+\tau)}{C_{p,0+\tau}}\right) & \longrightarrow 0 \\ & \Big\| & & \Big\downarrow{\scriptstyle H^n(\alpha)} & & \Big\downarrow{\scriptstyle H^n(\beta)} & \\ 0 \longrightarrow & H^n(C_{p,1+\tau}) & \longrightarrow & H^n(C([p+1]+\tau)) & \longrightarrow & H^n\left(\dfrac{C([p+1]+\tau)}{C_{p,1+\tau}}\right) & \longrightarrow 0 \end{array}$$

with the exact lines. This diagram shows that injectivity of $H^n(\alpha)$ is implied by injectivity of $H^n(\beta)$. In turn, the latter is a consequence of the fact that the complex $C([p+1]+\tau)/C([p+0]+\tau)$ is of the type n (as results from the exact cohomology sequence corresponding to (4.50)). But this follows from the exact cohomology sequence of $C([p]+0+\tau) \subset C([p+1]+\varepsilon) = C([p]+1+\varepsilon)$ and the observations which started the induction. □

REMARK. The proof shows that Proposition 4.4.11 remains valid for the complexes $C_{p,\tau}(R)$.

Comparing the proposition just proved with (4.46) and (4.41), we obtain the following result estimating the cohomology of $K^{p,*}$ and consequently the term $E_1(\mathcal{Y})$.

THEOREM. *The following estimates are valid*:

1. *Let* $h^{l,q}(K_0^{1,0}) = 0$ (*resp.*, $H^{l,q}(K_0^{1,0}) = 0$) *for* $q \le s$ ($s \le n$), *and let* $K_0^{1,0}$ *be a direct summand in* $\mathcal{C}\mathrm{Diff}([\varkappa], \mathcal{F}(\mathcal{Y}))$. *Then* $E_1^{1,q}(\mathcal{Y}) = H^q(K^{1,*})$ *for* $q < s$ *and* $E_1^{p,q}(\mathcal{Y}) = H^q(K^{p,*})$ *for* $q < s - 1$, $p > 1$.

2. *Assume that*

(a) $K_0^{1,0}$ *is a direct summand in* $\mathcal{C}\mathrm{Diff}([\varkappa], \mathcal{F}(\mathcal{Y}))$,
(b) *there exists a number* l_0 *such that* $h^{p,q}(K_0^{1,0}) = 0$ (*resp.*, $H^{p,q}(K_0^{1,0}) = 0$) *for* $(l, q) \ne (l_0, q)$, *and finally*,
(c) *the natural embedding* $C((0,0)) \to C((1,0))$ *induces a group monomorphism* $H^n(C((0,0))) \to H^n(C((1,0)))$.

Then for $p \ge 1$ *one has* $E_1^{p,q} = H^q(K^{p,*})$ *if* $q < n - 1$.

Essentially, this theorem establishes a direct relation between the δ-Spencer cohomology of the operator ∇ and the first term of the $\mathcal{C}$-spectral sequence. Its first assertion tells us that if the δ-Spencer complex associated with ∇ is acyclic in dimensions $\le s$, then the term E_1 is acyclic in "bi-dimensions" $q \le s - 2$, $p > 0$, i.e., all its eventually nontrivial components are situated in $n - s + 2$ horizontal lines of the standard (p, q)-diagram starting with the n-th and going down. This result, however, can be improved by one line (namely, the $(n - s + 2)$-nd), which is also trivial. The second assertion gives this improved version only for $s = n - 1$ just to avoid more sophisticated homological techniques. In full generality, this is done in Chapter 5 in terms of the *compatibility complex*. But we do not discuss here how the δ-Spencer cohomology is exactly related to this complex. Let us emphasize also that in the proof of Theorem 4.4.11 given before we were limited to use rather elementary tools of homological algebra.

4.4.12. It is worth paying special attention to nonoverdetermined systems of partial differential equations, which are of primary interest in what concerns applications. A rigorous concept of nonoverdeterminedness does not reduce to the commonly known requirement that the number of independent equations does not exceed the number of unknown functions, which is too rough to be useful. A correct approach is in the framework of the formal theory of partial differential equations (see [**44, 45, 98, 103**]). Apart from some regularity conditions, nonoverdeterminedness means that $\ker \mathcal{S}_\nabla = \ker v_\nabla^{(1)} = 0$ (see Subsection 4.4.8). In this case the complex $K_0^{1,*}$ is isomorphic to the complex $\widetilde{\mathcal{S}P}$, and the left filtration in it is the image of the standard filtration $\{\widetilde{\mathcal{S}_k P}\}$ in $\mathcal{S}P$. Therefore, one can take $-n$ for the number l_0 appearing in Theorem 4.4.11 (see Remark 4.4.10). Further, since $\ker \mathcal{S}_\nabla = 0$, one has

$$C((0,0)) = \mathcal{C}\mathrm{Diff}\{P, P\} = \mathcal{C}\mathrm{Diff}(P, \widetilde{\mathcal{S}P}),$$
$$C((1,0)) = \mathcal{C}\mathrm{Diff}\{[\varkappa], P\} = \mathcal{C}\mathrm{Diff}([\varkappa], \widetilde{\mathcal{S}P}),$$

and the natural embedding $C((0,0)) \to C((1,0))$ is identified with the mapping $\square \mapsto \square \circ \nabla$, where $\square \in \mathcal{C}\mathrm{Diff}(P, \widetilde{\mathcal{S}P})$. Consequently, the corresponding cohomology mapping $H^n(C((0,0))) \to H^n(C((1,0)))$ (see Subsection 4.3.5) is identified with

$$\mathcal{C}\mathrm{Diff}(\nabla, \mathrm{id}_{\hat{P}})\colon H^n(\mathcal{C}\mathrm{Diff}\{P,P\}) = \mathcal{C}\mathrm{Diff}(P,\hat{P}) \to H^n(\mathcal{C}\mathrm{Diff}\{[\varkappa],P\}) = \mathcal{C}\mathrm{Diff}([\varkappa],\hat{P}),$$

i.e., with the correspondence $\Delta \mapsto \Delta \circ \nabla$, $\Delta \in \mathcal{C}\mathrm{Diff}(P,\hat{P})$. Let us verify that this mapping is a monomorphism. Since $\mathcal{C}\mathrm{Diff}(Q,\hat{P}) = \mathcal{C}\mathrm{Diff}(Q,\mathcal{F}(\mathcal{Y})) \otimes \hat{P}$, it suffices to check that the mapping

$$\mathcal{C}\mathrm{Diff}(\nabla, \mathrm{id}_{\mathcal{F}})\colon \mathcal{C}\mathrm{Diff}(P,\mathcal{F}(\mathcal{Y})) \to \mathcal{C}\mathrm{Diff}([\varkappa],\mathcal{F}(\mathcal{Y}))$$

is a monomorphism. But this is obvious, since $\mathcal{C}\mathrm{Diff}(\nabla, \mathrm{id}_{\mathcal{F}}) = v_\nabla^{(1,0)}$.

Thus, we have obtained the following result:

THEOREM (Two Line Theorem). *Let $K_0^{1,0}$ be a direct summand in the complex $\mathcal{C}\mathrm{Diff}([\varkappa],\mathcal{F}(\mathcal{Y}))$ and assume that $\ker \mathcal{S}_\nabla = 0$ (this is the case when $\operatorname{char}\nabla$ is nowhere dense). Then for $p \geq 1$ one has $E_1^{p,q}(\mathcal{Y}) = H^q(K^{p,*}) = 0$ if $q \neq n-1, n$. In addition, $E_1^{1,n-1}(\mathcal{Y}) = \ker \nabla^*$ and $E_1^{1,n} = \operatorname{coker} \nabla^*$.*

PROOF. According to the above, the first part of the theorem follows from Theorem 4.4.11 (2). The second part is a corollary of (4.42) and Theorem 4.4.7. □

4.4.13. To conclude this section, let us describe a method of computing the term $E_2(\mathcal{Y})$. To do this, we need the notion of *homotopy* in the category DE.

Let $\mathcal{O} \in \mathrm{Ob(DE)}$ and $I = [0,1]$. The algebra $\mathcal{F}(\mathcal{O} \times I)$ is defined as an algebra of functions $f_t \in \mathcal{F}(\mathcal{O})$ depending smoothly on $t \in I$. Set

$$\mathcal{F}_p(\mathcal{O} \times I) = \{f_t \in \mathcal{F}_p(\mathcal{O}) \mid \mathcal{F}_p(\mathcal{O}) \text{ is a filtration in } \mathcal{F}(\mathcal{O})\}.$$

Thus an FG-category is defined over $\mathcal{F}(\mathcal{O} \times I)$.

We regard I as an object of DE by setting

$$\mathcal{C}\Phi(I) = \{\varphi \in \Phi \mid \varphi(t) = 0,\ \forall t\}$$

for any representing object $\Phi(I)$. In other words, the points $t \in I$ and only these points are integral manifolds of the Cartan distribution on I defined in this way. Let π_1, π_2 be the natural projections of $\mathcal{O} \times I$ onto $\mathcal{O}$ and I, respectively. Introduce the operation $\mathcal{C}$ on $\mathcal{O} \times I$ assuming that submodules $\mathcal{C}\Phi(\mathcal{O} \times I)$ in $\Phi(\mathcal{O} \times I)$ are generated by $\pi_1^*\mathcal{C}\Phi(\mathcal{O})$ and $\pi_2^*\mathcal{C}\Phi(I)$. Obviously, the operation $\mathcal{C}$ defined in such a way transforms $\mathcal{O} \times I$ to an object of DE. In this setting, the projections π_i, $i = 1,2$, as well as the mappings $\alpha_t\colon \mathcal{O} \to \mathcal{O} \times I$, $\alpha_t(x) = (x,t)$, $x \in \mathcal{O}$, are morphisms in the category DE.

REMARK. Changing I above to any object of DE, one gets the definition of the direct product in DE.

DEFINITION. Two morphisms $F_0, F_1\colon \mathcal{O}_1 \to \mathcal{O}_2$ in DE are said to be *homotopic* if there exists a morphism $F\colon \mathcal{O}_1 \times I \to \mathcal{O}_2$ such that $F_0 = F \circ \alpha_0$, $F_1 = F \circ \alpha_1$.

Since the term E_1 of the $\mathcal{C}$-spectral sequence admits the infinitesimal Stokes formula and the morphisms in DE induce homomorphisms of the corresponding $\mathcal{C}$-spectral sequences, one can use the same arguments as in the category of smooth manifolds to prove the following result.

THEOREM. *If two morphisms $F_0, F_1\colon \mathcal{O}_1 \to \mathcal{O}_2$ of the objects in DE are homotopic, then the homomorphisms $F_i\colon E_2^{p,q}(\mathcal{O}_2) \to E_2^{p,q}(\mathcal{O}_1)$, $i = 1,2$, coincide.*

This fact can be used for computing the term $E_2(\mathcal{O})$ along the same lines as in computations of the de Rham cohomology of smooth manifolds. As an illustration, we present one result which can be regarded as an analog of the Poincaré lemma in the category DE.

Let $\mathcal{Y} \subset J^k(\pi)$, π being a vector bundle. Assume that the fiberwise homothety group of π is a symmetry group for $\mathcal{Y}$. Such equations will be called *conic.* Note that linear equations are obviously conic.

PROPOSITION. *If an equation $\mathcal{Y}$ is conic, then $E_2^{p,q}(\mathcal{Y}) = 0$ for $p > 0$ and $E_2^{0,q}(\mathcal{Y}) = H^q(M)$, where M is the base of π.*

PROOF. If σ is the zero section of π, then $j(\sigma)\colon M \to \mathcal{Y}$ is a homotopy equivalence in DE. $\square$

COROLLARY. *If an equation $\mathcal{Y}$ is conic, then complexes* (4.43) *and* (4.44) *are acyclic in positive dimensions.*

4.5. Applications to the Lagrangian formalism with constraints and to the theory of conservation laws

Below we collect some consequences of the results obtained in Sections 4.3 and 4.4 by paying more attention to the terms $E_1^{\varepsilon,n}$, $E_1^{\varepsilon,n-1}$ and to the differentials $d_1^{\varepsilon,n}$, $d_1^{\varepsilon,n-1}$, $\varepsilon = 0, 1$.

4.5.1. In applications of the $\mathcal{C}$-spectral sequence, the knowledge of the cohomology $H^*(\mathcal{Y})$ to which it converges is important. We start with a number of remarks in this connection, restricting ourselves to the case $\mathcal{Y} \subset J^k(\pi)$. The general case reduces to this by standard topological techniques and contains nothing new.

First of all note that the projection $\pi_{\infty,k}\colon \mathcal{Y}_\infty \to \mathcal{Y}$ is not in general a fiber bundle and even not a surjection. Taking an apparently incompatible equation $\mathcal{Y}' \subset J^k(\pi)$ (e.g., one can take $\mathcal{Y}' \subset \pi_k^{-1}(M')$, where $M' \subset M$ is a submanifold, $\dim M' < \dim M$) and by "gluing" it to a "good" equation $\mathcal{Y}'' \subset J^k(\pi)$, one obtains examples of this kind. Therefore, in general, the cohomology of $\mathcal{Y}$ and that of $\mathcal{Y}_\infty$ are not related to each other in a reasonable way. On the other hand, if natural projections $\mathcal{Y}^{s+1} \to \mathcal{Y}^s$ are surjective, then under certain regularity conditions, $\mathcal{Y}^{s+1}$ and $\mathcal{Y}^s$ are of the same homotopic type and therefore $H^*(\mathcal{Y}^{s+1}) = H^*(\mathcal{Y}^s)$. Exact formulations of the corresponding conditions can be found in [**45, 103**]. Here we state a result by Goldschmidt obtained in [**45**].

PROPOSITION. *Let $\mathcal{Y} \subset J^k(\pi)$ be a subbundle of the bundle $\pi_k\colon J^k(\pi) \to M$, $\mathcal{Y}^1 \to \mathcal{Y}^0 = \mathcal{Y}$ a surjection, and let the symbol of the equation $\mathcal{Y}$ be 2-acyclic. Then for any $s \geq 0$ the projections $\mathcal{Y}^{s+1} \to \mathcal{Y}^s$ are affine bundles and therefore $H^*(\mathcal{Y}_\infty) = H^*(\mathcal{Y})$.*

In particular, if $\mathcal{Y}$ is not an overdetermined equation and is a subbundle in π_k, then $H^*(\mathcal{Y}_\infty) = H^*(\mathcal{Y})$.

COROLLARY. *Consider an equation $\mathcal{Y}$ such that the conditions of Theorem* 4.4.11 (2) *and of Proposition* 4.5.1 *hold. Then $E_2^{0,n-1} = H^{n-1}(\mathcal{Y})$.*

PROOF. Indeed, according to Corollary 4.4.7, one obtains from Theorem 4.4.11 that $E_2^{0,n-1} = H^{n-1}(\mathcal{Y})$. $\square$

Thus, under the above assumptions one has the exact sequence

$$0 \to H^{n-1}(\mathcal{Y}) \to \bar{H}^{n-1}(\mathcal{Y}_\infty) = E_1^{0,n-1} \xrightarrow{d_1^{0,n-1}} E_1^{1,n-1} = \ker \nabla^*. \tag{4.51}$$

Recall that the group $\bar{H}^{n-1}(\mathcal{Y}_\infty)$ is interpreted as the group of conservation laws for the equation $\mathcal{Y}$. Thus it is natural to say that a conservation law $\omega \in H^{n-1}(\mathcal{Y}) \subset \bar{H}^{n-1}(\mathcal{Y}_\infty)$ is *rough*, since it is determined by the topology of $\mathcal{Y}$ only and not by specifics of an individual solution. In particular, if $V \subset E$ is a solution of $\mathcal{Y}$, then the cohomology class $j(V)^*(\omega)$ does not change under deformations of this solution.

DEFINITION. The quotient group $\bar{\bar{H}}^{n-1}(\mathcal{Y}) = \bar{H}^{n-1}(\mathcal{Y}_\infty)/H^{n-1}(\mathcal{Y})$ (defined under the above assumptions) is called the group of *proper conservation laws* for the equation $\mathcal{Y}$.

In the sequel, the class of equations described by the following definition will be important.

DEFINITION. Let $\mathcal{Y}$ be a differential equation.

1. If $\mathcal{Y}$ satisfies the assumptions of Theorem 4.4.11 (2) and Proposition 4.5.1, then $\mathcal{Y}$ is called *normal with respect to* ∇.
2. An equation $\mathcal{Y} = \{\varphi = 0\}$ normal with respect to $\nabla = \ell_\varphi$ will be called *normal*.

THEOREM. *Let $\mathcal{Y}$ be normal with respect to ∇. Then $\bar{\bar{H}}^{n-1}(\mathcal{Y}) \subset \ker \nabla^*$. If in addition $H^n(\mathcal{Y}) \subset \bar{H}^n(\mathcal{Y}_\infty)$ (in particular, if $H^n(\mathcal{Y}) = 0$), then $\bar{\bar{H}}^{n-1}(\mathcal{Y}) = \ker d_1^{1,n-1}$.*

PROOF. The first part of the theorem follows from the exactness of (4.51). The second is a consequence of the equality $E_2^{1,n-1}(\mathcal{Y}) = E_\infty^{1,n-1}(\mathcal{Y})$ (see Corollary 4.4.7). □

This gives an efficient method for computing conservation laws (or for estimation of their amount). In fact, an upper bound is given by the kernel of the operator ∇^*, and in many cases it can be computed by a rather efficient procedure. The operator $d_1^{1,n-1}$ also admits a sufficiently simple description (see Subsection 4.5.5). It allows one to remove "unnecessary" elements from $\ker \nabla^*$ and to obtain $\bar{\bar{H}}^{n-1}(\mathcal{Y})$. For example, the scheme described easily produces conservation laws for many equations such as Korteweg–de Vries, Burgers, Klein–Gordon equation, and for a number of others. One of the methods of this type is described in [**140**]. See also [**127**], the collection [**139**], and the book [**19**].

REMARK (1). It is useful to keep in mind that for normal equations we have $\ker d_1^{1,n-1}/\bar{\bar{H}}^{n-1}(\mathcal{Y}) \subset H^n(\mathcal{Y})$, and consequently $\ker d_1^{1,n-1} = \bar{\bar{H}}^{n-1}(\mathcal{Y})$ modulo topological aspects controlled rather easily.

REMARK (2). Let $\mathcal{Y} \subset J^k(\pi)$. Then the equality $H^n(\mathcal{Y}) = (\pi_k|_{\mathcal{Y}})^*(H^n(M))$ implies $H^n(\mathcal{Y}) \subset \bar{H}^n(\mathcal{Y}_\infty)$.

4.5.2. Now let $\mathcal{Y} \subset J^k(\pi)$ be a *conformally selfadjoint equation*. This means that $\mathcal{Y} = \{\varphi = 0\}$ and $\ell_\varphi = \lambda \ell_\varphi^*$ (see Subsection 2.3.8). Suppose also that $\mathcal{Y}$ is a normal equation. Then according to Theorem 3.3.5 one has

$$\operatorname{sym} \mathcal{Y} = \ker \ell_\varphi = \ker \ell_\varphi^* = E_1^{1,n-1}. \tag{4.52}$$

This implies the following important result which can be interpreted as the *inverse Noether theorem*.

THEOREM (Inverse Noether theorem). *Let a normal equation* $\mathcal{Y} = \{\varphi = 0\} \subset J^k(\pi)$ *be conformally self-adjoint. Then* $d_1^{0,n-1}$ *takes* $\bar{H}^{n-1}(\mathcal{Y})$ *to* $E_1^{1,n-1}(\mathcal{Y})$ *and the corresponding quotient mapping embeds the group* $\bar{\bar{H}}^{n-1}(\mathcal{Y})$ *of proper conservation laws of the equation* $\mathcal{Y}$ *into the algebra* $\operatorname{sym}\mathcal{Y}$ *of its symmetries. In other words, to every conservation law of* $\mathcal{Y}$ *there corresponds a symmetry. The kernel of this correspondence coincides with* $H^{n-1}(\mathcal{Y})$.

COROLLARY. *If a normal equation* $\mathcal{Y} \subset J^k(\pi)$ *is in addition an Euler–Lagrange system and* $H^{n-1}(E_\pi) = 0$, *then the operator* $d_1^{0,n-1}$ *embeds the group of conservation laws of* $\mathcal{Y}$ *into the algebra* $\operatorname{sym}\mathcal{Y}$.

PROOF. To prove the result it suffices to compare Theorems 4.3.9 and 4.5.2. □

REMARK (1). Extensive literature is dedicated to the inverse Noether theorem for Euler–Lagrange equations. The fact that the Noether-like mechanism works for conformally selfadjoint equations seems to be quite unexpected.

REMARK (2). We emphasize that the results of this and previous subsections usually hold for nonoverdetermined equations one encounters in geometry and mathematical physics. The normality conditions which assure the validity of these results are not restrictive and can be easily verified.

4.5.3. We shall now describe explicitly the differential $d_1^{0,n-1}$ and the isomorphism $E_1^{1,n-1} = \ker \nabla^*$. keeping in mind their importance for the theory of conservation laws. Among other things, this description will clarify why Theorem 4.5.2 is the inverse to the classical Noether theorem.

We shall work in a special chart, or equivalently, we assume that $\mathcal{Y}_\infty \subset J^\infty(\pi)$. Restrictions of various objects over $J^\infty(\pi)$ to $\mathcal{Y}_\infty$ are denoted by adding brackets $[\bullet]$ to the original notation (cf. Section 4.4). We use the notation of Subsection 4.4.1 and write $e = \langle a \rangle$ if $e \in E_1^{p,q}$ and $a \in E_0^{p,q}$ is a d_0-cocycle representing e.

We identify E_0 (resp., $E_0(\mathcal{Y})$) with the bicomplex $\mathcal{C}\mathrm{Diff}^{\mathrm{alt}}\{\varkappa\}$ (resp., with $\mathcal{C}\mathrm{Diff}^{\mathrm{alt}}\{[\varkappa]\}/K_0$) and E_1 (resp., $E_1(\mathcal{Y})$) with the $\bar{d}$-cohomology of these complexes. Then for $e = \langle a \rangle$, $a \in E_0^{p,q}$, one has

$$d_1^{p,q} = \langle l_{p,q}(a) \rangle. \tag{4.53}$$

From Proposition 4.4.3 it follows that

$$d_1^{p,q}([e]) = [\langle l_{p,q}(a) \rangle] \bmod K_0. \tag{4.54}$$

According to (4.4), $l_{0,q}(\omega) = \ell_\omega$, $\omega \in \Lambda_0$, and hence

$$d_1^{0,q}\langle [\omega] \rangle = \langle [\ell_\omega] \rangle \bmod K_0. \tag{4.55}$$

This is the required formula.

Suppose further that $e = \square \bmod K_0 \in E_1^{1,n-1}(\mathcal{Y})$, where the operator $\square$ belongs to $\mathcal{C}\mathrm{Diff}([\varkappa], \Lambda^{n-1}(\mathcal{Y}_\infty))$. Then $\bar{d} \circ \square = 0 \bmod K_0$, i.e., $\bar{d} \circ \square = \Delta \circ \nabla$ (recall that $\mathcal{Y}$ is a regular equation with respect to ∇). Since $E_1^{n-1} = H^{n-1}(K^{1,*})$, the isomorphism $\ker \nabla^* = H^n(K_0^{1,*})$ is realized by the coboundary operator of the exact cohomology sequence corresponding to the exact sequence of complexes $0 \to K_0^{1,*} \to \widetilde{\mathcal{S}[\varkappa]} \to K^{1,*} \to 0$ and $d_{1,q} = -\bar{d}$ (see Subsection 4.3.4). This isomorphism takes the element e to $-\Delta^*(1) \in \ker \nabla^*$.

Conversely, let $\psi \in \ker \nabla^* \subset [\hat{P}] = \mathrm{Hom}_{\mathcal{F}(\mathcal{Y})}([P], \bar{\Lambda}^n(\mathcal{Y}_\infty))$. Then

$$\mu(\psi \circ \nabla) = \nabla^*(\psi) = 0$$

and therefore $\psi \circ \nabla = \bar{d} \circ \square$ (see Subsection 4.3.5). Here, according to the Green formula, $\square(\chi) = \mathcal{K}_\lambda(\psi \circ \nabla \circ \chi)$ and the isomorphism under consideration takes $\psi \in \ker \nabla^*$ to $e = -\square \bmod K_0 \in E_1^{1,n-1}$.

Let $e = d_1^{0,n-1}\langle[\omega]\rangle$ and $\nabla = [\ell_\varphi]$ (see Subsection 4.4.5). So, by (4.55), $e = [\ell_\omega]$. Hence, to find the element of $\ker \nabla^*$ corresponding to e one has to find an operator Δ satisfying the relation $\bar{d} \circ [\ell_\omega] = \Delta \circ [\ell_\varphi]$. But on $\mathcal{Y}_\infty$ one has $\bar{d} \circ [\ell_\omega] = \Delta \circ [\ell_{\bar{d}\omega}]$ and $\bar{d}\omega = 0$. Consequently, $\bar{d}\omega = \square(\varphi)$, where $\square \in \mathcal{C}\mathrm{Diff}(P, \bar{\Lambda}^n)$. Since, in addition, $[\ell_{\square(\varphi)}] = [\square] \circ [\ell_\varphi]$, one has $\Delta = [\square]$ and $\Delta^*(1) = [\square^*(1)]$. Thus we have proved the following result describing the image of $d_1^{0,n-1}$.

PROPOSITION. *Let $\mathcal{Y} = \{\varphi = 0\} \subset J^k(\pi)$ be a normal equation. We identify $E_1^{1,n-1}$ with $\ker \ell_\varphi^*$. Then for $h = \langle[\omega]\rangle \in \bar{H}^{n-1}(\mathcal{Y})$, $\omega \in \bar{\Lambda}^{n-1}$, one has $d_1^{0,n-1}(h) = [\square^*(1)]$, where $\square \in \mathcal{C}\mathrm{Diff}(P, \bar{\Lambda}^n)$ is such that $\bar{d}\omega = -\square(\varphi)$.*

COROLLARY. *Let $f \in \operatorname{sym} \mathcal{Y}$. Then under the assumptions of Proposition 4.5.3 one has*

$$\mathcal{E}_f\{h\} = (\mathcal{E}_f + [\Delta^*])(h), \qquad h \in \ker[\ell_\varphi^*],\ h = [\tilde{h}],$$

where the operator $\Delta \in \mathcal{C}\mathrm{Diff}(P, P)$ is such that $\mathcal{E}_f(\varphi) = \Delta(\varphi)$. In other words, the Lie derivative of an element $h \in \ker[\ell_\varphi^*]$, where h is viewed as an element of $E_1^{1,n-1}$, along the field $\mathcal{E}_f$ equals its "usual" Lie derivative $\mathcal{E}_f(h)$ plus the element $[\Delta^*](h)$.

PROOF. Note first that the operator Δ exists because $f \in \operatorname{sym} \mathcal{Y}$ and $\mathcal{Y}$ is a normal equation. Let us find an operator $\tilde{\square}$ such that $\bar{d}\mathcal{E}_f(\omega) = -\tilde{\square}\varphi$. We have $\bar{d}\mathcal{E}_f(\omega) = \mathcal{E}_f(\bar{d}\omega) = -\mathcal{E}_f(\square(\varphi))$. Using the Green formula for $\square(\varphi)$, one obtains

$$\mathcal{E}_f(\square(\varphi)) = \mathcal{E}_f\big(\langle\varphi, \square^*(1)\rangle + \bar{d}\mathcal{K}_\lambda(\square \circ \varphi)\big) = \mathcal{E}_f(\langle\varphi, \tilde{h}\rangle) + \bar{d}\mathcal{E}_f(\mathcal{K}(\square \circ \varphi)),$$

since $\square^*(1) = \tilde{h}$ and $\mathcal{E}_f \circ \bar{d} = \bar{d} \circ \mathcal{E}_f$. But

$$\begin{aligned}\mathcal{E}_f(\langle\varphi, \tilde{h}\rangle) &= \langle\mathcal{E}_f(\varphi), \tilde{h}\rangle + \langle\varphi, \mathcal{E}_f(\tilde{h})\rangle = \langle\Delta(\varphi), \tilde{h}\rangle + \langle\varphi, \mathcal{E}_f(\tilde{h})\rangle \\ &= \big(\tilde{h}^* \circ \Delta + \mathcal{E}_f(\tilde{h})^*\big)(\varphi).\end{aligned}$$

On the other hand, using the formula for infinitesimal transformation of $\mathcal{K}_\lambda$ along $\mathcal{E}_f$ (see Subsection 1.3.7), we get

$$\begin{aligned}\mathcal{E}_f\big(\mathcal{K}_\lambda(\square \circ \varphi)\big) &= \mathcal{E}_f(\mathcal{K}_\lambda)(\square \circ \varphi) + \mathcal{K}_\lambda([\mathcal{E}_f, \square \circ \varphi]) \\ &= \bar{d} \circ v(\square \circ \varphi) + \mathcal{K}\big(\square \circ \mathcal{E}_f(\varphi) + \mathcal{E}_f(\square) \circ \varphi\big) \\ &= \bar{d} \circ v(\square \circ \varphi) + \mathcal{K}_\lambda\big(\square \circ \Delta(\varphi) + \mathcal{E}_f(\square) \circ \varphi\big).\end{aligned}$$

Hence one can see that $\nabla\colon \psi \mapsto \mathcal{E}_f(\mathcal{K}_\lambda(\square \circ \psi))$ is a $\mathcal{C}$-differential operator. Thus

$$\mathcal{E}_f(\square(\varphi)) = \big(\tilde{h}^*\Delta + \mathcal{E}_f(\tilde{h})^* + \bar{d} \circ \nabla\big)(\varphi),$$

i.e., $\tilde{\square} = \tilde{h}^* \circ \Delta + \mathcal{E}_f(\tilde{h})^* + \bar{d} \circ \nabla$ and $\mathcal{E}\{h\} = [\tilde{\square}^*(1)] = \Delta^*(h) + \mathcal{E}_f(h)$. □

4.5.4. Let us establish relations between Theorems 4.2.4 and 4.5.2 (or, to be precise, Corollary 4.5.2), i.e., between the "direct" and "inverse" Noether theorems. To do this, we give a more adequate formulation of Theorem 4.2.4.

DEFINITION. Let $\chi \in \varkappa(\pi)$ and $\Omega \in \bar{H}^n(J^\infty(\pi))$.

1. The element χ is called a *symmetry* of a Lagrangian Ω if $\chi\{\Omega\} = 0$ on $\mathcal{Y}_\infty$, where $\mathcal{Y}$ is the Euler–Lagrange equation corresponding to Ω.

2. If $\chi\{\Omega\} = 0$, then χ is called a *Noetherian symmetry* of Ω.

Denote by $\operatorname{Sym}\Omega$ the set of all symmetries of a Lagrangian Ω and by $\operatorname{sym}\Omega$ the set of all its Noetherian symmetries. Obviously, $\operatorname{sym}\Omega \subset \operatorname{Sym}\Omega$.

If χ is a symmetry of Ω and ρ is its density, then $[\mathfrak{Э}_\chi(\rho)] = [\bar{d}\nu]$ on $\mathcal{Y}_\infty$, where $\nu \in \bar{\Lambda}^{n-1}(J^\infty(\pi))$. Let a field $Y \in \mathcal{C}D(\mathcal{F}(\pi))$ be such that $\nu = -Y \lrcorner\, \rho$. Then

$$(\mathfrak{Э}_\chi + Y)(\rho) = \mathfrak{Э}_\chi(\rho) + Y(\rho) - \mathfrak{Э}_\chi(\rho) - \bar{d}(Y \lrcorner\, \rho) = 0$$

on $\mathcal{Y}_\infty$. Hence, the field $X = \mathfrak{Э}_\chi + Y$ is a symmetry of the density ρ (see Section 4.2) and according to Theorem 4.2.4 the corresponding c-density is $c_\lambda(\rho, X) = \mathcal{K}_\lambda(\ell_\rho \circ \chi) - \nu$. The mapping

$$\mathrm{Nt}\colon \operatorname{Sym}\Omega \to \bar{H}^{n-1}(\mathcal{Y}), \quad \chi \mapsto \langle \mathcal{K}_\lambda(\ell_\rho \circ \chi) - \nu \rangle$$

arising in this way will be called the *Noether mapping*. Now let us find how this mapping depends on ρ and ν.

Let $\rho' = \rho + \bar{d}\psi$. Then, by the Green formula,

$$\bar{d}\mathcal{K}_\lambda(\ell_{\bar{d}\psi} \circ \chi) = \bar{d}\mathcal{K}_\lambda(\bar{d} \circ \ell_\psi \circ \chi) = \bar{d}(\ell_\psi(\chi))$$

so that $\mathcal{K}_\lambda(\ell_{\bar{d}\psi} \circ \chi) = \ell_\psi(\chi) + \bar{d}\alpha$ if $\bar{H}^{n-1}(J^\infty(\pi)) = 0$ (or locally). On the other hand,

$$[\ell_{\rho'}(\chi)] = [\ell_\rho(\chi)] + [\ell_{\bar{d}\psi}(\chi)] = [\bar{d}\nu] + [\bar{d}\ell_\psi(\chi)]$$

so that $\nu + \ell_\psi(\chi)$ can be taken as the element ν for ρ'. This shows that the Noether mapping does not depend locally on ρ but in general is multivalued. The degree of ambiguity coincides with the image of the homomorphism $\bar{H}^{n-1}(J^\infty(\pi)) \to \bar{H}^{n-1}(\mathcal{Y}_\infty)$. Freedom in the choice of ν leads obviously to the same ambiguity. Recalling that $\bar{H}^{n-1}(J^\infty(\pi)) = H^{n-1}(J^\infty(\pi))$, we obtain

PROPOSITION. *The Noether mapping is defined uniquely modulo the image of the natural homomorphism* $H^{n-1}(J^\infty(\pi)) \to \bar{H}^{n-1}(\mathcal{Y}_\infty)$.

If χ is a Noetherian symmetry, then $\mathfrak{Э}_\chi(\rho) = \bar{d}\nu$, so that by the Green formula, one has

$$0 = \ell_\rho(\chi) - \bar{d}\nu = \bar{d}\big(\mathcal{K}_\lambda(\ell_\rho \circ \chi) - \nu\big) + (\mathcal{E}(\Omega), \chi) = dc_\lambda(\rho, X) + (\mathcal{E}(\Omega), \chi).$$

Recall that $\ell^*_\varphi = \ell_\varphi$ if $\varphi = \mathcal{E}(\Omega)$ (Theorem 4.3.9). Due to Proposition 4.5.3, the element of $\ker \ell^*_\varphi = \ker \ell_\varphi$, where $\varphi = \mathcal{E}(\Omega)$, corresponding to the conservation law $h = \langle[\omega]\rangle$, $\omega = c_\lambda(\rho, X)$, is of the form $\square^*(1)$, where $\bar{d}\omega = -\square(\varphi)$. But $\bar{d}\omega = -(\mathcal{E}(\Omega), \chi) = -(\varphi, \chi)$. Hence, χ can be taken for the operator $\square$, where χ is understood as an element of $\operatorname{Hom}(\hat{\varkappa}, \bar{\Lambda}^n) \subset \mathcal{C}\mathrm{Diff}(\hat{\varkappa}, \bar{\Lambda}^n)$. Thus, identifying $E_1^{1,n-1}$ with $\ker \ell^*_\varphi = \ker \ell_\varphi = \operatorname{sym}\mathcal{Y}$, one gets

$$d_1^{0,n-1}\langle c_\lambda(\rho, X)\rangle = d_1^{0,n-1}(\mathrm{Nt}(\chi)) = \chi. \tag{4.56}$$

Thus we have proved the following

THEOREM. *Let a normal equation $\mathcal{Y} \subset J^k(\pi)$ be the Euler–Lagrange equation corresponding to a Lagrangian $\Omega \in \bar{H}^n(J^\infty(\pi))$. Then the Noether mapping considered on the set of Noetherian symmetries is invertible with $d_1^{0,n-1}$ being the inverse mapping.*

REMARK (1). The above arguments show that the theorem is not in general valid for non-Noetherian symmetries. The additional correcting summand is $\square^*(1)$, where $\ell_\rho(\chi) - \bar{d}\nu = \square(\varphi)$.

REMARK (2). Noetherian symmetries can be found by a constructive procedure. In fact, from the properties of the variational complex described in Subsection 4.3.8 it follows that under appropriate topological conditions (or locally) the procedure reduces to solving the equation $\mathcal{E}(\Im_\chi(\rho)) = 0$, $\Omega = \langle\rho\rangle$. In general, one needs information on zero cohomology of complex (4.44) or, which is the same, on the term $E_2^{0,n}(\mathcal{Y})$.

4.5.5. Since $\ker d_1^{1,n-1} = \bar{\bar{H}}^{n-1}(\mathcal{Y})$ (see Theorem 4.5.2), the operator $d_1^{1,n-1}$ turns out to be very important. Let us describe it in more details. To do this, additional information concerning the term $E_1^{2,n-1}$ is needed. In this subsection, the hypothesis of Theorem 4.4.2 is assumed to hold.

Keeping in mind (4.39) for $p = 2$ and noticing that the kernel of the skew-symmetrization mapping $\mathcal{C}\mathrm{Diff}_{(1)}\{[\varkappa];[\varkappa]\} \to \mathcal{C}\mathrm{Diff}_{(2)}^{\mathrm{alt}}\{[\varkappa]\}$ coincides with the module $\mathcal{C}\mathrm{Diff}_{(2)}^{\mathrm{sym}}\{[\varkappa]\}$, it is also easy to see that $\ker v_\nabla^{(2)}$ is the image of the complex $\mathcal{C}\mathrm{Diff}_{(2)}^{\mathrm{sym}}\{[P]\}$ under the composition of mappings

$$\mathcal{C}\mathrm{Diff}_{(2)}^{\mathrm{sym}}\{[P]\} \hookrightarrow \mathcal{C}\mathrm{Diff}_{(1)}\{[P];[P]\} \xrightarrow{\mathcal{C}\mathrm{Diff}\{\nabla;\mathrm{id}\}} \mathcal{C}\mathrm{Diff}_{(1)}\{[\varkappa];[P]\},$$

which is obviously a monomorphism under the above assumptions.

Taking into account that in the case under consideration $H^{n-1}(K_0^{p,*}) = 0$ and using the isomorphism between $\ker v_\nabla^{(2)}$ and $\mathcal{C}\mathrm{Diff}_{(2)}^{\mathrm{sym}}\{[P]\}$ (see (4.42) for $p = 2$), we obtain

$$0 \to \mathcal{C}\mathrm{Diff}^{\mathrm{sym}}([P],[\hat{P}]) \to \ker\nabla_{(2)} \to H^{n-1}(K^{2,*}) = E_1^{2,n-1} \to 0, \tag{4.57}$$

where $\mathcal{C}\mathrm{Diff}^{\mathrm{sym}}(Q,\hat{Q})$ denotes the set of all selfadjoint operators acting from Q to $\hat{Q}$ and, according to Subsection 4.3.6,

$$H^n(\ker v_\nabla^{(2)}) = H^n(\mathcal{C}\mathrm{Diff}_{(2)}^{\mathrm{sym}}\{[P]\}) = \mathcal{C}\mathrm{Diff}^{\mathrm{sym}}([P],[\hat{P}]).$$

Since $\ker\nabla_{(2)} \subset H^n(\mathcal{C}\mathrm{Diff}_{(1)}\{[\varkappa];[P]\}) = \mathcal{C}\mathrm{Diff}\{[\varkappa],[\hat{P}]\}$, one can use (4.57) and consider $E_1^{2,n-1}$ as a subgroup in $\mathcal{C}\mathrm{Diff}([\varkappa],[\hat{P}])/\Theta$, where Θ is the image of the embedding

$$\mathcal{C}\mathrm{Diff}^{\mathrm{sym}}([P],[\hat{P}]) \to \mathcal{C}\mathrm{Diff}\{[\varkappa],[\hat{P}]\}, \quad \square \mapsto \square\circ\nabla.$$

PROPOSITION. *Let $\mathcal{Y} = \{\varphi = 0\} \subset J^k(\pi)$ be a normal equation and* $[f] \in \ker[\ell_\varphi^*]$, $f \in \hat{P}$. *Then, identifying $E_1^{1,n-1}$ with* $\ker[\ell_\varphi^*]$, *one has*

$$d_1^{1,n-1}([f]) = [\ell_f - \Delta^*] \bmod \Theta,$$

where $\ell_\varphi^*(f) = \Delta(\varphi)$, $\Delta \in \mathcal{C}\mathrm{Diff}(P,\hat{\varkappa})$.

PROOF. Note first that $\ell_\varphi^*(f) = 0$ on $\mathcal{Y}_\infty$. Since $\mathcal{Y}$ is a normal and consequently regular equation, there exists an operator Δ such that $\ell_\varphi^*(f) = \Delta(\varphi)$.

Consider the Green formula for the operator $f\circ\ell_\varphi \in \mathcal{C}\mathrm{Diff}(\varkappa,\bar{\Lambda}^n)$ in the form

$$(f\circ\ell_\varphi)(\chi) + (\ell_\varphi^*(f),\chi) = \bar{d}\square, \quad \chi\in\varkappa(\pi). \tag{4.58}$$

Then the isomorphism $\ker[\ell_\varphi^*] \to E_1^{1,n-1}$ takes the element $[f]$ to $\langle[\square]\rangle \in E_1^{1,n-1}$ and according to (4.54) one has

$$d_1^{1,n-1}\langle[\square]\rangle = [\langle l_{1,n-1}(\square)\rangle] = [l_{1,n-1}(\square)] \bmod K_0.$$

The image of $d_1^{1,n-1}\langle[\square]\rangle$ under the inclusion

$$E_1^{2,n-1} = H^{n-1}(K^{2,*}) \to H^n(K_0^{2,*})$$

(see (4.41)) equals $\langle e\rangle$, where

$$e = \bar{d}_{2,n-1}[\ell_{1,n-1}(\square)] = -[\ell_{1,n}(\bar{d}\circ\square)] = [l_{1,n}(f\circ\ell_\varphi + \ell_\varphi^*(f))], \quad \ell_\varphi^*(f)\in\hat{\varkappa},$$

is understood as an operator acting from $\varkappa$ to $\bar{\Lambda}^n$. Note that the last equality follows from (4.58).

But from (4.4) we have

$$l_{1,n}(f\circ\ell_\varphi) = (\ell_f,\ell_\varphi) - (\ell_\varphi,\ell_f),$$

where (ℓ_f,ℓ_φ), $(\ell_\varphi,\ell_f)\in\mathcal{C}\mathrm{Diff}_{(2)}(\varkappa;\bar{\Lambda}^n)$ and

$$(\ell_f,\ell_\varphi)(\chi_1,\chi_2) = (\ell_f(\chi_1),\ell_\varphi(\chi_2))$$

(similarly for (ℓ_φ,ℓ_f)). The bracket on the right-hand side of the last equality denotes the natural pairing $\hat{P}\times P\to\bar{\Lambda}^n$.

Using (4.4) again, we see that

$$l_{1,n}(h) = (\ell_h,\mathrm{id}_\varkappa) - (\mathrm{id}_\varkappa,\ell_h), \quad h\in\hat{\varkappa}\subset\mathcal{C}\mathrm{Diff}(\varkappa,\bar{\Lambda}^n).$$

If in addition $h = \ell_\varphi^*(f) = \Delta(\varphi)$, then $\ell_h = \ell_{\Delta(\varphi)} = \Delta\circ\ell_\varphi + \varphi^*\circ\Delta'$ for some operator $\Delta'\in\mathcal{C}\mathrm{Diff}(\varkappa,\mathrm{Hom}(P,\hat{\varkappa}))$.

Putting together the above relations, one obtains

$$e = [l_{1,n}(f\circ\ell_\varphi + \ell_\varphi^*(f))] = [(\ell_f,\ell_\varphi) - (\ell_\varphi,\ell_f) + (\Delta\circ\ell_\varphi,\mathrm{id}_\varkappa) - (\mathrm{id}_\varkappa,\Delta\circ\ell_\varphi)].$$

Therefore, the inverse image $\bar{e}$ of the element e under the natural projection

$$\mathcal{C}\mathrm{Diff}_{(1)}\{[\varkappa],[P]\}\to K_0^{2,*}\subset\mathcal{C}\mathrm{Diff}_{(2)}^{\mathrm{alt}}\{[\varkappa]\}, \quad \partial\mapsto(\partial\circ[\ell_\varphi])_{\mathrm{alt}},$$

obviously equals $(\ell_\varphi,\mathrm{id}_P)-(\mathrm{id}_\varkappa,\Delta)$. This shows that the corresponding cohomology class $\langle e\rangle\in\mathcal{C}\mathrm{Diff}([\varkappa],[\hat{P}])$ equals $\mu(e) = \ell_\varphi - \Delta^*$ (see Subsection 4.3.5). □

REMARK. It is useful to keep in mind that Θ consists of all operators of the form $\square\circ[\ell_\varphi^*]$, where $\square = \square^*\in\mathcal{C}\mathrm{Diff}([P],[\hat{P}])$.

The following result illustrates how the developed formalism is related to the linear theory of Section 1.5.

COROLLARY. *Let $\mathcal{Y}$ be a linear equation satisfying the assumptions of the proposition proved above. If $f = (\pi_\infty|_{\mathcal{Y}_\infty})^*(g)$, then $d_1^{1,n-1}([f]) = 0$, i.e., $[f]$ determines a conservation law for the equation $\mathcal{Y}$.*

PROOF. Indeed, in this case $\ell_f = 0$. Moreover, linearity of $\mathcal{Y}$ implies that $\varphi = \hat{\square}(\varphi_{\mathrm{id}})$ (see Subsection 2.1.2) so that $\ell_\varphi = \ell_{\hat{\square}(\varphi_{\mathrm{id}})} = \hat{\square}\circ\ell_{\varphi_{\mathrm{id}}} = \hat{\square}$. Thus $\ell_\varphi^*(f) = \hat{\square}^*(f) = \pi_\infty^*(\square(g))$. Therefore, the filtration of $\ell_\varphi^*(f)$ equals -1, while that of $\Delta(\varphi)$ is nonnegative if $\Delta\neq 0$. This proves that $\Delta = 0$. □

This corollary shows that for any normal linear equation $\mathcal{Y}$, a solution of the adjoint equation determines a conservation law. Moreover, conservation laws of such a type are independent (cf. Subsection 1.5.4).

4.5.6. Lagrangian formalism with constraints. Recall that in Subsection 4.3.8 we established the coincidence of the differential $d_1^{0,n}$ with the Euler operator for the Lagrangian formalism without constraints. Being of a functorial nature, this fact should remain true in the formalism with constraints as well. Here we show that this is really the case. To be more precise, we consider the situation where the quantities under variation satisfy a differential equation $\mathcal{Y} \subset J^k(E,n)$. In the notation of Subsection 4.2.1, this means that the domain B' related to this variational problem is an open subset in $\mathcal{Y}_\infty$.

Recall that in accordance with Subsections 4.2.1 and 4.2.2, Lagrangian density should be understood as a form $\omega \in \bar{\Lambda}^n(B, \partial B)$, while the Lagrangian is the cohomology class $\mathcal{L} = \langle\omega\rangle \in \bar{H}^n(B, \partial B)$.

In contrast to Section 4.2, where the arguments were based on the first variation formula (which is a naive approach to variational problems), we shall use the Stokes formula (4.16) restricted to $L \in M^n_{\mathrm{adm}}(E)$ (or, resp., to $\sigma \in \Gamma_{\mathrm{adm}}(\pi)$). A special character of variational problems with constraints is revealed by the fact that an infinitesimal variation of a solution L of the constraint equation (i.e., a "tangent vector" to "the space of solutions of the constraint equation" at the "point" L) is not usually represented as a value at L of a global vector field (i.e., of an element of $\operatorname{sym}\mathcal{Y}$). Therefore, we start with clarifying the meaning of such a "tangent vector". To do this, assume that $\mathcal{Y} = \{\varphi = 0\}$, $\varphi \in P$, and consider (in a special chart) the restriction $\ell_{\varphi,\sigma}$ of the operator ℓ_φ to a section $\sigma \in \Gamma_{\mathrm{loc}}(\pi)$: $\ell_{\varphi,\sigma} \in \operatorname{Diff}(\Gamma(\pi), P_\sigma)$, $P_\sigma = j(\sigma)^*(P)$. If $P = \mathcal{F}(\pi,\xi)$, then $P_\sigma = \Gamma(\xi)$ and $\ell_{\varphi,\sigma} = j(\sigma)^* \circ \ell_\varphi \circ \pi_\infty^*$. If, in addition, σ is a solution of the equation $\mathcal{Y}$, then any solution χ_σ of the equation $\ell_{\varphi,\sigma}(\chi_\sigma) = 0$ should be understood as a tangent vector (or as an infinitesimal variation of the solution σ) to the space of solutions of $\mathcal{Y}$ at the "point" σ. In the generic situation, an element $\chi_L \in j(L)^*[\varkappa]$, $\varkappa = \varkappa(J^k(E,n))$, will be called a *tangent vector* to the space of solutions of $\mathcal{Y}$ at L if χ_L is a solution of the equation $\ell_{\varphi,\sigma} = 0$ in a special chart, where the section σ is such that the graphs of $j(\sigma)$ and $j(L)$ coincide in this chart. The set of such "vectors" will be denoted by $\tau_L(\mathcal{Y})$, or by $\tau_\sigma(\mathcal{Y})$ if $\mathcal{Y} \subset J^\infty(\pi)$.

The "infinitesimal Stokes formula" needed below corresponds to the value of (4.16) for $p = 0$ on L (resp., on σ) and is of the form

$$\chi_L\{e\} = \chi_L \lrcorner\, d_1^{0,q}(e), \quad \chi_L \in \tau_L(\mathcal{Y}), \tag{4.59}$$

or, respectively,

$$\chi_\sigma\{e\} = \chi_\sigma \lrcorner\, d_1^{0,q}(e), \quad \chi_\sigma \in \tau_\sigma(\mathcal{Y}).$$

Both the right- and the left-hand side of these formulas belong to $H^q(L)$ (resp., to $H^q(\mathcal{U})$, where $\mathcal{U}$ is the domain of σ) and are defined in an obvious way. For example, $\chi_\sigma\{e\} = \langle \ell_{\rho,\sigma}(\chi_\sigma)\rangle$, where $e = \langle[\rho]\rangle$, $\rho \in \bar{\Lambda}^q(J^\infty(\pi))$. Formulas (4.59) are proved in the same way as (4.16).

When studying a variational problem, for "infinitesimal variations" of the manifold $L \in M^n_{\mathrm{adm}}(E)$ (resp., of $\sigma \in \Gamma_{\mathrm{adm}}(\pi)$) one should take the elements $\chi_L \in \tau_L(\mathcal{Y})$ (resp., $\chi_\sigma \in \tau_\sigma(\mathcal{Y})$) consistent with the eventual additional conditions of the problem. In other words, one should distinguish an appropriate subspace of "tangent vectors" in $\tau_L(\mathcal{Y})$ (resp., in $\tau_\sigma(\mathcal{Y})$). Such subspaces will be denoted by $\operatorname{Adm}(L)$ (resp., by $\operatorname{Adm}(\sigma)$). Their exact description depends on the type of the problem under consideration and usually causes no difficulty. For example, the boundary conditions of the initial variational problem in an obvious way lead to some boundary conditions for the equation $\ell_{\varphi,\sigma} = 0$. These conditions are to be satisfied by the solutions belonging to $\mathrm{Adm}(\sigma)$.

To deduce the Euler–Lagrange equations it remains to note that when the variation is $\chi_L \in \mathrm{Adm}(L)$ (resp., $\chi_\sigma \in \mathrm{Adm}(\sigma)$), the rate of changes of the functional under consideration, by definition, equals $\chi_L(\mathcal{L})$ (resp., $\chi_\sigma(\mathcal{L})$).

DEFINITION. A variational problem is said to be *regular* if the condition $\chi_L \lrcorner d_1^{0,n}(\mathcal{L}) = 0$ (resp., $\chi_\sigma \lrcorner d_1^{0,n}(\mathcal{L}) = 0$) for all $\chi_L \in \mathrm{Adm}(L)$ (resp., $\chi_\sigma \in \mathrm{Adm}(L)$) implies $j(L)^* d_1^{0,n}(\mathcal{L}) = 0$ (resp., $j(\sigma)^* d_1^{0,n}(\mathcal{L}) = 0$).

It should be noted that, strictly speaking, the operators $d_1^{1,n}$ in the definition above must be regarded as differentials of the $\mathcal{C}$-spectral sequence for the pair $(B, \partial B)$.

Keeping in mind the last definition and using the above considerations together with formula (4.59), we get the following

THEOREM. *A necessary condition for a manifold L (resp., a section σ) to be an extremal for a regular variational problem is $j(L)^*(d_1^{0,n}(\mathcal{L})) = 0$ (resp., $j(\sigma)^*(d_1^{0,n}(\mathcal{L})) = 0$).*

In other words, the equation $d_1^{0,n}(\mathcal{L}) = 0$ is an analog of the Euler–Lagrange equations with constraints. It should be noted that the terms $E_1^{1,n}(\mathcal{Y}) \ni d_1^{0,n}(\mathcal{L})$ are not modules over the base algebra $\mathcal{F}(\mathcal{Y})$, and this is why a well-defined interpretation for the expressions like $j(L)^*(e), e \in E_1^{1,n}(\mathcal{Y})$ is needed. We give it in the next subsection. Also we show there that Theorem 4.5.6 may be understood as a general result concerning Lagrange multipliers.

4.5.7. The terms $E_0^{p,q} = E_0^{p,q}(\mathcal{O})$ of the $\mathcal{C}$-spectral sequence for an object $\mathcal{O}$ of the category DE are obviously $\mathcal{F}(\mathcal{O})$-modules, and the differentials $d_0^{p,q}$ are $\mathcal{C}$-differential operators (see (4.2)). Therefore, the term $\{E_0, d_0\}$ can be restricted to the graph of the mapping $j(L)$ (resp., $j(\sigma)$). To be more precise, we consider restrictions $E_{0,L}^{p,q} = j(L)^*(E_0^{p,q})$ (resp., $E_{0,\sigma}^{p,q} = j(\sigma)^*(E_0^{p,q})$). Since the operators $d_0^{p,q}$ are $\mathcal{C}$-differential, they induce the operators $d_{0,L}^{p,q}\colon E_{0,L}^{p,q} \to E_{0,L}^{p,q+1}$ (resp., $d_{0,\sigma}^{p,q}\colon E_{0,\sigma}^{p,q} \to E_{0,\sigma}^{p,q+1}$). Obviously, $d_{0,L}^{p,q+1} \circ d_{0,L}^{p,q} = 0$ (resp., $d_{0,\sigma}^{p,q+1} \circ d_{0,\sigma}^{p,q} = 0$), and $\{E_{0,L}^{p,q}, d_{0,L}^{p,q}\}_{p=\mathrm{const}}$ (resp., $\{E_{0,\sigma}^{p,q}, d_{0,\sigma}^{p,q}\}_{p=\mathrm{const}}$) are complexes. Cohomology of these complexes at the term $E_{0,L}^{p,q}$ (resp., $E_{0,\sigma}^{p,q}$) is denoted by $E_{1,L}^{p,q}$ (resp., $E_{1,\sigma}^{p,q}$). Since these complexes are quotients of the complexes $\{E_0^{p,q}, d_0^{p,q}\}_{p=\mathrm{const}}$, one has the natural cohomology mappings $E_1^{p,q} \to E_{1,L}^{p,q}$ (resp., $E_1^{p,q} \to E_{1,\sigma}^{p,q}$), which are denoted by $j(L)^*$ (resp., $j(\sigma)^*$). This notation was used in this sense in the theorem and definition above.

Let us now clarify the meaning of the equation $d_1^{0,n}(\mathcal{L}) = 0$. To do this, we assume the constraint equation $\mathcal{Y}$ to be regular with respect to the operator ∇ and use the relation $E_1^{1,n} = \operatorname{coker} \nabla^*$ (Theorem 4.4.7) together with the natural restriction mappings $E_r^{p,q}(J^\infty(E,n)) \to E_r^{p,q}(B)$ for $B \subset \mathcal{Y}_\infty \subset J^\infty(E,n)$ (resp., $E_r^{p,q}(J^\infty(\pi)) \to E_r^{p,q}(B)$ for $B \subset \mathcal{Y}_\infty \subset J^\infty(\pi)$). Thus, if $\mathcal{L} = \langle \omega \rangle$ and $\omega = [\rho]$, $\rho \in \bar{\Lambda}^n(J^\infty(\pi))$ (resp., $\rho \in \bar{\Lambda}^n(J^\infty(\pi))$), then

$$d_1^{0,n}(\mathcal{L}) = \{[d_1^{0,n}\langle \rho \rangle] \bmod \operatorname{im} \nabla^*\} \in [\hat{\varkappa}].$$

In the case $\mathcal{Y} = \{\varphi = 0\} \subset J^k(\pi)$ one has

$$d_1^{0,n}(\mathcal{L}) = [\ell_\rho^*(1)] \bmod \operatorname{im} \nabla^*.$$

Therefore, denoting by ∇_σ the restriction of the operator ∇ to the section σ, one sees that the equality $j(\sigma)^* d_1^{0,n}(\mathcal{L}) = 0$ is equivalent to

$$j(\sigma)^*(\ell_\rho^*(1)) + \nabla_\sigma^*(\hat{p}_\sigma) = 0$$

for some $\hat{p}_\sigma \in j(\sigma)^*(\hat{P})$ or to

$$j(\sigma)^*\big([\ell_\rho^*(1)] + \nabla^*([\hat{p}])\big) = 0$$

for some $\hat{p} \in \hat{P}$. If $\mathcal{Y} = \{\varphi = 0\}$ and $\nabla = \ell_\varphi$, then

$$[\ell_{(\hat{p},\varphi)}] = [(\ell_{\hat{p}}, \varphi) + (\hat{p}, \ell_\varphi] = [(\hat{p}, \ell_\varphi)] = [\hat{p}^* \circ \ell_\varphi].$$

Hence, $\nabla^*([\hat{p}]) = [\ell_\varphi^*(\hat{p})] = [(\hat{p}^* \circ \ell_\varphi)^*(1)] = [\ell_{(\hat{p},\varphi)}^*(1)]$. Consequently, the extremality condition for the section σ takes the form

$$j(\sigma)^*\big([\ell_\rho^*(1)] + [\ell_{(\hat{p},\varphi)}]^*(1)\big) = j(\sigma)^*\big(\ell_{\rho+(\hat{p},\varphi)}^*(1)\big) = 0.$$

Since $\ell_{\rho+(\hat{p},\varphi)}^*(1) = \mathcal{E}(\bar{\mathcal{L}})$, where $\bar{\mathcal{L}} = \langle \rho + (\hat{p}, \varphi) \rangle$, we obtain the following

THEOREM. *Any extremal of the original variational problem is an extremal for the "free" Lagrangian $\bar{\mathcal{L}} = \bar{\mathcal{L}}(\hat{p})$ for some $\hat{p} \in \hat{P}$.*

This theorem gives an interpretation of Theorem 4.5.6 as a theorem on Lagrange multipliers. In fact, the element $\hat{p} \in \hat{P}$ plays the role of a "Lagrange multiplier" λ in this situation.

Hence, in order to solve the original variational problem, one has to specify elements $\hat{p} \in \hat{P}$ in the theorem above. This can be done as follows. First, consider the system

$$\begin{cases} \varphi = 0, \\ \mathcal{E}(\bar{\mathcal{L}}(\hat{p})) = 0. \end{cases} \tag{4.60}$$

This system is incompatible for a generic $\hat{p}$. Hence, one has to solve the compatibility conditions in order to find admissible Lagrange multipliers $\hat{p}$. Further, the extremals of the initial problem are found as the solutions of system (4.60) for admissible $\hat{p}$ satisfying the corresponding additional conditions (e.g., boundary conditions).

4.5.8. The obtained results allow one to pose the inverse problem of the calculus of variations with constraints and to formulate criteria of its solvability. Namely, a nonlinear operator determined by the element $\mathcal{E}(\mathcal{L})$, where $\mathcal{L}$ is the same as in the previous two subsections, is called the Euler–Lagrange operator for problems with constraints. Then one has

THEOREM. *If $e \in E_1^{1,n}(B)$ is the Euler–Lagrange operator for some variational problem, then $d_1^{1,n}(e) = 0$. This condition is sufficient if the constraint equation is conic.*

PROOF. The result follos from the exactness of complex (4.44) and Corollary 4.4.13. □

Obviously, this fact is useful for an analysis of a more delicate problem: is an equation $e = 0$, $e \in E_1^{1,n}$ an Euler–Lagrange equation (cf. Subsection 4.3.9)?

Acyclicity of complex (4.44) for conic equations leads to the following result.

PROPOSITION. *Consider a nonconstraint variational problem with a Lagrangian $\mathcal{L}$ and let $\mathcal{Y} = \{\mathcal{E}(\mathcal{L}) = 0\}$. If the equation $\mathcal{Y}$ is normal and conic, and $\bar{H}^n(\mathcal{Y}) = 0$, then $\mathcal{L}|_{\mathcal{Y}_\infty} = 0$.*

COROLLARY ("Generalized Schwarz formula"). *Let $\mathcal{Y}$ be as in the proposition above. Then there exists $\theta \in \bar{\Lambda}^{n-1}(\mathcal{Y}_\infty)$ such that*

$$\int_{\mathcal{U}_\sigma} \sigma^*(\omega) = \int_{\partial\mathcal{U}_\sigma} \sigma^*(\theta)$$

for any extremal σ.

PROOF. Indeed, if $\mathcal{L} = \langle\omega\rangle$, then the equality $\mathcal{L}|_{\mathcal{Y}_\infty} = 0$ means that $\omega = \bar{d}\theta$ on $\mathcal{Y}_\infty$, i.e., if $\sigma\colon \mathcal{U}_\sigma \to E_\pi$ is an extremal, then $\int_{\mathcal{U}_\sigma} \sigma^*(\omega) = \int_{\partial\mathcal{U}_\sigma} \sigma^*(\theta)$. □

Note that the generalized Schwarz formula is valid if and only if $\mathcal{L}|_{\mathcal{Y}_\infty} = 0$, where $\mathcal{Y} = \{\mathcal{E}(\mathcal{L}) = 0\}$. In additioin, the form θ can be found explicitly via the infinitesimal Stokes formula in $E_1(\mathcal{Y}_\infty)$. For example, if the equation is conic, the answer is given by formula (4.34).

REMARK. Of course, the results of this subsection are valid not just for the class of conic equations but for a much larger class. Conic equations are distinguished by the fact that the required acyclicity of the complex (4.44) can be proved using the simplest homotopy.

4.6. Characteristic classes

Let $\# = \#_S\colon E_\# \to \mathbb{R}^n$ be the bundle of geometric structures of the type S over $\mathbb{R}^n$ (see Section 3.5). Consider a smooth manifold M, $\dim M = n$, and a point $x \in M$. Let $\pi\colon E \to M$ be the natural bundle of the type S. Then an arbitrary local diffeomorphism $H\colon \mathbb{R}^n \to M$ satisfying $H(0) = x$ allows one to identify $J_x^\infty(\pi)$ with $J_0^\infty(\#)$ and $G_x^\infty M$ with $\mathbf{G}^\infty(n)$, where

$$\mathbf{G}^\infty(n) = \{\, [\Phi]_0^\infty \mid \Phi \in \operatorname{Diffeo}_0 \mathbb{R}^n \,\}$$

is the differential group of infinite order and

$$G_x^\infty M = \{\, [\Phi]_x^\infty \mid \Phi \in \operatorname{Diffeo}_x M \,\}.$$

These identifications lead to the diffeomorphism $J_x^\infty(\pi)/G_x^\infty M = J_0^\infty(\#)/\mathbf{G}^\infty(n)$, which is obviously independent of H. As a result, we obtain a canonical diffeomorphism $J^\infty(\pi)/\operatorname{Diffeo} M = J_0^\infty(\#)/\mathbf{G}(n)$. The diffiety $J_0^\infty(\#)/\mathbf{G}(n)$ is called the *characteristic diffiety* of geometric structures of the type S and is denoted by $\operatorname{Ch}(S)$. For a given structure $\omega\colon M \to E$, the composition

$$M \xrightarrow{j_\infty(\omega)} J^\infty(\pi) \to J^\infty(\pi)/\operatorname{Diffeo} M = \operatorname{Ch}(S)$$

is called the *characteristic mapping* and is denoted by $\zeta(\omega)\colon M \to \operatorname{Ch}(S)$. Using this mapping, it seems to be natural to define the characteristic class of a geometric structure ω corresponding to a cohomology class $\nu \in H^*(\operatorname{Ch}(S))$ as the inverse image $\zeta(\omega)^*(\nu) \in H^*(M)$. But this definition is void, since the diffiety $\operatorname{Ch}(S)$ is usually contractible (as in the case of pseudo-Riemannian metrics, for example). The following considerations show how to improve this definition.

Note that the mapping $\operatorname{Diffeo} M \ni F \mapsto \hat{F} \in \operatorname{Diffeo} E$ (see p. 122 in Section 3.5) is a differential operator. Its order p is called the *order of a geometric structure* of the type S. If a structure of order p is homogeneous, the group

$$\mathbf{G}^p(n) = \{ [\Phi]_0^p \mid \Phi \in \operatorname{Diffeo}_0 \mathbb{R}^n \}$$

acts transitively on the fiber $\#^{-1}(0)$ of the bundle $\#$. This implies, first, that, up to an isomorphism, the stabilizer $\mathbf{G}_y^p(n) \subset \mathbf{G}^p(n)$ of a point $y \in \#^{-1}(0)$ is independent of the choice of y. Let us denote this stabilizer by $\mathbf{G}(S)$. Secondly, the diffiety $\operatorname{Ch}(S)$ is canonically identified with the quotient space $\#_{\infty,0}^{-1}(y)/\mathbf{G}_y^\infty(n)$, where $\mathbf{G}_y^\infty(n)$ is the inverse image of the stabilizer $\mathbf{G}_y^p(n) \subset \mathbf{G}^p(n)$ under the natural mapping $\mathbf{G}^\infty(n) \to \mathbf{G}^p(n)$. Further, let us temporarily assume that the group $\mathbf{G}^\infty(n)$ acts freely on the space $J_0^\infty(\#)$. This assumption implies, in particular, that the projection

$$\mu\colon \#_{\infty,0}^{-1}(y) \to \#_{\infty,0}^{-1}(y)/\mathbf{G}_y^\infty(n) = \operatorname{Ch}(S)$$

is a principal $\mathbf{G}_y^\infty(n)$-bundle. Since the space $\#_{\infty,0}^{-1}(y)$ is contractible, $\operatorname{Ch}(S)$ has the homotopy type of the classifying space of the group $\mathbf{G}_y^\infty(n)$. But the group $\mathbf{G}^\infty(n)$ is obtained from $\mathbf{G}^p(n)$ as a result of extensions by groups of the form $\mathbb{R}^r$. Therefore, the groups $\mathbf{G}_y^\infty(n)$ and $\mathbf{G}_y^p(n)$ are of the same homotopy type and consequently have the homotopy type of $\mathbf{G}(S)$. Thus, our assumption leads to the conclusion that the characteristic diffiety $\operatorname{Ch}(S)$ is of the same homotopy type as the classifying space $\mathrm{B}\mathbf{G}(S)$ of the group $\mathbf{G}(S)$. From this we would obtain a natural theory of characteristic classes for homogeneous geometric structures.

Of course, this assumption does not always hold, because, for example, the automorphism group of a geometric structure may be nontrivial. Nevertheless, the following statement is valid:

LEMMA. *The subset $J_{0,\mathrm{sing}}^\infty(\#)$ of $J_0^\infty(\#)$ where the action of $\mathbf{G}^\infty(n)$ is not free, has infinite codimension.*

We shall prove this result a bit later. Now we note that it saves the above arguments. Namely, it suffices to pass from the space $J_0^\infty(\#)$ to its "regular part" $J_{0,\mathrm{reg}}^\infty(\#) = J_0^\infty(\#) \setminus J_{0,\mathrm{sing}}^\infty(\#)$. In this case, $\#_{\infty,0}^{-1}$ must be replaced by $\#_{\infty,0}^{-1} \cap J_{0,\mathrm{reg}}^\infty(\#)$. This will lead to the *regular characteristic diffiety* $\operatorname{Ch}_{\mathrm{reg}}(S) = J_{0,\mathrm{reg}}^\infty(\#)/\mathbf{G}^\infty(n)$. Hence we obtain

PROPOSITION. *The diffiety $\operatorname{Ch}_{\mathrm{reg}}(S)$ has the homotopy type of the classifying space $\mathrm{B}\mathbf{G}(S)$ of the group $\mathbf{G}(S)$.*

Let $\nu \in H^*(\operatorname{Ch}_{\mathrm{reg}}(S))$ be a de Rham cohomology class of the diffiety $\operatorname{Ch}_{\mathrm{reg}}(S)$ and ω a geometric structure of the type S on a manifold M, $\dim M = n$. We define the ν-characteristic class of the structure ω as the element $\bar{\zeta}(\omega)^*(\nu) \in H^*(M)$, where $\bar{\zeta}(\omega)\colon M \to \operatorname{Ch}_{\mathrm{reg}}(S)$ is a *regularization* of the characteristic mapping $\zeta(\omega)$. To be more precise, $\bar{\zeta}(\omega) = \mu \circ j_\infty(\omega')$, where ω' is a small deformation of the structure such that $j_\infty(\omega') \cap J_{0,\mathrm{sing}}^\infty(\#) = \emptyset$. Since $\operatorname{codim} J_{0,\mathrm{sing}}^\infty(\#) = \infty$, such deformations exist and they are homotopic to each other in $J_{0,\mathrm{reg}}^\infty(\#)$.

REMARK. The regular characteristic diffiety is an n-dimensional stratum of the characteristic orbifold diffiety $\operatorname{Ch}(S)$.

EXAMPLE. Within this approach, Pontryagin's characteristic classes can be obtained as characteristic classes of Riemannian metrics. The characteristic classes of homotopic metrics obviously coincide. Since on a given manifold, there exists only one homotopy class of Riemannian metrics, the corresponding characteristic classes are independent of the choice of a metric. Therefore, such characteristic classes are invariants of the manifold itself. This is a paraphrase of the classical theory of characteristic classes based on the sequence

$$\boxed{\text{Riemannian metric}} \Rightarrow \boxed{\text{Levi-Civitta connection}}$$

$$\Rightarrow \boxed{\text{Curvature tensor}} \Rightarrow \boxed{\text{Chern–Weyl homomorphism}}$$

EXAMPLE. Characteristic classes of 3-webs on two-dimensional manifolds are trivial, and the diffiety $\mathrm{Ch}_{\mathrm{reg}}(S)$ (as well as $\mathrm{Ch}(S)$) is homotopy trivial.

The above lemma is a direct consequence of the following statement.

PROPOSITION. *There exists an integer k such that $J^\infty_{0,\mathrm{sing}}(\#) \subset \mathcal{E}_\infty$, where $\mathcal{E} \subset J^{k+1}(\#)$ is an analytic subset of nonzero codimension.*

PROOF. Let $Q \subset J^k_0(\#)$, $0 \le k \le \infty$, be a subset and $r(Q) = \bigcup_{x\in\mathbb{R}^n} \hat{r}^{(k)}_x(Q)$, where $r_x\colon \mathbb{R}^n \to \mathbb{R}^n$ is a translation by the vector $x \in \mathbb{R}^n$. The correspondence $Q \mapsto r(Q)$ obviously establishes a bijection between $\mathbf{G}^{p+k}(n)$-invariant subsets in $J^k_0(\#)$ and $\mathrm{Diffeo}(M)$-invariant subsets in $J^k(\#)$ and preserves their codimension, if the latter is defined.

We denote by $\bar{\mathbf{G}}^p(n)$, $1 \le p \le \infty$, a natural extension of the group $\mathbf{G}^p(n)$ by the group of translations of the space $\mathbb{R}^n$. It is easy to see that the orbits of the group $\mathrm{Diffeo}\, M$ in $J^k(\#)$ coincide with the orbits of the natural action of $\bar{\mathbf{G}}^{p+k}(n)$ in $J^k(\#)$. Therefore, the invariants of the group $\bar{\mathbf{G}}^\infty(n)$ in $J^\infty(\#)$ coincide with the differential invariants of geometric structures of the type S. Denote by $J^\infty_{\mathrm{sing}}(\#) \subset J^\infty(\#)$ the subset where the action of $\bar{\mathbf{G}}^\infty(n)$ is not free. Then obviously

$$J^\infty_{0,\mathrm{sing}}(\#) \subset r\Big(J^\infty_{0,\mathrm{sing}}(\#)\Big) \subset J^\infty_{\mathrm{sing}}(\#),$$

and thus it suffices to show that $J^\infty_{\mathrm{sing}}(\#) \subset \mathcal{E}_\infty$ for an appropriate $\mathcal{E}$.

We now recall some general facts. Let a (real) Lie group G act analytically on a (real) analytic manifold W. Then generic (and of maximal dimension) orbits span everywhere dense set in W. Further, any (real) analytic subset is stratified by (perhaps open) analytic manifolds.

We shall apply these results to the action of the group $G = \bar{\mathbf{G}}^{p+k}(n)$ on the manifold $W = J^k(\#)$, which is analytic. Indeed, if $k = 0$, the manifold $W = J^0(\#)$ is identified with the homogeneous space $\bar{\mathbf{G}}^p(n)/(\mathbf{G}(S) \times \mathbb{R}^n)$ of the group $\bar{\mathbf{G}}^p(n)$. Hence, the action is analytic (see [**31**]). For $k > 0$, it is necessary to use the fact that the action of the group $\bar{\mathbf{G}}^{p+k}(n)$ on $J^k_0(\#)$ is obtained from the action of $\mathbf{G}^p(n) \subset \bar{\mathbf{G}}^p(n)$ on $J^0_0(\#) = \#^{-1}(0) \subset J^0(\#)$ by a purely algebraic procedure of extension (see [**2**] and also [**58**, Chapter 1]). We denote by Σ_k the singular set of the $\bar{\mathbf{G}}^{p+k}(n)$-action on $J^k(\#)$. By the above-said, this set is a stratified analytic submanifold.

Further, the condition for differential invariants $I_1, \dots, I_n \in C^\infty(J^k(\#))$ to be independent of a structure ω in the vicinity of a point $x \in \mathbb{R}^n$ is equivalent to $\bar{d}I_1 \wedge \dots \wedge \bar{d}I_n\big|_\rho \neq 0$, where $\rho = [\omega]^{k+1}_x$ and $\bar{d}$ denotes the horizontal de Rham differential. Let us define the set $\Delta_{k+1} \subset J^{k+1}(\#)$ by the following condition: $\rho \in \Delta_{k+1}$ if and only if for any set $I_1, \ldots, I_k$ of analytic differential invariants defined in the vicinity of the point $\#_{k+1,k}(\rho) \in J^k(\#)$ one has $\bar{d}I_1 \wedge \cdots \wedge \bar{d}I_n\big|_\rho = 0$. Obviously, Δ_{k+1} is an analytic set in $J^{k+1}(\#)$ invariant under the action of Diffeo$(\mathbb{R}^n)$. Moreover, if the number k is such that our differential invariants are realized by functions on $J^k(\#)$, then $\operatorname{codim} \Delta_{k+1} \geq 1$. Indeed, let $\rho \in \Delta_{k+1}$ and $\rho' = \#_{k+1,k}(\rho) \notin \Sigma_k$. Then generic orbits of the group $\bar{\mathbf{G}}^{p+k}(n)$ in $J^k(\#)$ are of codimension $\geq n$. Therefore, in the vicinity of the point ρ' one can construct n analytic invariants $I_1, \ldots, I_n$ satisfying $\bar{d}I_1 \wedge \cdots \wedge \bar{d}I_n \neq 0$. This implies that the points $\tau \in J^{k+1}(\#)$ for which $\bar{d}I_1 \wedge \cdots \wedge \bar{d}I_n\big|_\tau = 0$ constitute an analytic hypersurface in $J^{k+1}(\#)$ and the latter contains a neighborhood of the point ρ in Δ_{k+1}. This proves the desired statement, since the set Σ_k is analytic and $\operatorname{codim} \Sigma_k \geq 1$.

It follows directly from definitions that a point θ lies in $J^\infty_{\text{sing}}(\#)$ if and only if there exists a vector field X on $\mathbb{R}^n$ such that

$$\hat{X}^{(\infty)}_\theta \in \mathcal{C}_\theta \tag{4.61}$$

(see Section 3.5 for the definition of $\hat{X}$) and $[X]^\infty_a \neq 0$ for $a = \#_\infty(\theta)$. We represent the field $\hat{X}^{(\infty)}_\theta$ in the form $\hat{X}^{(\infty)}_\theta = \mathcal{Э}_\varphi + X_\infty$. Then the condition (4.61) is equivalent to $\mathcal{Э}_\varphi|_\theta = 0$, or, which is the same, to $D_\sigma(\varphi) = 0$ for any σ. This implies that $\theta \in \mathcal{E}'_\infty$, where $\mathcal{E}'$ is the equation $\varphi = 0$. Thus, we obtain

$$J^\infty_{\text{sing}}(\#) \subset \mathcal{E}'_\infty. \tag{4.62}$$

Let us now choose the number k as above. Then $\theta_{k+1} = \#_{\infty,k+1}(\theta) \in \Delta_{k+1}$. Indeed, if $\theta_{k+1} \notin \Delta_{k+1}$ and $\theta = [\omega]^\infty_a$, then in the vicinity of $a \in \mathbb{R}^n$ there exists a local coordinate system (y_i) composed of differential invariants of the structure ω. If Γ_t is a local one-parameter group generated by the field X, the functions $y_i \circ \Gamma_t$ also form a local coordinate system in a neighborhood of the point $\Gamma_t(a)$. For small t, this system differs from the system (y_i) on the level of l-jets, where l is such that $[X]^l_a \neq 0$. But the functions $y_i \circ \Gamma_t$ are differential invariants of the structure ω_t obtained from ω by the action of Γ_t. Consequently, $[X(\omega)]^l_a \neq 0$, where $X(\omega) = d\omega/dt|_{t=0}$ is the Lie derivative of the structure ω along the field X. Since

$$[X(\omega)]^\infty_a = \big(j_\infty(\omega)^*(\varphi)\big)(a) = \varphi(\theta) = 0,$$

we arrive at a contradiction.

Thus, $\#_{\infty,k+1}(\mathcal{E}'_\infty) \subset \Delta_{k+1}$. Therefore $\mathcal{E}'_\infty \subset \mathcal{E}_\infty$, where $\mathcal{E} = \Delta_{k+1}$. Together with (4.62) this implies $J^\infty_{\text{sing}}(\#) \subset \mathcal{E}_\infty$. □

Now we compute the groups $E_1^{p,q}(\mathrm{Ch}_{\mathrm{reg}}(S))$. This computation is based on the following property of the $\mathcal{C}$-spectral sequence: for any diffiety $\mathcal{O}$, an element $\rho \in E_1^{p,q}(\mathcal{O})$, $p > 0$, vanishes if and only if all its restrictions $\rho|_{\mathcal{U}_\alpha} \in E_1^{p,q}(\mathcal{U}_\alpha)$ vanish for any open cover $\{\mathcal{U}_\alpha\}$ of $\mathcal{O}$, $\mathcal{O} = \bigcup_\alpha \mathcal{U}_\alpha$. Thus, we start with a local description of $\mathrm{Ch}_{\mathrm{reg}}(S)$.

We recall the construction in Section 3.5. Let $x_1, \ldots, x_n$ be independent scalar invariants of geometric structures of the type S on n-dimensional manifolds. We define new invariants p^j_σ in the following way. Let a structure ω on a manifold M be generic with respect to the invariants x_i, i.e., the functions $x_1(\omega), \ldots, x_n(\omega)$ form a local coordinate system on M. Define the value of the invariant p^j_σ on ω by

$$p^j_\sigma(\omega) = \frac{\partial^{|\sigma|}\omega^j}{\partial x_\sigma(\omega)}(x_1(\omega), \ldots, x_n(\omega)),$$

where ω^j are the components of ω in the local chart $x_1(\omega), \dots, x_n(\omega)$. It is easy to see [**2**, Chapter 7] that any differential invariant is a function of x_i and p^j_σ. On the other hand, these invariants are related by equations

$$\psi_i(\dots, p^j_\sigma, \dots) = x_i, \qquad i = 1, \dots, n, \tag{4.63}$$

where ψ_i is a local expression for the invariant x_i. Hence the algebra $\mathcal{A}$ of all scalar differential invariants coincides locally with the function algebra $\mathcal{F}(\mathcal{Y}_\infty)$, where the equation $\mathcal{Y}$ is given by system (4.63). In this way, we construct a cover of the diffiety $\mathrm{Ch}_{\mathrm{reg}}(S)$ by open charts of the form $\mathcal{Y}_\infty$.

Consider an arbitrary chart $\mathcal{Y}_\infty$ in $\mathrm{Ch}_{\mathrm{reg}}(S)$. Let $\mathcal{U}$ be the inverse image of $\mathcal{Y}_\infty$ under the natural projection $\xi\colon J^\infty(\pi) \to \mathrm{Ch}(S)$, where π is the bundle of geometric structures of the type S on some manifold M. Choose a special coordinate system on $J^\infty(\pi)$ and define a mapping $\gamma\colon \mathcal{Y}_\infty \to \mathcal{U}$ by the following condition: the coordinates of the point $\gamma(\theta)$, $\theta \in \mathcal{Y}_\infty$, coincide with the coordinates (x_i, p^j_σ) of the point θ. We note that γ is a morphism in the category of differential equations and $\xi \circ \gamma = \mathrm{id}_{\mathcal{Y}_\infty}$.[1]

Hence the homomorphism $\xi^{p,q}_r\colon E^{p,q}_r(\mathcal{Y}_\infty) \to E^{p,q}_r(\mathcal{U})$ of the $\mathcal{C}$-spectral sequences possesses a left inverse, $\gamma^{p,q}_r$. Consequently, $\xi^{p,q}_r$ is an embedding. Since $\mathcal{U}$ is an open set in $J^\infty(\pi)$, one has $E^{p,q}_1(\mathcal{U}) = 0$ for $p > 0$ and $q \neq n$. Hence, $E^{p,q}_1(\mathcal{Y}_\infty) = 0$ for $p > 0$ and $q \neq n$ as well. This is valid for any chart $\mathcal{Y}_\infty$ in $\mathrm{Ch}_{\mathrm{reg}}(S)$, and therefore for the diffiety $\mathrm{Ch}_{\mathrm{reg}}(S)$ itself.

PROPOSITION. *One has $E^{p,q}_1(\mathrm{Ch}_{\mathrm{reg}}(S)) = 0$ for all $p > 0$ and $q \neq n$.*

This result implies that $E^{p,q}_2(\mathrm{Ch}_{\mathrm{reg}}(S)) = 0$ for $p > 0$ and $q \neq n$, which means that $E^{p,n}_2(\mathrm{Ch}_{\mathrm{reg}}(S)) = E^{p,n}_\infty(\mathrm{Ch}_{\mathrm{reg}}(S)) = H^{p+n}(\mathrm{Ch}_{\mathrm{reg}}(S))$. As a consequence, we obtain

PROPOSITION. *All characteristic classes of bordisms of geometric structures of the type S reduce to standard characteristic classes of $\mathbf{G}(S)$-bundles.*

Further, for $q < n$ one has

$$\bar{H}^q(\mathrm{Ch}_{\mathrm{reg}}(S)) = E^{0,q}_1(\mathrm{Ch}_{\mathrm{reg}}(S)) = E^{0,q}_2(\mathrm{Ch}_{\mathrm{reg}}(S)) = H^q(\mathrm{Ch}_{\mathrm{reg}}(S)),$$

which gives us the following result:

PROPOSITION. *Consider geometric structures ω of the type S on a manifold M, $\dim M = n$. Then*

1. *For $i < n$, all i-dimensional characteristic classes are inverse images*

$$\xi(\omega)^*(\eta) \in H^i(M)$$

of horizontal cohomology classes $\eta \in \bar{H}^i(\mathrm{Ch}_{\mathrm{reg}}(S))$.

2. *For $i = n$, they are of the form $\xi(\omega)^*(\eta) \in H^n(M)$, where*

$$\eta \in E^{0,n}_2(\mathrm{Ch}_{\mathrm{reg}}(S)) \subset \bar{H}^n(\mathrm{Ch}_{\mathrm{reg}}(S))$$

are trivial Lagrangians on $\mathrm{Ch}_{\mathrm{reg}}(S)$.

REMARK. For the case of pseudo-Riemannian metrics, similar results were obtained in different terms by Gilkey [**41**]. Following Tsujishita [**110, 112**], these results can be formulated as follows. Consider the manifold $J^\infty(\pi)$, where π is the bundle of pseudo-Riemannian metrics of a given type, and reduce the algebra

[1]The mapping γ realizes the so-called *slice* commonly used in physics.

of smooth functions on $J^\infty(\pi)$ to the algebra of functions polynomial in jet variables. In the corresponding "polynomial" $\mathcal{C}$-spectral sequence, consider the part $E_r^{p,q}(\pi)_{\mathrm{inv}}$ invariant with respect to diffeomorphisms of the base of π. Then the groups $E_1^{0,q}(\pi)_{\mathrm{inv}}$, $q < n$, and $E_2^{0,n}(\pi)_{\mathrm{inv}}$ are generated by the Pontryagin and Euler classes. Within this approach, characteristic classes are treated as cohomology of the invariant part of the de Rham complex on $J^\infty(\pi)$.

4.7. Concluding remarks

The problem of a *conceptual* definition of the Euler operator is by no means trivial, and this is demonstrated by its solution, the $\mathcal{C}$-spectral sequence. For this reason, it is no wonder that after discovering this sequence in 1977 [**129, 121**], the author was able to solve some old problems related to the Euler operator and presented in this chapter. The history of these problems, e.g., of the inverse problem of the calculus of variations, is rather instructive from many points of view. The main lesson is that one correct definition may bring a number of, sometimes nontrivial, theorems. For more than 100 years since this problem was posed by Helmholtz, numerous results of various generality were obtained (see, for example, [**33, 1, 102, 75, 35, 48, 108, 109, 52, 100**]). The work by Tulczyjew [**114**] and Kupershmidt [**77**] (an unpublished original version of the latter circulated in Moscow since 1977, while the English version on the differential algebra language appeared in Manin's survey [**88**]) are culminating points of this period. It is these papers where a construction of the complex starting with the Euler operator was given (similar to the construction of the de Rham complex starting with the differential of functions) and local exactness of this complex was proved. The only global result on the inverse problem preceding the discovery of the $\mathcal{C}$-spectral sequence, is due to Takens [**107**] (his preprint appeared a few months before [**121**]).

Thus the "one-line theorem" [**121**] (see Subsection 4.3.7), being in essence a direct consequence of the construction of the $\mathcal{C}$-spectral sequence on the manifold of infinite jets, not only delivers a complete synthesis of the earlier local results, but also gives a complete global solution to the inverse problem itself and to all its higher analogs. It is worth noting that in the constrained situation (see Subsection 4.5.6) no results on the inverse problem were obtained before [**121**].

The method of the $\mathcal{C}$-spectral sequence, as shown in this chapter, allows one to construct a systematic theory of conservation laws for arbitrary systems of partial differential equations. One rather particular case is the method given by the first Noether variational theorem. Until [**121**], it was practically the only method of finding conservation laws of particular differential equations. In [**19**], the reader will find numerous applications of the $\mathcal{C}$-spectral sequence theory. In this respect, it is interesting to note that the strikingly simple method of finding conservation laws (by solving the equation $\ell^*_{\mathcal{Y}}\varphi = 0$) is little known to people working in mathematical physics and applications. This relates not only to conservation laws. For example, anti-fields in the anti-field formalism are just the entities appearing in the first column of the $\mathcal{C}$-spectral sequence for the Yang–Mills equations; see [**10, 49**].

The above results concerning the structure of the $\mathcal{C}$-spectral sequence show that its complexity, i.e., the number of nontrivial lines, is proportional to the complexity of the δ-Spencer cohomology of the universal linearization operator. In fact, on the basis of the method developed in this chapter, it is possible to obtain much more precise results for equations of a particular interest. Since the discovery of the $\mathcal{C}$-spectral sequence, much attention was paid to the term $E_1^{1,n-1} = \ker \ell_\varphi^*$ responsible for conservation laws. The reader will find many concrete results and basic techniques used so far in computations of conservation laws in [**127, 139**] and [**19**]. Some efficient methods developed recently and presented in the next chapter allow one to get other results (see [**39, 40, 118**]). By contrast, just a few works were dedicated to complete computations of the terms $E_1^{p,q}$ different from $E_1^{1,n-1}$ (see, for instance, [**4, 56**]).

Some equivariant versions of the $\mathcal{C}$-spectral sequence were approached by several authors [**5, 112, 113, 119**]. Results equivalent to computation of some term $E_1^{p,q}$ were obtained by M. Henneaux and his collaborators using a different approach. We mention the works [**10, 49, 23**] as typical ones.

Results equivalent to those of this chapter were obtained by Bryant and Griffiths in the context of exterior differential systems; see [**25, 26, 27**].

CHAPTER 5

Introduction to Secondary Calculus

The theory of higher symmetries and the $\mathcal{C}$-spectral sequence demonstrate a remarkable analogy with the calculus of vector fields and differential forms on a smooth manifold. Analyzing the nature of this analogy, one can observe that this is just a small piece of a much larger picture that involves not only vector fields and differential forms, but the entire differential calculus. A counterpart of classical calculus arising from this analogy is called *Secondary Calculus*. In this chapter, we give an introduction to Secondary Calculus by exhibiting the key ideas and main results obtained in this field up to now. Our presentation is expository, and in particular, no proofs are given. We start with a discussion of the above-mentioned analogy in more details.

A remark concerning the notation used in this chapter is necessary. First, we denote the terms of the $\mathcal{C}$-spectral sequence by $\mathcal{C}E_r^{p,q}$ to distinguish them from the terms of other spectral sequences considered simultaneously. Second, sometimes it is convenient to denote the i-th cohomology of a complex by $H^i(\partial)$, where ∂ stands for its differential. For instance, $H^i(d)$ is the alternative notation for the i-th de Rham cohomology. Third, the following convention concerning signs is adopted:

the symbol of a graded object appearing in the exponent of (-1)
means the grading of this object.

It should also be emphasized that there are two kinds of $\mathcal{F}(\mathcal{O})$-modules we will consider. "Passive" modules belong to the FG-category over $\mathcal{O}$, as before, whereas "active" modules acting on passive ones belong, in a sense, to the dual category composed of geometric *co-filtered* modules. This is rather important for a rigorous treatment of the material. Nevertheless, this aspect is almost irrelevant in our exposition here, and we will not pay attention to it.

5.1. First idea of Secondary Calculus

5.1.1. First, we note that there are two natural operations on differential forms associated with a vector field X on a manifold M. Namely, these are

$$i_X\colon \Lambda^p(M) \to \Lambda^{p-1}(M) \quad \text{(insertion)}$$

and

$$L_X\colon \Lambda^p(M) \to \Lambda^p(M) \quad \text{(Lie derivative)}.$$

They are related to one another by means of the "infinitesimal Stokes formula" (or "Cartan's formula"):

$$L_X = i_X \circ d + d \circ i_X.$$

On the other hand, as we have already seen in Subsection 4.3.10, the following two operations are associated with a higher symmetry $\chi \in \operatorname{sym} \mathcal{Y}$:

$$i_\chi \colon \mathcal{C}E_1^{p,*}(\mathcal{Y}_\infty) \to \mathcal{C}E_1^{p-1,*}(\mathcal{Y}_\infty),$$

also denoted by $\chi \lrcorner$, and

$$L_\chi \colon \mathcal{C}E_1^{p,*}(\mathcal{Y}_\infty) \to \mathcal{C}E_1^{p,*}(\mathcal{Y}_\infty)$$

with $\mathcal{C}E_1^{p,*}(\mathcal{Y}_\infty) = \sum_{q\geq 0} \mathcal{C}E_1^{p,*}(\mathcal{Y}_\infty)$. Moreover, in this notation formula (4.16) becomes

$$L_\chi = i_\chi \circ d_1 + d_1 \circ i_\chi$$

with $d_1 = d_1(\mathcal{Y}_\infty)$. In fact, all the above-said is valid for an arbitrary object $\mathcal{O}$ of the category DE as well, and the analogy can be summarized as follows:

$$\begin{aligned} M &\longleftrightarrow \mathcal{O}, \\ \Lambda^p(M) &\longleftrightarrow \mathcal{C}E_1^{p,*}(\mathcal{O}), \\ D(M) &\longleftrightarrow \operatorname{sym} \mathcal{O}, \\ i_X &\longleftrightarrow i_\chi, \\ L_X &\longleftrightarrow L_\chi, \end{aligned}$$

where $\operatorname{sym} \mathcal{O}$ stands for $D_\mathcal{C}(\mathcal{O})/\mathcal{C}D(\mathcal{O})$.

Continuing this analogy, we recall that the differential

$$\mathcal{E} = d_1^{0,n} \colon \mathcal{C}E_1^{0,n}(\mathcal{O}) = \bar{H}^n(\mathcal{O}) \to \mathcal{C}E_1^{1,n}(\mathcal{O}) \tag{5.1}$$

is the Euler operator, i.e., it assigns to a functional (action) the corresponding Euler–Lagrange equation. Since it plays the same role in searching for extremals of functionals as the standard differential

$$d \colon C^\infty(M) \to \Lambda^1(M)$$

does for extremals of functions, we can extend the above analogy list with

$$d = d_0 \longleftrightarrow \mathcal{E} = \mathcal{E}_0.$$

These and other similar facts lead to a suspicion that this list can be, in fact, extended to all components of the classical differential calculus, i.e., to differential operators, jets, tensors, etc. In other words, one can expect the existence of a new calculus composed of such analogs. We call this, at the moment hypothetical, calculus *secondary* for two reasons. One is that Secondary Calculus is "programmed" in the language of the classical Calculus, which is "primary" in this respect. The other reason comes from quantum physics.

Namely, the Schrödinger approach to quantum mechanics looks as

Hamilton equations (ordinary differential equations)	$\xrightarrow{\text{quantization}}$	Schrödinger equation (partial differential equation)

Thus, partial differential equations may be viewed as "quantized" ordinary equations. In quantum field theory, the starting point is classical fields, which are to be quantized. From the mathematical point of view, we start with partial differential equations describing classical fields, i.e., "quantized" ordinary equations, and obtain "quantized" partial differential equations, i.e., "twice quantized" ordinary ones. Assuming that these "quantized" partial differential equations belong to the list of above-mentioned analogs, we find an additional motivation to use the word "secondary" for yet unknown natural analogs of partial differential equations.

We note that the term "natural" is used here in physical context, and as such must be distinguished from its mathematical usage. While the latter is in the sense of category theory, the former is in its direct sense, i.e., as a fact of Nature. Coincidence of these two meanings may be seen in the fact that (primary) calculus is the only known (or "natural") language of classical physics.

Of course, these considerations are too vague and general, and the reader is referred to Chapter 0 for further details.

For the convenience of terminology, further on we use the adjective "secondary" in combination with a term of the classical calculus to refer to its analog in Secondary Calculus. For instance, higher symmetries of an equation $\mathcal{Y}$ are secondary vector fields on $\mathcal{Y}_\infty$. Secondary differential p-forms on $\mathcal{O}$ are elements of $\mathcal{C}E_1^{p,*}(\mathcal{O})$. More generally, the right column of the above list of analogs is composed of secondary objects corresponding to the standard ones in the left column.

5.1.2. The basic problem in Secondary Calculus is

> *to define the secondary analog of a given object of the classical calculus and then* ***to compute it.***

We call this problem the *secondarization problem.* One feature of Secondary Calculus to be emphasized from the very beginning is that the concepts to operate with must not only be defined but *computed* as well. For instance, in order to work with secondary differential forms one needs to compute the first term of the $\mathcal{C}$-spectral sequence, i.e., to compute a certain type of cohomology. So, in contrast to the standard calculus, we no longer have the possibility to represent a secondary differential form as an analytic expression of a certain kind, established once and for all.

At first sight, it seems hardly possible to formalize the secondarization problem so that it would admit a general solution. It can be shown, however, that the required formalization is reduced essentially to a formalization of *differential calculus.* In its turn, the latter problem admits a solution along the lines sketched in [**136**] and [**133**] (see also [**64, 71, 116, 145**]). In what follows, we are not going to discuss the exact meaning of the words "differential calculus" as we use them now and further on. It would require a separate, long enough exposition. We hope, however, that particular cases of the secondarization problem considered below will give the reader a rather precise idea of what we have in mind. One of our main goals in the subsequent exposition is to show informally what is the *secondarization algorithm.*

5.1.3. Before proceeding in this direction, it is convenient to develop a proper geometric language.

First, note that an object of the category DE (see Subsection 3.4.9) can be viewed as a (generally, infinite-dimensional) manifold $\mathcal{O}$ supplied with a smooth function algebra $A = \mathcal{F}(\mathcal{O})$ and a finite-dimensional distribution $\mathcal{C} = \mathcal{C}(\mathcal{O})$ such that the triple $(\mathcal{O}, \mathcal{F}(\mathcal{O}), \mathcal{C}(\mathcal{O}))$ is locally of the form $(\mathcal{Y}_\infty, \mathcal{F}(\mathcal{Y}_\infty), \mathcal{C}(\mathcal{Y}_\infty))$ for a certain partial differential equation (PDE) $\mathcal{Y}$. Here "locally" means a suitable localization with respect to the Zariski topology corresponding to the algebra $\mathcal{F}(\mathcal{O})$.

DEFINITION. A pair of the form $(\mathcal{Y}_\infty, \mathcal{C}(\mathcal{Y}_\infty))$ is called an *elementary diffiety.*

Elementary diffieties are geometric objects that play the same role in the theory of partial differential equations as affine algebraic varieties do in the theory of algebraic equations.

We now define the *diffiety dimension* $\operatorname{Dim}\mathcal{O}$ of an elementary diffiety $\mathcal{O} = (\mathcal{Y}_\infty, \mathcal{C}(\mathcal{Y}_\infty))$ as the dimension of the structure distribution $\mathcal{C}(\mathcal{Y}_\infty)$. In other words, this is the number of "independent variables" in $\mathcal{Y}$.

Hence, a general diffiety, as understood above, is obtained by gluing together elementary diffieties of the same diffiety dimension.

It is also natural to introduce the *diffiety morphisms*, which will be called *smaps*. If $\mathcal{O} = (\mathcal{O}, \mathcal{F}(\mathcal{O}), \mathcal{C}(\mathcal{O}))$ and $\mathcal{O}' = (\mathcal{O}', \mathcal{F}(\mathcal{O}'), \mathcal{C}(\mathcal{O}'))$ are two diffieties, then a smap from $\mathcal{O}$ to $\mathcal{O}'$ is a smooth mapping $\Phi\colon \mathcal{O} \to \mathcal{O}'$ which respects the Cartan distribution. More exactly, "smooth" means that $\Phi \circ f \in \mathcal{F}(\mathcal{O})$ if $f \in \mathcal{F}(\mathcal{O}')$. "Respects" means that the image of the distribution $\mathcal{C}(\mathcal{O})$ under the differential of Φ belongs to $\mathcal{C}(\mathcal{O}')$, i.e.,

$$d_\theta\Phi\colon \mathcal{C}_\theta \to \mathcal{C}_{\Phi(\theta)}, \quad \forall\, \theta \in \mathcal{O},$$

where $d_\theta\Phi$ stands for the differential of Φ at θ.

Written in coordinates, smaps from $\mathcal{O}$ to $\mathcal{O}'$ can be viewed as infinite prolongations of differential operators, taking solutions of $\mathcal{Y}$ to those of $\mathcal{Y}'$, where $\mathcal{Y}_\infty$ and $\mathcal{Y}'_\infty$ represent $\mathcal{O}$ and $\mathcal{O}'$ locally. More exactly, if Δ is such an operator and $u = f(x)$ is the local description of L, then the derivatives of $\Delta(f)$ evaluated at the corresponding point y are jet coordinates of $\Phi(\theta) \in \mathcal{Y}'_\infty$, $\theta = [L]^\infty_{x_0} \in \mathcal{Y}_\infty$.

The following class of smaps is of special interest. A smap $\Phi\colon \mathcal{O}' \to \mathcal{O}$ is called a *covering* (in the category DE) if Φ is surjective and $d_\theta\Phi\colon \mathcal{C}_\theta \to \mathcal{C}_{\Phi(\theta)}$ is an isomorphism for every $\theta \in \mathcal{O}'$. The concept of a covering allows us to formalize the idea of a nonlocal symmetry (conservation law, etc.) for a given PDE $\mathcal{Y}$. Namely, if $\Phi\colon \mathcal{Y}'_\infty \to \mathcal{Y}_\infty$ is a covering, then a symmetry (a conservation law, etc.) of $\mathcal{Y}'_\infty$ is called a (Φ-) nonlocal symmetry (resp., conservation law, etc.) of $\mathcal{Y}$. The objects of differential calculus, i.e., smooth functions, etc., on $\mathcal{Y}'_\infty$ contain, in general, integrals and other "nonlocalities" when being expressed in terms of the corresponding quantities on $\mathcal{Y}$. This is why it is natural to call them "nonlocal".

Various well-known constructions in the theory of PDE make implicit use of special kinds of coverings. For example, such are Bäcklund transformations, the procedure of factorization of PDE, Wahlquist–Estabrook prolongation structures, etc. This shows the importance of this concept. A Bäcklund transformation connecting solutions of equations $\mathcal{Y}'$ and $\mathcal{Y}''$ can be viewed as a diagram of coverings

$$\mathcal{O}' \xleftarrow{\Phi'} \mathcal{O} \xrightarrow{\Phi''} \mathcal{O}''$$

with $\mathcal{O}' = (\mathcal{Y}'_\infty, \mathcal{C}\mathcal{Y}'_\infty))$, $\mathcal{O}'' = (\mathcal{Y}''_\infty, \mathcal{C}(\mathcal{Y}''_\infty))$, and $\mathcal{O} = (\mathcal{Y}, \mathcal{C}(\mathcal{Y}_\infty))$ for a suitable equation $\mathcal{Y}$.

It is remarkable that nonlocal quantities of a certain kind, say, symmetries of a given equation, form a "community" (for example, a Lie algebra in the case of symmetries) in spite of the fact that they live on different coverings (see, for instance, [**74**]). This makes it possible to develop a common calculus, which makes the methods of symmetry analysis much more powerful. Thus, it looks very attractive to extend the category DE in order to include nonlocal quantities as well. This goal can be achieved by an appropriate generalization of the notion of diffiety.

DEFINITION. A *(general) diffiety* is locally the inverse limit of a (generally, infinite) sequence of coverings:

$$\mathcal{O}_1 \xleftarrow{\Phi_1} \mathcal{O}_2 \xleftarrow{\Phi_2} \cdots \longleftarrow \mathcal{O}_i \xleftarrow{\Phi_i} \cdots$$

with $\mathcal{O}_i$ being elementary diffieties.

A general diffiety $\mathcal{O}$ is supplied in a natural way with an n-dimensional distribution $\mathcal{C}(\mathcal{O})$ with $n = \operatorname{Dim} \mathcal{O}_i$, $i = 1, 2, \dots$, which is the inverse limit of distributions $\mathcal{C}(\mathcal{O}_i)$. We continue to call $\mathcal{C}(\mathcal{O})$ the *Cartan distribution* (= the *infinite-order contact structure*) on $\mathcal{O}$. It is important to emphasize that $\mathcal{C}(\mathcal{O})$ is a *Frobenius* distribution. Similarly, the algebra $\mathcal{F}(\mathcal{O})$ of smooth functions on $\mathcal{O}$, defined as the direct limit of the algebras $\mathcal{F}(\mathcal{O}_i)$, has a distinguished class of equivalent, i.e., inscribed one into another, filtrations. This makes it possible to define the FG-category of $\mathcal{F}(\mathcal{O})$-modules and to develop a differential calculus in it along the lines of Section 3.1. For example, filtered derivations of the algebra $\mathcal{F}(\mathcal{O})$ form an $\mathcal{F}(\mathcal{O})$-module denoted by $D(\mathcal{O})$ whose elements are interpreted as (*primary*) vector fields on $\mathcal{O}$. The standard commutator operation endows it with a Lie algebra structure. The submodule $\mathcal{C}D(\mathcal{O})$ of $D(\mathcal{O})$ composed of the vector fields belonging to the Cartan distribution $\mathcal{C}(\mathcal{O})$ is a Lie algebra ideal in it. This is equivalent to the fact that $\mathcal{C}(\mathcal{O})$ is a Frobenius distribution. Thus, the Lie algebra

$$\operatorname{sym}(\mathcal{O}) = D_{\mathcal{C}}(\mathcal{O})/\mathcal{C}D(\mathcal{O})$$

with $D_{\mathcal{C}}(\mathcal{O}) = \{X \in D(\mathcal{O}) \mid [X, \mathcal{C}D(\mathcal{O})] \subset \mathcal{C}D(\mathcal{O})\}$ is defined and its elements are *secondary vector fields* on $\mathcal{O}$.

The $\mathcal{F}(\mathcal{O})$-module $\Lambda^p(\mathcal{O})$ of (*primary*) differential p-forms on $\mathcal{O}$ is defined as in Subsection 3.1.5 for elementary diffieties, and the corresponding Cartan submodule $\mathcal{C}\Lambda^p(\mathcal{O}) \subset \Lambda^p(\mathcal{O})$ is defined as in Subsection 3.1.7. Thus, the construction of the $\mathcal{C}$-spectral sequence given in Section 4.3 remains valid for an arbitrary diffiety $\mathcal{O}$ as well. We denote it by

$$\{\mathcal{C}E_r^{p,q}(\mathcal{O}), d_r^{p,q}\}$$

and call the elements of its first term *secondary differential forms* on the diffiety $\mathcal{O}$.

The mapping $\Phi^* \colon \Lambda^*(\mathcal{O}') \to \Lambda^*(\mathcal{O})$ associated with a smap $\Phi \colon \mathcal{O} \to \mathcal{O}'$ obviously takes $\mathcal{C}\Lambda^*(\mathcal{O}')$ to $\mathcal{C}\Lambda^*(\mathcal{O})$, and therefore, $\mathcal{C}^p\Lambda^*(\mathcal{O}')$ to $\mathcal{C}^p\Lambda^*(\mathcal{O})$ for any p. In other words, Φ^* is a chain mapping preserving the $\mathcal{C}$-filtration. For this reason, it generates a mapping of $\mathcal{C}$-spectral sequences

$$\Phi_r^* \colon \mathcal{C}E_r^{p,q}(\mathcal{O}') \to \mathcal{C}E_r^{p,q}(\mathcal{O}), \qquad \forall r \geq 0,$$

which is a cochain mapping of the complex $\{\mathcal{C}E_r(\mathcal{O}'), d_r\}$ to $\{\mathcal{C}E_r(\mathcal{O}), d_r\}$ such that the induced mapping in cohomology coincides with Φ_{r+1}^*. The mapping Φ_1^* is naturally interpreted as the secondary analog of the mapping of the differential form algebras induced by a smooth mapping. Similarly, Φ_2^* is the secondary analog of the corresponding mapping of de Rham cohomology algebras.

These examples show that diffieties are objects on which Secondary Calculus grows naturally. They also clearly demonstrate that the *primary* calculus on a diffiety $\mathcal{O}$ understood as the differential calculus in the FG category over $\mathcal{F}(\mathcal{O})$ is the raw material used to build Secondary Calculus on $\mathcal{O}$.

The study of diffieties in the most general form has started just recently, and so, not very much is known about them at present. Later we shall discuss their structure and concrete methods of work with them. These topics are touched upon, for instance, in [**74**] and [**19**].

5.1.4. It is very instructive to specify the already known secondary constructions to 0-Dimensional (not to be confuseed with 0-dimensional!) diffieties.

An n-dimensional manifold M can be viewed as a diffiety in two different but natural ways, according to the choice of the Cartan distribution on it. By defining the Cartan distribution on M to be the 0-dimensional distribution

$$M \ni a \mapsto \{0\} = \mathcal{C}_a \subset T_a M,$$

we get a 0-Dimensional diffiety. Its integral submanifold is formed by points of M.

Another possibility to endow M with a Cartan distribution is

$$M \ni a \mapsto \mathcal{C}_a = T_a M.$$

In this way, one gets an n-Dimensional diffiety whose integral submanifolds are open domains of M. Therefore, n-dimensional manifolds are the smallest possible n-Dimensional diffieties, and as such they arise as solutions of PDE.

Thus, an n-dimensional manifold M viewed as a 0-Dimensional diffiety is a "society" whose "members", i.e., points, have a trivial inner structure. On the other hand, M viewed as an n-Dimensional diffiety, is not a society but a singleton, i.e., a "point" from the social point of view, having a rich inner structure. In the case of a generic diffiety, we have a mixture of these extreme options.

Denote by $\mathcal{O}_M$ (resp., $\mathcal{O}_M^\bullet$) the 0-Dimensional (resp., n-Dimensional) diffiety associated with M as explained before. Then, as is easy to see, $\mathcal{C}D(\mathcal{O}_M) = 0$ and

$$\operatorname{sym}(\mathcal{O}_M) = D(M).$$

Thus, secondary vector fields in this case are identical to the standard (= primary) ones. On the other hand,

$$\operatorname{sym}(\mathcal{O}_M^\bullet) = 0.$$

The $\mathcal{C}$-spectral sequence for $\mathcal{O}_M$ looks as follows. Put

$$\Lambda_+^k(M) = \sum_{i \geq k} \Lambda^i(M).$$

Then, obviously, $\mathcal{C}\Lambda(\mathcal{O}_M) = \Lambda_+^1(M)$ and therefore $\mathcal{C}^k\Lambda(\mathcal{O}_M) = \Lambda_+^k(M)$. Hence

$$\mathcal{C}E_0^{p,0}(\mathcal{O}_M) = \Lambda^p(M) \quad \text{and} \quad \mathcal{C}E_0^{p,q}(\mathcal{O}_M) = 0 \text{ if } q \neq 0.$$

Since $d_0 \colon \mathcal{C}E_0^{p,q} \to \mathcal{C}E_0^{p,q+1}$, we have $d_0(\mathcal{O}_M) = 0$. This implies that

$$\mathcal{C}E_1^{p,q}(\mathcal{O}_M) = \mathcal{C}E_0^{p,q}(\mathcal{O}_M)$$

and, in particular,

$$\mathcal{C}E_1^{p,0}(\mathcal{O}_M) = \Lambda^p(M) \quad \text{and} \quad d_1^{p,0} = d \quad \text{(exterior differential)}.$$

Hence, the first term of the $\mathcal{C}$-spectral sequence of a manifold M interpreted as a 0-Dimensional diffiety coincides with the de Rham complex of M.

On the other hand, the $\mathcal{C}$-spectral sequence for $\mathcal{O}_M^\bullet$ is quite different. In fact, $\mathcal{C}\Lambda(\mathcal{O}_M^\bullet) = 0$, and therefore $\mathcal{C}^k\Lambda(\mathcal{O}_M^\bullet) = 0$ for $k > 0$. Consequently,

$$\mathcal{C}E_0^{0,q}(\mathcal{O}_M^\bullet) = \Lambda^q(M), \quad \text{and} \quad \mathcal{C}E_0^{p,q}(\mathcal{O}_M^\bullet) = 0 \quad \text{for } p \neq 0.$$

This implies that $d_0^{0,q} = d$ (exterior differential), while $d_0^{p,q} = 0$ for $p \neq 0$. Thus,

$$\mathcal{C}E_1^{0,q}(\mathcal{O}_M^\bullet) = H^q(M) \quad \text{(de Rham cohomology of } M\text{)}$$

and

$$\mathcal{C}E_1^{p,q}(\mathcal{O}_M^\bullet) = 0 \quad \text{for } p \neq 0.$$

The above considerations show that secondary vector fields and differential forms turn out to be identical to standard ones in the class of 0-Dimensional diffieties. It is natural to generalize this observation to the following general principle:

> *Any concept of Secondary Calculus must be identical to its classical (primary) counterpart when restricted to the category of* 0*-Dimensional diffieties.*

There is a number of other indications in favor of this principle. We call it the *correspondence principle* and take it as a guide in solving the secondarization problem. It is worth noting that this principle is parallel to the *Bohr correspondence principle* in quantum mechanics, which controls the quantization problem. The latter requires that the classical mechanics should be the "limit" of quantum mechanics as "$\hbar \to 0$". The analog of "$\hbar \to 0$" in the context of secondarization problem is "$\operatorname{Dim} \mathcal{O} \to 0$". Since the Planck constant $\hbar$ is a true constant, the expression "$\hbar \to 0$" has only a heuristic meaning, which cannot be formalized in a rigorous way. On the other hand, some facts indicate the existence of a more sophisticated version of the diffiety dimension which takes values in $\mathbb{R}$. In other words, it is plausible that a rigorous meaning to "$\operatorname{Dim} \mathcal{O} \to 0$" can be given.

5.1.5. The discussion in the previous subsection makes it natural to regard informally a generic n-Dimensional diffiety $\mathcal{O}$ as a "society" composed of "individuals" of the form $\mathcal{O}_M^\bullet$, M being an n-dimensional integral submanifold of $\mathcal{C}(\mathcal{O})$. These submanifolds play the role of points composing a smooth finite-dimensional manifold. Hence

> *"secondary points" are integral submanifolds of the Cartan distribution.*

In particular, secondary points of an elementary diffiety $\mathcal{Y}_\infty$ are identified with solutions of $\mathcal{Y}$.

Now it is not difficult to understand what is the secondary analog of the evaluation mapping

$$e_a \colon C^\infty(M) \to C^\infty(M)/\mathcal{I}_a = C^\infty(\{a\}) = \mathbb{R}$$

with $\mathcal{I}_a = \{f \in C^\infty(M) \mid f(a) = 0\}$ being the ideal of the point $a \in M$. To do this, one has just to replace all components of the above formula with their secondary analogs:

$$\begin{aligned} M &\Longleftrightarrow \mathcal{O}, \\ M \ni a &\Longleftrightarrow L \in \mathcal{O}, \\ C^\infty(M) &\Longleftrightarrow \mathcal{C}E_1^{0,*}(\mathcal{O}), \\ C^\infty(\{a\}) = \mathbb{R} &\Longleftrightarrow \mathcal{C}E_1^{0,*}(\mathcal{O}_L^\bullet) = H^*(L), \end{aligned}$$

with L being an integral submanifold of $\mathcal{C}(\mathcal{O})$. As a result, we see that the "secondary evaluation mapping" e_L at the "secondary point L" looks as

$$e_L \colon \mathcal{C}E_1^{0,*}(\mathcal{O}) \to H^*(L).$$

This is the mapping between the first terms of the $\mathcal{C}$-spectral sequences associated with $\mathcal{O}$ and $\mathcal{O}_L$, generated by the canonical inclusion smap $L \hookrightarrow \mathcal{O}$.

A consequence of the previous considerations is that the de Rham cohomology algebra $H^*(L)$ of a "secondary point" L is the secondary analog of the algebra of scalars $\mathbb{R}$ attached to a point of a manifold. Thus, in Secondary Calculus, scalars are not absolute and vary from one secondary point to another. It is worth noting that "classical scalars" do not depend on points to which they are attached properly. This is due to the fact that such points have trivial inner structure and cannot be distinguished from one another as "individuals". However, variability of scalars from point to point is not surprising. For instance, such a situation is common in algebraic geometry. But we encounter a really new phenomenon when trying to secondarize the concept of a vector space over a given algebra of scalars. There are delicate moments in solving this problem. For instance, one of them is how to interpret properly the multiplication of scalars:

$$\mu\colon \mathbb{R}\times\mathbb{R}\to\mathbb{R}. \tag{5.2}$$

The obvious solution, exterior multiplication in $H^*(L)$, is not entirely correct.

5.2. Secondary linear algebra

In this section, basic notions of secondary linear algebra are introduced. It would hardly be possible to motivate the choices we make, from the very beginning. Their justification comes *a posteriori* in the form of an elegant self-consistency of the results.

5.2.1. To proceed, we need some additional constructions. We fix a manifold M, $\dim M = n$. A *differential complex* $(K,\partial) = \{(K_i,\partial_i)\}$ on M is a complex

$$0\to K_0 \xrightarrow{\partial_0} K_1 \xrightarrow{\partial_1} \cdots \xrightarrow{\partial_{i-1}} K_i \xrightarrow{\partial_i} \cdots$$

composed of $C^\infty(M)$-modules K_i and differential operators ∂_i connecting them.

Let $(K,\partial) = \{(K_i,\partial_i)\}$ and $(\tilde K,\tilde\partial) = \{(\tilde K_i,\tilde\partial_i)\}$ be two differential complexes on M. We set

$$\mathcal{G}\,\mathrm{Diff}(K,\tilde K) = \sum_\alpha \mathcal{G}\,\mathrm{Diff}^\alpha(K,\tilde K),$$

where

$$\mathcal{G}\,\mathrm{Diff}^\alpha(K,\tilde K) = \{\Box\colon K\to\tilde K \mid \Box(K_i)\subset \tilde K_{i+\alpha},\ \Box|_{K_i}\in \mathrm{Diff}(K_i,\tilde K_{i+\alpha}),\ \forall i\}.$$

The $C^\infty(M)$-module $\mathcal{G}\,\mathrm{Diff}(K,\tilde K)$ becomes a complex with respect to the differential L:

$$L(\Box) = \tilde\partial\circ\Box - (-1)^\alpha\Box\circ\partial \quad\text{for}\quad \Box\in\mathcal{G}\,\mathrm{Diff}^\alpha(K,\tilde K). \tag{5.3}$$

Denote by $\mathcal{L}^\alpha(K,\tilde K)$ the α-th cohomology of the complex $(\mathcal{G}\,\mathrm{Diff}(K,\tilde K), L)$ and put

$$\mathcal{L}(K,\tilde K) = \sum_\alpha \mathcal{L}^\alpha(K,\tilde K).$$

Then the elements of $\mathcal{L}(K,\tilde K)$ are homotopy classes of graded cochain mappings from K to $\tilde K$. Therefore, any $h\in\mathcal{L}(K,\tilde K)$ generates a mapping in cohomology denoted also by h:

$$h\colon H^*(K)\to H^*(\tilde K). \tag{5.4}$$

DEFINITION. The cocycles of the complex $(\mathcal{G}\,\mathrm{Diff}(K, \tilde{K}), L)$ are called *differential (graded cochain) mappings* from (K, ∂) to $(\tilde{K}, \tilde{\partial})$. The same names are given to the induced mappings in cohomology.

In other words, only mappings of the form (5.4) are considered to be differential in cohomology.

These definitions emphasize that we deal with the category whose objects are differential complexes on M and morphisms are differential graded cochain mappings. The corresponding notion of homotopy equivalence is as follows.

DEFINITION. Two differential complexes K and $\tilde{K}$ are said to be (*differentially*) *homotopy equivalent* if there exist $h \in \mathcal{L}^0(K, \tilde{K})$ and $\tilde{h} \in \mathcal{L}^0(\tilde{K}, K)$ such that

$$\tilde{h}h \sim \mathrm{id}_K, \qquad h\tilde{h} \sim \mathrm{id}_{\tilde{K}}\,.$$

In the context under consideration the cohomology of differential complexes is the only quantity we are interested in. On the other hand, their cohomological origin must not be ignored, and the concept of differential homotopy type puts these two aspects together. Thus, the differential homotopy types are of principal interest for our purposes. Speaking informally, a differential cohomology space itself can be thought of as a "set", while the homotopy class of the original differential complex, as a *differential linear* structure on it. In turn, the elements of $\mathcal{L}(K, \tilde{K})$ are interpreted as *differential linear* mappings from $H^*(K)$ to $H^*(\tilde{K})$. The first question in this connection is: what are linear differential mappings of secondary scalars into themselves? The answer to this question consists in computing the cohomology of the complex $(\mathcal{G}\,\mathrm{Diff}(\Lambda^*(M), \Lambda^*(M)), L)$, i.e., $\mathcal{L}^\alpha(\Lambda^*(M), \Lambda^*(M))$.

THEOREM. *There is a natural isomorphism*

$$\mathcal{L}(\Lambda^*(M), \Lambda^*(M)) = H^*(M).$$

Moreover, the differential linear mapping corresponding to $h \in H^*(M)$ under this isomorphism is the operator of left multiplication by h.

This theorem is the secondary analog of the following classical triviality:

$$\mathrm{Hom}_{\mathbb{R}}(\mathbb{R}, \mathbb{R}) = \mathbb{R}.$$

Indeed, if M is a one point manifold, then $\Lambda^*(M) = C^\infty(M) = \mathbb{R}$, and linear differential operators over 0-dimensional manifolds are just linear mappings.

We adopt a special notation to distinguish the differential graded algebra $\mathcal{G}\,\mathrm{Diff}(\Lambda^*(M), \Lambda^*(M))$, which will play an important role in what follows:

$$\mathcal{S}(M) = \mathcal{G}\,\mathrm{Diff}(\Lambda^*(M), \Lambda^*(M)).$$

Note also that the standard exterior differential $d = d_M$ belongs to $\mathcal{S}(M)$.

5.2.2. The following notion is crucial for our purposes.

DEFINITION. A (left) *$\mathcal{S}$-module* (over M) is a differential complex (K, ∂) supplied with a (left) $\mathcal{S}(M)$-module structure

$$\mu\colon \mathcal{S}(M) \to \mathcal{G}\,\mathrm{Diff}(K, K)$$

such that

(i) if $f \in \mathcal{C}^\infty(M)$, then $\mu(f) = f$ (the operator of multiplication by f);
(ii) $\mu(d) = \partial$.

$\mathcal{S}$-modules possess a number of special properties. First of all, they are unitary $\Lambda^*(M)$-modules. This follows from (i) and the natural inclusion $\Lambda^*(M) \hookrightarrow \mathcal{S}(M)$ associating with a differential form ω the operator of (left) multiplication by ω. Next, by (ii), ∂ is a first-order differential operator. In addition, also by (ii), μ, being a homomorphism of associative algebras, is a cochain mapping of the complex $(\mathcal{S}(M), L)$ to $(\mathcal{G}\,\mathrm{Diff}(K,K), L)$. By Theorem 5.2.1, the corresponding mapping in cohomology can be viewed as the mapping

$$\mu^*\colon H^*(M) \to \mathcal{L}(K,K).$$

This defines an action of $H^*(M)$ on $H^*(K)$. In other words, the cohomology of an $\mathcal{S}$-module over M is an $H^*(M)$-module in a natural way.

Example (1). The algebra $\mathcal{S}(M)$ regarded as the canonical left $\mathcal{S}(M)$-module and supplied with the differential $\partial\colon \square \mapsto d\circ\square$, $\square \in \mathcal{S}(M)$, becomes an $\mathcal{S}$-module. Its k-th cohomology space is canonically isomorphic to $\Lambda^n(M) \otimes_{C^\infty(M)} D_{n-k}(M)$, where $D_{n-k}(M)$ is the $C^\infty(M)$-module of $(n-k)$-multivector fields on M. According to this isomorphism, an element

$$\omega \otimes Z \in \Lambda^n(M) \otimes_{C^\infty(M)} D_{n-k}(M), \quad Z = X_1 \wedge \cdots \wedge X_{n-k},$$

acts on $\Lambda^*(M)$ by taking $\rho \in \Lambda^*(M)$ to $\omega \wedge i_{X_1}(\dots(i_{X_{n-k}}(\rho))\dots)$. Thus, such an operator acts trivially on $\Lambda^j(M)$ for $j \neq n-k$ and coincides with the right multiplication by $i_{X_1}(\dots(i_{X_{n-k}}(\omega))\dots) \in \Lambda^k(M)$ if $j = n-k$. This also shows that the above cohomology action of $H^i(M)$ is trivial for $i > 0$ and reduces to the multiplication by locally constant functions for $i = 0$.

Example (2). The de Rham complex $(\Lambda^*(M), d)$ becomes an $\mathcal{S}$-module with respect to the canonical action of $\mathcal{S}(M)$ on $\Lambda^*(M)$. The cohomology action of $H^*(M)$ in this case is identified with the multiplication $H^*(M)\times H^*(M) \to H^*(M)$.

Definition. Given $\mathcal{S}$-modules $\mathcal{K} = (K, \partial, \mu)$ and $\tilde{\mathcal{K}} = (\tilde{K}, \tilde{\partial}, \tilde{\mu})$, a differential mapping $\Delta\colon K \to \tilde{K}$ is called an *$\mathcal{S}$-homomorphism* if

$$\Delta \circ \mu(\square) = (-1)^{\Delta\square}\tilde{\mu}(\square)\circ\Delta, \quad \forall \square \in \mathcal{S}(M).$$

The $\mathcal{S}$-homomorphisms are automatically graded cochain mappings. Indeed,

$$\Delta\circ\partial = \Delta\circ\mu(d) = (-1)^{\Delta}\tilde{\mu}(d)\circ\Delta = (-1)^{\Delta}\tilde{\partial}\circ\Delta.$$

Consequently, any $\mathcal{S}$-homomorphism Δ generates a mapping of cohomology groups $H(\Delta)\colon H^*(K) \to H^*(\tilde{K})$, which, as is easy to see, is also a homomorphism of $H^*(M)$-modules.

Example (3). The mapping $\mathcal{S}(M) \to \Lambda^*(M)$ taking $\square \in \mathcal{S}(M)$ to $\square(1) \in \Lambda^*(M)$ is an $\mathcal{S}$-homomorphism of the $\mathcal{S}$-modules described in Examples 5.2.2 (1) and 5.2.2 (2), respectively.

Denote by $\mathrm{Hom}_{\mathcal{S}}(\mathcal{K}, \tilde{\mathcal{K}})$ the vector space of all $\mathcal{S}$-homomorphisms from $\mathcal{K}$ to $\tilde{\mathcal{K}}$. The following result is easily deduced from Definition 5.2.2.

Example. $\mathrm{Hom}_{\mathcal{S}}(\Lambda^*(M), \Lambda^*(M)) = \mathbb{R}$. This example shows that $\mathcal{S}$-homomorphisms are very rigid as compared with generic differential cochain mappings. Below we use this property of $\mathcal{S}$-homomorphisms to describe the structure of "good" $\mathcal{S}$-modules.

The importance of $\mathcal{S}$-modules for our purposes is that they are *secondary vector spaces* (over the "secondary point" M). More precisely, when saying that the cohomology $H^*(K)$ with $\mathcal{K} = (K, \partial, \mu)$ is a secondary vector space we use the adjective "secondary" exactly in the same sense as was done earlier with secondary functions, differential forms, etc. But at the same time it should be emphasized that the cohomological origin of this vector space (in the standard sense of this word) cannot be ignored. In other words, $H^*(K)$ understood as a ("normal") abstract vector space plays no role in Secondary Calculus.

Given $\mathcal{S}$-modules $\mathcal{K} = (K, \partial, \mu)$ and $\tilde{\mathcal{K}} = (\tilde{K}, \tilde{\partial}, \tilde{\mu})$, a *secondary linear mapping* from $\mathcal{K}$ to $\tilde{\mathcal{K}}$ is the homotopy class of a differential cochain mapping from K to $\tilde{K}$. This means that we deal only with mappings from $H^*(K)$ to $H^*(\tilde{K})$ arising from $\mathcal{L}(K, \tilde{K})$.

These definitions allow us to develop the *secondary linear algebra* in a systematic way. For instance, the tensor product of two $\mathcal{S}$-modules $\mathcal{K} = (K, \partial, \mu)$ and $\tilde{\mathcal{K}} = (\tilde{K}, \tilde{\partial}, \tilde{\mu})$ is defined as follows:

$$\mathcal{K} \otimes^{\mathcal{S}} \tilde{\mathcal{K}} = (K \otimes_{\Lambda^*(M)} \tilde{K}, \partial \otimes^{\mathcal{S}} \tilde{\partial}, \mu \otimes^{\mathcal{S}} \tilde{\mu}),$$

where

$$(\partial \otimes^{\mathcal{S}} \tilde{\partial})(\kappa \otimes \tilde{\kappa}) = \partial(\kappa) \otimes \tilde{\kappa} + (-1)^{\kappa} \kappa \otimes \tilde{\partial}(\tilde{\kappa})$$

and the $\mathcal{S}$-module structure $\mu \otimes^{\mathcal{S}} \tilde{\mu} \colon \mathcal{S}(M) \to K \otimes_{\Lambda^*(M)} \tilde{K}$ is defined uniquely by the condition

$$((\mu \otimes^{\mathcal{S}} \tilde{\mu})(i_X))(\kappa \otimes \tilde{\kappa}) = \mu(i_X)(\kappa) \otimes \tilde{\kappa} + (-1)^{\kappa} \kappa \otimes \tilde{\mu}(i_X)(\tilde{\kappa}).$$

Similarly, $\mathrm{Hom}_{\Lambda^*(M)}(K, \tilde{K})$ can be supplied with a natural $\mathcal{S}$-module structure. Denote this $\mathcal{S}$-module by $\mathrm{Hom}^{\mathcal{S}}(\mathcal{K}, \tilde{\mathcal{K}})$. If $\Delta \in \mathcal{S}(M)$ is a derivation of $\Lambda^*(M)$, then its action Δ^H on $\mathrm{Hom}^{\mathcal{S}}(\mathcal{K}, \tilde{\mathcal{K}})$ is defined as

$$\Delta^H(h) = \mu(\Delta) \circ h - (-1)^{h\Delta} h \circ \tilde{\mu}(\Delta), \quad h \in \mathrm{Hom}_{\Lambda^*(M)}(K, \tilde{K}).$$

In particular, d^H is a differential (it might also be denoted by $\mathrm{Hom}^{\mathcal{S}}(\partial, \tilde{\partial})$) in $\mathrm{Hom}_{\Lambda^*(M)}(K, \tilde{K})$.

REMARK. Conceptually, it is more satisfactory to understand *secondary linear mappings* as cohomology classes of the complex $\mathrm{Hom}^{\mathcal{S}}(\mathcal{K}, \tilde{\mathcal{K}})$ rather than elements of $\mathcal{L}(K, \tilde{K})$. However, in the next subsection it will be shown that these approaches coincide on the class of *natural* $\mathcal{S}$-modules.

5.2.3. Now we introduce a special class of $\mathcal{S}$-modules, which are, in a "cohomological" sense, projective objects in a suitable category. They deserve to be called secondary vector spaces even more than generic $\mathcal{S}$-modules if we recall that the usual vector spaces are always projective modules over the ground field.

First we recall the standard notion of a connection in a suitable algebraic setting.

DEFINITION. Let P be a $C^\infty(M)$-module. A homomorphism of $C^\infty(M)$-modules, $\nabla \colon D(M) \to \mathrm{Diff}_1(P, P)$, is called a *connection* in P if for any $f \in C^\infty(M)$, $X \in D(M)$, and $p \in P$ we have

$$\nabla(X)(fp) = f(\nabla(X)(p)) + X(f)p.$$

In the sequel, we write ∇_X instead of $\nabla(X)$.

Let us associate with a $C^\infty(M)$-module P the $\Lambda^*(M)$-module ΛP defined by

$$\Lambda P = \sum_{i\geq 0} \Lambda^i P, \quad \Lambda^i P = \Lambda^i(M) \otimes_{C^\infty(M)} P.$$

Elements of ΛP are called *P-valued differential forms*. The P-valued differential forms are identified in an obvious way with skew-symmetric $C^\infty(M)$-multilinear functions on $D(M)$. As usual, the value of the form $\omega \in \Lambda^k P$ on $X_1, \dots, X_k \in D(M)$ is denoted by $\omega(X_1, \dots, X_k)$.

The $\Lambda^*(M)$-module structure in ΛP is natural:

$$\rho \wedge (\omega \otimes p) = (\rho \wedge \omega) \otimes p, \quad \rho, \omega \in \Lambda^*(M), \; p \in P.$$

For a vector field $X \in D(M)$ the insertion operator

$$i_X \colon \Lambda P^k \to \Lambda P^{k-1}$$

is defined by

$$i_X(\rho \otimes p) = i_X(\rho) \otimes p, \quad \rho \in \Lambda^k(M), \; p \in P.$$

With a connection ∇ in P one naturally associates a first-order differential operator d_∇ on ΛP taking $\Lambda^k P$ to $\Lambda^{k+1} P$:

$$\begin{aligned} d_\nabla(\omega)(X_1, \dots, X_{k+1}) &= \sum_i (-1)^{i-1} \nabla_{X_i}(\omega(X_1, \dots, \hat{X}_i, \dots, X_{k+1})) \\ &+ \sum_{i<j} (-1)^{i+j} \omega([X_i, X_j], X_1, \dots, \hat{X}_i, \dots, \hat{X}_j, \dots, X_{k+1}) \end{aligned}$$

with $\omega \in \Lambda^k P$.

A connection ∇ is said to be *flat* if

$$[\nabla_X, \nabla_Y] = \nabla_{[X,Y]}$$

for all $X, Y \in D(M)$. It is easy to see that ∇ is flat if and only if $d_\nabla^2 = 0$. Therefore, a flat connection in P transforms ΛP to a differential complex.

PROPOSITION. *Let ∇ be a flat connection in the module P. Then the complex $(\Lambda P, d_\nabla)$ admits a unique $\mathcal{S}$-module structure μ such that*

$$\mu(\omega) = \omega^P, \quad \mu(i_X) = i_X^P,$$

for all $\omega \in \Lambda^*(M)$, $X \in D(M)$, where ω^P is the operator of left multiplication by ω in ΛP, and i_X^P is the insertion of X into operators with values in P-valued forms.

This assertion results directly from the fact that the algebra $\mathcal{S}(M)$ is generated by d, i_X, $X \in D(M)$, and the operators of multiplication by $\omega \in \Lambda^*(M)$.

COROLLARY. *A flat connection ∇ generates a unique* $\mathrm{Diff}(C^\infty(M))$*-module structure ν in P such that*

(i) $\nu(f) = f$ *(the operator of multiplication by f, $f \in C^\infty(M)$);*
(ii) $\nu(X) = \nabla_X$, $X \in D(M)$;
(iii) $[f, \nu(\square)] = \nu([f, \square])$, $\square \in \mathcal{C}\,\mathrm{Diff}(C^\infty(M))$.

We distinguish the $\mathcal{S}$-module structures in ΛP described in Proposition 5.2.3 and the $\mathrm{Diff}(C^\infty(M))$-module structures described in Corollary 5.2.3 by calling them *natural*. Hence, for a projective $C^\infty(M)$-module P we have the following natural identifications:

Flat connections in P	$\Longleftrightarrow$	Natural Diff $C^\infty(M)$-module structures in P	$\Longleftrightarrow$	Natural $\mathcal{S}$-module structures in ΛP

5.2.4. Now we characterize natural $\mathcal{S}$-modules to distinguish them from others. Recall some elementary facts from the theory of connections.

We associate with a vector space V the $C^\infty(M)$-module $C^\infty(M)\otimes_{\mathbb{R}} V$ endowed with the connection $\nabla^V = \nabla$:

$$\nabla_X = X \otimes \mathrm{id}_V \in \mathrm{Diff}_1(C^\infty(M) \otimes_{\mathbb{R}} V, C^\infty(M) \otimes_{\mathbb{R}} V), \quad X \in D(M),$$

with

$$(X \otimes \mathrm{id}_V)(f \otimes v) = X(f) \otimes v, \quad f \in C^\infty(M),\ v \in V.$$

This connection is flat. The corresponding $\Lambda^*(M)$-module $\Lambda(C^\infty(M) \otimes_{\mathbb{R}} V)$ is naturally identified with $\Lambda^*(M)\otimes_{\mathbb{R}} V$. Respectively, the differential d_{∇^V} is identified with

$$d \otimes \mathrm{id}_V \colon \Lambda^k(M) \otimes_{\mathbb{R}} V \to \Lambda^{k+1}(M) \otimes_{\mathbb{R}} V.$$

It is easy to see that

$$H^0(d_{\nabla^V}) = \{\varphi \in C^\infty(M) \otimes_{\mathbb{R}} V \mid d_{\nabla^V}(\varphi) = 0\} = V.$$

The $\Lambda^*(M)$-module $\Lambda^*(M) \otimes_{\mathbb{R}} V$ carries a natural $\mathcal{S}$-module structure μ defined by $\mu(\square) = \square \otimes \mathrm{id}_V$. In other words, the $\mathcal{S}$-module

$$\mathcal{K}^0 = (\Lambda^*(M) \otimes_{\mathbb{R}} V, d_{\nabla^V})$$

is of the form ΛQ with $Q = C^\infty(M) \otimes_{\mathbb{R}} V$.

Now let ∇ be a flat connection in a $C^\infty(M)$-module P, and let $\mathcal{K} = (\Lambda P, d_\nabla)$ be the associated natural $\mathcal{S}$-module (Proposition 5.2.3). If $V = H^0(d_\nabla)$, where $p \in H^0(d_\nabla)$ if and only if $d_\nabla(p) = 0$, then the mapping

$$\Delta\colon \Lambda^*(M) \otimes_{\mathbb{R}} V \to \Lambda P, \quad \Delta(\omega \otimes p) = \omega^P(p), \tag{5.5}$$

is an $\mathcal{S}$-module monomorphism of $\mathcal{K}^0$ to $\mathcal{K}$. If, in addition, M is simply connected, then Δ is an isomorphism. This shows that Δ is a *local* isomorphism, i.e., $\Delta_{\mathcal{U}}\colon \mathcal{K}^0_{\mathcal{U}} \to \mathcal{K}_{\mathcal{U}}$ is an isomorphism of $\mathcal{S}$-modules over $\mathcal{U}$ for a suitable neighborhood $\mathcal{U}$ of a given point of M. Here the subscript $\mathcal{U}$ refers to the localization to $\mathcal{U}$.

Note also another natural identification.

PROPOSITION. *Let $\mathcal{K} = (\Lambda P, d_\nabla)$ be the $\mathcal{S}$-module associated with the flat connection ∇ according to Proposition* 5.2.3. *Then there is a natural isomorphism*

$$H^0(d_\nabla) = \mathrm{Hom}^{\mathcal{S}}(\Lambda^*(M), \mathcal{K}).$$

Note that an analog of $\mathcal{S}$-homomorphism (5.5) can be constructed for any $\mathcal{S}$-module $\mathcal{K} = (K, \partial, \mu)$ such that $K_i = 0$, $i < 0$. Indeed, in such a case, $H^0(K) = \ker \partial|_{K_0} \subset K_0$ and one can put

$$\Delta\colon \Lambda^*(M) \otimes_{\mathbb{R}} H^0(K) \to K, \quad \Delta(\omega \otimes \kappa) = \mu(\omega)(\kappa), \tag{5.6}$$

with $\omega \in \Lambda^*(M)$ and $\kappa \in K_0$, $\partial(\kappa) = 0$.

Summarizing the previous considerations, we finally get

PROPOSITION. *An $\mathcal{S}$-module $\mathcal{K} = (K, \partial, \mu)$ is natural if and only if $K_i = 0$, $i < 0$, and the $\mathcal{S}$-homomorphism* (5.6) *is a local isomorphism.*

5.2.5. Operations in "secondary linear algebra" become particularly transparent in the case of natural modules. For instance, if $\mathcal{K} = (\Lambda P, d_\nabla)$ and $\tilde{\mathcal{K}} = (\Lambda \tilde{P}, d_{\tilde{\nabla}})$, then a straightforward verification shows that

$$\mathcal{K} \otimes^{\mathcal{S}} \tilde{\mathcal{K}} = (\Lambda(P \otimes_{C^\infty(M)} \tilde{P}), d_{\nabla \otimes \tilde{\nabla}}),$$

where $\nabla \otimes \tilde{\nabla}$ is the (flat) connection in $P \otimes_{C^\infty(M)} \tilde{P}$ defined by

$$(\nabla \otimes \tilde{\nabla})_X = \nabla_X \otimes \mathrm{id}_{\tilde{K}} + \mathrm{id}_K \otimes \tilde{\nabla}_X.$$

Similarly, one has the following canonical identification:

$$\mathrm{Hom}^{\mathcal{S}}(\mathcal{K}, \tilde{\mathcal{K}}) = \left\{ \Lambda \left(\mathrm{Hom}_{C^\infty(M)}(P, \tilde{P}) \right), d_{\mathrm{Hom}(\nabla, \tilde{\nabla})} \right\},$$

where $\mathrm{Hom}(\nabla, \tilde{\nabla})$ is the (flat) connection in $\mathrm{Hom}_{C^\infty(M)}(P, \tilde{P})$ defined by

$$\left[\mathrm{Hom}(\nabla, \tilde{\nabla}) \right]_X (\varphi) = \tilde{\nabla}_X \circ \varphi - \varphi \circ \nabla_X.$$

The problem of the description of "secondary linear mappings" between natural $\mathcal{S}$-modules admits a solution that is, in a sense, universal. This problem consists in computing the cohomology space $\mathcal{L}(\mathcal{K}, \tilde{\mathcal{K}})$ for $\mathcal{K} = (\Lambda P, d_\nabla)$ and $\tilde{\mathcal{K}} = (\Lambda \tilde{P}, d_{\tilde{\nabla}})$. In this case, $\mathcal{L}(\mathcal{K}, \tilde{\mathcal{K}})$ is the cohomology of the complex $(\mathcal{G}\,\mathrm{Diff}(\Lambda P, \Lambda \tilde{P}), L)$ with the differential $L = L_{\nabla, \tilde{\nabla}}$ defined (see (5.3)) by

$$L_{\nabla, \tilde{\nabla}}(\square) = d_{\tilde{\nabla}} \circ \square - (-1)^{\square} \square \circ d_\nabla.$$

THEOREM. *The cohomology of the complex* $(\mathcal{G}\,\mathrm{Diff}(\Lambda P, \Lambda \tilde{P}), L_{\nabla, \tilde{\nabla}})$ *is canonically isomorphic to that of* $(\Lambda(\mathrm{Hom}_{C^\infty(M)}(P, \tilde{P})), d_{\mathrm{Hom}(\nabla, \tilde{\nabla})})$:

$$0 \to \mathrm{Hom}_{C^\infty(M)}(P, \tilde{P}) \xrightarrow{d_{\mathrm{Hom}(\nabla, \tilde{\nabla})}} \Lambda^1(M) \otimes_{C^\infty(M)} \mathrm{Hom}_{C^\infty(M)}(P, \tilde{P})$$
$$\xrightarrow{d_{\mathrm{Hom}(\nabla, \tilde{\nabla})}} \cdots \xrightarrow{d_{\mathrm{Hom}(\nabla, \tilde{\nabla})}} \Lambda^n(M) \otimes_{C^\infty(M)} \mathrm{Hom}_{C^\infty(M)}(P, \tilde{P}) \to 0.$$

In other words, we showed that secondary linear mappings of the natural $\mathcal{S}$-module $\mathcal{K} = (\Lambda P, d_\nabla)$ to the natural $\mathcal{S}$-module $\tilde{\mathcal{K}} = (\Lambda \tilde{P}, d_{\tilde{\nabla}})$ are cohomology classes of the complex $\mathrm{Hom}^{\mathcal{S}}(\mathcal{K}, \tilde{\mathcal{K}})$. See also the remark at the end of the previous subsection.

If $P = \tilde{P}$, $\nabla = \tilde{\nabla}$, then $\mathrm{End}\, P = \mathrm{Hom}_{C^\infty(M)}(P, \tilde{P})$ and we write $d_{\mathrm{End}\,\nabla}$ for $d_{\mathrm{Hom}(\nabla,\nabla)}$. In this particular case, Theorem 5.2.5 gives

$$H^*(L_{\nabla,\nabla}) = H^*(d_{\mathrm{End}\,\nabla}). \tag{5.7}$$

Note that (5.7) generalizes Theorem 5.2.1 to arbitrary flat connections. Indeed, this theorem is a particular case of (5.7) for $P = C^\infty(M)$ and for the canonical flat connection ∇ in $C^\infty(M)$, for which $\nabla_X = X$.

Now let μ be the natural $\mathcal{S}$-module structure in $\{\Lambda P, d_\nabla\}$ (Proposition 5.2.3). Then by (5.7) the cohomology mapping

$$\mu^* \colon H^*(\mathcal{S}(M)) \to H^*(L_{\nabla,\nabla}) \tag{5.8}$$

is identified with the mapping

$$H^*(M) = H^*(d) \to H^*(d_{\mathrm{End}\,\nabla}) \tag{5.9}$$

induced by the canonical cochain mapping

$$(\Lambda^*(M), d) \to (\Lambda(\mathrm{End}\, P), d_{\mathrm{End}\,\nabla})$$

with $\omega \mapsto \omega \otimes \mathrm{id}_P$. Map (5.8) (and therefore (5.9)) is the secondary analog of the mapping $\mathbb{R} \to \operatorname{End} V$ (where V is a real vector space) which associates with a scalar the corresponding scalar linear operator. The above considerations lead to the result that might have been predicted from the very beginning:

THEOREM. *The action of $H^*(M)$ on $H^*(d_\nabla)$ induced by the natural $\mathcal{S}$-module structure on $(\Lambda P, d_\nabla)$ coincides with the canonical action of $H^*(M)$ on $H^*(d_\nabla)$.*

5.3. Secondary modules

5.3.1. Now we are ready to introduce the notion of a *secondary module* and related constructions. This is achieved by repeating, in the context of $\mathcal{C}$-theory, the fundamentals of secondary linear algebra developed earlier. For this reason, here we can restrict ourselves to the list of the corresponding definitions, constructions, and results. What is done below can be interpreted geometrically as elements of the theory of *secondary vector bundles.*

Fix a diffiety $\mathcal{O}$. We work in the framework of the category of FG-modules over the filtered algebra $\mathcal{F}(\mathcal{O})$ (see Section 3.1) without mentioning it explicitly. Thus, "$\mathcal{F}(\mathcal{O})$-module" means "filtered geometric module over the filtered algebra $\mathcal{F}(\mathcal{O})$", "differential operator" means "filtered differential operator over $\mathcal{F}(\mathcal{O})$", and so on. We start with repeating the basic definitions of Subsection 5.2.5 in the context of $\mathcal{C}$-differential calculus over $\mathcal{O}$.

A *$\mathcal{C}$-complex* $K = \{K_i, \Delta_i\}$ over $\mathcal{O}$ is a complex

$$0 \to K_0 \xrightarrow{\Delta_0} K_1 \xrightarrow{\Delta_1} \cdots \xrightarrow{\Delta_{i-1}} K_i \xrightarrow{\Delta_i} \cdots$$

of $\mathcal{F}(\mathcal{O})$-modules and $\mathcal{C}$-differential operators. With the $\mathcal{C}$-complexes $(K, \Delta) = \{K_i, \Delta_i\}$ and $(\tilde{K}, \tilde{\Delta}) = \{\tilde{K}_i, \tilde{\Delta}_i\}$ we associate the complex

$$(\mathcal{GC}\operatorname{Diff}(K, \tilde{K}), L)\colon\ \mathcal{GC}\operatorname{Diff}(K, \tilde{K}) = \sum_\alpha \mathcal{GC}\operatorname{Diff}^\alpha(K, \tilde{K}),$$

where

$$\mathcal{GC}\operatorname{Diff}^\alpha(K, \tilde{K}) = \{\square\colon K \to \tilde{K} \mid \square(K_i) \subset \tilde{K}_{i+\alpha},\ \square|_{K_i} \in \mathcal{C}\operatorname{Diff}(K_i, \tilde{K}_{i+\alpha}),\ \forall i\}.$$

The differential L is defined as

$$L(\square) = \tilde{\Delta} \circ \square - (-1)^\alpha \square \circ \Delta \quad \text{for } \square \in \mathcal{GC}\operatorname{Diff}^\alpha(K, \tilde{K}).$$

Denote by $\mathcal{CL}^\alpha(K, \tilde{K})$ its α-th cohomology space and put

$$\mathcal{CL}(K, \tilde{K}) = \sum_\alpha \mathcal{CL}^\alpha(K, \tilde{K}).$$

Hence, the elements of $\mathcal{CL}(K, \tilde{K})$ are homotopy classes of graded cochain mappings from K to $\tilde{K}$. Therefore, any $h \in \mathcal{CL}(K, \tilde{K})$ generates naturally a mapping in cohomology, which is also denoted by h:

$$h\colon H^*(K) \to H^*(\tilde{K}).$$

Two $\mathcal{C}$-complexes K and $\tilde{K}$ are said to be *$\mathcal{C}$-homotopy equivalent* if there exist $h \in \mathcal{CL}^0(K, \tilde{K})$ and $\tilde{h} \in \mathcal{CL}(\tilde{K}, K)$ such that $\tilde{h} \circ h$ and $h \circ \tilde{h}$ are $\mathcal{C}$-homotopy equivalent to id_K and $\mathrm{id}_{\tilde{K}}$, respectively.

Note that for $K = \tilde{K}$ the complex $(\mathcal{GC}\operatorname{Diff}(K, K), L)$ is, in addition, an associative algebra with respect to the composition operation. More precisely, it is a differential graded algebra (DG-algebra) (see, for instance, [**87**]). The case $K = \bar{\Lambda}^*(\mathcal{O})$ deserves a special notation:

$$\mathcal{S}(\mathcal{O}) = \mathcal{GC}\,\mathrm{Diff}(\bar{\Lambda}^*(\mathcal{O}), \bar{\Lambda}^*(\mathcal{O})).$$

Note that $\bar{d} = \bar{d}(\mathcal{O}) \in \mathcal{S}(\mathcal{O})$.

Similarly to secondary vector spaces, secondary modules are $\mathcal{C}$-homotopy equivalence classes of a special type of $\mathcal{C}$-complexes. The following $\mathcal{C}$-homotopy equivalence is very important.

THEOREM. *The natural inclusion of $\mathcal{C}$-complexes*

$$\{\bar{\Lambda}^*(\mathcal{O}), \bar{d}\} \hookrightarrow \{\mathcal{S}(\mathcal{O}), L\}$$

taking a horizontal differential form ω on $\mathcal{O}$ to the operator of left multiplication by ω is a $\mathcal{C}$-homotopy equivalence. In particular,

$$\mathcal{CL}(\bar{\Lambda}^*(\mathcal{O}), \bar{\Lambda}^*(\mathcal{O})) = \bar{H}^*(\mathcal{O}).$$

As we have already noticed in Subsection 5.1.4, the horizontal de Rham cohomology algebra $\bar{H}^*(\mathcal{O}) = \mathcal{C}E_1^{0,*}(\mathcal{O})$ is interpreted as the "secondary smooth function algebra" on $\mathcal{O}$. Now this idea can be made more precise.

DEFINITION. The *secondary smooth function algebra* of a diffiety $\mathcal{O}$ is the $\mathcal{C}$-homotopy type of the horizontal de Rham complex $(\bar{\Lambda}^*(\mathcal{O}), \bar{d})$ of $\mathcal{O}$, or, equivalently, that of the complex $\mathcal{S}(\mathcal{O}) = (\mathcal{GC}\,\mathrm{Diff}(\bar{\Lambda}^*(\mathcal{O}), \bar{\Lambda}^*(\mathcal{O})), L)$.

5.3.2. The next step towards secondary modules is to distinguish a suitable class of $\mathcal{C}$-complexes.

DEFINITION. An *$\mathcal{S}$-module* (over $\mathcal{O}$) is a $\mathcal{C}$-complex (K, Δ) supplied with a left $\mathcal{S}(\mathcal{O})$-module structure

$$\mu\colon \mathcal{S}(\mathcal{O}) \to \mathcal{GC}\,\mathrm{Diff}(K, K)$$

such that

(i) if $f \in \mathcal{F}(\mathcal{O})$, then $\mu(f) = f$ (multiplication by f),
(ii) $\mu(\bar{d}) = \Delta$.

We emphasize the following consequences of the given definition:

1. An $\mathcal{S}$-module is, in particular, a unitary $\bar{\Lambda}^*(\mathcal{O})$-module (due to the natural inclusion $\bar{\Lambda}^*(\mathcal{O}) \hookrightarrow \mathcal{S}(\mathcal{O})$).
2. The structure homomorphism μ is a homomorphism of graded associative algebras.
3. The structure homomorphism μ is a cochain mapping of $(\mathcal{S}(\mathcal{O}), L)$ to the complex $(\mathcal{GC}\,\mathrm{Diff}(K, K), L)$ (due to 2 and (ii)) and as such generates a mapping in cohomology $\mu^*\colon \bar{H}^*(\mathcal{O}) \to \mathcal{CL}(K, K)$.
4. By 3, a natural action of $\bar{H}^*(\mathcal{O})$ on $H^*(K)$ is defined, i.e., $H^*(K)$ is an $\bar{H}^*(\mathcal{O})$-module.
5. The structure homomorphism μ is of degree zero (due to (i)), while Δ is a first-order differential operator (due to (i) and (ii)).

DEFINITION. Let $\mathcal{K} = (K, \Delta, \mu)$ and $\tilde{\mathcal{K}} = (\tilde{K}, \tilde{\Delta}, \tilde{\mu})$ be $\mathcal{S}$-modules. A graded $\mathcal{C}$-differential operator $\nabla\colon K \to \tilde{K}$ is called an *$\mathcal{S}$-module homomorphism* (or, shorter, *$\mathcal{S}$-homomorphism*) if and only if

$$\nabla \circ \mu(\square) = (-1)^{\nabla\square}\tilde{\mu}(\square) \circ \nabla \quad \text{for each } \square \in \mathcal{S}(\mathcal{O}).$$

Note (see Subsection 5.2.2) that an $\mathcal{S}$-homomorphism is automatically a graded cochain mapping, and the corresponding mapping $H(\nabla)\colon H^*(K) \to H^*(\tilde{K})$ in cohomology is also a homomorphism of $\bar{H}^*(\mathcal{O})$-modules. We denote by $\mathrm{Hom}_{\mathcal{S}}(\mathcal{K}, \tilde{\mathcal{K}})$ the linear space of all $\mathcal{S}$-homomorphisms from $\mathcal{K}$ to $\tilde{\mathcal{K}}$ in the context of $\mathcal{C}$-theory.

The $\mathcal{C}$-analogs of Examples 5.2.2 (1) and 5.2.2 (2) are the simplest $\mathcal{S}$-modules over a diffiety $\mathcal{O}$. Namely, the first of them is the algebra $\mathcal{S}(\mathcal{O})$ understood as the canonical left $\mathcal{S}(\mathcal{O})$-module, whereas the second is $\bar{\Lambda}^*(\mathcal{O})$. The evaluation mapping $\mathcal{S}(\mathcal{O}) \to \bar{\Lambda}^*(\mathcal{O})$, $\square \mapsto \square(1)$, is an $\mathcal{S}$-homomorphisms (see Example 5.2.2 (3)).

EXAMPLE. It is not difficult to verify that

$$\mathrm{Hom}_{\mathcal{S}}(\mathcal{S}(\mathcal{O}), \mathcal{S}(\mathcal{O})) = \mathcal{S}(\mathcal{O}), \qquad \mathrm{Hom}_{\mathcal{S}}(\bar{\Lambda} * (\mathcal{O}), \mathcal{S}(\mathcal{O})) = 0,$$
$$\mathrm{Hom}_{\mathcal{S}}(\mathcal{S}(\mathcal{O}), \bar{\Lambda}^*(\mathcal{O})) = \bar{\Lambda}^*(\mathcal{O}), \qquad \mathrm{Hom}_{\mathcal{S}}(\bar{\Lambda}^*(\mathcal{O}), \bar{\Lambda}^*(c\mathcal{O})) = \mathbb{R}.$$

DEFINITION. A *secondary module* over a diffiety $\mathcal{O}$ is the $\mathcal{C}$-homotopy type of an $\mathcal{S}$-module.

Any $\mathcal{S}$-module (K, Δ) may be viewed as a *secondary vector bundle*. Indeed, given a "secondary point" $M \hookrightarrow \mathcal{O}$, i.e., an integral submanifold M of $\mathcal{O}$, the restriction (K_M, Δ_M) of (K, Δ) to M is well defined due to the fact that the differential of the latter complex is a $\mathcal{C}$-differential operator. In particular, the restriction of $(\bar{\Lambda}^*(\mathcal{O}), \bar{d})$ to M is $(\Lambda^*(M), d)$. For the same reason, the restriction of $\mathcal{GC}\,\mathrm{Diff}(K, K)$ to M is $\mathcal{G}\,\mathrm{Diff}(K_M, K_M)$. In particular, $\mathcal{S}(\mathcal{O})$ restricts to $\mathcal{S}(M)$. In addition, the structure homomorphism μ can also be restricted to M by supplying the complex (K_M, Δ_M) with an $\mathcal{S}$-module structure in this way. In other words, (K_M, Δ_M) becomes a "secondary vector space" over M. This is the "secondary fiber" attached to the "secondary point" of M. It is easy to see also that an $\mathcal{S}$-homomorphism $\Delta\colon K \to \tilde{K}$ restricts naturally to an $\mathcal{S}$-homomorphism $\Delta_M\colon K_M \to \tilde{K}_M$, i.e., to a "secondary linear mapping" of "secondary fibers" over M, etc.

It is natural to call the restriction operation described above the *secondary evaluation mapping* at the *secondary point* M. Denote it by ϵ_M. It is important to note that

the secondary evaluation mapping ϵ_M takes $\mathcal{C}$-homotopy types over $\mathcal{O}$ to differential homotopy classes over M,

in full agreement with the fact that secondary modules and secondary vector spaces are the corresponding homotopy types.

5.3.3. Basic constructions of linear algebra with secondary modules are defined in parallel with those of secondary vector spaces. For instance, given $\mathcal{S}$-modules $\mathcal{K} = (K, \partial, \mu)$ and $(\tilde{K}, \tilde{\partial}, \tilde{\mu})$ over a diffiety $\mathcal{O}$, their tensor product $\mathcal{K} \otimes^{\mathcal{S}} \tilde{\mathcal{K}}$ is defined as the tensor product $K \otimes_{\bar{\Lambda}^*(\mathcal{O})} \tilde{K}$ of $\bar{\Lambda}^*(\mathcal{O})$-modules supplied canonically with the $\mathcal{S}$-module structure $\mu \otimes^{\mathcal{S}} \tilde{\mu}$. This structure is uniquely characterized by the actions of $\mathcal{C}$-derivations of the algebra $\bar{\Lambda}^*(\mathcal{O})$ on $K \otimes_{\bar{\Lambda}^*(\mathcal{O})} \tilde{K}$. If δ is such a derivation, then

$$[(\mu \otimes^{\mathcal{S}} \tilde{\mu})(\delta)](\kappa \otimes \tilde{\kappa}) = [\mu(\delta)](\kappa) \otimes \tilde{\kappa} + (-1)^{\delta\kappa} \kappa \otimes [\tilde{\mu}(\delta)](\tilde{\kappa}), \quad \kappa \otimes \tilde{\kappa} \in K \otimes_{\bar{\Lambda}^*(\mathcal{O})} \tilde{K}.$$

Similarly, the module $\mathrm{Hom}^{\mathcal{S}}(\mathcal{K}, \tilde{\mathcal{K}})$ of *secondary pre-homomorphisms* of the secondary module $\mathcal{K}$ to $\tilde{\mathcal{K}}$ coincides with $\mathrm{Hom}_{\bar{\Lambda}^*(\mathcal{O})}(\mathcal{K}, \tilde{\mathcal{K}})$ as $\bar{\Lambda}^*(\mathcal{O})$-module, with the canonical $\mathcal{S}$-module structure $\mathrm{Hom}^{\mathcal{S}}(\mu, \tilde{\mu})$ uniquely defined by

$$[\mathrm{Hom}^{\mathcal{S}}(\mu, \tilde{\mu})(\delta)](h) = \tilde{\mu}(\delta) \circ h - (-1)^{h\delta} h \circ \mu(\delta), \quad h \in \mathrm{Hom}^{\mathcal{S}}(\mathcal{K}, \tilde{\mathcal{K}}),$$

for any $\mathcal{C}$-derivation δ of the algebra $\bar{\Lambda}^*(\mathcal{O})$. We emphasize that it is important not to confuse $\mathrm{Hom}^{\mathcal{S}}(\mathcal{K}, \tilde{\mathcal{K}})$ with $\mathrm{Hom}_{\mathcal{S}}(\mathcal{K}, \tilde{\mathcal{K}})$.

DEFINITION. The $\mathcal{C}$-homotopy types of the $\mathcal{S}$-modules $\mathcal{K} \otimes^{\mathcal{S}} \tilde{\mathcal{K}}$ and $\mathrm{Hom}^{\mathcal{S}}(\mathcal{K}, \tilde{\mathcal{K}})$ are called the *secondary tensor product* and the *secondary module of homomorphisms* of the secondary modules that are the $\mathcal{C}$-homotopy types of the $\mathcal{S}$-modules $\mathcal{K}$ and $\tilde{\mathcal{K}}$, respectively.

This definition immediately brings up the problem of computing the cohomology of the complexes $\mathcal{K} \otimes^{\mathcal{S}} \tilde{\mathcal{K}}$ and $\mathrm{Hom}^{\mathcal{S}}(\mathcal{K}, \tilde{\mathcal{K}})$. A similar problem can be formuated for any secondary concept, and a general method to solve it will be presented below. This is essentially the same technique as that we have used earlier to compute the first term of the $\mathcal{C}$-spectral sequence.

As we have already noted when discussing secondary vector spaces, there exist two different ways to introduce the notion of a secondary linear mapping. The one that is more naive is given by

DEFINITION (1). *Secondary linear mappings* from a secondary module $\mathcal{M}$ represented by an $\mathcal{S}$-module $\mathcal{K} = (K, \partial)$ to another one represented by $\tilde{\mathcal{K}} = (\tilde{K}, \tilde{\partial})$ are cohomology classes of the complex $(\mathcal{GC}\,\mathrm{Diff}(K, \tilde{K}), L)$. The *secondary module of secondary homomorphisms* from $\mathcal{M}$ to $\tilde{\mathcal{M}}$ is the $\mathcal{C}$-homotopy type of the complex $(\mathcal{GC}\,\mathrm{Diff}(K, \tilde{K}), L)$.

A more sophisticated but more satisfactory from the conceptual point of view definition is obtained by substituting the $\mathcal{S}$-module $\mathrm{Hom}^{\mathcal{S}}(\mathcal{K}, \tilde{\mathcal{K}})$ for the complex $(\mathcal{GC}\,\mathrm{Diff}(K, \tilde{K}), L)$ in the preceding definition.

DEFINITION (2). *Secondary homomorphisms* from a secondary module $\mathcal{M}$ to $\tilde{\mathcal{M}}$ are cohomology classes of the complex $\mathrm{Hom}^{\mathcal{S}}(\mathcal{K}, \tilde{\mathcal{K}})$. The *secondary module of secondary homomorphisms* from $\mathcal{M}$ to $\tilde{\mathcal{M}}$ is the $\mathcal{C}$-homotopy type of the complex $\mathrm{Hom}^{\mathcal{S}}(\mathcal{K}, \tilde{\mathcal{K}})$.

In the next subsection we shall see that these definitions coincide on the class of natural secondary modules.

5.3.4. Now we are going to introduce *natural secondary modules* over a diffiety $\mathcal{O}$, which are in the same relation with general secondary modules as natural secondary vector spaces are with general ones. This class of modules plays a principal role in solving the secondarization problem; the reasons for that become clear later. In what follows, the definitions and constructions of Subsection 5.2.3 are placed into the context of $\mathcal{C}$-theory.

The starting point is the notion of a $\mathcal{C}$-connection, the $\mathcal{C}$-analog of the standard notion.

DEFINITION. Let P be an $\mathcal{F}(\mathcal{O})$-module. An $\mathcal{F}(\mathcal{O})$-module homomorphism

$$\nabla \colon \mathcal{C}D(\mathcal{O}) \to \mathcal{C}\,\mathrm{Diff}_1(P, P)$$

is called a *$\mathcal{C}$-connection* in P if

$$\nabla(X)(fp) = f\nabla(X)(p) + X(f)p$$

for any $f \in \mathcal{F}(\mathcal{O})$, $X \in \mathcal{C}D(\mathcal{O})$, $p \in P$. Below we follow the standard notation and write ∇_X instead of $\nabla(X)$.

A $\mathcal{C}$-connection is said to be *flat* if $[\nabla_X, \nabla_Y] = \nabla_{[X,Y]}$ for any vector fields $X, Y \in \mathcal{C}D(\mathcal{O})$. An equivalent formulation is to say that its *curvature tensor* defined as

$$R_\nabla(X,Y)\colon \stackrel{\text{def}}{=} [\nabla_X, \nabla_Y] - \nabla_{[X,Y]}, \quad X, Y \in \mathcal{C}D(\mathcal{O}),$$

vanishes.

Since ∇_X is a $\mathcal{C}$-differential operator, the $\mathcal{C}$-connection ∇ can be restricted to any integral submanifold M of $\mathcal{O}$ by supplying it with a connection, say ∇^M. More exactly,

$$\nabla^M_{X|_M}(p|_M) = \nabla_X(p)|_M, \quad X \in \mathcal{C}D(\mathcal{O}),\ p \in P.$$

Let us associate with an $\mathcal{F}(\mathcal{O})$-module P the graded $\bar{\Lambda}^*(\mathcal{O})$-module $\bar{\Lambda}P = \sum_i \bar{\Lambda}^i P$, where

$$\bar{\Lambda}P = \bar{\Lambda}^*(\mathcal{O}) \otimes_{\mathcal{F}(\mathcal{O})} P, \qquad \bar{\Lambda}^i P = \bar{\Lambda}^i(\mathcal{O}) \otimes_{\mathcal{F}(\mathcal{O})} P.$$

The elements of $\bar{\Lambda}P$ are called *P-valued horizontal differential forms.*

The operator $d_\nabla\colon \bar{\Lambda}^p(\mathcal{O}) \to \bar{\Lambda}^{p+1}(\mathcal{O})$, $i = 0, 1, 2, \ldots$, defined as

$$\begin{aligned} d_\nabla(\omega)(X_1, \ldots, X_{p+1}) &= \sum_i (-1)^{i-1} \nabla_{X_i}(\omega(X_1, \ldots, \hat{X}_i, \ldots, X_{p+1})) \\ &\quad + \sum_{i<j} (-1)^{i+j} \omega([X_i, X_j], X_1, \ldots, \hat{X}_i, \ldots, \hat{X}_j, \ldots, X_{p+1}) \end{aligned}$$

with $\omega \in \bar{\Lambda}^p P$ and $X_1, \ldots, X_{p+1} \in \mathcal{C}D(\mathcal{O})$, is naturally associated with a $\mathcal{C}$-connection ∇ in P. It is easy to see that d_∇ is a $\mathcal{C}$-differential operator and $d_\nabla^2 = 0$ if and only if ∇ is flat. Thus, in this case $(\bar{\Lambda}P, d_\nabla)$ is a $\mathcal{C}$-complex. This complex carries an $\mathcal{S}$-module structure in a natural way, because the algebra $\mathcal{S}(\mathcal{O})$ is generated by the operators $\bar{d}$, i_X, $X \in \mathcal{C}D(\mathcal{O})$, and ω^P, where ω^P, $\omega \in \bar{\Lambda}^*(\mathcal{O})$, stands for the operator of left multiplication by ω in $\bar{\Lambda}P$.

PROPOSITION. *Consider the $\mathcal{C}$-complex $(\bar{\Lambda}P, d_\nabla)$. It admits a unique $\mathcal{S}$-module structure μ such that*

$$\mu(\omega) = \omega^P, \ \mu(i_X) = i_X^P, \quad \omega \in \bar{\Lambda}^*(\mathcal{O}),\ X \in \mathcal{C}D(\mathcal{O}),$$

where i_X^P denotes the operator of insertion of X into horizontal P-valued forms.

COROLLARY. *A flat connection ∇ generates a unique $\mathcal{C}\operatorname{Diff}(\mathcal{F}(\mathcal{O}), \mathcal{F}(\mathcal{O}))$-module structure ν in P such that*

(i) $\nu(f) = f$ *(multiplication by $f \in \mathcal{F}(\mathcal{O})$)*;
(ii) $\nu(X) = \nabla_X$, $X \in \mathcal{C}D(\mathcal{O})$;
(iii) $[f, \nu(\Box)] = \nu([f, \Box])$, $\Box \in \mathcal{C}\operatorname{Diff}(\mathcal{F}(\mathcal{O}), \mathcal{F}(\mathcal{O}))$.

DEFINITION. The $\mathcal{S}$-modules described in Proposition 5.3.4 are said to be *natural.* The $\mathcal{C}$-homotopy types of the natural $\mathcal{S}$-modules are called *natural secondary modules.*

DEFINITION. The $\mathcal{C}\operatorname{Diff}(\mathcal{F}(\mathcal{O}), \mathcal{F}(\mathcal{O}))$-modules described in Corollary 5.3.4 are said to be *horizontal.*

Hence, for a projective $\mathcal{F}(\mathcal{O})$-module P, we have the following natural identifications:

Flat $\mathcal{C}$-connections in P	$\Longleftrightarrow$	Horizontal module structures in P	$\Longleftrightarrow$	Natural $\mathcal{S}$-module structures in $\bar{\Lambda}P$

Let us note that the secondary evaluation mapping ϵ_M takes a natural $\mathcal{S}$-module over $\mathcal{O}$ to a natural $\mathcal{S}$-module over M. Namely, ϵ_M takes the natural $\mathcal{S}$-module associated with a flat $\mathcal{C}$-connection ∇ in a $\mathcal{F}(\mathcal{O})$-module P to the natural $\mathcal{S}$-module (over M) associated with the flat connection ∇^M in the $C^\infty(M)$-module $P|_M$.

One of the most important examples of natural modules is provided by the theory of $\mathcal{C}$-spectral sequences. Namely, these are columns $(\mathcal{C}E_0^{p,*}(\mathcal{O}), d_0)$ of the 0-th term of the $\mathcal{C}$-spectral sequence. Indeed,

$$\mathcal{C}E_0^{p,q}(\mathcal{O}) = \bar{\Lambda}^q(\mathcal{O}) \otimes_{\mathcal{F}(\mathcal{O})} \mathcal{C}^p\Lambda^p(\mathcal{O}).$$

Therefore, $\mathcal{C}E_0^{p,*}(\mathcal{O}) = \bar{\Lambda}P$, with $P = \mathcal{C}^p\Lambda^p(\mathcal{O})$. The $\mathcal{F}(\mathcal{O})$-module $\mathcal{C}^p\Lambda^p(\mathcal{O})$ carries a canonical $\mathcal{C}$-connection $\mathcal{C}\nabla^p$ defined by

$$(\mathcal{C}\nabla^p)_X(\omega) = L_X(\omega), \quad X \in \mathcal{C}D(\mathcal{O}),\ \omega \in \mathcal{C}^p\Lambda^p(\mathcal{O}),$$

where L_X stands for the operator of Lie derivative along X. This definition is well-posed due to the fact that $L_{fX}(\omega) = fL_X(\omega)$ for $f \in \mathcal{F}(\mathcal{O})$, $\omega \in \mathcal{C}^p\Lambda^p$, $X \in \mathcal{C}D(\mathcal{O})$. In other words, the correspondence

$$X \mapsto L_X|_{\mathcal{C}^p\Lambda^p(\mathcal{O})}, \quad X \in \mathcal{C}D(\mathcal{O}),$$

is a homomorphism of $\mathcal{F}(\mathcal{O})$-modules. It can be easily verified that $d_0|_{\mathcal{C}E_0^{p,*}(\mathcal{O})} = d_{\mathcal{C}\nabla^p}$. Moreover, the standard formula $[L_X, L_Y] = L_{[X,Y]}$ shows that the $\mathcal{C}$-connection $\mathcal{C}\nabla^p$ is flat. With these observations, we get a more satisfactory version of the notion of secondary differential forms.

DEFINITION. The *secondary module* of *secondary differential p-forms* over a diffiety $\mathcal{O}$ is the secondary module that is represented by the natural $\mathcal{S}$-module $(\mathcal{C}E_0^{p,*}(\mathcal{O}), d_0)$.

The simplest class of natural $\mathcal{S}$-modules corresponds to free $\mathcal{F}(\mathcal{O})$-modules. Such modules are of the form $\mathcal{F}(\mathcal{O}) \otimes_{\mathbb{R}} V$, where V is an $\mathbb{R}$-vector space. The canonical $\mathcal{C}$-connection ∇^V in it is defined as

$$\nabla_X^V(f \otimes v) = X(f) \otimes v, \quad X \in \mathcal{C}D(\mathcal{O}),\ v \in V.$$

This connection is flat, and the natural $\mathcal{S}$-module $\mathcal{K}^V$ corresponding to it, is identified with the free $\bar{\Lambda}^*(\mathcal{O})$-module $\bar{\Lambda}(\mathcal{O}) \otimes V$ supplied with the $\mathcal{S}(\mathcal{O})$-action μ^V defined as $\mu^V(\square)(\omega \otimes v) = \square(\omega) \otimes v$, $\square \in \mathcal{S}(\mathcal{O})$. The secondary module corresponding to $\mathcal{K}^V$ is interpreted as the module of secondary V-valued functions on $\mathcal{O}$.

The following assertion gives an interpretation of the 0-th cohomology of a natural secondary module.

PROPOSITION. *Let $\mathcal{K} = (\Lambda P, d_\nabla)$ be the natural $\mathcal{S}$-module associated with a flat $\mathcal{C}$-connection ∇. Then there is a natural isomorphism*

$$H^0(d_\nabla) = \mathrm{Hom}_{\mathcal{S}}(\Lambda^*(M), \mathcal{K}).$$

For an arbitrary $\mathcal{S}$-module $\mathcal{K} = (K, \partial, \mu)$ such that $K_i = 0$, $i < 0$, we have a natural $\mathcal{S}$-homomorphism $I_{\mathcal{K}} \colon \mathcal{K}^V \to \mathcal{K}$ with $V = H^0(K)$,

$$I_{\mathcal{K}}(\omega \otimes \kappa) = \mu(\omega)(\kappa), \quad \omega \in \bar{\Lambda}^*(\mathcal{O}),\ \kappa \in K_0,\ \partial(\kappa) = 0.$$

The $\mathcal{S}$-homomorphism $I_{\mathcal{K}}$ gives a characterization of natural $\mathcal{S}$-modules.

PROPOSITION. *An $\mathcal{S}$-module $\mathcal{K} = (K, \partial, \mu)$ is natural if and only if $K_i = 0$, $i < 0$, and the $\mathcal{S}$-homomorphism $I_{\mathcal{K}}$ is a local isomorphism.*

5.3.5. As in the case of secondary vector spaces (see Subsection 5.2.5), operations of "secondary linear algebra" on natural $\mathcal{S}$-modules over a diffiety $\mathcal{O}$ can be described directly as operations on the corresponding $\mathcal{C}$-connections. Namely, the tensor product $\mathcal{K} \otimes^{\mathcal{S}} \tilde{\mathcal{K}}$ of natural $\mathcal{S}$-modules $\mathcal{K} = (\bar{\Lambda}P, \bar{d}_\nabla)$ and $\tilde{\mathcal{K}} = (\bar{\Lambda}\tilde{P}, \bar{d}_{\tilde{\nabla}})$ is identified canonically with the natural $\mathcal{S}$-module corresponding to the $\mathcal{C}$-connection $\nabla \otimes \tilde{\nabla}$ in $P \otimes_{C^\infty(M)} \tilde{P}$, where

$$(\nabla \otimes \tilde{\nabla})_X = \nabla_X \otimes \mathrm{id}_{\tilde{K}} + \mathrm{id}_K \otimes \tilde{\nabla}_X, \quad X \in \mathcal{C}D(\mathcal{O}).$$

In other words,

$$\mathcal{K} \otimes^{\mathcal{S}} \tilde{\mathcal{K}} = \left(\bar{\Lambda}(P \otimes_{\mathcal{F}(\mathcal{O})} \tilde{P}), \bar{d}_{\nabla \otimes \tilde{\nabla}}\right).$$

Similarly,

$$\mathrm{Hom}^{\mathcal{S}}(\mathcal{K}, \tilde{\mathcal{K}}) = \left(\bar{\Lambda}\left(\mathrm{Hom}_{\mathcal{F}(\mathcal{O})}(P, \tilde{P})\right), \bar{d}_{\mathrm{Hom}(\nabla, \tilde{\nabla})}\right),$$

where the flat $\mathcal{C}$-connection $\mathrm{Hom}(\nabla, \tilde{\nabla})$ in $\mathrm{Hom}_{\mathcal{F}(\mathcal{O})}(P, \tilde{P})$ is defined by

$$\left[\mathrm{Hom}(\nabla, \tilde{\nabla})\right]_X (h) = \tilde{\nabla}_X \circ h - h \circ \nabla_X, \quad h \in \mathrm{Hom}_{\mathcal{F}(\mathcal{O})}(P, \tilde{P}).$$

According to Definition 5.3.3 (1), the problem of describing secondary homomorphisms between natural secondary modules $\mathcal{M}$ and $\tilde{\mathcal{M}}$ requires computation of the cohomology $\mathcal{L}(\mathcal{K}, \tilde{\mathcal{K}})$ of the complex $(\mathcal{GC}\,\mathrm{Diff}(\bar{\Lambda}P, \bar{\Lambda}\tilde{P}), L)$ for the natural $\mathcal{S}$-modules $\mathcal{K} = (\bar{\Lambda}P, \bar{d}_\nabla)$ and $\tilde{\mathcal{K}} = (\bar{\Lambda}\tilde{P}, \bar{d}_{\tilde{\nabla}})$ representing them.

The differential L of this complex is defined (see (5.3)) as

$$L = L_{\nabla, \tilde{\nabla}}(\square) = \bar{d}_{\tilde{\nabla}} \circ \square - (-1)^{\square} \square \circ \bar{d}_\nabla.$$

THEOREM. *The canonical inclusion*

$$\left(\bar{\Lambda}(\mathrm{Hom}_{\mathcal{F}(\mathcal{O})}(P, \tilde{P})), \bar{d}_{\mathrm{Hom}(\nabla, \tilde{\nabla})}\right) \hookrightarrow \left(\mathcal{GC}\,\mathrm{Diff}(\bar{\Lambda}P, \bar{\Lambda}\tilde{P}), L_{\nabla, \tilde{\nabla}}\right)$$

is a $\mathcal{C}$-homotopy equivalence. In particular,

$$H^*(L_{\nabla, \tilde{\nabla}}) = H^*(\bar{d}_{\mathrm{Hom}(\nabla, \tilde{\nabla})}). \tag{5.10}$$

Note that Theorem 5.3.1 is a particular case of this theorem with $P = \mathcal{F}(\mathcal{O})$ and flat $\mathcal{C}$-connection ∇ defined as $\nabla_X = X$, $X \in \mathcal{C}D(\mathcal{O})$.

COROLLARY. *Definitions* 5.3.3 (1) *and* (2) *of secondary homomorphisms are equivalent in the class of natural secondary modules.*

The particular case $P = \tilde{P}$, $\nabla = \tilde{\nabla}$ deserves to be mentioned separately. Put $\mathrm{End}_{\mathcal{F}(\mathcal{O})} P = \mathrm{Hom}_{\mathcal{F}(\mathcal{O})}(P, P)$. Then Theorem 5.3.5 shows that

$$H^*(L_{\nabla, \nabla}) = H^*(\bar{d}_{\mathrm{End}\,\nabla}), \tag{5.11}$$

where $\bar{d}_{\mathrm{End}\,\nabla}$ stands for $\bar{d}_{\mathrm{Hom}(\nabla, \nabla)}$. Let us note in this connection that, in accordance with the chosen approach, either the secondary module corresponding to $(\mathcal{GC}\,\mathrm{Diff}(\bar{\Lambda}P, \bar{\Lambda}P), L_{\nabla, \nabla})$, or that corresponding to $(\bar{\Lambda}(\mathrm{End}_{\mathcal{F}(\mathcal{O})} P), \bar{d}_{\mathrm{End}\,\nabla})$ is interpreted as the *algebra of secondary endomorphisms* of the secondary module $\mathcal{M}$ represented by $(\bar{\Lambda}P, \bar{d}_\nabla)$. Evidently, the action of this algebra on $\mathcal{M}$ is just the action of $\mathcal{CL}(K, K)$ on $H^*(K)$, $K = (\bar{\Lambda}P, d_\nabla)$, discussed earlier. It is worth emphasizing that "algebra of secondary endomorphism" is to be understood as a *secondary algebra* over the "secondary smooth function algebra" over the diffiety $\mathcal{O}$ in question. More precisely, this assumes the existence of a homomorphism of DG-algebras representing the $\mathcal{C}$-homotopy types of secondary algebras in question. In the case we are interested in, this is the $\mathcal{S}$-module structure mapping

$$\mu\colon \mathcal{S}(\mathcal{O}) \to \mathcal{GC}\,\mathrm{Diff}(\bar{\Lambda}P, \bar{\Lambda}P).$$

It induces the mapping in cohomology

$$\mu^*: H^*(\mathcal{S}(M)) \to H^*(L_{\nabla,\nabla}), \tag{5.12}$$

which is identified, due to Theorems 5.3.1 and 5.3.5, with the mapping

$$\bar{H}^*(\mathcal{O}) = H^*(\bar{d}) \to H^*(\bar{d}_{\mathrm{End}\,\nabla}) \tag{5.13}$$

induced by the canonical cochain mapping

$$(\bar{\Lambda}^*(\mathcal{O}), \bar{d}) \to (\bar{\Lambda}(\mathrm{End}_{\mathcal{F}(\mathcal{O})} P), \bar{d}_{\mathrm{End}\,\nabla})$$

taking $\omega \in \bar{\Lambda}(\mathcal{O})$ to $\omega \otimes \mathrm{id}_P$.

These observations yield the following result.

THEOREM. *The action of $\bar{H}^*(\mathcal{O})$ on $\bar{H}^*(\bar{d}_\nabla)$ induced by the natural $\mathcal{S}$-module structure in $\{\bar{\Lambda}P, \bar{d}_\nabla\}$ coincides with the canonical action of $\bar{H}^*(\mathcal{O})$ on $\bar{H}^*(\bar{d}_\nabla)$.*

5.4. Secondary linear differential operators

In this section, secondary linear differential operators are discussed. There are different approaches to the subject, and we present here the one that does not require additional cohomological constructions and therefore is simpler to expose. However, it is not the best from the conceptual point of view. In fact, it is clear from the preceding discussion of secondary homomorphisms that secondary differential operators should be cohomology classes of suitable secondary modules, and this is really the case.

5.4.1. Let (K, Δ, μ) and $(\tilde{K}, \tilde{\Delta}, \tilde{\mu})$ be $\mathcal{S}$-modules connected by a graded differential operator $\Xi\colon K \to \tilde{K}$, which is no longer supposed to be $\mathcal{C}$-differential. For an operator $\Box \in \mathcal{S}(\mathcal{O})$ put

$$\delta_\Box(\Xi) = \Xi \circ \mu(\Box) - (-1)^{\Box \cdot \Xi} \tilde{\mu}(\Box) \circ \Xi.$$

DEFINITION. Let $\mathcal{M}$ and $\tilde{\mathcal{M}}$ be secondary modules associated with $\mathcal{S}$-modules (K, Δ, μ) and $(\tilde{K}, \tilde{\Delta}, \tilde{\mu})$, respectively. A *secondary differential operator* of order $\le k$ from $\mathcal{M}$ to $\tilde{\mathcal{M}}$ is a mapping $H^*(K) \to H^*(\tilde{K})$ induced by a cochain mapping $\Xi\colon K \to \tilde{K}$ which is a graded differential operator over $\mathcal{F}(\mathcal{O})$ such that

$$\delta_{\Box_1}(\delta_{\Box_2} \cdots (\delta_{\Box_k}(\Xi)) \dots) \in \mathcal{GC}\,\mathrm{Diff}(K, \tilde{K})$$

for any $\Box_1, \dots, \Box_k \in \mathcal{S}(\mathcal{O})$.

Evidently, this definition imitates the standard algebraic Definition 1.1.2 of usual differential operators (in the notation of Subsection 1.1.2) in the equivalent form $\delta_{a_1,\dots,a_s}(\Box) \in \mathrm{Hom}_A(P, Q)$.

In the following two examples Ω_p stands for the secondary module of secondary differential p-forms (see Definition 5.3.4).

EXAMPLE (1). The differential $d_1\colon \mathcal{C}E_1^{p,*}(\mathcal{O}) \to \mathcal{C}E_1^{p+1,*}(\mathcal{O})$ is a first-order secondary differential operator. More precisely, this is the mapping in cohomology corresponding to a first-order secondary differential operator acting from Ω_p to Ω_{p+1}.

By the One Line Theorem (see Subsection 4.3.7), for $\mathcal{O} = J^\infty(\pi)$ the first differential $d_1\colon \mathcal{C}E_1^{0,*}(\mathcal{O}) \to \mathcal{C}E_1^{1,*}(\mathcal{O})$ reduces to the Euler operator $\mathcal{E} = d_1^{0,n}\colon \mathcal{C}E_1^{0,n}(\mathcal{O}) \to \mathcal{C}E_1^{1,n}(\mathcal{O})$ assigning to an action the corresponding Euler–Lagrange equation. The well-known expression for the Euler operator in terms of local jet coordinates presents it as a differential operator of infinite order. But now we see that it is a first-order secondary differential operator, just as the secondary analog of the classical de Rham differential $d\colon C^\infty(M) \to \Lambda^1(M)$ should be.

The next example tells us that higher symmetries, interpreted earlier for quite different reasons as secondary vector fields, are in fact secondary differential operators of the first order.

EXAMPLE (2). If $\mathcal{O} = \mathcal{Y}_\infty$ and $\chi \in \operatorname{Sym}\mathcal{O}$, then the secondary Lie derivative L_χ along χ is a first-order secondary differential operator. This assertion is equivalent to Theorem 3.3.7.

These examples provide an additional confirmation of the list of analogs we started in Subsection 5.1.1 and make it substantially more precise.

Now we show what secondary scalar differential operators, i.e., those acting on the "secondary smooth function algebra" (Definition 5.3.1), look like for $\mathcal{O} = J^\infty(\pi)$. Let us fix a local jet chart (x_i, u_σ^j). A k-th order differential operator $\square\colon \mathcal{F} \to \mathcal{F}$, $\mathcal{F} = \mathcal{F}(J^\infty(\pi))$, is said to be *vertical* (with respect to the chosen chart) if its local expression does not contain derivations with respect to x_i. In other words, it looks as

$$\square = \sum_{0<s\le k} \sum_{\substack{i_1,\dots,i_s \\ \sigma_1,\dots,\sigma_s}} a_{\sigma_1,\dots,\sigma_s}^{i_1,\dots,i_s} \frac{\partial^s}{\partial u_{\sigma_1}^{i_1} \cdots \partial u_{\sigma_s}^{i_s}}.$$

PROPOSITION. *A vertical differential operator represents a secondary differential operator if and only if* $[D_i, \square] = 0$, $i = 1, \dots, n$, *and vice versa. Any operator of this kind is of the form*

$$\ni_\nabla = \sum_{i,\sigma} \mathcal{L}_\sigma(\nabla_i) \circ \frac{\partial}{\partial u_\sigma^i}$$

with $\nabla = (\nabla_1, \dots, \nabla_m)$ *being a vertical differential operator and*

$$\mathcal{L}_\sigma(\nabla_i) = [D_{i_1}, [D_{i_2}, \dots, [D_{i_s}, \nabla_i] \dots]] \quad \text{for} \quad \sigma = (i_1, \dots, i_s).$$

Vertical operators ∇ play the same role as generating functions for evolutionary derivations. However, ∇ is not defined uniquely if $\operatorname{ord}\nabla > 0$. In fact, $\ni^2 = 0$, i.e., $\ni_{\ni_\Delta} = 0$, for any vertical Δ.

Secondary differential operators are different from the usual ones in some respects. For instance, a scalar secondary operator of order greater than one is not, generally, a sum of compositions of lower order secondary operators. However, this standard property remains valid "asymptotically" in the sense that it is still true for all its restrictions to secondary functions of a given filtration.

5.4.2. In this subsection, we give an example of how some *natural* ideas can be programmed in the language of Secondary Calculus. Namely, the notion of a secondary Poisson structure on $J^\infty(\pi)$ is defined by copying verbatim the classical pattern and giving the "secondary" meaning to all its ingredients. This way one gets the Hamiltonian formalism in field theory.

In the classical situation, one of several ways to define a Poisson bracket $\{\cdot,\cdot\}$ on a manifold M is by means of a $C^\infty(M)$-homomorphism

$$\Gamma\colon \Lambda^1(M) \to D(M) \quad \text{such that} \quad \{f,g\} = \Gamma(df)(g). \tag{5.14}$$

Such a homomorphism Γ is called *Hamiltonian* if the resuting bracket is skew-symmetric and satisfies the Jacobi identity. Thus, the Poisson structures on M are in one-to-one correspondence with the Hamiltonian homomorphisms. The advantage of approach (5.14) for our purposes is that all its components except $D(M)$ are already secondarized. As for $D(M)$, we show later (Theorem 5.7.3) that the secondarization $D(M) \Rightarrow \operatorname{sym}\mathcal{O}$, which was so important in capturing the first idea of Secondary Calculus, is not, in fact, completely satisfactory. More precisely, $\operatorname{sym}\mathcal{O}$ is only the 0-th component of the whole algebra of secondary vector fields, which is graded according to the general notion of a secondary differential operator introduced above. Nevertheless, for $\mathcal{O} = J^\infty(\pi)$ only the 0-th component is not trivial, and therefore, in this case the concept of a Poisson structure is directly secondarizable on the basis of what is already known. The definition that follows is obtained from (5.14) by replacing all the objects appearing in it by their secondary analogs and taking into account the special form they acquire on $\mathcal{O} = J^\infty(\pi)$.

DEFINITION. A *secondary Poisson structure* on $J^\infty(\pi)$ is a Lie algebra structure on $\bar{H}^n(J^\infty(\pi)) = \mathcal{C}E_1^{0,n}(J^\infty(\pi))$:

$$\{\mathcal{L}_1, \mathcal{L}_2\} = L_{\Delta(d_1(\mathcal{L}_1))}(\mathcal{L}_2), \quad \mathcal{L}_i \in \bar{H}^n(J^\infty),\ d_1 = d_1^{0,n}, \tag{5.15}$$

given by a $\mathcal{C}$-differential operator

$$\Delta\colon \mathcal{C}E_1^{1,n}(\mathcal{O}) = \hat{\kappa} \to \kappa = \operatorname{sym}(J^\infty(\pi)),$$

where $d_1^{0,n}$ is the differential in the first term of the $\mathcal{C}$-spectral sequence (= the Euler operator) on $J^\infty(\pi)$.

The fact that the operator Δ in the above definition is assumed to be $\mathcal{C}$-differential deserves a special comment. Indeed, according to the general theory of secondary differential operators, it should be the mapping in cohomology corresponding to a secondary homomorphism connecting secondary first-order differential forms and vector fields. This means that on the level of $\mathcal{S}$-modules representing them, the operator, say Δ^{pre}, representing Δ, should be $\mathcal{C}$-differential. Therefore, in the case $\mathcal{O} = J^\infty(\pi)$ the $\mathcal{C}$-differential operator Δ^{pre} takes the n-th order component of the $\mathcal{S}$-module $(\mathcal{C}E_0^{1,n}(J^\infty(\pi)), d_0)$, i.e., $\mathcal{C}\operatorname{Diff}(\kappa, \bar{\Lambda}^n)$, to the 0-th order component of the complex $(\bar{\mathcal{J}}(\kappa) \otimes_{\mathcal{F}(J^\infty(\pi))} \bar{\Lambda}^*(\mathcal{O}), S)$ with S being the Spencer operator, i.e., to $\bar{\mathcal{J}}(\kappa)$. It is not difficult to verify that the operator Δ in this case coincides with the composition

$$\hat{\kappa} = \mathcal{C}\operatorname{Diff}_0(\kappa, \bar{\Lambda}^n) \hookrightarrow \mathcal{C}\operatorname{Diff}(\kappa, \bar{\Lambda}^n) \xrightarrow{\Delta^{\mathrm{pre}}} \mathcal{J}(\kappa) \xrightarrow{\pi_{\infty,0}} \bar{\mathcal{J}}^0(\kappa) = \kappa,$$

and therefore is $\mathcal{C}$-differential as well.

As in the classical case, we call a $\mathcal{C}$-differential operator $\Delta\colon \hat{\kappa} \to \kappa$ *Hamiltonian* if the corresponding bracket (5.15) is a Lie bracket, i.e., is skew-symmetric and satisfies the Jacobi identity. The following result gives a useful characterization of Hamiltonian operators over $J^\infty(\pi)$.

THEOREM. *A $\mathcal{C}$-differential operator $\Delta\colon \hat{\kappa} \to \kappa$ (over $J^\infty(\pi)$) is Hamiltonian if and only if it is skew-adjoint, i.e., $\Delta^* = -\Delta$, and the identity*

$$[\mathcal{Э}_{\Delta\varphi}, \Delta] = \ell_{\Delta\varphi} \circ \Delta + \Delta \circ l^*_{\Delta\varphi} \tag{5.16}$$

holds for an arbitrary polynomial $\varphi = \varphi(x)$ in x of order $\leq \deg\Delta + \Phi(\Delta)$ with $\Phi(\Delta)$ being the highest jet order of the coefficients of Δ.

EXAMPLE. Any skew-symmetric operator Δ with constant coefficients, i.e., of the form $\Delta = \sum a_\sigma D_\sigma$ with a_σ being constant matrices, is Hamiltonian.
Indeed, it is easy to see that for such an operator Δ both sides of (5.16) vanish.

Due to Theorem 5.4.2, a complete classification of secondary Poisson structures, i.e., Poisson structures in field theory, with respect to the group of Lie transformations can be obtained for small values of

$$n = (\text{the number of independent variables}),$$
$$m = (\text{the number of dependent variables}),$$

and

$$\operatorname{ord}\Delta = (\text{the order of the Hamiltonian operator } \Delta).$$

A typical result of this kind is given below. We call a scalar $\mathcal{C}$-differential operator nondegenerate if its symbol is different from zero everywhere.

PROPOSITION. *Let $m = n = 1$ and let (x, u, p) be a standard contact chart in $J^1(\pi)$. Then, locally, any nondegenerate Hamiltonian operator Δ of order 3 can be brought by a contact transformation to the normal form*

$$\Delta = \pm D^3 + 2\lambda u D + \lambda p$$

with λ being a nonnegative constant. Such a form is unique.

This proposition as well as Example 5.4.2 shows that in contrast with the classical situation no analog of the Darboux lemma is possible for Poisson structures in field theory if one takes $\mathcal{C}$-transformations (see Section 3.4) as the classifying group. In other words, the Darboux lemma fails to be valid in Secondary Calculus if $\mathcal{C}$-transformations are interpreted to be "secondary diffeomorphisms". However, it looks plausible that the Darboux lemma can be restored in Secondary Calculus if coverings (see Subsection 5.1.3) are taken for the secondarization of local diffeomorphisms. In fact, all known classification results concerning Hamiltonian operators are in favor of this conjecture (see, for instance, [**9**]). For example, we have:

EXAMPLE. In the notation of Example 5.4.2, the covering corresponding to the operator $-\partial/\partial x + (\lambda/2)u^2$ takes the Hamiltonian operator D to $-D^3 + 2\lambda u D + \lambda p$.

5.5. $\mathcal{C}_\nabla$-spectral sequence

The passage from the 0-th term of the $\mathcal{C}$-spectral sequence to the first one comprises, in particular, the passage from an $\mathcal{S}$-module to its cohomology. Our intention now is to describe a similar construction for any natural $\mathcal{S}$-module. To do that, we need a suitable generalization of the notion of adjoint operator, to which the beginning of this section is devoted.

5.5.1. Below we use a direct cohomological method to introduce the necessary generalization of the notion of adjoint operator. It has some advantages in comparison with the "elementary" approach of Section 1.2. One of them is that it fits Secondary Calculus very naturally.

Let P and Q be $\mathcal{F}(\mathcal{O})$-modules and ∇ a flat connection in P. The operator

$$S_{\nabla}^{Q}\colon \mathcal{C}\,\mathrm{Diff}(Q,\bar{\Lambda}^{s}P)\to \mathcal{C}\,\mathrm{Diff}(Q,\bar{\Lambda}^{s+1}P),\quad S_{\nabla}^{Q}(\square)=\bar{d}_{\nabla}\circ\square,$$

is a differential in $\mathcal{C}\,\mathrm{Diff}(Q,\bar{\Lambda}P)$. The following result is fundamental for the cohomological approach to the theory of adjoint operators.

THEOREM. *If P and Q are projective $\mathcal{F}(\mathcal{O})$-modules, then $H^{i}(S_{\nabla}^{Q})=0$ for $i\neq n$ and $H^{n}(S_{\nabla}^{Q})=\hat{Q}\otimes_{\mathcal{F}(\mathcal{O})}P$.*

Any operator $\square\in\mathcal{C}\,\mathrm{Diff}(Q,R)$ generates a cochain mapping

$$(\mathcal{C}\,\mathrm{Diff}(R,\bar{\Lambda}P),S_{\nabla}^{R})\to(\mathcal{C}\,\mathrm{Diff}(Q,\bar{\Lambda}P),S_{\nabla}^{Q}),\quad \Delta\mapsto\Delta\circ\square,$$

with $\Delta\in\mathcal{C}\,\mathrm{Diff}(R,\bar{\Lambda}P)$. If the modules P, Q and R are projective, Theorem 5.5.1 shows that the corresponding mapping in cohomology, denoted by

$$\square^{*\nabla}\colon H^{*}(S_{\nabla}^{R})\to H^{*}(S_{\nabla}^{Q}),$$

can be viewed as a mapping from $\hat{R}\otimes_{\mathcal{F}(\mathcal{O})}P$ to $\hat{Q}\otimes_{\mathcal{F}(\mathcal{O})}P$:

$$\square^{*\nabla}\colon \hat{R}\otimes_{\mathcal{F}(\mathcal{O})}P\to\hat{Q}\otimes_{\mathcal{F}(\mathcal{O})}P. \tag{5.17}$$

We say that the operator $\square^{*\nabla}$ is *left ∇-adjoint* to $\square$.

The concept of the ∇-adjoint operator turns out to be identical to the standard one in the case where $P=\mathcal{F}(\mathcal{O})$ and ∇ is the canonical flat connection in $\mathcal{F}(\mathcal{O})$. Commutativity of diagram (1.17) in Subsection 1.5.3 demonstrates it clearly.

The definition implies that the standard property

$$(\square_1\circ\square_2)^{*\nabla}=\square_2^{*\nabla}\circ\square_1^{*\nabla}$$

of adjoint operators still holds for left ∇-adjoint operators. However, for obvious reasons the double left ∇-adjoint operator

$$(\square^{*\nabla})^{*\nabla}\colon \widehat{\hat{Q}\otimes_{\mathcal{F}(\mathcal{O})}P}\otimes_{\mathcal{F}(\mathcal{O})}P\to\widehat{\hat{R}\otimes_{\mathcal{F}(\mathcal{O})}P}\otimes_{\mathcal{F}(\mathcal{O})}P$$

does not coincide with the original operator $\square$.

Note the canonical identification for any projective $\mathcal{F}(\mathcal{O})$-modules P and T:

$$\widehat{\hat{T}\otimes_{\mathcal{F}(\mathcal{O})}P}\otimes_{\mathcal{F}(\mathcal{O})}P=T\otimes_{\mathcal{F}(\mathcal{O})}\mathrm{End}_{\mathcal{F}(\mathcal{O})}P.$$

This allows one to view the double left ∇-adjoint operator as

$$(\square^{*\nabla})^{*\nabla}\colon Q\otimes_{\mathcal{F}(\mathcal{O})}\mathrm{End}_{\mathcal{F}(\mathcal{O})}P\to R\otimes_{\mathcal{F}(\mathcal{O})}\mathrm{End}_{\mathcal{F}(\mathcal{O})}P$$

and to relate the original operator $\square$ with its double ∇-adjoint by the following commutative diagram:

$$\begin{array}{ccc}
Q\otimes_{\mathcal{F}(\mathcal{O})}\mathrm{End}_{\mathcal{F}(\mathcal{O})}P & \xrightarrow{(\square^{*\nabla})^{*\nabla}} & R\otimes_{\mathcal{F}(\mathcal{O})}\mathrm{End}_{\mathcal{F}(\mathcal{O})}P\\
{\scriptstyle \mathrm{id}_Q\otimes\mathbf{1}_P}\uparrow & & \uparrow{\scriptstyle \mathrm{id}_R\otimes\mathbf{1}_P}\\
Q & \xrightarrow{\quad\square\quad} & P
\end{array}$$

in which $\mathbf{1}_P$ stands for the identity endomorphism of P, etc.

Combining the operation of taking the ∇-adjoint operator with that of passing to the standard adjoint operator, we obtain two different operators. One of them is

$$\square^\nabla \stackrel{\text{def}}{=} (\square^*)^{*\nabla} : Q \otimes_{\mathcal{F}(\mathcal{O})} P \to R \otimes_{\mathcal{F}(\mathcal{O})} P.$$

We call it the ∇-*extension* of $\square$. The other one, $\square^{*\nabla *} = (\square^{*\nabla})^*$, takes $\hat{Q} \widehat{\otimes_{\mathcal{F}(\mathcal{O})}} P$ to $\hat{R} \widehat{\otimes_{\mathcal{F}(\mathcal{O})}} P$. The canonical identification

$$\hat{T} \widehat{\otimes_{\mathcal{F}(\mathcal{O})}} S = \mathrm{Hom}_{\mathcal{F}(\mathcal{O})}(S, T),$$

valid for any projective $\mathcal{F}(\mathcal{O})$-modules S and T, allows one to regard this operator as

$$\square^{*\nabla *} \colon \mathrm{Hom}_{\mathcal{F}(\mathcal{O})}(P, Q) \to \mathrm{Hom}_{\mathcal{F}(\mathcal{O})}(P, R).$$

5.5.2. For a differential operator of the form

$$\Delta \colon Q \to R \otimes_{\mathcal{F}(\mathcal{O})} P$$

with P and ∇ as above, there exists another kind of adjoint operator, defined as follows. Consider the operator

$$*(Q, R)_k \colon \mathcal{C}\,\mathrm{Diff}_k(Q, R) \to \mathcal{C}\,\mathrm{Diff}_k(\hat{R}, \hat{Q}), \quad \square \mapsto \square^*.$$

It is not difficult to see that it is a $\mathcal{C}$-differential operator of order $\le k$. Its ∇-extension $*(Q,R)_k^\nabla \stackrel{\text{def}}{=} [*(Q,R)_k]^\nabla$ is

$$*(Q,R)_k^\nabla : \mathcal{C}\,\mathrm{Diff}_k(Q, R) \otimes_{\mathcal{F}(\mathcal{O})} P \to \mathcal{C}\,\mathrm{Diff}_k(\hat{R}, \hat{Q}) \otimes_{\mathcal{F}(\mathcal{O})} P,$$

or

$$*(Q,R)_k^\nabla : \mathcal{C}\,\mathrm{Diff}_k(Q, R \otimes_{\mathcal{F}(\mathcal{O})} P) \to \mathcal{C}\,\mathrm{Diff}_k(\hat{R}, \hat{Q} \otimes_{\mathcal{F}(\mathcal{O})} P),$$

due to the canonical identifications

$$\mathcal{C}\,\mathrm{Diff}_k(Q, R) \otimes_{\mathcal{F}(\mathcal{O})} P \to \mathcal{C}\,\mathrm{Diff}_k(Q, R \otimes_{\mathcal{F}(\mathcal{O})} P),$$

etc. If $l \le k$, then, obviously,

$$*(Q,R)_k^\nabla|_{\mathcal{C}\,\mathrm{Diff}_l(Q, R\otimes_{\mathcal{F}(\mathcal{O})} P)} = *(Q,R)_l^\nabla.$$

For this reason, the direct limit of the mappings $*(Q,R)_k^\nabla$, $k \to \infty$,

$$*(Q,R)^\nabla : \mathcal{C}\,\mathrm{Diff}(Q, R \otimes_{\mathcal{F}(\mathcal{O})} P) \to \mathcal{C}\,\mathrm{Diff}(\hat{R}, \hat{Q} \otimes_{\mathcal{F}(\mathcal{O})} P)$$

is well defined. In other words, this mapping associates with a $\mathcal{C}$-differential operator $\Delta \colon Q \to R \otimes_{\mathcal{F}(\mathcal{O})} P$ another $\mathcal{C}$-differential operator taking $\hat{R}$ to $\hat{Q} \otimes_{\mathcal{F}(\mathcal{O})} P$. We denote the latter by $\Delta^{\nabla *}$, so that

$$\Delta^{\nabla *} = *(Q,R)^\nabla(\Delta) \colon \hat{R} \to \hat{Q} \otimes_{\mathcal{F}(\mathcal{O})} P.$$

It is easy to see that

$$(\Delta^{*\nabla})^{*\nabla} = \Delta.$$

The operator $\Delta^{*\nabla}$ is called the *right adjoint* to Δ. As above, $\Delta^{*\nabla} = \Delta^*$ if $P = \mathcal{F}(\mathcal{O})$ and ∇ is the standard connection in $\mathcal{F}(\mathcal{O})$.

5.5.3. Now we show that a natural secondary module can be interpreted as the zero term of a generalization of the $\mathcal{C}$-spectral sequence.

We start with a flat connection ∇ in an $\mathcal{F}(\mathcal{O})$-module P and put, as before,

$$\Lambda P = \Lambda^*(\mathcal{O}) \otimes_{\mathcal{F}(\mathcal{O})} P, \qquad \Lambda^i P = \Lambda^i(\mathcal{O}) \otimes_{\mathcal{F}(\mathcal{O})} P,$$

and

$$d_\nabla \colon \Lambda^i P \to \Lambda^{i+1} P$$

for the associated differential in ΛP (see Subsection 5.2.3). The *$\mathcal{C}$-filtration* in ΛP is defined in a natural way:

$$\mathcal{C}^k \Lambda P = \mathcal{C}^k \Lambda^*(\mathcal{O}) \otimes_{\mathcal{F}(\mathcal{O})} P, \qquad \mathcal{C}^k \Lambda^i P = \mathcal{C}^k \Lambda^i(\mathcal{O}) \otimes_{\mathcal{F}(\mathcal{O})} P.$$

The spectral sequence corresponding to this filtration is called the *$\mathcal{C}_\nabla$-spectral sequence* and its terms are denoted by $(\mathcal{C}_\nabla E_r^{p,q}(\mathcal{O}), d_r^\nabla)$. Note that

$$\mathcal{C}_\nabla E_0^{p,q}(\mathcal{O}) = \mathcal{C}E_0^{p,q}(\mathcal{O}) \otimes_{\mathcal{F}(\mathcal{O})} P \tag{5.18}$$

and that the $\mathcal{C}_\nabla$-spectral sequence is a *module* over $\mathcal{C}$-spectral one, i.e., an action

$$\mathcal{C}E_r^{p,q}(\mathcal{O}) \times \mathcal{C}_\nabla E_r^{s,t}(\mathcal{O}) \to \mathcal{C}_\nabla E_r^{p+s,q+t}(\mathcal{O})$$

is naturally defined. More exactly, the term $\mathcal{C}_\nabla E_r(\mathcal{O})$ is a DG-module over the DG-algebra $\mathcal{C}E_r(\mathcal{O})$. By (5.18), all nontrivial terms $\mathcal{C}_\nabla E_0^{p,q}(\mathcal{O})$ of the $\mathcal{C}_\nabla$-spectral sequence are situated in the region $p \geq 0$, $0 \leq q \leq n$. Note that the $\mathcal{C}$-spectral sequence becomes identical to the $\mathcal{C}_\nabla$-spectral one in the case when $P = \mathcal{F}(\mathcal{O})$ and ∇ is the standard flat connection in $\mathcal{F}(\mathcal{O})$ for which $\nabla_X = X$. In this sense, the results concerning the structure of the $\mathcal{C}_\nabla$-spectral sequence presented below generalize those of Chapter 4 concerning the $\mathcal{C}$-spectral sequence.

The connection ∇ canonically determines a $\mathcal{C}$-connection $\bar{\nabla}$ in P:

$$\bar{\nabla}_X \colon = \nabla_X, \quad X \in \mathcal{C}D(\mathcal{O}).$$

The covariant differential $\bar{d}_{\bar{\nabla}} \colon \bar{\Lambda}^i P \to \bar{\Lambda}^{i+1} P$ is obviously the quotient of d_∇ due to the fact that $\bar{\Lambda}^i P = \Lambda^i P / \mathcal{C}\Lambda^i P$. Therefore, the complex $\{\mathcal{C}_\nabla E_0^{0,*}, d_0^\nabla\}$ is identified with $\{\bar{\Lambda} P, \bar{d}_{\bar{\nabla}}\}$ and hence the cohomology $H^*(\bar{d}_{\bar{\nabla}})$ is identified with the term $\mathcal{C}_\nabla E_1^{0,*}$. This relates the cohomology $H^*(\bar{d}_{\bar{\nabla}})$ with $H^*(d_\nabla)$, to which the $\mathcal{C}_\nabla$-spectral sequence converges. In this context, it is worth noting that diffieties of principal interest have homotopy types of finite-dimensional manifolds. For instance, the homotopy type of the diffiety $\mathcal{O} = \mathcal{Y}_\infty$ is the same as that of $\mathcal{Y}_k$, $k < \infty$, if k is sufficiently big. In such a situation, computation of the cohomology $H^*(d_\nabla)$ is reduced to a finite-dimensional model, to which all standard techniques of algebraic topology can be applied. In particular, this approach allows us to prove some vanishing results for the second term of the $\mathcal{C}_\nabla$-spectral sequence similarly to what has already been done for the $\mathcal{C}$-spectral sequence in Chapter 4.

In order to describe the first term of the $\mathcal{C}_\nabla$-spectral sequence, a ∇-analog of the fundamental isomorphism η (see Subsection 4.3.3) is needed.

PROPOSITION. *If $\mathcal{O} = J^\infty$, then for $p \geq 1$ we have the following natural isomorphism:*

$$(\mathcal{C}_\nabla E_0^{p,q}(\mathcal{O}), d_0^\nabla) = (\mathcal{C}\operatorname{Diff}_{(p)}^{\mathrm{alt}}(\kappa; \bar{\Lambda} P), S_\nabla^{(p)})$$

with $S_\nabla^{(p)}(\square) = (-1)^p \bar{d}_{\bar{\nabla}} \circ \square$.

Put now

$$\partial_\chi(\Delta)(\chi_1,\dots,\chi_p) = \partial_\chi(\Delta(\chi_1,\dots,\chi_p)) + \sum_i \Delta(\chi_1,\dots,\{\chi_i,\chi\},\dots,\chi_p)$$

for $\chi,\chi_i \in \kappa$ and $\Delta \in \mathcal{C}\operatorname{Diff}^{\mathrm{alt}}_{(p)}(\kappa;Q)$. Obviously, $\partial_\chi(\Delta)$ is a $\mathcal{C}$-differential operator.

THEOREM. *If $p \geq 1$, $\mathcal{O} = J^\infty(\pi)$, and P is a projective $\mathcal{F}(\mathcal{O})$-module, then*

1. $\mathcal{C}_\nabla E_1^{p,q}(J^\infty(\pi)) = 0$ *if* $q \neq n$.
2. $\mathcal{C}_\nabla E_1^{p,n}(J^\infty(\pi)) = L_p^\nabla(\kappa)$ *with*

$$L_p^\nabla(\kappa) = \{\square \in \mathcal{C}\operatorname{Diff}^{\mathrm{alt}}_{(p-1)}(\kappa;\hat\kappa \otimes_{\mathcal{F}} P) \mid \square^{\bar\nabla *}_{\chi_1,\dots,\chi_{p-2}} = -\square_{\chi_1,\dots,\chi_{p-2}}\}.$$

3. *The P-valued ∇-Euler operator* $d_1^{\nabla 0,n} = \mathcal{E}_0^\nabla : H^n(\bar d_\nabla) \to \hat\kappa \otimes_{\mathcal{F}} P$ *is given by*

$$\mathcal{E}_0^\nabla([\omega]) = \ell_\omega^{\bar\nabla *}(1), \quad \omega \in \bar\Lambda^n \otimes_{\mathcal{F}} P.$$

4. *The operators* $d_1^{\nabla p,n} = \mathcal{E}_p^\nabla$ *for* $p \geq 1$ *are given by*

$$\mathcal{E}_p^\nabla(\Delta)(\chi_1,\dots,\chi_p) = \sum_{i=1}^p (-1)^i \partial_{\chi_i}(\Delta)(\chi_1,\dots,\hat\chi_i,\dots,\chi_p)$$
$$+ \Big(\sum_{i=1}^{p-1} (-1)^{s-1} \ell^{*\bar\nabla}_{\chi_s} \circ \Delta_{\chi_1},\dots,\hat\chi_s,\dots,\chi_{p-1} + (-1)^p \ell^{\bar\nabla *}_{\Delta(\chi_1,\dots,\chi_{p-1})}\Big)(\chi_p).$$

In particular,

$$\mathcal{E}_1^\nabla(\varphi) = \ell_\varphi - \ell_\varphi^{\bar\nabla *}.$$

5. *The ∇-Lie derivative along $\chi \in \kappa$ in $\mathcal{C}_\nabla E_1(\nabla)$ is given by*

$$\chi\{\Delta\} = \partial_\chi(\Delta) + \ell_\chi^{*\bar\nabla}.$$

REMARK. Formulas for $\mathcal{E}_p^\nabla = d_1^{\nabla p,n}$ and the ∇-Lie derivative are obtained from the corresponding formulas for $d_1^{p,n}$ and L_χ in Subsections 4.3.8 and 4.3.10 by the substitutions $\ell_\chi^* \mapsto \ell_\chi^{*\bar\nabla}$ for $\chi \in \kappa$, $\ell_\varphi^* \mapsto \ell_\varphi^{\bar\nabla *}$ for $\varphi \in \mathcal{C}E_1^{1,n}$, and $L_\chi \mapsto L_\chi^\nabla \overset{\mathrm{def}}{=} [d_1^\nabla, i_\chi]$.

5.5.4. Now we are in a position to describe the first term of the $\mathcal{C}_\nabla$-spectral sequence for $\mathcal{O} = \mathcal{Y}_\infty$. The results presented below generalize those of Section 4.4 and, moreover, make them more precise. The key observation is the following

PROPOSITION. *The p-th column of the zero term of the $\mathcal{C}_\nabla$-spectral sequence is canonically identified with the natural $\mathcal{S}$-module*

$$(\bar\Lambda(P \otimes_{\mathcal{F}(\mathcal{O})} \mathcal{C}^p\Lambda^p(\mathcal{O})), \bar d_{\bar\nabla \otimes \mathcal{C}\nabla^p}),$$

where $\mathcal{C}\nabla^p$ is the canonical $\mathcal{C}$-connection in $\mathcal{C}^p\Lambda^p(\mathcal{O})$ (see Subsection 5.3.4).

Another important observation is that the standard skew-symmetrization operator

(5.19) $\mathcal{C}\Lambda^1(\mathcal{O}) \otimes_{\mathcal{F}(\mathcal{O})} \mathcal{C}^{p-1}\Lambda^{p-1}(\mathcal{O}) \xrightarrow{\mathrm{alt}} \mathcal{C}^p\Lambda^p(\mathcal{O}) \subset \mathcal{C}\Lambda^1(\mathcal{O}) \otimes_{\mathcal{F}(\mathcal{O})} \mathcal{C}^{p-1}\Lambda^{p-1}(\mathcal{O})$

canonically generates an $\mathcal{S}$-module homomorphism:

$$\Big(\bar\Lambda(P \otimes_{\mathcal{F}(\mathcal{O})} \mathcal{C}\Lambda^1(\mathcal{O}) \otimes_{\mathcal{F}(\mathcal{O})} \mathcal{C}^{p-1}\Lambda^{p-1}(\mathcal{O})), \bar d_{\bar\nabla \otimes \mathcal{C}\nabla^1 \otimes \mathcal{C}\nabla^{p-1}}\Big)$$
$$\to \Big(\bar\Lambda(P \otimes_{\mathcal{F}(\mathcal{O})} \mathcal{C}^p\Lambda^p(\mathcal{O})), \bar d_{\bar\nabla \otimes \mathcal{C}\nabla^p}\Big),$$

which, in addition, is a cochain mapping. The corresponding mapping in cohomology may be regarded, in view of Proposition 5.5.4, as a mapping

$$H^*(\bar{d}_{\bar{\nabla}\otimes \mathcal{C}\nabla^1\otimes\mathcal{C}\nabla^{p-1}}) \to H^*(\bar{d}_{\bar{\nabla}\otimes\mathcal{C}\nabla^p}) = \mathcal{C}_\nabla E_1^{p,*}.$$

Since the skew-symmetrization operator "alt" is a projection, i.e., $\mathrm{alt}^2 = \mathrm{alt}$, the term $\mathcal{C}_\nabla E_1^{p,*}$ is identified with the subspace of $H^*(\bar{d}_{\bar{\nabla}\otimes \mathcal{C}\nabla^1\otimes\mathcal{C}\nabla^{p-1}})$ on which the permutation group acts skew-symmetrically.

If $\mathcal{O} = \mathcal{Y}_\infty$ and the equation $\mathcal{Y}$ is given by $F = 0$ with $F \in Q$, Q being an $\mathcal{F}(J^\infty(\pi))$-module, then $\ell_F\colon \kappa \to Q$ and, as we have already seen in Subsection 4.4.4, $\mathcal{C}\Lambda^1(\mathcal{O})$ is identified with the cokernel of the mapping

$$\ell_{[F]}^{\mathrm{Diff}}\colon \mathcal{C}\,\mathrm{Diff}([Q], \mathcal{F}(\mathcal{O})) \to \mathcal{C}\,\mathrm{Diff}([\kappa], \mathcal{F}(\mathcal{O})), \tag{5.20}$$

which takes $\Delta \in \mathcal{C}\,\mathrm{Diff}([Q], \mathcal{F}(\mathcal{O}))$ to $\Delta\circ\ell_{[F]}$. Consequently, the $\mathcal{F}(\mathcal{O})$-module

$$\bar{\Lambda}(P\otimes_{\mathcal{F}(\mathcal{O})}\mathcal{C}\Lambda^1(\mathcal{O})\otimes_{\mathcal{F}(\mathcal{O})}\mathcal{C}^{p-1}\Lambda^{p-1}(\mathcal{O}))$$

is identified with the cokernel of the mapping

$$\ell_{[F]}^{\mathrm{Diff}}\otimes \mathrm{id}_{P\otimes \mathcal{C}^{p-1}\Lambda^{p-1}(\mathcal{O})}\colon \mathcal{C}\,\mathrm{Diff}([Q],\mathcal{F}(\mathcal{O}))\otimes_{\mathcal{F}(\mathcal{O})}\bar{\Lambda}\Big(P\otimes_{\mathcal{F}(\mathcal{O})}\mathcal{C}^{p-1}\Lambda^{p-1}(\mathcal{O})\Big)$$
$$\to \mathcal{C}\,\mathrm{Diff}([\kappa],\mathcal{F}(\mathcal{O}))\otimes_{\mathcal{F}(\mathcal{O})}\bar{\Lambda}\Big(P\otimes_{\mathcal{F}(\mathcal{O})}\mathcal{C}^{p-1}\Lambda^{p-1}(\mathcal{O})\Big).$$

Finally, in view of the canonical identification

$$\mathcal{C}\,\mathrm{Diff}(A,B)\otimes_{\mathcal{F}(\mathcal{O})} C \to \mathcal{C}\,\mathrm{Diff}(A, B\otimes_{\mathcal{F}(\mathcal{O})} C)$$

valid for arbitrary $\mathcal{F}(\mathcal{O})$-modules A, B, C, with the module C being projective, the mapping $\ell_{[F]}^{\mathrm{Diff}}\otimes \mathrm{id}_{P\otimes \mathcal{C}^{p-1}\Lambda^{p-1}(\mathcal{O})}$ is identified with the mapping

$$\ell_{[F]}^{(p)}\colon \mathcal{C}\,\mathrm{Diff}([Q],\bar{\Lambda}(P\otimes_{\mathcal{F}(\mathcal{O})}\mathcal{C}^{p-1}\Lambda^{p-1}(\mathcal{O})))$$
$$\to \mathcal{C}\,\mathrm{Diff}([\kappa],\bar{\Lambda}(P\otimes_{\mathcal{F}(\mathcal{O})}\mathcal{C}^{p-1}\Lambda^{p-1}(\mathcal{O})))$$

that takes an operator $\Delta \in \mathcal{C}\,\mathrm{Diff}([Q],\bar{\Lambda}(P\otimes_{\mathcal{F}(\mathcal{O})}\mathcal{C}^{p-1}\Lambda^{p-1}(\mathcal{O})))$ to $\Delta\circ\ell_{[F]}$.

It is worth mentioning that a complex of the form

$$(\mathcal{C}\,\mathrm{Diff}(R,\bar{\Lambda}S), S_{\nabla}^{R}),$$

where R and S are $\mathcal{F}(\mathcal{O})$-modules and ∇ is a flat connection in S, is an $\mathcal{S}$-module with respect to the $\mathcal{S}(\mathcal{O})$-action $\square \mapsto \mu(\Delta)\circ\square$, where $\square \in \mathcal{C}\,\mathrm{Diff}(R,\bar{\Lambda}S)$, $\Delta\in\mathcal{S}(\mathcal{O})$ and μ is the natural $\mathcal{S}$-module structure in $\bar{\Lambda}S$ associated with ∇. Recall also that $S_{\nabla}^{R}(\square) = \bar{d}_\nabla\circ\square$. Now it is easy to see that $\ell_{[F]}^{(p)}$ is an $\mathcal{S}$-homomorphism. This proves the following key assertion.

PROPOSITION. *For $\mathcal{O} = \mathcal{Y}_\infty$, the $\mathcal{S}$-module*

$$\Big(\bar{\Lambda}(P\otimes_{\mathcal{F}(\mathcal{O})}\mathcal{C}\Lambda^1(\mathcal{O})\otimes_{\mathcal{F}(\mathcal{O})}\mathcal{C}^{p-1}\Lambda^{p-1}(\mathcal{O})), \bar{d}_{\bar{\nabla}\otimes\mathcal{C}\nabla^1\otimes\mathcal{C}\nabla^{p-1}}\Big)$$

is canonically isomorphic to the cokernel of the $\mathcal{S}$-homomorphism $\ell_{[F]}^{(p)}$, and therefore, the $\mathcal{S}$-module $(\mathcal{C}_\nabla E_0^{p,*}, d_0^\nabla)$ is isomorphic to the skew-symmetric part of this cokernel.

REMARK. The skew-symmetric part of the cokernel of $\ell_{[F]}^{(p)}$ means the image of the alternator (5.19) under the isomorphism of Proposition 5.5.4.

Now we can show the isomorphism of Proposition 5.5.4 in action by using it to prove the Two Line Theorem for the $\mathcal{C}_\nabla$-spectral sequence. Recall that the "normality" assumption for the equation $\mathcal{Y}$ in question guarantees injectivity of the mapping (5.20) and hence injectivity of any mapping of the form

$$\mathcal{C}\operatorname{Diff}([Q],R)\to\mathcal{C}\operatorname{Diff}([\kappa],R),\quad \square\mapsto\square\circ\ell_{[F]},$$

for any projective $\mathcal{F}(\mathcal{O})$-module R. In particular, in this case all mappings $\ell_{[F]}^{(p)}$, $p\geq 1$, are injective and we get the following short exact sequences of $\mathcal{S}$-modules:

$$0\to\mathcal{C}\operatorname{Diff}([Q],\bar{\Lambda}(P\otimes_{\mathcal{F}(\mathcal{O})}\mathcal{C}^{p-1}\Lambda^{p-1}(\mathcal{O})))$$
$$\xrightarrow{\ell_{[F]}^{(p)}}\mathcal{C}\operatorname{Diff}([\kappa],\bar{\Lambda}(P\otimes_{\mathcal{F}(\mathcal{O})}\mathcal{C}^{p-1}\Lambda^{p-1}(\mathcal{O})))\to\operatorname{coker}\ell_{[F]}^{(p)}\to 0.$$

The corresponding long exact sequence in cohomology is

$$\cdots\to H^i(S^{[Q]}_{\bar{\nabla}\otimes\mathcal{C}\nabla^{p-1}})\to H^i(S^{[\kappa]}_{\bar{\nabla}\otimes\mathcal{C}\nabla^{p-1}})$$
$$\to H^i(\operatorname{coker}\ell_{[F]}^{(p)})\to H^{i+1}(S^{[Q]}_{\bar{\nabla}\otimes\mathcal{C}\nabla^{p-1}})\to\cdots.$$

By virtue of Theorem 5.5.1 and Definition 5.17 of the ∇-adjoint operator, this sequence reduces to

$$0\to\ker\ell_{[F]}^{*(\bar{\nabla}\otimes\mathcal{C}\nabla^{p-1})}\to[\hat{Q}]\otimes_{\mathcal{F}(\mathcal{O})}\mathcal{C}^{p-1}\Lambda^{p-1}(\mathcal{Y}_\infty)$$
$$\xrightarrow{\ell_{[F]}^{*(\bar{\nabla}\otimes\mathcal{C}\nabla^{p-1})}}[\hat{\kappa}]\otimes_{\mathcal{F}(\mathcal{O})}\mathcal{C}^{p-1}\Lambda^{p-1}(\mathcal{Y}_\infty)\to\operatorname{coker}\ell_{[F]}^{*(\bar{\nabla}\otimes\mathcal{C}\nabla^{p-1})}\to 0$$

This proves the desired result.

THEOREM (∇-Two Line Theorem). *If ∇ is a flat connection in a projective $\mathcal{F}(\mathcal{O})$-module Q, then, under the hypothesis of the ordinary Two Line Theorem, $\mathcal{C}_\nabla E_1^{p,n}=0$ for $p>0$, $q\neq n-1,n$, and $\mathcal{C}_\nabla E_1^{p,n-1}$ (resp., $\mathcal{C}_n a E_1^{p,n}$) is isomorphic to the skew-symmetric part of* $\ker\ell_{[F]}^{*(\bar{\nabla}\otimes\mathcal{C}\nabla^{p-1})}$ *(resp.,* $\operatorname{coker}\ell_{[F]}^{*(\bar{\nabla}\otimes\mathcal{C}\nabla^{p-1})}$*).*

REMARK. By substituting $\ell_{[F]}^{*\mathcal{C}\nabla^p}$ for $\ell_{[F]}^{*(\bar{\nabla}\otimes\mathcal{C}\nabla^{p-1})}$ in the statement of the last theorem one gets an improved version of the "ordinary" Two Line Theorem.

5.5.5. To prove a k-analog of the above theorem, we need two prerequisites. One of them is the following fact of elementary homological algebra, which can be easily proved by induction.

PROPOSITION. *Let*

$$0\to A_1\xrightarrow{\alpha_1}A_2\xrightarrow{\alpha_2}\cdots\xrightarrow{\alpha_{k-1}}A_k\xrightarrow{\beta}B\to 0$$

be an exact sequence of cochain complexes and $H^i(A_s)=0$ for $i\neq n$, $s=1,\dots,k$. Then $H^i(B)=H^{i-n+k}(K)$, where K is the complex

$$0\to H^n(A_1)\xrightarrow{H^n(\alpha_1)}H^n(A_2)\xrightarrow{H^n(\alpha_2)}\cdots\xrightarrow{H^n(\alpha_{k-1})}H^n(A_k)\to 0$$

and $H^i(K)$ stands for its cohomology at the term $H^n(A_i)$. In particular, if $H^i(B)\neq 0$, then $n-k<i\leq n$ and $H^{n-k+1}(B)=\ker H^n(\alpha_1)$, $H^n(B)=\operatorname{coker}H^n(\alpha_{k-1})$.

On the other hand, it is known from the formal theory of partial differential equations that under a week regularity condition any linear $\mathcal{C}$-differential operator $\Delta\colon S \to T$ can be extended to a $\mathcal{C}$-complex

$$0 \to S \xrightarrow{\Delta} T = T_1 \xrightarrow{\Delta_1} T_2 \xrightarrow{\Delta_2} \cdots \xrightarrow{\Delta_{k-2}} T_{k-1} \to 0 \tag{5.21}$$

such that the sequence

$$0 \to \mathcal{C}\,\mathrm{Diff}(T_{k-1}, \mathcal{F}(\mathcal{O})) \xrightarrow{S_{k-2}} \cdots \xrightarrow{S_1} \mathcal{C}\,\mathrm{Diff}(T_1, \mathcal{F}(\mathcal{O})) \xrightarrow{S_0} \mathcal{C}\,\mathrm{Diff}(S, \mathcal{F}(\mathcal{O}))$$

with $S_i(\square) = \square \circ \Delta_i$ for $\square \in \mathcal{C}\,\mathrm{Diff}(T_{i+1}, \mathcal{F}(\mathcal{O}))$ is exact.

DEFINITION. The complex (5.21) is called a *compatibility complex* for Δ of length $k-1$.

Among compatibility complexes of a given operator there exist some of length $\leq n$. The operator is said to be of *length* k if the minimal length of compatibility complexes it admits is equal to $k-1$.

Suppose now that the operator $\ell_{[F]}\colon [\kappa] \to [Q]$ is of length k and let

$$0 \to [\kappa] = Q_0 \xrightarrow{\ell_{[F]}=\Delta_0} [Q] = Q_1 \xrightarrow{\Delta_1} \cdots \xrightarrow{\Delta_{k-2}} Q_{k-1} \to 0$$

be its compatibility complex of length $k-1$. So, the sequence

$$0 \to \mathcal{C}\,\mathrm{Diff}(Q_{k-1}, \mathcal{F}(\mathcal{O})) \xrightarrow{S_{k-2}} \cdots \xrightarrow{S_1} \mathcal{C}\,\mathrm{Diff}([Q], \mathcal{F}(\mathcal{O})) \xrightarrow{S_0} \mathcal{C}\,\mathrm{Diff}([\kappa], \mathcal{F}(\mathcal{O}))$$

is exact. Its tensor product with the natural $\mathcal{S}$-module

$$\Omega_p = \bar{\Lambda}\Big(P \otimes_{\mathcal{F}(\mathcal{O})} \mathcal{C}^{p-1}\Lambda^{p-1}(\mathcal{O})\Big)$$

may be viewed as the following sequence of cochain mappings (and, moreover, of $\mathcal{S}$-homomorphisms):

(5.22)

$$0 \to \mathcal{C}\,\mathrm{Diff}(Q_{k-1}, \Omega_p) \xrightarrow{S^{(p)}_{k-2}} \cdots \xrightarrow{S^{(p)}_1} \mathcal{C}\,\mathrm{Diff}([Q], \Omega_p) \xrightarrow{S^{(p)}_0 = \ell^{(p)}_{[F]}} \mathcal{C}\,\mathrm{Diff}([\kappa], \Omega_p),$$

which is also exact. Then Propositions 5.5.1 and 5.5.5 show that the cohomology of the complex $\operatorname{coker} \ell^{(p)}_{[F]}$ is isomorphic to that of the complex

$$\begin{aligned}
0 \to \hat{Q}_{k-1} \otimes_{\mathcal{F}(\mathcal{O})} P \otimes_{\mathcal{F}(\mathcal{O})} \mathcal{C}^{p-1}\Lambda^{p-1}(\mathcal{O}) &\xrightarrow{\Delta^{*(\bar\nabla\otimes\mathcal{C}\nabla^{p-1})}_{k-2}} \cdots \\
\xrightarrow{\Delta^{*(\bar\nabla\otimes\mathcal{C}\nabla^{p-1})}_{1}} &[\hat{Q}] \otimes_{\mathcal{F}(\mathcal{O})} P \otimes_{\mathcal{F}(\mathcal{O})} \mathcal{C}^{p-1}\Lambda^{p-1}(\mathcal{O}) \\
\xrightarrow{\ell^{*(\bar\nabla\otimes\mathcal{C}\nabla^{p-1})}_{[F]}} &[\hat{\kappa}] \otimes_{\mathcal{F}(\mathcal{O})} P \otimes_{\mathcal{F}(\mathcal{O})} \mathcal{C}^{p-1}\Lambda^{p-1}(\mathcal{O}) \to 0.
\end{aligned} \tag{5.23}$$

This complex is composed of n-th cohomology of complexes forming (5.22). More precisely,

$$H^i(\operatorname{coker} \ell^{(p)}_{[F]}) = \ker \Delta^{*(\bar\nabla\otimes\mathcal{C}\nabla^{p-1})}_{n-i-1} / \operatorname{im} \Delta^{*(\bar\nabla\otimes\mathcal{C}\nabla^{p-1})}_{n-i}.$$

Now the main result on the structure of the first term of the $\mathcal{C}_\nabla$-spectral sequence follows from this and Proposition 5.5.4.

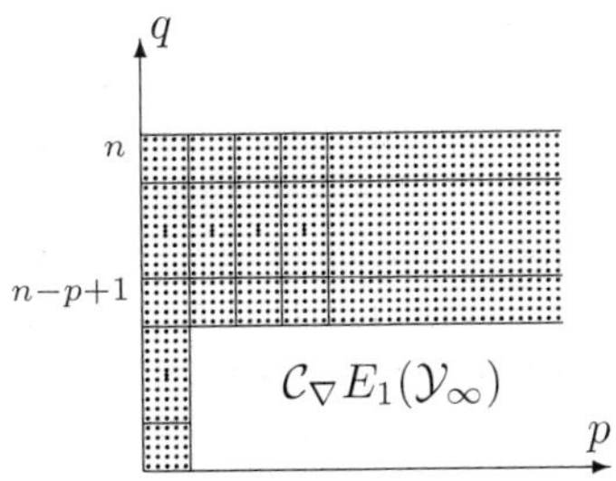

FIGURE 5.1

THEOREM (∇-k-Line Theorem). *If $\mathcal{O} = \mathcal{Y}_\infty$, $\mathcal{Y} = \{F = 0\}$, and $\ell_{[F]}$ is of length k, then $\mathcal{C}_\nabla E_1^{p,q} = 0$ for $p > 0$ and $q \le n - k$. If $n - k < q \le n$, then the term $\mathcal{C}_\nabla E_1^{p,q}$ is isomorphic to the skew-symmetric part of the module* $\ker \Delta_{n-i-1}^{*(\bar\nabla\otimes\mathcal{C}\nabla^{p-1})} / \operatorname{im} \Delta_{n-i}^{*(\bar\nabla\otimes\mathcal{C}\nabla^{p-1})}$.

The region in the standard (p, q)-diagram for the first term of the $\mathcal{C}_\nabla$-spectral sequence in which all its nontrivial components are located is shaded in Figure 5.1.

It is instructive to compare the Two Line and k-Line Theorems with those considered in Section 4.4. A result of the formal theory of partial differential equations asserts that if the stable δ-Spencer cohomology of an operator Δ vanishes in dimensions greater than k, then the length of Δ does not exceed k. In Section 4.4, the Two Line and k-Line Theorems were proved by using the acyclicity of the δ-Spencer cohomology. This approach is practically equivalent to the approach discussed above and does not require more sophisticated constructions related to $\mathcal{S}$-modules, etc. However, its serious shortcoming is that it hides the natural structure of computations.

5.6. Secondary multivector-valued differential forms

In this subsection, we show how to secondarize multivector-valued differential forms (= mixed skew-symmetric tensor fields). The result, which is interesting by itself, is not the only goal of this and subsequent sections. The procedure described below sheds light on some key points that are common in solving any concrete secondarization problem, and therefore, gives an idea of what the general secondarization algorithm is.

More precisely, the problem we are concerned with now is to find the secondary analog of the $C^\infty(M)$-module

$$\Lambda^i D_j(M) \stackrel{\text{def}}{=} \Lambda^i(M) \otimes_{C^\infty(M)} D_j(M),$$

elements of which are called *multivector-valued differential forms*. Further on, the following notation is used:

$$\Lambda^* D_j(M) \stackrel{\text{def}}{=} \Lambda^*(M) \otimes_{C^\infty(M)} D_j(M),$$

$$\Lambda^i D_*(M) \stackrel{\text{def}}{=} \Lambda^i(M) \otimes_{C^\infty(M)} D_*(M),$$

$$\Lambda^* D_*(M) \stackrel{\text{def}}{=} \Lambda^*(M) \otimes_{C^\infty(M)} D_*(M).$$

5.6.1. The first step in secondarization of a primary concept is to bring it to a "secondarizable" form. As a rule, there are several natural and equivalent ways to do this. For instance, in the case we are interested in now, it would be possible first to secondarize differential forms, which is already done, then to do the same with multivector fields, and finally to use secondary linear algebra (tensor product in the current context) to get the general result. Secondary multivector fields can also be constructed using the secondary wedge product operation. However, a more direct approach is chosen here, which works in "indecomposable" situations as well, such as, for instance, vector fields.

Our starting point is the following secondarizable interpretation of multivector-valued differential forms.

PROPOSITION. *The $C^\infty(M)$-module of $C^\infty(M)$-linear graded differential operators of order i acting on the graded algebra $\Lambda^*(M)$ coincides with $\Lambda^* D_i(M)$.*

Recall that a multivector-valued differential form $\omega \otimes V \in \Lambda^* D_*(M)$ acts on $\Lambda^*(M)$ by $\varrho \mapsto \omega \wedge i_V(\varrho)$, $\rho \in \Lambda^*(M)$. Thus, in what follows, multivector-valued differential forms are secondarized as differential operators of this kind.

The second step in the secondarization procedure is to consider operators of the same kind on the diffiety $\mathcal{O}$ in question, by paying attention to the Cartan distribution $\mathcal{C}(\mathcal{O})$, which should be "respected" by this procedure. We put

$$\mathcal{C}_r D_i(\mathcal{O}) = \{V \in D_i(\mathcal{O}) \mid i_V(\mathcal{C}^p\Lambda^*(\mathcal{O})) \subset \mathcal{C}^{p-i+r}\Lambda^*(\mathcal{O})\}$$

and obtain in this way the filtration

$$0 \subset \mathcal{C}_i D_i \subset \cdots \subset \mathcal{C}_r D_i \subset \mathcal{C}_{r-1} D_i \subset \cdots \subset \mathcal{C}_0 D_i = D_i, \tag{5.24}$$

which is dual to the $\mathcal{C}$-filtration in $\Lambda^*(\mathcal{O})$. The graded algebra associated with filtration (5.24) acts naturally on the graded algebra associated with the $\mathcal{C}$-filtration, i.e., on $\mathcal{C}E_0(\mathcal{O})$. More precisely, let us put

$$D_{r,s}(\mathcal{O}) = \mathcal{C}_s D_{r+s}(\mathcal{O}) / \mathcal{C}_{s+1} D_{r+s}(\mathcal{O}).$$

Then any element $\bar{H} \in D_{r,s}(\mathcal{O})$ takes $\mathcal{C}E_0^{p,q}(\mathcal{O})$ to $\mathcal{C}E_0^{p-r,q-s}(\mathcal{O})$.

The elements of $D_{r,s}(J^\infty(\pi))$ are represented in terms of jet coordinates by formal series of the form

$$\sum a^{i_1\dots i_r}_{\sigma_1\dots\sigma_r,j_1\dots j_s} \frac{\partial}{\partial u^{i_1}_{\sigma_1}} \wedge \cdots \wedge \frac{\partial}{\partial u^{i_r}_{\sigma_r}} \wedge D_{j_1} \wedge \cdots \wedge D_{j_s}$$

with $a^{i_1\dots i_r}_{\sigma_1\dots\sigma_r,j_1\dots j_s} \in \mathcal{F}(\mathcal{O})$. It is important to note that the $\mathcal{F}(\mathcal{O})$-module $D_{p,0}(\mathcal{O})$ carries in a natural way a flat $\mathcal{C}$-connection, which will be denoted by $\mathcal{C}\nabla_p$. It is defined as follows:

$$(\mathcal{C}\nabla_p)_X([V]) = [L_X(V)],$$

where $X \in \mathcal{C}D(\mathcal{O})$, $V \in D_p(\mathcal{O})$ and $[\cdot,\cdot]$ stands for cosets modulo $\mathcal{C}_1 D_p(\mathcal{O})$.

The following assertion is an analog of Proposition 5.6.1.

PROPOSITION. *The $\mathcal{F}(\mathcal{O})$-module of $\mathcal{F}(\mathcal{O})$-linear graded differential operators acting on $\mathcal{C}E_0(\mathcal{O})$ coincides with $\mathcal{C}E_0(\mathcal{O}) \otimes_{\mathcal{F}(\mathcal{O})} D_{*,*}(\mathcal{O})$, where $D_{*,*} = \sum_{p,q} D_{p,q}$.*

5.6.2. According to the notion of a secondary module, secondary operators must be homotopy classes of graded cochain mappings. A graded operator $\Delta\colon \mathcal{C}E_0(\mathcal{O}) \to \mathcal{C}E_0(\mathcal{O})$ is a cochain mapping of $(\mathcal{C}E_0(\mathcal{O}), d_0)$ into itself if and only if $L_0(\Delta) \overset{\text{def}}{=} [d_0, \Delta] = 0$. The following assertion substantially reduces the number of candidates for secondary multivector-valued differential forms.

PROPOSITION. *If* $0 \neq \Delta \in \mathcal{C}E_0(\mathcal{O}) \otimes_{\mathcal{F}(\mathcal{O})} D_{*,q}(\mathcal{O})$ *with* $q > 0$, *then* $L_0(\Delta) \neq 0$.

On the other hand, L_0 takes $\mathcal{C}E_0^{s,t}(\mathcal{O}) \otimes D_{p,0}(\mathcal{O})$ to $\mathcal{C}E_0^{s,t+1}(\mathcal{O}) \otimes D_{p,0}(\mathcal{O})$ and $L_0^2 = 0$. Thus, we obtain the complex

$$
\begin{aligned}
0 \to \mathcal{C}E_0^{s,0}(\mathcal{O}) \otimes_{\mathcal{F}(\mathcal{O})} D_{p,0}(\mathcal{O}) \xrightarrow{L_0} \cdots \xrightarrow{L_0} \mathcal{C}E_0^{s,t}(\mathcal{O}) \otimes_{\mathcal{F}(\mathcal{O})} D_{p,0}(\mathcal{O})& \\
\xrightarrow{L_0} \cdots \xrightarrow{L_0} \mathcal{C}E_0^{s,n}(\mathcal{O}) \otimes_{\mathcal{F}(\mathcal{O})} D_{p,0}(\mathcal{O}) \to 0.&
\end{aligned} \tag{5.25}
$$

It is not difficult to see that this complex coincides with

$$\left(\bar{\Lambda}(\mathcal{C}^s\Lambda^s(\mathcal{O}) \otimes_{\mathcal{F}(\mathcal{O})} D_{p,0}(\mathcal{O})), \bar{d}_{\mathcal{C}\nabla^s \otimes \mathcal{C}\nabla_p}\right).$$

In other words, the complex associated with the flat $\mathcal{C}$-connection $\mathcal{C}\nabla^s \otimes \mathcal{C}\nabla_p$ is canonically isomorphic to complex (5.25). Therefore, the latter is an $\mathcal{S}$-module.

DEFINITION. The $\mathcal{C}$-homotopy type of $\mathcal{S}$-module (5.25) is called the *secondary module of multivector-valued differential forms.* Thus, the secondary p-vector-valued differential s-forms on $\mathcal{O}$ are cohomology classes of the complex $(\mathcal{C}E_0^{s,*}(\mathcal{O}) \otimes_{\mathcal{F}(\mathcal{O})} D_{p,0}(\mathcal{O}), L_0)$.

In particular, *secondary p-vector fields* on $\mathcal{O}$ are cohomology classes of the complex $(\bar{\Lambda}^*(\mathcal{O}) \otimes_{\mathcal{F}(\mathcal{O})} D_{p,0}(\mathcal{O}), L_0)$ and *secondary vector-valued differential s-forms* on $\mathcal{O}$ are cohomology classes of the complex $(\mathcal{C}E_0^{s,*}(\mathcal{O}) \otimes_{\mathcal{F}(\mathcal{O})} D_{1,0}(\mathcal{O}), L_0)$.

Denote by $H^*_{\Lambda^s D_p}(\mathcal{O})$ the cohomology of complex (5.25). By definition, secondary p-vector-valued differential s-forms are elements of this space. The notation $H^*_{\Lambda^* D}, H^*_{D_*}$, etc., is interpreted similarly. For instance, the elements of the cohomology space $H^*_{D_p}(\mathcal{O})$ are secondary p-vector fields on $\mathcal{O}$ with $H^*_{D_p}(\mathcal{O})$ being the cohomology of the complex $(\bar{\Lambda}^*(\mathcal{O}) \otimes_{\mathcal{F}(\mathcal{O})} D_{p,0}(\mathcal{O}), L_0)$. The notation $H^q_{\Lambda^s D_p}(\mathcal{O})$ refers to the q-th cohomology of complex (5.25), so that

$$H^*_{\Lambda^s D_p}(\mathcal{O}) = \sum_q H^q_{\Lambda^s D_p}(\mathcal{O}).$$

Symbols like $H^q_{\Lambda^s D}(\mathcal{O})$, $H^*_{D_*}(\mathcal{O})$, etc. have obvious meaning. For instance,

$$H^*_{D_*}(\mathcal{O}) = \sum_{p,q} H^q_{D_p}(\mathcal{O}) \quad \text{with} \quad H^q_{D_p}(\mathcal{O}) = H^q_{\Lambda^0 D_p}(\mathcal{O}).$$

Note also that the natural embeddings like

$$D_*(M) \hookrightarrow \Lambda^* D_*(M), \quad \Lambda^* D(M) \hookrightarrow \Lambda^* D_*(M),$$

etc., are reproduced on the secondary level as

$$H^*_{D_*}(\mathcal{O}) \hookrightarrow H^*_{\Lambda^* D_*}(\mathcal{O}), \quad H^*_{\Lambda^* D}(\mathcal{O}) \hookrightarrow H^*_{\Lambda^* D_*}(\mathcal{O}),$$

etc., respectively. The preceding discussion gives us a chance to make the notion of a secondary vector field more precise.

DEFINITION. The $\mathcal{C}$-homotopy type of the complex $(\bar{\Lambda}^*(\mathcal{O}) \otimes_{\mathcal{F}(\mathcal{O})} D_{1,0}(\mathcal{O}), L_0)$ is called the *secondary module of secondary vector fields* on $\mathcal{O}$. Respectively, elements of the cohomology space $H_D^*(\mathcal{O})$ are called *secondary vector fields* on $\mathcal{O}$.

It is easy to see that $\operatorname{sym}\mathcal{O} = H_D^0(\mathcal{O})$. Thus, the interpretation of higher symmetries as secondary vector fields, which was so important as an heuristic argument in capturing the first idea of Secondary Calculus, must be corrected. Namely, higher symmetries form only the zero-degree component of the true concept. The structure of higher components will be described later.

Secondary $(1,1)$*-type tensors*, or, which is the same, *secondary tangent bundle homomorphisms*, also deserve a special mention. They are, obviously, elements of the cohomology space $H^*_{\Lambda^1 D}(\mathcal{O})$, and moreover, in the case $\mathcal{O} = \mathcal{Y}_\infty$, they are just *recursion operators* for the equation $\mathcal{Y}$. It will be shown below that $H^*_{\Lambda^1 D}(\mathcal{O})$ is an associative algebra acting naturally on $H_D^*(\mathcal{O})$ and, in particular, on higher symmetries of $\mathcal{Y}$.

5.6.3. As is well known, multivector-valued differential forms are subject of a number of *natural operations* such as various products, insertions, brackets, etc. It is natural to expect that they all can be secondarized as well, and now we will show how it can be done in details. The following two simple rules will help to make this procedure almost automatic in many situations.

The first rule is based on the observation that L_0-cohomology classes are homotopy classes of cochain mappings of the complex $(\mathcal{C}E_0(\mathcal{O}), d_0)$ into itself. Since the composition of two cochain mappings is also a cochain mapping whose homotopy class is determined completely by the homotopy classes of the factors, we get:

The composition rule: *Any operation on multivector-valued differential forms understood as* $C^\infty(M)$*-linear differential operators over the graded algebra* $\Lambda^*(M)$*, which is a combination of their compositions, is secondarizable.*

Further, we note that the homogeneous components of a nonhomogeneous cochain mapping are obviously also cochain mappings whose homotopy classes are uniquely determined by that of the original mapping. Therefore, we have:

The homogeneity rule: *If a nonhomogeneous operation is secondarizable, then each of its homogeneous components is secondarizable as well.*

Application of this rule in the current context is based on the fact that a $C^\infty(M)$-linear graded differential operator acting on $\Lambda^*(M)$ decomposes canonically into a sum of its homogeneous components, which are also $C^\infty(M)$-linear differential operators. Recall that such an operator of order k is called *homogeneous* if it vanishes on forms of degree less than k.

Now we pass to the secondarization of concrete operations. In what follows, the classical pattern of the operation to be secondarized precedes the corresponding secondarization procedure. Greek letters refer to differential forms, while capital Latin letters denote multivectors.

Multiplication by a form: $\alpha \otimes V \mapsto \omega \wedge \alpha \otimes V$. Multiplication by an element $\theta \subset \mathcal{C}E_0(\mathcal{O})$ can be interpreted as a zero-degree $\mathcal{C}$-differential operator acting on $\mathcal{C}E_0(\mathcal{O})$. Since $[d_0, \theta] = \pm d_0\theta$, the L_0-cohomology class of such an operator coincides with its d_0-cohomology class, which is an element of $\mathcal{C}E_1(\mathcal{O})$. Thus, the following pairing is well defined:

$$\mathcal{C}E_1^{p,q}(\mathcal{O}) \times H^t_{\Lambda^s D_r}(\mathcal{O}) \to H^{q+t}_{\Lambda^{p+s} D_r}(\mathcal{O}), \quad [\theta] \times [\Xi] \mapsto [\theta \wedge \Xi],$$

where $\theta \in \mathcal{C}E_0^{p,q}(\mathcal{O})$, $\Xi \in \mathcal{C}E_0^{s,t}(\mathcal{O}) \otimes D_{p,0}(\mathcal{O})$ and the bracket $[\cdot]$ stands for the corresponding cohomology classes. In this way, $H^*_{\Lambda^* D_*}(\mathcal{O})$ becomes a $\mathcal{C}E_1(\mathcal{O})$-module and $H^*_{\Lambda^* D_r}(\mathcal{O})$ is its submodule, $r \geq 1$. In particular, $H^*_{\Lambda^* D}(\mathcal{O}) = H^*_{\Lambda^* D_1}(\mathcal{O})$ is a $\mathcal{C}E_1(\mathcal{O})$-submodule of $H^*_{\Lambda^* D_*}(\mathcal{O})$. Similarly, an $\bar{H}^*(\mathcal{O})$-module structure on $H^*_{D_*}(\mathcal{O})$ is defined. We keep the usual wedge notation "$\wedge$" for these multiplications.

Exterior product of multivectors: $(P, Q) \mapsto P \wedge Q$. Note that the wedge product of two multivectors on a manifold M coincides with their composition if they are interpreted as $C^\infty(M)$-linear differential operators on $\Lambda^*(M)$. Thus, this product structure can be secondarized by applying the composition rule to the following "pre-secondarized" wedge product in $\bar{\Lambda}^*(\mathcal{O}) \otimes_{\mathcal{F}(\mathcal{O})} D_{*,0}(\mathcal{O})$:

$$(\omega \otimes V) \wedge (\rho \otimes W) = (-1)^{\rho V} (\omega \wedge \rho) \otimes (V \wedge W)$$

for $\omega, \rho \in \bar{\Lambda}^*(\mathcal{O})$, $V, W \in D_{*,0}$. The obtained secondary wedge product

$$\wedge\colon H^p_{D_q}(\mathcal{O}) \times H^r_{D_s}(\mathcal{O}) \to H^{p+r}_{D_{q+s}}(\mathcal{O})$$

is obviously compatible with the $\bar{H}^*(\mathcal{O})$-module structure in $H^*_{D_*}(\mathcal{O})$.

Exterior product of multivector-valued differential forms and the insertion operation: $(\alpha \otimes V) \wedge (\beta \otimes W) = (-1)^{\beta V} (\alpha \wedge \beta) \otimes (V \wedge W)$ and $(\alpha \otimes V) \lrcorner (\beta \otimes W) = \alpha \wedge (V \lrcorner \beta) \otimes W$. To secondarize these two operations, we note that the composition $(\alpha \otimes V) \circ (\beta \otimes W)$ of two forms understood as operators on $\Lambda^*(M)$ is also a $C^\infty(M)$-linear differential operator on $\Lambda^*(M)$ of order $\leq \deg V + \deg W$, whose homogeneous component of order $\deg V + \deg W$ is exactly the wedge product of these forms, while the result of the insertion of the first of them into the second is the $\deg W$-order component. Thus, the composition rule together with homogeneity allows one to secondarize both operations simultaneously. It remains to define precisely what are the corresponding "pre-secondarizations". For $\alpha, \beta \in \bar{\Lambda}^*(\mathcal{O})$, $\omega, \rho \in E_0^{*,0}(\mathcal{O})$, and $V, W \in D_{*,0}(\mathcal{O})$, we put

$$(\alpha \otimes \omega \otimes V) \wedge (\beta \otimes \rho \otimes W) = (-1)^{\beta(\omega + V) + \rho V} (\alpha \wedge \beta) \otimes (\omega \wedge \rho) \otimes (V \wedge W)$$

and

$$(\alpha \otimes \omega \otimes V) \lrcorner (\beta \otimes \rho \otimes W) = (-1)^{\beta(\omega + V)} (\alpha \wedge \beta) \otimes (\omega \wedge (V \lrcorner \rho)) \otimes W.$$

Consider the composition $(\alpha \otimes \omega \otimes V) \circ (\beta \otimes \rho \otimes W)$ of $\bar{\Lambda}^*(\mathcal{O})$-linear operators. Its homogeneous component of biorder $(\deg V + \deg W, 0)$ is the "pre-secondarized" wedge product of the original forms, and the corresponding operation $\lrcorner$ is its $(\deg W, 0)$-order component. Hence, the secondarized operations $\wedge$ and $\lrcorner$ on multivector-valued differential forms are pairings of the form

$$H^p_{\Lambda^r D_s}(\mathcal{O}) \times H^k_{\Lambda^l D_m}(\mathcal{O}) \xrightarrow{\ \wedge\ } H^{p+k}_{\Lambda^{r+l} D_{s+m}}(\mathcal{O}),$$

$$H^p_{\Lambda^r D_s}(\mathcal{O}) \otimes H^k_{\Lambda^l D_m}(\mathcal{O}) \xrightarrow{\ \lrcorner\ } H^{p+k}_{\Lambda^{r+l-s} D_m}(\mathcal{O}).$$

It is worth noting that the pairing

$$H^*_{\Lambda^p D}(\mathcal{O}) \times H^*_{\Lambda^1 D}(\mathcal{O}) \xrightarrow{\ \lrcorner\ } H^*_{\Lambda^p D}(\mathcal{O})$$

determines an associative algebra structure in $H^*_{\Lambda^1 D}(\mathcal{O})$ for $p = 1$ and a right $H^*_{\Lambda^1 D}(\mathcal{O})$-module structure in $H^*_{\Lambda^p D}(\mathcal{O})$.

The Richardson–Nijenhuis bracket: $[\Xi, \Phi]_{\mathrm{RN}} = \Xi \lrcorner \Phi - (-1)^{\Xi \cdot \Phi} \Phi \lrcorner \Xi$, $\Xi, \Phi \in \Lambda^* D(M)$. It is easy to see that

$$[\Xi, \Phi]_{\mathrm{RN}} = [\Xi, \Phi],$$

where $[\cdot,\cdot]$ is the graded commutator of operators acting on $\Lambda^*(M)$. Thus, the Richardson–Nijenhuis bracket can be secondarized either by using the already secondarized operation $\lrcorner$, or by secondarizing directly the commutator with the help of the composition rule. Obviously, both approaches lead to the same pairing

$$H^q_{\Lambda^p D}(\mathcal{O}) \times H^s_{\Lambda^r D}(\mathcal{O}) \xrightarrow{[\cdot,\cdot]_{\mathrm{RN}}} H^{q+s}_{\Lambda^{p+r-1}D}(\mathcal{O}).$$

In this way, $H_{\Lambda^* D}(\mathcal{O})$ becomes a Lie algebra with respect to the Richardson–Nijenhuis bracket and $H_{\Lambda^1 D}(\mathcal{O})$ is a Lie subalgebra of it. The pairing

$$H^*_{\Lambda^1 D}(\mathcal{O}) \times H^*_{\Lambda^p D}(\mathcal{O}) \xrightarrow{[\cdot,\cdot]_{\mathrm{RN}}} H^*_{\Lambda^p D}(\mathcal{O})$$

defines a Lie module structure in $H^*_{\Lambda^p D}(\mathcal{O})$ over $H^*_{\Lambda^1 D}(\mathcal{O})$.

Now we pass to secondarization of the Schouten–Nijenhuis bracket and the Frölicher–Nijenhuis one, which are first-order bidifferential operators over $C^\infty(M)$, in contrast to the operations considered above, which are $C^\infty(M)$-linear. There exist a number of alternative definitions of these brackets, but the secondarizable version we need is not very common and is described below.

The Schouten–Nijenhuis bracket. The L-commutator of graded differential operators acting on $\Lambda^*(M)$ is defined as

$$[\![\Delta,\nabla]\!] = \frac{1}{2}\Big([[\Delta,d],\nabla] - (-1)^{\nabla}[\Delta,[\nabla,d]]\Big).$$

Its restriction to $D_*(M)$ gives the Schouten–Nijenhuis bracket. This suggests a definition of the secondary Schouten–Nijenhuis bracket in $H^*_{D_*}(\mathcal{O})$ by the formula

$$[\![u,v]\!] = \frac{1}{2}\Big([[u,d_1],v] - (-1)^{v}[u,[v,d_1]]\Big), \tag{5.26}$$

where $[\cdot,\cdot]$ denotes the standard commutator of linear operators acting on $\mathcal{C}E_1(\mathcal{O})$, d_1 is the differential in the first term of the $\mathcal{C}$-spectral sequence and $(-1)^v = (-1)^{p+q}$ for $v \in H^q_{D_p}(\mathcal{O})$. It is not difficult to verify that $[\![u,v]\!] \in H^*_{D_*}(\mathcal{O})$. Therefore, the bracket (5.26) is a pairing

$$H^q_{D_p}(\mathcal{O}) \times H^s_{D_r}(\mathcal{O}) \xrightarrow{[\![\cdot,\cdot]\!]} H^{q+s}_{D_{p+r-1}}(\mathcal{O}).$$

It defines a Lie algebra structure on $H^*_{D_*}(\mathcal{O})$. Moreover,

$$[\![\varphi u,v]\!] = \varphi \wedge [\![u,v]\!] - (-1)^{u(v+1)+\varphi}[d_1\varphi,v]\circ u$$

with $\varphi \in \bar{H}^*(\mathcal{O})$. This formula shows that the secondary Schouten–Nijenhuis bracket is a first-order secondary bidifferential operator on $H^*_{D_*}(\mathcal{O})$.

The Frölicher–Nijenhuis bracket. In terms of the L-commutator, this bracket is defined by the formula

$$[\Xi,\Phi]_{FN} = [\![\Xi,\Phi]\!] + \frac{1}{2}(-1)^{\Phi}\big[\Xi \lrcorner \Phi + (-1)^{\Xi\Phi}\Xi \lrcorner \Phi, d\big]$$

with $\Xi,\Phi \in \Lambda^* D(M)$. Its secondary version is obtained directly by giving "secondary" meaning to all its terms:

$$[\![\theta,\varphi]\!]_{FN} = [\![\theta,\varphi]\!] + \frac{1}{2}(-1)^{\varphi}\big[\theta \lrcorner \varphi + (-1)^{\theta\varphi}\varphi \lrcorner \theta, d_1\big]$$

for $\theta,\varphi \in H^*_{\Lambda^* D}(\mathcal{O})$. The pairing obtained in this way,

$$H^q_{\Lambda^p D}(\mathcal{O}) \times H^s_{\Lambda^r D}(\mathcal{O}) \xrightarrow{[\cdot,\cdot]_{\mathrm{FN}}} H^{q+s}_{\Lambda^{p+r}D}(\mathcal{O}),$$

is a Lie algebra structure on $H^*_{\Lambda^* D}(\mathcal{O})$.

Thus, all basic operations on multivector-valued differential forms are secondarizable. Moreover, there exists a general method allowing one to avoid discussion of particular cases. But this requires some new technical tools and terminology.

5.7. The secondary Frölicher–Nijenhuis formalism

The final step in any secondarization procedure is the computation of the corresponding cohomology. In the case of secondary multivector-valued differential forms this is $H^*_{\Lambda^s D_p}(\mathcal{O})$. It turns out that the method developed earlier with the same purpose for the first term of the $\mathcal{C}$-spectral sequence works in the general case as well. It is remarkable that the only point to be modified when passing from one particular case to another is the *fundamental isomorphism* η, which should be described properly. Such a description for general multivector-valued differential forms is rather cumbersome (see [**132**] for a systematic exposition). Thus, in what follows we consider only the case $p = 1$, for which all necessary ingredients are already at our disposal and which, in addition, is rather instructive.

5.7.1. Thus, we pass to computation of the cohomology $H^*_{\Lambda^* D}(\mathcal{O})$. As in the case of the $\mathcal{C}$-spectral sequence, it is convenient to start with the particular case $\mathcal{O} = J^\infty(\pi)$. Consider the $\mathcal{F}(J^\infty(\pi))$-module $\bar{\mathcal{J}}(\kappa)$ of infinite horizontal jets of κ and the corresponding horizontal Spencer jet-complex (see Section 3.1):

$$0 \to \bar{\mathcal{J}}(\kappa) \xrightarrow{S} \bar{\Lambda}^1(J^\infty(\pi)) \otimes_{\mathcal{F}} \bar{\mathcal{J}}(\kappa) \xrightarrow{S} \cdots \xrightarrow{S} \bar{\Lambda}^n(J^\infty(\pi)) \otimes_{\mathcal{F}} \bar{\mathcal{J}}(\kappa) \to 0$$

with $S\colon \omega \otimes \bar{\jmath}_\infty(\chi) \mapsto \bar{d}\omega \otimes \bar{\jmath}_\infty(\chi)$, $\omega \in \bar{\Lambda}^*(J^\infty(\pi))$, $\chi \in \kappa$.

PROPOSITION. *There exists a natural isomorphism of $\mathcal{F}(J^\infty(\pi))$-modules*

$$\eta\colon D_{1,0}(J^\infty(\pi)) \to \bar{\mathcal{J}}(\kappa)$$

such that $\eta(\partial_\chi) = \bar{\jmath}_\infty(\chi)$. Moreover, this isomorphism together with that of Proposition 4.3.3 *yields the isomorphism*

$$\eta\colon \mathcal{C}E_0^{p,q}(J^\infty(\pi)) \otimes_{\mathcal{F}} D_{1,0}(J^\infty(\pi)) \to \mathcal{C}\operatorname{Diff}^{\mathrm{alt}}_{(p)}(\kappa; \bar{\Lambda}^q(J^\infty(\pi)) \otimes_{\mathcal{F}} \bar{\mathcal{J}}(\kappa)).$$

The Spencer differential S induces the differential $S\colon \square \mapsto S \circ \square$ in the modules

$$\mathcal{C}\operatorname{Diff}^{\mathrm{alt}}_{(p)}(\kappa; \bar{\Lambda}^q(J^\infty(\pi)) \otimes_{\mathcal{F}} \bar{\mathcal{J}}(\kappa)).$$

Thus, $H^*_{\Lambda^* D}(J^\infty(\pi))$ is isomorphic to the cohomology of the complex obtained in such a way. Its cohomology can be computed easily due to the fact that the Spencer jet-complex is acyclic in positive dimensions and all its 0-cocycles are of the form $\bar{\jmath}_\infty(\chi)$. In this way, one gets the following result.

THEOREM (One Line Theorem). *One has the isomorphisms:*

(i) $H^0_{\Lambda^p D}(J^\infty(\pi)) = \mathcal{C}\operatorname{Diff}^{\mathrm{alt}}_{(p)}(\kappa; \kappa)$;
(ii) $H^q_{\Lambda^p D}(J^\infty(\pi)) = 0$ *for* $q > 0$.

The secondary vector-valued differential p-forms on $J^\infty(\pi)$ are thereby identified with the skew-symmetric p-ary $\mathcal{C}$-differential operators on κ with values in κ.

The natural action

$$H^*_{\Lambda^* D_*}(\mathcal{O}) \times \mathcal{C}E_1(\mathcal{O}) \to \mathcal{C}E_1(\mathcal{O})$$

restricted to secondary vector-valued differential forms looks as follows:

$$H^s_{\Lambda^r D}(\mathcal{O}) \times \mathcal{C}E_1^{p,q}(\mathcal{O}) \to \mathcal{C}E_1^{p+r-1,q+s}(\mathcal{O}).$$

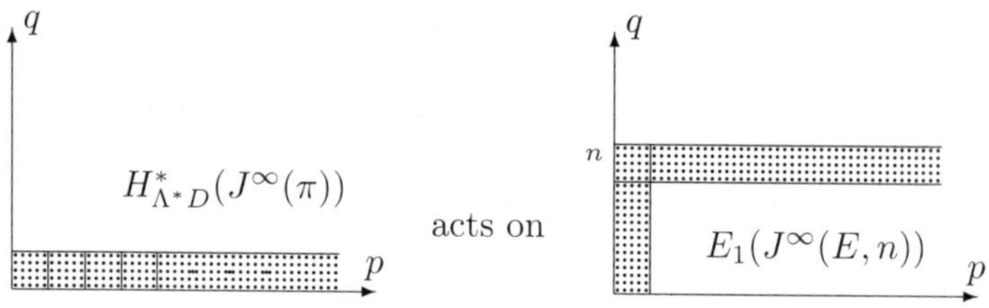

FIGURE 5.2

According to One Line Theorems 4.3.7 and 5.7.1, on $J^\infty(\pi)$ it reduces to

$$\mathcal{C}\,\mathrm{Diff}^{\mathrm{alt}}_{(r)}(\kappa;\kappa)\times L_p(\kappa)\to L_{p+r-1}(\kappa).$$

Denote the result of the action of $\Delta\in\mathcal{C}\,\mathrm{Diff}^{\mathrm{alt}}_{(r)}(\kappa;\kappa)$ on $\square\in L_p(\kappa)$ by $\Delta\lrcorner\,\square$. Then the following formula holds:

$$\begin{aligned}(\Delta\lrcorner\,\square)(\chi_1,\dots,\chi_{p+r-2})&=\sum_{|I|=r}(-1)^{(I,J)}\square(\Delta(\chi_I),\chi_J)\\&\quad+\sum_{|I|=r-1}(-1)^{(I,J)}\Delta^*_{\chi_I}(\square(\chi_J)),\end{aligned}\tag{5.27}$$

where χ_I is $\chi_1,\dots,\chi_{i_r}$ for $I=(i_1,\dots,i_r)$, $1\le i_1<\dots<i_r\le p+r-2$, and J is composed of remaining χ_i, etc.; here $(-1)^{(I,J)}$ denotes the sign of the permutation (I,J) of $(1,2,\dots,p+r-2)$. In this way, the lines of the $H^*_{\Lambda^*D}(J^\infty(\pi))$-diagram act on the lines of the $E_1(J^\infty(E,n))$ term as shown in Figure 5.2.

5.7.2. In its turn, an explicit description of secondary vector-valued differential forms on $J^\infty(\pi)$ obtained in the previous subsection allows us to describe explicitly the basic natural operations on them defined in Subsection 5.6.3. It can be done in two different ways, and each of them has its own advantages. One deals, in the spirit of formula (5.27), directly with multidifferential operators representing secondary vector-valued differential forms, while the other refers to jet coordinates and Cartan forms. Both descriptions are given in parallel below.

We start with some conventions concerning the notation that will be used. First, χ_I stands for $\chi_{i_1},\dots,\chi_{i_s}$ if $I=(i_1,\dots,i_s)$, and $|I|$ means the length of I, i.e., s. If arguments in a formula are numbered by $1,2,\dots,m$, then $(\alpha,\beta,I,\bar I)$ stands for the permutation of $(1,2,\dots,m)$ formed by two singlets α and β, and subsets I and $\bar I$, with $\bar I$ being the complement of (α,β,I) in $(1,2,\dots,m)$. Both I and $\bar I$ are supposed to be increasing. Finally, $(-1)^{(\alpha,\beta,I,\bar I)}$ stands for the sign of the permutation $(\alpha,\beta,I,\bar I)$. The similar convention is adopted in the cases where one or both singlets are absent.

It is convenient also to introduce a shorter notation for the Cartan forms:

$$\xi^i_\sigma=U_1(p^i_\sigma)=dp^i_\sigma-\sum_j p^i_{\sigma j}\,dx_j$$

and to identify locally the algebra of Cartan forms

$$\mathcal{C}E_0^{*,0}(J^\infty(\pi))=\sum_{p\ge0}\mathcal{C}^p\Lambda^p(J^\infty(\pi))$$

with the graded algebra of polynomials

$$\sum a^{i_1\dots i_p}_{\sigma_1\dots\sigma_p}\xi^{i_1}_{\sigma_1}\cdots\xi^{i_p}_{\sigma_p}, \quad a^{i_1\dots i_p}_{\sigma_1\dots\sigma_p}\in\mathcal{F}(J^\infty(\pi)), \tag{5.28}$$

in anti-commuting variables ξ^i_σ. Recall also that the algebra $\mathcal{C}\operatorname{Diff}\mathcal{F}(J^\infty(\pi))$ acts naturally on the algebra of Cartan forms

$$(\square,\omega)\mapsto\square^{\mathcal{C}\nabla^p}(\omega), \quad \square\in\mathcal{C}\operatorname{Diff}\mathcal{F}(J^\infty(\pi)),\ \omega\in\mathcal{C}^p\Lambda^p(J^\infty(\pi))$$

due to the natural $\mathcal{C}$-connection $\mathcal{C}\nabla^p$ in $\mathcal{C}^p\Lambda^p(J^\infty(\pi))$. To simplify the notation, we shall write $\square(\omega)$ instead of $\square^{\mathcal{C}\nabla^p}(\omega)$. For instance, with this convention the following evident formula holds:

$$D_\tau(\xi^k_\sigma)=\xi^k_{\sigma+\tau}.$$

This formula, together with the Leibniz rule, describes completely the action of an operator $\square=\sum a_\sigma D_\sigma$ on a polynomial of the form (5.28).

The $\mathcal{C}$-connection $\mathcal{C}\nabla^p\otimes\mathcal{C}\nabla_1$ in $\mathcal{C}^p\Lambda^p(J^\infty(\pi))\otimes_{\mathcal{F}(J^\infty(\pi))}\varkappa$ allows one to define evolutionary derivations whose generating functions belong to $\mathcal{C}^p\Lambda^p(J^\infty(\pi))$. Namely, if P is an $\mathcal{F}(J^\infty(\pi))$-module and $f\in P$, then we put

$$\mathcal{Э}_\omega(f)=\ell_f^{\mathcal{C}\nabla^p\otimes\mathcal{C}\nabla_1}(\omega), \quad \omega\in\mathcal{C}^p\Lambda^p(J^\infty(\pi))\otimes_{\mathcal{F}(J^\infty(\pi))}\varkappa.$$

If, in a chosen jet chart, $\omega=(\omega_1,\dots,\omega_p)$ with $\omega_i\in\mathcal{C}^p\Lambda^p(J^\infty(\pi))$, then

$$\mathcal{Э}_\omega=\sum_{\sigma,i}D_\sigma(\omega_i)\frac{\partial}{\partial p^i_\sigma}.$$

These generalized evolutionary derivations together with the operator

$$U\colon\mathcal{C}^p\Lambda^p(J^\infty(\pi))\to\mathcal{C}^{p+1}\Lambda^{p+1}(J^\infty(\pi)), \quad U(\theta)=\sum_{i,\tau}\xi^i_\tau\wedge\frac{\partial(\theta)}{\partial p^i_\tau},$$

will be used further on.

The fundamental isomorphism η identifies the modules

$$\mathcal{C}^p\Lambda^p(J^\infty(\pi))\otimes_{\mathcal{F}(J^\infty(\pi))}\varkappa \quad\text{and}\quad \mathcal{C}\operatorname{Diff}^{\mathrm{alt}}_{(p)}(\varkappa;\varkappa).$$

Therefore, according to Theorem 5.7.1,

$$H^0_{\Lambda^pD}(J^\infty(\pi))=\mathcal{C}^p\Lambda^p(J^\infty(\pi))\otimes_{\mathcal{F}(J^\infty(\pi))}\varkappa.$$

This isomorphism allows us to interpret secondary vector-valued differential forms as polynomials in ξ_i with coefficients in $\varkappa$. This coordinatewise description will be used in the subsequent list of basic operations. This list, containing only nontrivial pairings, is parallel to that of Subsection 5.6.3, where the vector multiplicity should be taken equal to 1.

Multiplication by a secondary form:

$$\mathcal{C}E_1^{*,0}(J^\infty(\pi))\times H^0_{\Lambda^*D}(J^\infty(\pi))\xrightarrow{\mu}H^0_{\Lambda^*D}(J^\infty(\pi)).$$

If $n=0$, this operation is the standard multiplication of vector-valued forms on a manifold by a differential form. If $n>0$, then

$$\mathcal{C}E_1^{*,0}(J^\infty(\pi))=\mathcal{C}E_1^{0,0}(J^\infty(\pi))=\{\text{locally constant functions on } J^\infty(\pi)\}$$

and the operation is just the multiplication by a local constant.

Insertion into secondary vector-valued differential forms:

$$H^0_{\Lambda^pD}(J^\infty(\pi))\times H^0_{\Lambda^rD}(J^\infty(\pi))\xrightarrow{\lrcorner}H^0_{\Lambda^{p+r-1}D}(J^\infty(\pi)).$$

It is defined as follows:

$$\text{(i)}\quad (\Delta \lrcorner \square)(\chi_1, \dots, \chi_{p+r-1}) = \sum_{|I|=p} (-1)^{(I,\bar I)} \square(\Delta(\chi_I), \chi_{\bar I})$$

with $\Delta \in \mathcal{C}\operatorname{Diff}^{\mathrm{alt}}_{(p)}(\kappa;\kappa)$, $\square \in \mathcal{C}\operatorname{Diff}^{\mathrm{alt}}_{(r)}(\kappa;\kappa)$.

$$\text{(ii)}\quad \omega \lrcorner \rho = \sum_k \left(\sum_{i,\sigma} D_\sigma(\omega^i) \wedge \frac{\partial \rho^k}{\partial \xi^i_\sigma} \right) \frac{\partial}{\partial u^k}$$

with $\omega \in \mathcal{C}^p\Lambda^p(J^\infty(\pi)) \otimes_{\mathcal{F}(J^\infty(\pi))} \kappa$, $\omega = \sum_i \omega^i \partial/\partial u^i$, $\omega^i \in \mathcal{C}^p\Lambda^p(J^\infty(\pi))$, and similarly for $\rho \in \mathcal{C}^p\Lambda^p(J^\infty(\pi))$.

The secondary Richardson–Nijenhuis bracket:

$$H^0_{\Lambda^p D}(J^\infty(\pi)) \times H^0_{\Lambda^r D}(J^\infty(\pi)) \xrightarrow{[\cdot,\cdot]_{\mathrm{RN}}} H^0_{\Lambda^{p+r-1} D}(J^\infty(\pi))$$

reduces to the preceding formulas in an obvious way.

The secondary Frölicher–Nijenhuis bracket:

$$H^0_{\Lambda^k D}(J^\infty(\pi)) \times H^0_{\Lambda^l D}(J^\infty(\pi)) \xrightarrow{[\cdot,\cdot]_{\mathrm{FN}}} H^0_{\Lambda^{k+l} D}(J^\infty(\pi)).$$

One has

$$\begin{aligned}
\text{(i)}\quad [\Delta, \square]_{\mathrm{FN}}(\chi_1, \dots, \chi_{k+l}) &= \sum_{|I|=k} (-1)^{(I,\bar I)} \{\Delta(\chi_I), \square(\chi_{\bar I})\} \\
&+ \sum_{\alpha, |I|=k} (-1)^{(\alpha, I, \bar I)+k} \square(\{\chi_\alpha, \Delta(\chi_I)\}, \chi_{\bar I}) \\
&- \sum_{\alpha, |I|=l} (-1)^{(\alpha, I, \bar I)+(k-1)l} \Delta(\{\chi_\alpha, \square(\chi_I)\}, \chi_{\bar I}) \\
&- \sum_{\alpha<\beta, |I|=k-1} (-1)^{(\alpha,\beta,I,\bar I)+k} \square(\Delta(\{\chi_\alpha, \chi_\beta\}, \chi_I), \chi_{\bar I}) \\
&+ \sum_{\alpha<\beta, |I|=l-1} (-1)^{(\alpha,\beta,\bar I)+(k-1)l} \Delta(\square(\{\chi_\alpha, \chi_\beta\}, \chi_I), \chi_{\bar I})
\end{aligned}$$

with $\Delta \in \mathcal{C}\operatorname{Diff}^{\mathrm{alt}}_{(k)}(\kappa;\kappa)$, $\square \in \mathcal{C}\operatorname{Diff}^{\mathrm{alt}}_{(l)}(\kappa;\kappa)$.

$$\text{(ii)}\qquad \{\omega, \rho\}^i = \sum_i (\ni_\omega[\rho^i] - (-1)^{(\omega+1)(\rho+1)} \ni_\rho[\omega^i])$$

with $\omega \in \mathcal{C}^k\Lambda^k(J^\infty(\pi)) \otimes_{\mathcal{F}(J^\infty(\pi))} \kappa$, $\omega = \sum_i \omega^i \partial/\partial u^i$, $\omega^i \in \mathcal{C}^k\Lambda^k(J^\infty(\pi))$ and similarly for $\rho \in \mathcal{C}^l\Lambda^l(J^\infty(\pi))$, where $\ni_\omega[\theta]$, $\theta \in \mathcal{C}^s\Lambda^s(J^\infty(\pi))$, is defined as

$$\ni_\omega[\theta] = \sum_{i,\sigma} \left(D_\sigma(\omega^i) \wedge \frac{\partial \theta}{\partial p^i_\sigma} + D_\sigma(U_1(\omega^i)) \wedge \frac{\partial \theta}{\partial \xi^i_\sigma} \right).$$

Lie derivatives of secondary differential forms along secondary form-valued vector fields:

$$H^0_{\Lambda^s D}(J^\infty(\pi)) \times \mathcal{C}E_1^{p,n} \xrightarrow{L} \mathcal{C}E_1^{p+s,n}.$$

The fundamental isomorphism η allows one to represent this derivation as a pairing

$$\mathcal{C}\operatorname{Diff}^{\mathrm{alt}}_{(s)}(\kappa;\kappa) \times L_p(\kappa) \to L_{p+s}(\kappa).$$

The exact formula is as follows:

$$
\begin{aligned}
L_\Delta(\Box)(\chi_1,\dots,\chi_{s+p-1}) = &\sum_{|I|=s}(-1)^{(I,\bar I)}\Delta(\chi_I)\{\Box(\chi_{\bar I})\}\\
&+\sum_{\alpha,|I|=s-1}(-1)^{(\alpha,I,\bar I)+s}\chi_\alpha\{\Delta^*_{\chi_I}(\Box(\chi_{\bar I}))\}\\
&-\sum_{\alpha,|I|=s-1}(-1)^{(\alpha,I,\bar I)+s}\Delta^*_{\chi_I}(\chi_\alpha \lrcorner\, \mathcal{E}_1(\Box(\chi_{\bar I})))\\
&-\sum_{\alpha<\beta,|I|=s-2}(-1)^{(\alpha,\beta,I,\bar I)+s}\Delta_{\{\chi_\alpha,\chi_\beta\},\chi_I}(\Box(\chi_{\bar I}))\\
&+\sum_{\alpha,|I|=s}(-1)^{(\alpha,I,\bar I)+s+1}\Box(\{\chi_\alpha,\Delta(\chi_I)\},\chi_{\bar I})\\
&-\sum_{\alpha<\beta,|I|=s-1}(-1)^{(\alpha,\beta,I,\bar I)+s}\Box(\Delta(\{\chi_\alpha,\chi_\beta\},\chi_I),\chi_{\bar I})
\end{aligned}
$$

with $\Delta \in \mathcal{C}\,\mathrm{Diff}^{\mathrm{alt}}_{(s)}(\kappa;\kappa)$, $\Box \in L_p(\kappa)$. Elements of $L_p(\kappa)$ in this formula are interpreted as skew-symmetric $\hat\kappa$-valued $(p-1)$-differential operators on κ, i.e., $L_p(\kappa) \subset \mathcal{C}\,\mathrm{Diff}^{\mathrm{alt}}_{(p-1)}(\kappa;\hat\kappa)$.

5.7.3. To compute secondary vector-valued differential forms on a diffiety $\mathcal{O} = \mathcal{Y}_\infty$, we can apply essentially the same method as the one used in Subsection 5.5.4 for the $\mathcal{C}_\nabla$-spectral sequence. The only necessary modification is to replace the Diff-*compatibility complex* used there by its dual, the Jet-*compatibility complex*.

Thus, consider an equation $\mathcal{Y} = \{F = 0\}$ with $F \in P$ and the corresponding universal linearization operator $\ell_F\colon \kappa \to P$. Let $\ell_{[F]}\colon Q_0 \to Q_1$ (with $Q_0 = [\kappa] = \kappa|_{\mathcal{Y}_\infty}$, $Q_1 = P|_{\mathcal{Y}_\infty}$) be its restriction to $\mathcal{Y}_\infty$. Consider its Jet-prolongation (see Subsection 1.1.10):

$$\ell^{\mathrm{Jet}}_{[F]}\colon \bar{\mathcal{J}}(Q_0) \to \bar{\mathcal{J}}(Q_1), \quad f\bar{\jmath}(q) \mapsto f\bar{\jmath}(\ell_{[F]}(q)),\ f \in \mathcal{F}(\mathcal{O}),\ q \in Q.$$

Then

$$
\begin{aligned}
\mathrm{id}_{\mathcal{C}^p\Lambda^p(\mathcal{Y}_\infty)\otimes\bar\Lambda^*(\mathcal{Y}_\infty)} \otimes \ell^{\mathrm{Jet}}_{[F]}\colon &\mathcal{C}^p\Lambda^p(\mathcal{Y}_\infty) \otimes_{\mathcal{F}(\mathcal{Y}_\infty)} \bar\Lambda^*(\mathcal{Y}_\infty) \otimes_{\mathcal{F}(\mathcal{Y}_\infty)} \bar{\mathcal{J}}(Q_0)\\
&\to \mathcal{C}^p\Lambda^p(\mathcal{Y}_\infty) \otimes_{\mathcal{F}(\mathcal{Y}_\infty)} \bar\Lambda^*(\mathcal{Y}_\infty) \otimes_{\mathcal{F}(\mathcal{Y}_\infty)} \bar{\mathcal{J}}(Q_1)
\end{aligned}
$$

is a cochain mapping of the horizontal Jet-Spencer complexes with coefficients in $\mathcal{C}^p\Lambda^p(\mathcal{Y}_\infty)$.

PROPOSITION. *The fundamental isomorphism identifies $D_{1,0}(\mathcal{Y}_\infty)$ and $\ker \ell^{\mathrm{Jet}}_{[F]}$. If the $\mathcal{F}(\mathcal{Y}_\infty)$-module $\mathcal{C}^p\Lambda^p(\mathcal{Y}_\infty) \otimes_{\mathcal{F}(\mathcal{Y}_\infty)} \bar\Lambda^*(\mathcal{Y}_\infty)$ is projective, then*

$$\mathcal{C}^p\Lambda^p(\mathcal{Y}_\infty) \otimes_{\mathcal{F}(\mathcal{Y}_\infty)} \bar\Lambda^*(\mathcal{Y}_\infty) \otimes_{\mathcal{F}(\mathcal{Y}_\infty)} D_{1,0}(\mathcal{Y}_\infty) = \ker(\mathrm{id}_{\mathcal{C}^p\Lambda^p(\mathcal{Y}_\infty)\otimes\bar\Lambda^*(\mathcal{Y}_\infty)} \otimes \ell^{\mathrm{Jet}}_{[F]}).$$

A $\mathcal{C}$-complex

$$0 \to Q_0 \xrightarrow{\ell_{[F]}} Q_1 \xrightarrow{\Delta_1} \cdots \xrightarrow{\Delta_{k-2}} Q_{k-1} \to 0$$

is called a Jet-*compatibility complex* for $\ell_{[F]}$ if the corresponding sequence of Jet-prolonged operators,

$$\bar{\mathcal{J}}(Q_0) \xrightarrow{\ell^{\mathrm{Jet}}_{[F]}} \bar{\mathcal{J}}(Q_1) \xrightarrow{\Delta^{\mathrm{Jet}}_1} \cdots \xrightarrow{\Delta^{\mathrm{Jet}}_{k-2}} \bar{\mathcal{J}}(Q_{k-1}) \to 0, \tag{5.29}$$

is exact. Under a weak regularity condition, the formal theory of differential equations guarantees the existence of a Jet-compatibility complex of length $\le n$, i.e., with $k-1 \le n$. It is a Diff-compatibility complex as well (see Subsection 5.5.5).

Recall that the horizontal Jet-Spencer complex $(\bar{\Lambda}^*(\mathcal{O}) \otimes_{\mathcal{F}(\mathcal{O})} \bar{\mathcal{J}}(Q), \Sigma)$ with

$$\Sigma(\omega \otimes \bar{\jmath}(\chi)) = d\omega \otimes \bar{\jmath}(\chi), \quad \omega \in \bar{\Lambda}(\mathcal{O}),\ \chi \in \kappa,$$

has trivial cohomology in positive dimensions, and its zero cohomology is canonically isomorphic to Q. Its tensor product with a horizontal $\mathcal{F}(\mathcal{O})$-module T is called the *horizontal* Jet-*Spencer complex with coefficients in* T.

PROPOSITION. *Let T be a horizontal module. Then the zero-dimensional cohomology of the* Jet-*Spencer complex with coefficients in T is isomorphic to $T \otimes_{\mathcal{F}(\mathcal{O})} Q$, while all other cohomology is trivial.*

Exact sequence (5.29) generates the following exact sequence of Jet-Spencer complexes with coefficients in T:

$$(5.30)\quad T \otimes_{\mathcal{F}(\mathcal{O})} \bar{\Lambda}^*(\mathcal{O}) \otimes_{\mathcal{F}(\mathcal{O})} \bar{\mathcal{J}}(Q_0) \xrightarrow{\mathrm{id}_{T\otimes\bar{\Lambda}^*(\mathcal{O})} \otimes \ell^{\mathrm{Jet}}_{[F]}} \cdots \xrightarrow{\mathrm{id}_{T\otimes\bar{\Lambda}^*(\mathcal{O})} \otimes \Delta^{\mathrm{Jet}}_{k-2}} T \otimes_{\mathcal{F}(\mathcal{O})} \bar{\Lambda}^*(\mathcal{O}) \otimes_{\mathcal{F}(\mathcal{O})} \bar{\mathcal{J}}(Q_{k-1}) \to 0.$$

According to Proposition 5.7.3, the standard arguments of homological algebra applied to complex (5.30) prove the following key result:

PROPOSITION. *The cohomology $H^i(\ker \mathrm{id}_{T\otimes\bar{\Lambda}^*(\mathcal{O})} \otimes \ell^{\mathrm{Jet}}_{[F]})$ is canonically isomorphic to the i-th cohomology of the complex*

$$0 \to T \otimes_{\mathcal{F}(\mathcal{O})} Q_0 \xrightarrow{\ell^{\nabla^T}_{[F]}} T \otimes_{\mathcal{F}(\mathcal{O})} Q_1 \xrightarrow{\Delta^{\nabla^T}_1} \cdots \xrightarrow{\Delta^{\nabla^T}_{k-1}} T \otimes_{\mathcal{F}(\mathcal{O})} Q_{k-1} \to 0,$$

where ∇^T denotes the structure connection of the horizontal module T.

Combining this assertion specified for the horizontal module $T = \mathcal{C}^p\Lambda^p(J^\infty(\pi))$ and Proposition 5.7.3 one gets the following main result.

THEOREM (k-Line Theorem from the Bottom). *If the universal linearization operator of the equation $\mathcal{Y} = \{F = 0\}$ is of length k, then $H^i_{\Lambda^p D}(\mathcal{Y}_\infty) = 0$ for $i \ge k$. If $0 \le i < k$, then*

$$H^i_{\Lambda^p D}(\mathcal{Y}_\infty) = \ker \Delta_i^{\mathcal{C}\nabla^p} / \operatorname{im} \Delta_{i-1}^{\mathcal{C}\nabla^{p-1}}$$

under the assumption $\Delta_0 = \ell_{[F]}$. In particular, $H^0_{\Lambda^p D}(\mathcal{Y}_\infty) = \ker \ell^{\mathcal{C}\nabla^p}_{[F]}$.

Under the hypothesis of the Two Line Theorem, we have additionally the isomorphism $H^1_{\Lambda^p D}(\mathcal{Y}_\infty) = \operatorname{coker} \ell^{\mathcal{C}\nabla^p}_{[F]}$. In particular, this shows that higher symmetries of normal differential equations possess, in general, nontrivial components of degree 1.

It is remarkable that the k-Line Theorem from the Bottom is perfectly complementary to that "from the Top", like the two coils forming the DNA double helix. Namely, secondary operators acting on secondary differential forms are nontrivial in only those places where the nontrivial forms on which they act are nontrivial. This is illustrated in Figure 5.3.

Under the same assumptions as before, the k-Line Theorem from the Bottom remains valid for the general cohomology $H_{\mathcal{C}^p D_s}(\mathcal{Y}_\infty)$, i.e., for arbitrary secondary multivector-valued differential forms. Indeed, all k-line theorems presented in this

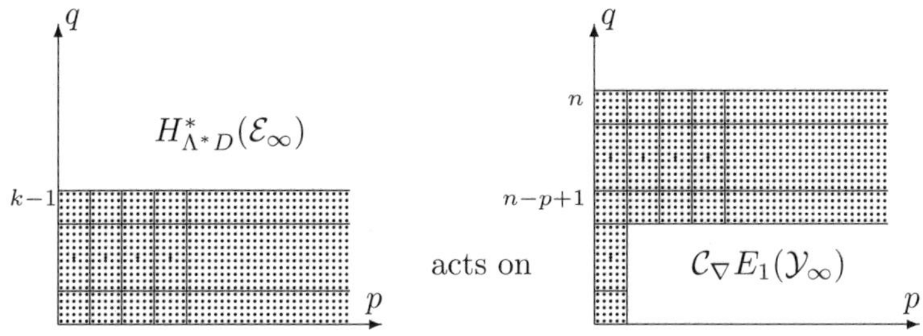

FIGURE 5.3

book are particular cases of a much more general result that concerns any secondary concept.

5.8. Concluding remarks

Brief bibliographical and historical remarks below are intended to give a general view of what has happened and is happening in Secondary Calculus. The first idea of the theory was formed at the end of the 1970s using the results in geometry of partial differential equations found earlier and described in Chapters 1–4 of the present book. The first systematic summary of them was given in the survey article [**124**]. The term Secondary Calculus was introduced in [**135**].

The secondarization problem arose naturally as soon as the idea of Secondary Calculus had acquired a sufficiently clear shape. During the 1980s, three attempts of secondarization have been made. First of them concerns the Hamiltonian formalism in field theory (see [**125**], where Definition 5.4.2 was proposed). At that time, it was rather intuitive and based of some examples from the theory of integrable systems; it was not obtained as a solution of the formally stated secondarization problem. Nevertheless, further development confirmed this guess to be right. In particular, first satisfactory classification results were obtained and a secondary analog of the classical Darboux lemma was proved for some small values of

$$n = (\text{number of independent variables}), \quad m = (\text{number of dependent variables})$$

in [**7, 8, 9**]. A serious shortcoming of this approach is that it does not lead directly to the corresponding $\mathcal{S}$-module, whose $\mathcal{C}$-homotopy type contains all secondary bivectors and among them the Poisson ones. Much later, Krasil′shchik constructed (see [**66**] and [**68**]), for a wide class of diffieties, a pair of complexes for which cochain mappings are secondary Poisson structures. For a systematic account see [**132**].

The second concept secondarized consciously in [**122**] was that of a covering (see Section 5.1). In the classical situation, a covering is, in particular, a mapping whose fibers have no internal structure. In the context of Secondary Calculus "no internal structure" means "Dim = 0". Therefore, the substitutions

$$\text{smap} \Rightarrow \text{mapping} \qquad \text{and} \qquad \text{Dim} = 0 \Rightarrow \dim = 0$$

in the standard definition of a topological covering bring naturally the secondary version of this notion. The latter allowed one to combine in a common theory many well-known constructions in the theory of partial differential equations and integrable systems such as Bäcklund transformations, Wahlquist–Estabrook prolongation structures, factorization of partial differential equations, nonlocal recursion operators, and the theory of nonlocal symmetries and conservation laws [**19, 74, 139**].

Factorization of PDE in connection with characteristic classes and scalar differential invariants was considered in [**119**] and [**131**] (see also [**2**] and [**106**]).

It is worth noting that the concept of a local diffeomorphism and that of a local covering coincide in the classical situation, but differ in the secondary context. It seems that the standard classification results, like the Darboux lemma, can be reasonably secondarized by interpreting the classical "local diffeomorphism" as the "secondary covering". In [**9**], this idea was shown to work well for Poisson structures (see Subsection 5.4.2).

Finally, the third concept, partially secondarized in the early 1980s, was that of a higher-order linear differential operator [**47**]. More precisely, this was done only for scalar differential operators on the basis of two different approaches: "geometric" and "functional". They both lead to the same result for $\mathcal{O} = J^\infty(\pi)$ reported in Subsection 5.4.1, which looks sufficiently satisfactory. However, for diffieties of a more general nature, these methods lead to different results. The functional approach is essentially the same as the one presented in Subsection 5.4.1. It is of cohomological nature as it should be, but it originates from a generally faulty interpretation of what are secondary functions: they were understood to be elements of $\bar{H}^n(\mathcal{O})$. This explains why this "archaic" approach works well if $\mathcal{O} = J^\infty(\pi)$ and fails otherwise.

Thus, these three "old" attempts of secondarization led to confidence in the existence of Secondary Calculus, and the problem of a satisfactory formalization of this concept arose.

Independently of Secondary Calculus, studies in geometry of partial differential equations along the lines exposed in [**124**] and [**71**] revealed various kinds of cohomology naturally associated with partial differential equations. In [**67**], Krasil′shchik defined and studied U-cohomology, later called $\mathcal{C}$-cohomology, by developing and generalizing the concept of higher symmetries. Later it turned out [**126**] that the U-cohomology is just secondary vector-valued differential forms subject to secondary Frölicher–Nijenhuis bracket formalism (see Section 5.6). In [**65, 66, 67, 68, 69**], Krasil′shchik, partially in collaboration with Kersten, initiated the study of graded ("super") generalizations of some basic constructions in geometry of partial differential equations. The results they obtained are interpreted in terms of Secondary Calculus as well.

As we have already seen, efficient computations of various kinds of horizontal cohomology is crucial for Secondary Calculus. The original author's method based on the use of Spencer's δ-cohomology as presented in Section 4.4 was subsequently gradually improved in works by Marvan [**90**], Tsujishita [**111**], Gessler [**39**], and Verbovetsky [**118**]. In its final form, due to Verbovetsky, it is presented in Subsections 5.5.4 and 5.5.5. A transparent and elegant form it acquired shows that one can hardly wish more on the level of general theory. Operators of the form ℓ_F^Φ are of key importance in applications of this method. This simple but rather subtle point was clearly understood in the course of evolution that starts with the observation due to Krasil′shchik that $\mathcal{C}$-differential operators act naturally on the algebra of Cartan forms.

It seems to be a little paradoxical that the notion of a module over the smooth function algebra of a manifold was secondarized last (see [**126**]). The results reported in Sections 5.3 and 5.4 show that it could hardly be predicted directly from the very beginning. It should be emphasized once again that just the notion of secondary module makes a solid foundation for Secondary Calculus.

Recent developments in quantum field theory related to the BRST transform method and anti-field formalism (see [**12, 13, 15, 14, 42, 46, 115**] and especially [**51**]) confirmed the expectations ([**135**]) concerning a fundamental role of Secondary Calculus in quantum field theory and its generalizations. A series of works by Henneaux and his collaborators (see, for instance, [**10**]) may be regarded as a direct bridge connecting Secondary Calculus and what was recently called Cohomological Physics. The collection [**50**] gives a rather complete view of the state of the art in this area as of 1997.

Bibliography

[1] V. Aldaya and J. A. de Azcarraga, *Variational principle on rth order jets of fibre bundles in field theory*, J. Math. Phys. **19** (1978) no. 9, 1869–1875.

[2] D. V. Alekseevski, V. V. Lychagin, and A. M. Vinogradov, *Basic ideas and concepts of differential geometry*, Geometry I, Encycl. Math. Sci., vol. 28, Springer-Verlag, Berlin, 1991.

[3] I. M. Anderson, *The variational bicomplex*, Academic Press, Boston, to appear.

[4] I. M. Anderson and N. Kamran, *The variational bicomplex for hyperbolic second-order scalar partial differential equations in the plane*, Duke Math. J. **87** (1997), 265–319.

[5] I. M. Anderson and J. Pohjanpelto, *The cohomology of invariant variational bicomplexes*. In: P. H. M. Kersten and I. S. Krasil′shchik (eds.), Geometric and algebraic structures in differential equations, Acta Appl. Math. **41** (1995), no. 1–3, 3–19.

[6] R. L. Anderson and N. H. Ibragimov, *Lie–Bäcklund transformations in applications*, SIAM, Philadelphia, 1979.

[7] A. M. Astashov, *Normal forms for Hamiltonian operators in field theory*, Dokl. AN SSSR **270** (1983), 1033–1037. (Russian)

[8] ———, *On generalization of the Darboux theorem to Hamiltonian field theory*, Mat. Zametki **36** (1984), 537–547; English transl. in Math. Notes **36** (1984).

[9] A. M. Astashov and A. M. Vinogradov, *On the structure of Hamiltonian operator in field theory*, J. Geom. and Phys. **3** (1986) no. 2, 263–287.

[10] G. Barnich, F. Brandt, and M. Henneaux, *Local BRST cohomology in the antifield formalism*: I. *General theorems*, Comm. Math. Phys. **174** (1995), 57–92; `arXiv:hep-th/9405109`.

[11] A. V. Bäcklund, *Zur Theorie der Flachentransformationen*, Math. Ann. Bd. **19** (1882), 387–422.

[12] I. A. Batalin and G. A. Vilkovisky, *Gauge algebra and quantization*, Phys. Lett. **102** (1981), 27–31.

[13] ———, *Quantization of gauge theories with linearly depended generators*, Phys. Rev. **D 28** (1983), 2567–2582.

[14] C. Becchi, A. Rouet, and R. Stora, *Renormalization of gauge theories*, Annals of Phys. (N.Y.) **98** (1976), 287–321.

[15] ———, *Renormalization of the abelian Higgs–Kibble model*, Comm. Math. Phys. **42** (1975), 127–162.

[16] G. Birkhoff, *Hydrodynamics: A study in logic, fact and similitude*, Revised Ed., Princeton Univ. Press, Princeton, NJ, 1960.

[17] G. W. Bluman and S. Kumei, *Symmetries and differential equations*, Appl. Math. Sci., vol. 81, Springer-Verlag, New York, 1989.

[18] A. V. Bocharov, *On Spencer cohomologies of affine algebraic varieties*, Acta Appl. Math., **49** (1997) 281–291.

[19] A. V. Bocharov, V. N. Chetverikov, S. V. Duzhin, N. G. Khor′kova, I. S. Krasil′shchik, A. V. Samokhin, Yu. N. Torkhov, A. M. Verbovetsky, and A. M. Vinogradov, *Symmetries and conservation laws for differential equations of mathematical physics*, Edited and with a preface by I. S. Krasil′shchik and A. M. Vinogradov, Amer. Math. Soc., Providence, RI, 1999.

[20] R. Bott and L. W. Tu, *Differential forms in algebraic topology*, Graduate Texts in Math., vol. 82, Springer-Verlag, New York Berlin, 1982.

[21] J. Braconnier, *Algèbres de Poisson*, C. R. Acad. Sci. Paris, Sér. A-B **284** (1977), no. 21, A1345–A1348.

[22] ———, *Sur les crochets généralisés*, C. R. Acad. Sci. Paris, Sér. A-B **286** (1978), no. 18, A763–A766.

[23] F. Brandt, *Gauge covariant algebra and local BRST cohomology*, Contemp. Math. vol. 219, Amer. Math. Soc., Providence, RI, 1998, pp. 53–68.

[24] R. L. Bryant, S. S. Chern, R. B. Gardner, H. L. Goldschmidt, and Ph. A. Griffiths, *Exterior differential systems*, MSRI Publ., vol. 18, Springer-Verlag, New York, 1991.

[25] R. L. Bryant and Ph. A. Griffiths, *Characteristic cohomology of differential systems,* I: *General theory*, J. Amer. Math. Soc. **8** (1995), 507–596; URL `http://www.math.duke.edu/~bryant`.

[26] ———, *Characteristic cohomology of differential systems,* II: *Conservation laws for a class of parabolic equations*, Duke Math. J. **78** (1995), 531–676.

[27] R. L. Bryant, Ph. A. Griffiths, and L. Hsu, *Hyperbolic exterior differential systems and their conservation laws, Part* I, Selecta Math. (N.S.) **1** (1995), 21–112.

[28] ———, *Hyperbolic exterior differential systems and their conservation laws, Part* II, Selecta Math. (N.S.) **1** (1995), 265–323.

[29] E. Cartan, *Les systèmes extérieurs et leur applications géométriques*, Hermann, Paris, 1946.

[30] V. N. Chetverikov, *On the structure of integrable $\mathcal{C}$-fields*, Diff. Geom. Appl. **1** (1991), 309–325.

[31] C. Chevalley, *Theory of Lie groups*, I, Princeton Univ. Press, Princeton, NJ, 1957.

[32] A. Connes, *Noncommutative geometry*, Academic Press, San Diego, CA, 1994.

[33] P. Dedecker, *Calcul des variations et topologie algébrique*, Mat. Soc. Roy. Sci. Liege **XIX** (1957), 1–216.

[34] ———, *On the generalization of symplectic geometry to multiple integrals in the calculus of variations*, Lecture Notes Math., vol. 580, Springer-Verlag, Berlin–New York, pp. 395–456.

[35] J. Douglas, *Solution of the inverse problem for the calculus of variations*, Trans. Amer. Math. Soc. **50** (1941), 71–128.

[36] W. I. Fushchich and A. G. Nikitin, *Symmetries of Maxwell's equations*, Reidel, Dordrecht, 1987.

[37] C. S. Gardner, *Korteweg-de Vries equation and generalizations,* IV. *The Korteweg-de Vries equation as a Hamiltonian system*, J. Math. Phys. **12** (1971), 1548–1551.

[38] C. S. Gardner, J. M. Greene, M. D. Kruskal, and R. M. Miura, *Korteweg-de Vries equation and generalization,* VI. *Methods for exact solution*, Comm. Pure Appl. Math. **27** (1974), 97–133.

[39] D. M. Gessler, *On the Vinogradov $\mathcal{C}$-spectral sequence for determined systems of differential equations*, Diff. Geom. Appl. **7** (1997), 303–324; URL `http://diffiety.ac.ru/preprint/`.

[40] ———, *The "three-line" theorem for the Vinogradov $\mathcal{C}$-spectral sequence of the Yang–Mills equation*, Acta Appl. Math. **56** (1999), 139–153; URL `http://diffiety.ac.ru/preprint/`.

[41] P. B. Gilkey, *Local invariants of a pseudo-Riemannian manifold*, Math. Scand. **36** (1975), 109–130.

[42] D. M. Gitman and I. V. Tyutin, *Quantization of fields with constraints*, Springer-Verlag, Berlin, 1990.

[43] C. Godbillon, *Géométrie différentielle et mécanique analytique*, Hermann, Paris, 1969.

[44] H. Goldschmidt, *Existence theorems for analytic linear partial differential equations*. Ann. Math. **86** (1967), 246–270.

[45] ———, *Integrability criteria for systems of non-linear partial differential equations*, J. Diff. Geom. **1** (1967), 269–307.

[46] J. Gomis, J. Paris, and S. Samuel, *Antibracket, antifields and gauge theory quantization*, Phys. Rep. **259** (1995), 1–145; `arXiv:hep-th/9412228`.

[47] V. N. Gusyatnikova, A. M. Vinogradov, and V. A. Yumaguzhin, *Secondary differential operators*, J. Geom. and Phys. **2** (1985), no 2, 23–65.

[48] P. Havas, *The range of application of Lagrange formalism*, I, Nuovo Cimento (Suppl.) **5** (1957), 363–388.

[49] M. Henneaux, B. Knaepen, and C. Schomblond, *Characteristic cohomology of p-form gauge theories*, Comm. Math. Phys. **186** (1997), 137–165; `arXiv:hep-th/9606181`.

[50] M. Henneaux, I. S. Krasil′shchik, and A. M. Vinogradov (eds.) *Secondary calculus and cohomological physics*, Contemp. Math. vol. 219, Amer. Math. Soc., Providence, RI, 1998.

[51] M. Henneaux and C. Teitelboim, *Quantization of gauge systems*, Princeton Univ. Press, Princeton, NJ, 1992.

[52] G. W. Horndeski, *Differential operators associated with Euler–Lagrange operator*, Tensor **28** (1974), 303–318.

[53] N. H. Ibragimov, *On the theory of Lie–Bäcklund transformation groups*, Mat. Sb. **109** (1979), no. 2, 229–253; English transl. in Math. USSR-Sb. **37** (1980).

[54] ———, *Transformation groups applied to mathematical physics*, Riedel, Boston, 1985.

[55] N. H. Ibragimov and A. B. Shabat, *Evolution equations with nontrivial Lie-Bäcklund group*, Funktsional. Anal. i Prilozhen. **14** (1980), no. 1, 25–36; English transl. in Functional Anal. Appl. **14** (1980).

[56] N. G. Khor′kova, *On the C-spectral sequence of differential equations*, Diff. Geom. Appl. **3** (1993), 219–243.

[57] F. Klein, *Gesammelte Mathematische Abhandlungen*, Springer-Verlag, Berlin, 1973. Erster Band: Liniengeometrie, Grundlegung der Geometrie, zum Erlanger Programm, Herausgegeben von R. Fricke und A. Ostrowski (von F. Klein mit ergänzenden Zusätsen verzehen), Reprint der Erstauflagen [Verlag von Julius Springer, Berlin, 1921].

[58] S. Kobayashi, *Transformation groups in differential geometry*, Ergebnisse der Mathematik und ihrer Grenzgebiete, Band 70, Springer-Verlag, New York–Heidelberg, 1972.

[59] I. Kolář, J. Slovák, and P. W. Michor, *Natural operations in differential geometry*, Springer-Verlag, Berlin–Heidelberg–New York, 1993.

[60] B. G. Konopelchenko, *Nonlinear integrable equations (Recursion operators, group-theoretical and Hamiltonian structures of soliton equations)*, Lecture Notes Phys., vol. 270, Springer-Verlag, Berlin, 1987.

[61] J.-L. Koszul, *Crochet de Schouten–Nijenhuis et cohomologie*. The mathematical heritage of E. Cartan (Lyon, 1984), Astérisque 1985, Numéro Hors Série, 257–271.

[62] I. S. Krasil′shchik, *A supersymmetry approach to Poisson structures over differential equations*, Proc. Conf. Diff. Appl., Brno, 1996, 14 pp. URL `http://www.ras.ru/EMIS/proceedings/index.htm`.

[63] ———, *Algebras with flat connections and symmetries of differential equations*, Lie Groups and Lie Algebras: Their Representations, Generalizations and Applications (B. P. Komrakov, I. S. Krasil′shchik, G. L. Litvinov, and A. B. Sossinsky, eds.), Kluwer, Dordrecht, 1998, pp. 425–434.

[64] ———, *Calculus over commutative algebras: a concise user guide*, Acta Appl. Math. **49** (1997), 235–248.

[65] ———, *Cohomology background in the geometry of PDE*, Contemp. Math., vol. 219, Amer. Math. Soc., Providence, RI, 1998, pp. 121–140.

[66] ———, *Hamiltonian formalism and supersymmetry for nonlinear differential equations*, Erwin Schrödinger International Inst. for Math. Physics, Wien, Preprint no. 257, 1995.

[67] ———, *Some new cohomological invariants for nonlinear differential equations*, Diff. Geom. Appl. **2** (1992), 307–350.

[68] I. S. Krasil′shchik and P. H. M. Kersten, *Deformations and recursion operators for evolution equations*, Geometry in Partial Differential Equations, A. Prastaro and Th. M. Rassias (eds.), World Scientific, Singapore, 1994, pp. 114–154.

[69] ———, *Graded differential equations and their deformations: A computational theory for recursion operators*, Acta Appl. Math. **41** (1994), 167–191.

[70] ———, *Symmetries and recursion operators for classical and supersymmetric differential equations*, Kluwer, Dordrecht, 2000.

[71] I. S. Krasil′shchik, V. V. Lychagin and A. M. Vinogradov, *Geometry of jet spaces and nonlinear partial differential rquations*, Gordon and Breach, New York, 1986.

[72] I. S. Krasil′shchik and A. M. Verbovetsky, *Homological methods in equations of mathematical physics*, Open Education, Opava, 1998.

[73] I. S. Krasil′shchik and A. M. Vinogradov (Eds.), *Algebraic aspects of differential calculus*, A special issue of Acta Applicandae Mathematicae **49** (1997), no. 3.

[74] ———, *Nonlocal trends in the geometry of differential equations: Symmetries, conservation laws, and Bäcklund transformations*, Acta Appl. Math. **15** (1989), 161–209.

[75] D. Krupka, *On the structure of the Euler mapping*, Arch. Math. (Brno), **X** (1974), no. 1, 55–62.

[76] M. D. Kruskal, R. M. Miura, C. S. Gardner, and N. J. Zabusky, *Korteweg-de Vries equation and generalizations, V. Uniqueness and nonexistence of polynomial conservation laws*, J. Math. Phys. **11** (1970), 952–960.

[77] B. A. Kupershmidt, *Geometry of jet bundles and the structure of Lagrangian and Hamiltonian formalism*, Lecture Notes Math., vol. 775, Springer-Verlag, Berlin–Heidelberg–New York, 1980, pp. 162–218.

[78] ———, *On the geometry of jet manifolds*, Uspekhi Mat. Nauk **30** (1975), 211–212. (Russian)

[79] M. Kuranishi, *Lectures on exterior differential systems*, Tata Institute Fundamental Research, Bombay, 1962.

[80] A. N. Leznov and M. V. Saveliev, *Group methods for integration of nonlinear dynamical systems*, Birkhäuser, Boston, 1987.

[81] S. Lie, *Gesammelte Abhandlungen*, Bd. 1–4, Teubner, Leipzig, and Aschelon, Kristiania, 1929.

[82] F. Lizzi, G. Marmo, G. Sparano, and A. Vinogradov, *Eikonal type equations for geometrical singularities of solutions in field theory*, J. Geom. Phys. **14** (1994), 211–235.

[83] V. V. Lychagin, *Geometric theory of singularities of solutions of nonlinear differential equations*, Itogi Nauk. Tekh. VINITI Ser. Probl. Geom. **20** (1988), 207–247 [Engl. transl.: J. Soviet Math. **51** (1990), 2735–2757].

[84] ———, *Geometry and topology of shock waves*, Dokl. AN SSSR **264** (1982), 551–555; English transl., Soviet Math. Dokl. **25** (1982), 685–689.

[85] ———, *Local classification of first order nonlinear partial differential equations*, Uspekhi Mat. Nauk **30** (1975), no. 1(181), 101–171; English transl., Russian Math. Surveys **30** (1975), no. 1, 105–175.

[86] ———, *Singularities of multivalued solutions of nonlinear differential equations and nonlinear phenomena*, Acta Appl. Math. **3** (1985), 135–173.

[87] S. Mac Lane, *Homology*, Springer-Verlag, Berlin, 1963.

[88] Yu. I. Manin, *Algebraic aspects of non-linear differential equations*, Itogi Nauki i Tekhniki. Sovremennye Problemy Matematiki, vol. 11. VINITI, Moscow, 1978, pp. 3–152; English transl., J. Soviet Math. **11** (1979), 1–122.

[89] M. Marvan, *A note on the category of partial differential equations*, Differential Geometry and its Applications, Proc. Conf. (Brno, Czechoslovakia, Aug. 24–30, 1986). URL `http://www.emis.de/proceedings/`.

[90] M. Marvan, *On the $\mathcal{C}$-spectral sequence with "general" coefficients*, Differential Geometry and its Applications, Proc. Conf. (Brno, Czechoslovakia, 1989), World Scientific, Singapore, 1990, pp. 361–371.

[91] J. McCleary, *A user's guide to spectral sequences*, Second Ed., Cambridge Univ. Press, Cambridge, 2001.

[92] A. V. Mikhailov, A. B. Shabat, and R. I. Yamilov, *The symmetry approach to a classification of nonlinear equations. Complete lists of integrable systems*, Uspekhi Mat. Nauk **42** (1987), no. 4(256), 3–53; English transl., Russian Math. Surveys **42** (1987), 1–63.

[93] R. M. Miura, *Korteweg-de Vries equation and generalizations,* I. *A remarkable explicit nonlinear transformation,* J. Math. Phys. **9** (1968), 1202–1204.

[94] R. M. Miura, C. S. Gardner, and M. D. Kruskal, *Korteweg-de Vries equation and generalizations,* II. *Existence of conservation laws and constants of motion*, J. Math. Phys. **9** (1968), 1204–1209.

[95] Jet Nestruev, *Smooth manifolds and observables*, Mosk. Tsentr Nepr. Mat. Obraz., Moscow, 2000; English transl. to appear.

[96] P. J. Olver, *Applications of Lie groups to differential equations*, Second Ed., Graduate Texts in Math., vol. 107, Springer-Verlag, New York, 1993.

[97] L. V. Ovsiannikov, *Group analysis of differential equations*, Academic Press, Boston, 1982.

[98] J.-F. Pommaret, *Systems of partial differential equations and Lie pseudogroups*, Gordon and Breach, New York, 1978.

[99] P. K. Rashevskii, *Riemannische Geometrie und Tensoranalysis*, Deutscher Verlag der Wissenschaften, Berlin, 1959.

[100] R. Santilli, *Necessary and sufficient conditions for the existence of Lagrangian in field theory*, I, II, Ann. Phys. **103** (1977), 354–468; III, ibid., **105** (1977), 227–258.

[101] L. I. Sedov, *Similarity and dimensional methods in mechanics*, Academic Press, Boston, 1959.

[102] J. Sniatycki, *On geometric structure of classical field theory in Lagrangian formalism*, Proc. Cambridge Phil. Soc. **68** (1970) no. 2, 475–484.

[103] D. C. Spencer, *Overdetermined systems of linear partial differential equations*, Bull. Amer. Math. Soc. **75** (1969), 179–239.
[104] S. Sternberg, *Lectures on differential geometry*, Chelsea, New York, 1983.
[105] C. H. Su and C. S. Gardner, *Korteweg-de Vries equation and generalizations,* III. *Derivation of the Korteweg-de Vries equation and Burgers equation*, J. Math. Phys. **10** (1969), 536–539.
[106] S. I. Svinolupov and V. V. Sokolov, *Factorization of evolution equations*, Uspekhi Mat. Nauk **47** (1992), 115–146; English transl. in Russian Math. Surveys **47** (1992).
[107] F. Takens, *A global version of the inverse problem of the calculus of variations*, J. Diff. Geom. **14** (1979) no. 9, 543–562.
[108] E. Tonti, *Variational formulation of non-linear differential equations*, I, Acad. Roy. Belg. Bull. Cl. Sci. **55** (1969), 137–165.
[109] ———, *Variational formulation of non-linear differential equations*, II, Acad. Roy. Belg. Bull. Cl. Sci. **55** (1969), 262–278.
[110] T. Tsujishita, *Formal geometry of systems of differential equations*, Sugaku **2** (1989), 1–40.
[111] ———, *Homological method of computing invariants of systems of differential equations*, Diff. Geom. Appl. **1** (1991), 3–34.
[112] ———, *On variation bicomplexes associated to differential equations*, Osaka J. Math. **19** (1982), 311–363.
[113] W. M. Tulczyjew, *Cohomology of the Lagrange complex*, Ann. Scuola Norm. Sup. Pisa (1988), 217–227.
[114] ———, *The Lagrange complex*, Bull. Soc. Math. France **105** (1977), 419–431.
[115] I. V. Tyutin, *Gauge invariance in field theory and statistical mechanics*, Preprint of the Lebedev FIAN, no. 39 (1975).
[116] A. M. Verbovetsky, *Differential operators over quantum spaces*, Acta Appl. Math. **49** (1997), 339–361; URL `http://diffiety.ac.ru/preprint/`.
[117] ———, *Lagrangian formalism over graded algebras*, J. Geom. And Phys. **18** (1996), 195–214; `arXiv:hep-th/9407037`.
[118] ———, *Notes on the horizontal cohomology*, Contemp. Math., vol. 219, Amer. Math. Soc., Providence, RI, 1998, pp. 211–232.
[119] A. M. Verbovetsky, A. M. Vinogradov, and D. M. Gessler, *Scalar differential invariants and characteristic classes of homogeneous geometric structures*, Math. Notes **51** (1992), 543–549.
[120] G. Vezzosi and A. M. Vinogradov, *On higher order analogues of the de Rham cohomology* (to appear).
[121] A. M. Vinogradov, *A spectral sequence associated with a nonlinear differential equation and algebro-geometric foundations of Lagrangian field theory with constraints*, Soviet Math. Dokl. **19** (1978), 144–148.
[122] ———, *Category of partial differential equations*, Lecture Notes Math., vol. 1108, Springer-Verlag, Berlin, 1984, pp. 77–102.
[123] ———, *Geometric singularities of solutions of partial differential equations*, Differential Geometry and its Applications, Proc. Conf. (Brno, 1986), Purkyně Univ. and Reidel, Dordrecht, 1987, pp. 359–379.
[124] ———, *Geometry of nonlinear differential equations*, J. Soviet Math. **17** (1981), 1624–1649.
[125] ———, *Hamiltonian structures in field theory*, Soviet Math. Dokl., **19** (1978), 790–794.
[126] ———, *Introduction to secondary calculus*, Contemp. Math., vol. 219, Amer. Math. Soc., Providence, RI, 1998, pp. 241–272.
[127] ———, *Local symmetries and conservation laws*, Acta Appl. Math. **3** (1984), 21–78.
[128] ———, *Multivalued solutions and a principle of classification of nonlinear partial differential equations*, Soviet Math. Dokl. **14** (1973), no. 3, 661–665.
[129] ———, *On the algebro-geometric foundations of Lagrangian field theory*, Soviet Math. Dokl. **18** (1978), 1200–1204.
[130] ———, *On geometric solution singularities* (in preparation).
[131] ———, *Scalar differential invariants, diffieties, and characteristic classes*, Mechanics, Analysis and Geometry: 200 Years after Lagrange, M. Francaviglia (Ed.), Elsevier, Amsterdam, 1991, pp. 379–416.
[132] ———, *Secondary Calculus: multivector-valued differential forms* (to appear).
[133] ———, *Some homological systems associated with differential calculus over commutative algebras*, Russian Math. Surveys **34** (1979), no. 6, 250–255.

[134] ———, *The C-spectral sequence, Lagrangian formalism, and conservation laws,* I. *The linear theory,* II. *The nonlinear theory*, J. Math. Anal. Appl. **100** (1984), 1–129.

[135] ———, *The category of differential equations and its significance for physics*, Proc. Conf. Diff. Geom. Appl. (Brno, 1984), D. Krupka (Ed.), J. E. Purkyně Univ., Brno, Czechoslovakia, 1984, pp. 289–301.

[136] ———, *The logic algebra for the theory of linear differential operators*, Soviet Math. Dokl. **13** (1972), 1058–1062.

[137] ———, *The theory of higher infinitesimal symmetries of nonlinear partial differential equations*, Soviet Math. Dokl. **20** (1979), 985–990.

[138] ———, *The union of the Schouten and Nijenhuis brackets, cohomology, and superdifferential operators*, Mat. Zametki **47** (1990), no. 6, 138–140; English transl. in Math. Notes **47** (1990).

[139] A. M. Vinogradov (Ed.), *Symmetries of partial differential equations*, Kluwer, Dordrecht, 1989 (reprint from Acta Appl. Math. **15** (1989), no. 1–2, and **16** (1989), no. 1–2).

[140] A. M. Vinogradov and I. S. Krasil′shchik, *A method of computing higher symmetries of nonlinear evolution equations and nonlocal symmetries*, Soviet Math. Dokl. **22** (1980), 235–239.

[141] ———, *What is the Hamiltonian formalism?*, Uspekhi Mat. Nauk **30** (1975), no. 1, 173–198; English transl., Russian Math. Surveys **30** (1975), no. 1, 177–202; see also London Math. Soc. Lecture Notes Ser., vol. 60, Cambridge Univ. Press, London, 1981, pp. 241–266.

[142] A. M. Vinogradov, I. S. Krasil′shchik, and V. V. Lychagin, *Applications of nonlinear equations in civil aviation*, Moscow Inst. of Civil Aviation, Moscow, 1977.

[143] ———, *Introduction to the geometry of nonlinear differential equations*, "Nauka", Moscow, 1986. (Russian)

[144] A. M. Vinogradov and V. A. Yumaguzhin, *Differential invariants of webs on 2-dimensional manifolds*, Mat. Zametki **48** (1990), no. 1, 26–37; English transl., Math. Notes **48** (1991), no. 1–2, 639–647.

[145] M. M. Vinogradov, *Principal functors of differential calculus in graded algebras*, Uspekhi Mat. Nauk. **44** (1989), no. 3 (267), 20–29.

[146] V. V. Zharinov, *Geometrical aspects of partial differential equations*, World Scientific, Singapore, 1992.

Index

adjoint complex, 41
adjoint operator, 32, 33
  in the $\mathcal{C}$-theory, 128
  left, 214
  right, 215
admissible submanifold of a variational problem, 133
algebra of symbols, 29
alternative Lagrangians, 22

Bohr correspondence principle, 195
bordisms of geometric structures, 186
  characteristic classes, 186
brackets
  Frölicher–Nijenhuis, 226
  higher Jacobi, 75
  Poisson, 10, 30, 106
  Richardson–Nijenhuis, 225
  Schouten–Nijenhuis, 226
bundle of jets, 58

Cartan distribution, 5, 62, 96, 193
Cartan plane, 96
Cartan submodule, 77, 99
category DE, *see* category of differential equations
category of differential equations, 9, 121
  homotopic morphisms, 171
  homotopy, 171
  morphisms, 121
  objects, 121
  Poincaré lemma, 172
$\mathcal{C}$-complete equation, 88
$\mathcal{C}$-complex, 203
$\mathcal{C}$-connection, 206
c-density, 47, 48, 136
  equivalence, 50
  nonlinear, 50
$\mathcal{C}$-differential operator, 100
$\mathcal{C}$-field, 72, 77, 107
$\mathcal{C}$-filtration, 138, 216
$\mathcal{C}$-general equation, 88
$\mathcal{C}$-Green formula, 130
characteristic, 76, 82
characteristic classes, 182
  Euler, 187
  of 3-webs, 184
  of bordisms, 24
  Pontryagin, 183, 187
characteristic covector, 30
characteristic diffiety, 182
  regular, 183
characteristic mapping, 182
  regularization, 183
$\mathcal{C}$-homotopy equivalence, 203
$\mathcal{C}_{(k)}$-field, 115
classical contact structure, 5
classical situation, 26
$\mathcal{C}_\nabla$-spectral sequence, 216
  vanishing theorems
    ∇-k-Line Theorem, 221
    ∇-Two Line Theorem, 219
cobweb of coverings, 14
common equation, 87
compatibility complex, 170, 220
  Diff-compatibility, 231
  Jet-compatibility, 231
conformally selfadjoint equation, 173
conic equation, 172
connection in a module, 199
conservation law, 23, 50
  linear, 48
  local, 136
  nonlocal, 192
  proper, 173
  rigid, 23
  rough, 173
contact structure
  classical, 5
  generalized, 61
  infinite-order, 193
  k-th order, 5, 62
correspondence principle, 195
covering, 13
  cobweb, 14
  in the category DE, 192
$\mathcal{C}$-spectral sequence, 138
  of a diffiety, 20
  vanishing theorems
    k-Line Theorem from the Bottom, 232
    One Line Theorem, 227
    Two Line Theorem, 161, 171
$\mathcal{C}$-transformation, 118

degree of a functor, 101
density of a linear conservation law, *see* c-density
de Rham complex, 28
Diff-compatibility complex, 231
diffeomorphism of domains in J^∞, 118
differential calculus functors, 27
  degree, 27
differential complex, 196
differential graded cochain mapping, 197
differential linear mapping, 197
differential linear structure, 197
diffiety, 7, 122
  $\mathcal{C}$-fields, 9
  contact fields, 9
  $\mathcal{C}$-spectral sequence, 20
  elementary, 191
  general, 193
  horizontal de Rham cohomology, 15
  horizontal differential, 15
  horizontal differential forms, 15
  secondary differential forms, 20
  symmetries, 9
  symmetry algebra, 9
  trivial contact fields, 9
diffiety dimension, 7, 192
diffiety morphisms, 192
divergence, 33
divided local chart, 4

elementary diffiety, 191
equation normal with respect to an operator, 173
equivalent c-densities, 50
equivalent equations, 92
equivalent filtrations, 121
Euler operator, 46, 136, 145
Euler–Lagrange equation, 46, 136
Euler–Lagrange operator for a problem with constraints, 181
Euler–Lagrange theorem, 136
evolutionary derivation, 10, 74
exceptional case, 62
exterior differentiation operator, 28
exterior product of differential forms, 28
extrinsic infinitesimal symmetry, 87
extrinsic symmetry, 87

factorization, 122
FG-category, 98
filtered differential operator, 97
filtered homomorphism, 97
first variation formula, 135
Φ-symmetry, *see* nonlocal symmetry
flat $\mathcal{C}$-connection, 207
flat connection, 200
formally integrable equation, 62
FP-equation, 47
Frölicher–Nijenhuis bracket, 226

functorial fiber bundle, 68
fundamental isomorphism, 227

g-connected domain, 67
general diffiety, 193
generalized contact structure, 61
generalized Noether theorem, 137
generalized Schwarz formula, 182
generalized symmetry, 13
generating function, 10
  of a conservation law, 23
  of a Lie field, 81, 84
  of an evolutionary derivation, 10
generating operator of a secondary differential operator, 18
geometric filtered module, 97
geometric modules, 28
geometric structure, 122, 182
  scalar differential invariant, 123
  type of, 122
geometrization functor, 28
gluing operator, 27
Green $\mathcal{L}$-formula, 46
Green formula, 41
  $\mathcal{C}$-formula, 130
  for a differential operator, 42
  general, 42
  $\mathcal{L}$-formula, 46

Hamiltonian formalism in the $\mathcal{C}$-theory, 106
Hamiltonian homomorphism, 212
Hamiltonian operator, 212
higher extrinsic infinitesimal symmetry, 109
higher infinitesimal symmetry, 107
higher Jacobi bracket, 75
homogeneous splitting, 42
homotopic morphisms in DE, 171
homotopy equivalence of differential complexes, 197
homotopy in the category DE, 171
horizontal de Rham cohomology of a diffiety, 15
horizontal de Rham complex, 100
horizontal differential form on a diffiety, 15
horizontal differential on a diffiety, 15
horizontal element, 101
horizontal Jet-Spencer complex with coefficients in a module, 232
horizontal module, 207
horizontalization operation, 100

infinite-order contact structure, 193
infinitesimal automorphism of the Cartan distribution, 107
infinitesimal Lie transformation, 81
infinitesimal Stokes formula, 150, 151
infinitesimal symmetry of a density, 136
infinitesimal symmetry of a Lagrangian density, 52

inner product, 84
insertion, 84
integrable $\mathcal{C}$-field, 121
integral manifold of the Cartan distribution, 8, 62
intrinsic infinitesimal symmetry, 87
inverse Noether theorem, 173
inverse problem in the calculus of variations, 23, 149
  strong, 149
  weak, 149
invertible differential operator, 120

jet of a submanifold, 57
jet operator, 27
jet order reduction, 28
Jet-compatibility complex, 231

k-Line Theorem from the Bottom, 232
k-th order contact structure, 5, 62

label, 3
label of a system of partial differential equations, 5
Lagrange multiplier, 181
Lagrangian, 134
Lagrangian density, 134
Lagrangian of variational problem, 134
Laplace equation, 122
left adjoint operator, 214
left filtration, 166
length of a differential operator, 220
Lie derivative, 150
Lie field, 81
Lie transformation, 80, 118
Lie–Bäcklund transformation, 13
lifting of a Lie field, 81
linear conservation law, 48
linear differential operator over an algebra, 26
linear Noether theorem, 54
linearization on a section, 73
l-integral manifold, 111
local conservation law, 136
locally maximal integral manifold of the Cartan distribution, 62
locally regular equation, 158

manifold of jets, 57, 58
module of differential forms, 27
module of jets, 27
module of symbols, 29
module-valued differential form, 200
module-valued horizontal differential forms, 207
morphism in the DE category, 121
morphism of equations, 50
multidifferential operator, 45
multivector-valued differential form, 221

∇-extension of a differential operator, 215
∇-k-Line Theorem, 221
∇-Two Line Theorem, 219
natural $\mathcal{S}$-module, 199, 207
natural $\mathcal{S}$-module structure, 201
natural bundle, 122
  type of, 122
natural secondary module, 206, 207
Noether current, 52
Noether mapping, 176
Noether theorem, 52
  generalized, 137
  inverse, 173
  linear, 54
Noetherian symmetry, 176
nondegenerate mapping, 113
nonexceptional case, 65
nonlinear c-density, 50
nonlinear differential equation, *see* system of nonlinear differential equations
nonlinear differential operator, 59
nonlocal symmetry, 13, 192
nonoverdetermined equation, 170
normal equation, 89, 173

object of the DE category, 121
One Line Theorem, 227
operator of universal linearization, 72
order of a geometric structure, 183

Poincaré duality, 39, 130
Poincaré lemma in the category DE, 172
Poisson bracket, 10, 30, 106
prolongation of a differential equation, 61
proper conservation law, 173

quadratic differential operator, 45
quadratic Lagrangian, 45
quadratic Lagrangian density, 45
quasi-integrable equation, 96
quotient equation, 122

recursion operator, 224
regular characteristic diffiety, 183
regular equation, 158
regular equation with respect to an operator, 158
regular variational problem, 180
relative vector field, 84
representing object, 27
$\rho_k(\pi)$-transformation, 68
Richardson–Nijenhuis bracket, 225
right adjoint operator, 215
right filtration, 166
rigid conservation law, 23
rigid equation, 87
R-manifold, 65, 97
rough conservation law, 173
R-plane, 5, 62

scalar differential invariant, 123
  value of, 123
scalar secondary differential operator, 17
Schouten–Nijenhuis bracket, 226
secondarization algorithm, 191
secondarization problem, 191
secondary p-vector field, 223
secondary algebra, 209
secondary differential form, 193, 208
  on a diffiety, 20
secondary differential operator, 210
secondary evaluation mapping, 205
secondary homomorphism, 206
secondary linear algebra, 199
secondary linear mapping, 199, 206
secondary module, 203, 205
secondary multivector-valued differential form, 223
secondary Poisson structure, 212
secondary pre-homomorphism, 205
secondary smooth function, 204
secondary tensor product, 206
secondary vector bundle, 203, 205
secondary vector field, 193, 224
secondary vector space, 199
secondary vector-valued differential s-form, 223
selfadjoint nonlinear system, 149
$\mathcal{S}$-homomorphism, 198, 204
skew-symmetric multiderivation, 27
smap, 9, 122, 192
$\mathcal{S}$-module, 197, 204
$\mathcal{S}$-module homomorphism, 204
smooth category, 28
smooth function on $J^\infty(E, n)$, 113
solution of a differential equation, 61
special characteristic classes, 24
special chart, 58
special coordinates, 58
Spencer $\mathcal{C}$Diff-complex, 129
Spencer Diff-complex, 38
Spencer Diff-operator, 38
Spencer Jet-complex, 31
s-solvable equation, 62
strong inverse problem, 149
strongly formally integrable equation, 62
symbol of a linear differential operator, 29
symmetry
  extrinsic, 87
  extrinsic infinitesimal, 87
  generalized, 13
  higher extrinsic infinitesimal, 109
  higher infinitesimal, 107
  infinitesimal of a density, 136
  infinitesimal of a Lagrangian density, 52
  intrinsic, 87
  intrinsic infinitesimal, 87
  Noetherian, 176
  nonlocal, 192
  of a diffiety, 9
  of a Lagrangian, 175
  of a Lagrangian density, 52
symmetry algebra of a diffiety, 9
system of nonlinear differential equations, 61
  $\mathcal{C}$-complete, 88
  $\mathcal{C}$-general, 88
  common, 87
  conformally selfadjoint, 173
  conic, 172
  equivalence, 92
  Euler–Lagrange, 46, 136
  formally integrable, 62
    strongly, 62
  label, 5
  locally regular, 158
  morphisms, 50
  nonoverdetermined, 170
  normal, 89, 173
  normal with respect to an operator, 173
  prolongation, 61
  quasi-integrable, 96
  regular, 158
    with respect to an operator, 158
  rigid, 87
  selfadjoint, 149
  solutions, 61
  s-solvable, 62
  symmetries, *see* symmetry

tangent vector to the space of solutions, 179
θ-integral manifold, 96
total derivative, 6, 60
transformation of a density, 51
transformations (including infinitesimal)
  $\mathcal{C}$-fields, 72, 77, 107
  $\mathcal{C}$-transformations, 118
  evolutionary derivations, 10, 74
  infinitesimal automorphisms of the Cartan distribution, 107
  infinitesimal Lie transformations, 81
  integrable $\mathcal{C}$-fields, 121
  Lie fields, 81
  Lie transformations, 80, 118
  Lie–Bäcklund transformations, 13
  translations by a section, 71
  trivial $\mathcal{C}$-fields, 79
  U-field, 85
  $U_k(\pi)$-transformation, 68
  U-transformation, 66
translation by a section, 71
trivial $\mathcal{C}$-field, 79
trivial density, 134
trivial Lagrangian, 149
trivial sequence, 167
triviality problem for Lagrangians, 22
Two Line Theorem, 161, 171
type of a U-manifold, 65

type of a geometric structure, 122
type of a natural bundle, 122

U-field, 85
U-geometry, 64
$U_k(\pi)$-transformation, 68
U-manifold, 64
universal linearization operator, 73
U-transformation, 66

variation of a density, 51
variational complex, 149
vertical differential operator, 105, 211

weak inverse problem, 149
web, 124